Postcranial Adaptation in Nonhuman Primates

Postcranial Adaptation in Nonhuman Primates

Edited by

DANIEL L. GEBO

NORTHERN ILLINOIS UNIVERSITY PRESS

DeKalb, 1993

Published by the Northern Illinois University Press,
DeKalb, Illinois 60115
Manufactured in the United States using acid-free paper ♾
Design by Julia Fauci

Library of Congress Cataloging-in-Publication Data
Postcranial adaptation in nonhuman primates / edited by
Daniel L. Gebo.
p. cm.
Includes bibliographical references and index.
ISBN 0-87580-179-X. —ISBN 0-87580-559-0 (pbk.)
1. Primates—Adaptation. 2. Primates—Anatomy.
3. Primates,
Fossil. I. Gebo, Daniel Lee, 1955-
QL737.P9P65 1993
599.8'044—dc20 93-16550
CIP

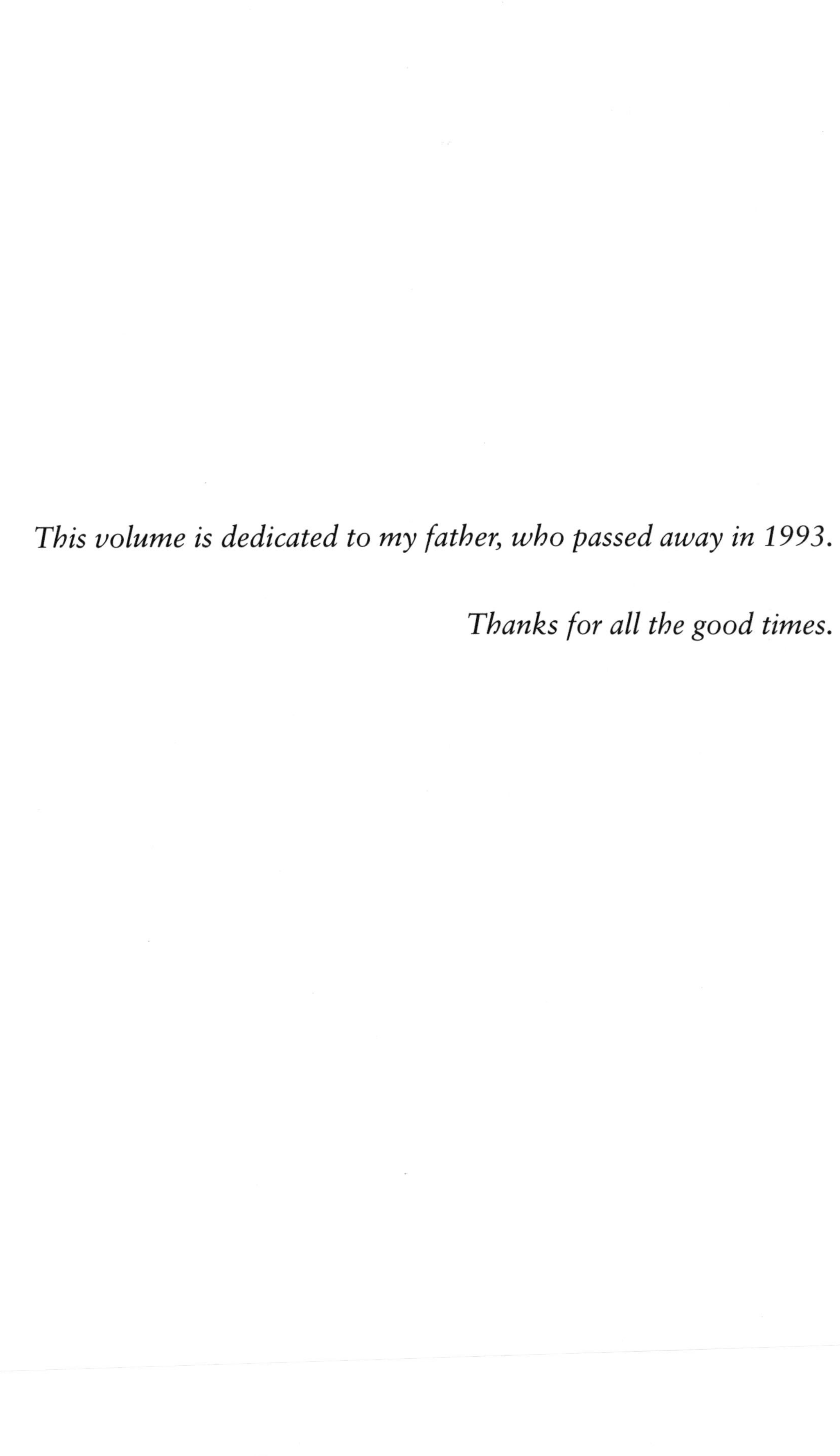

This volume is dedicated to my father, who passed away in 1993.

Thanks for all the good times.

CONTENTS

PREFACE

This volume is the result of an inspiration that came to me in 1990, and that I put to the test at the American Association of Physical Anthropology Conference in Miami, Florida, that same year. Thanks to the hard work of the contributors, the book is now a reality. Most anatomical textbooks are written on human anatomy and are chiefly for the use of medical students. This volume—like F. Jenkins's *Primate Locomotion* (1974) and the more recent volume *An Introduction to Human Evolutionary Anatomy* (Aiello and Dean, 1990)—tries to update students of primatology on our current understanding of primate functional anatomy and our application of this information to the primate fossil record. Compared to most other mammals, primates are known for their wide spectrum of postcranial adaptations related to their positional behavior (locomotion and postures). It is hardly surprising that these novel adaptations in positional behavior have led to the evolutionary origin of new groups of primates. Thus, the importance of functional anatomy to evolutionary studies is as great as ever.

Functional and comparative anatomical studies are also at the forefront of analysis when trying to understand both how an anatomical system works and (for the evolutionist) how likely it is that an anatomical system will duplicate itself in another lineage (convergence). The problem of convergence is the bane of evolutionary studies. It is here that functional and comparative studies can have significant impact on character analysis and on the refinement of better phylogenies. In the process, such studies can fuel an evolutionary debate or two (for example, bipedalism in early hominids). But it is not a one-way street. Evolutionary studies help functional morphologists better test their comparative samples for why and when their anatomical novelties evolved; they are also necessary when "current use" mechanical models, which never quite explain the comparative sample, are inadequate. The "adaptive past" should be as important to a functional morphologist in understanding form-function relationships as are the high-tech tools in use today. Overall, this combination of functional morphology with historical analysis is a powerful explanatory tool in primate evolutionary studies.

This volume is intended for upper-division undergraduates and those in the early graduate years. It is broken into three topical parts. In Part I, chapter 1 stands alone as our only chapter on biomechanics; it explains why biological materials and biomechanics are important considerations in understanding how primate limbs work. Part II is a systemic treatment of the primate body, although it is not exhaustive. Chapters 2 through 4 cover the forelimb, chapter 5 the spine, and chapters 6 and 7 the hindlimb. The comparative breadth varies from chapter to chapter. In Part III, chapters 8 through 11 discuss postcranial adaptations and locomotor behavior in extinct primates, from the beginning of the Eocene (55 mya) through the Miocene (from 22.5 to 5 mya). Each of the contributors was asked to synthesize and explain an anatomical system in living primates or postcranial adaptation in a group of extinct primates. All of the contributors performed their tasks admirably, especially considering that each contributor had to pass along the basic anatomical features (i.e., muscles, bones, and joints) in a functional perspective, make their chapter readable to upper-division undergraduates, and still add some of their current work for more provocative thinking. I thank all of them for their hard work. Writing these chapters turned out to be much harder than we thought it would be. Thanks again.

I would also like to acknowledge the help I received from Northern Illinois University Press. I thank also our outside reviewers for their many helpful comments and for the time they put into this volume. I take full blame for any mistakes, omissions, or other problems not foreseen by our reviewers. Finally, I would especially like to thank Marian Dagosto for her great patience in putting up with all of my mischief.

LITERATURE CITED

Aiello L and Dean C (1990) An Introduction to Human Evolutionary Anatomy. London: Academic Press.

Jenkins F (1974) Primate Locomotion. New York: Academic Press.

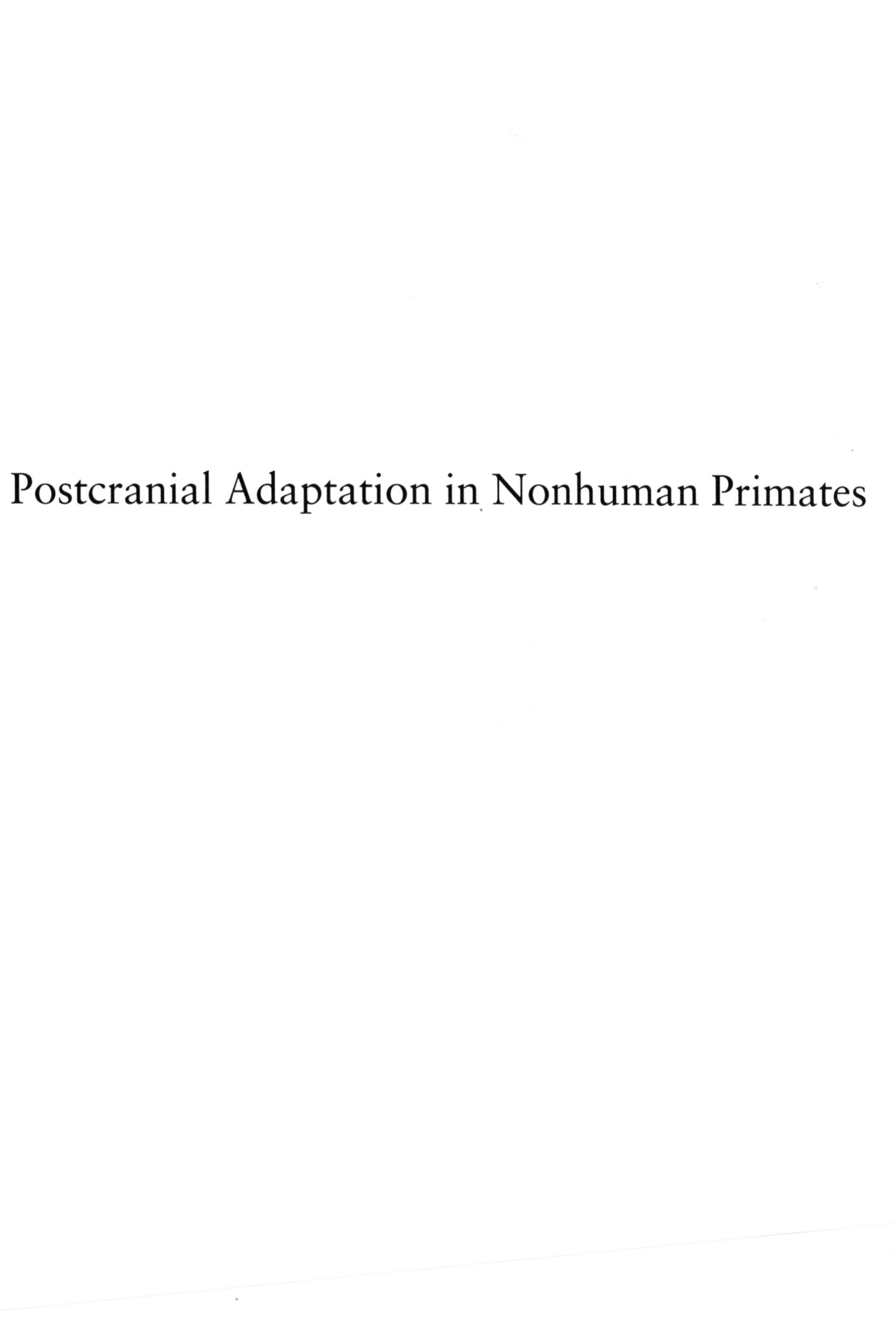

Postcranial Adaptation in Nonhuman Primates

Part I Biomechanics

C H A P T E R O N E

Biomechanics of Primate Limbs

Sharon M. Swartz

The designer of a bridge needs to know the strength of his steel or concrete and he needs to know how forces are transmitted through structures; he must draw on the expertise of materials science and structural engineering. A biologist studying an animal or a plant structure cannot understand it fully without the same sort of knowledge.

—R. McNeill Alexander

In this chapter, I hope to convey some fundamentals of a solid, mechanics-oriented approach to the study of animal structure. One might well ask why students of primate postcranial design should care about mechanics. The biomechanical approach emphasizes that the ability of an organism to deal with the mechanical forces that impinge upon it is a minimum requirement for life. This requirement sets the boundaries on the possible patterns of spatial organization of biological structures and determines what kinds of anatomies—drawn from the large realm of possible designs—will actually work. Animals devote a tremendous amount of biomass and energy to dealing with mechanical forces, particularly the postcranium. These forces arise from a variety of sources, in particular, the activity of the muscles active during normal movement; the accelerations and decelerations of the entire body and body segments with respect to one another as the position of the body changes; and the force of gravity. Physiological studies of biological beings regularly analyze how organisms cope with thermal and/or chemical parameters in their environments; biomechanics adds another kind of physical variable, force, to our arena of interest.

From a mechanical perspective, the properties of the tissues comprising the postcranium and the deployment of these tissues in building anatomical structures determine the ways in which organisms perform in their environments. Understanding how a specific skeletal element with a particular morphology functions in a particular taxon requires some knowledge of the behavior of bone as a tissue, as well as an understanding of the basic properties of the tissues directly associated with the bone (i.e., cartilage, muscle, ligaments). The anatomy of combinations of these different building blocks will then reflect both material composition and structure or architecture. I will consider these topics in turn, then consider some of the methods through which these subjects are studied, and conclude with a case study demonstrating how this kind of approach may be able to illuminate our understanding of the design of the primate postcranium.

MATERIALS

Stress, Strain, and Failure

When a force or a load is applied to a solid object, Newton's Second Law ($F = ma$ or force = mass × acceleration) tells us that the structure will accelerate. If, however, the object is somehow restrained from moving, it must still accommodate the force, even if acceleration is impossible. Rather than changing velocity, the structure will deform; it will undergo a change in length in the same direction as the applied force (Fig. 1.1). Even when an applied force causes motion, it may also cause a small

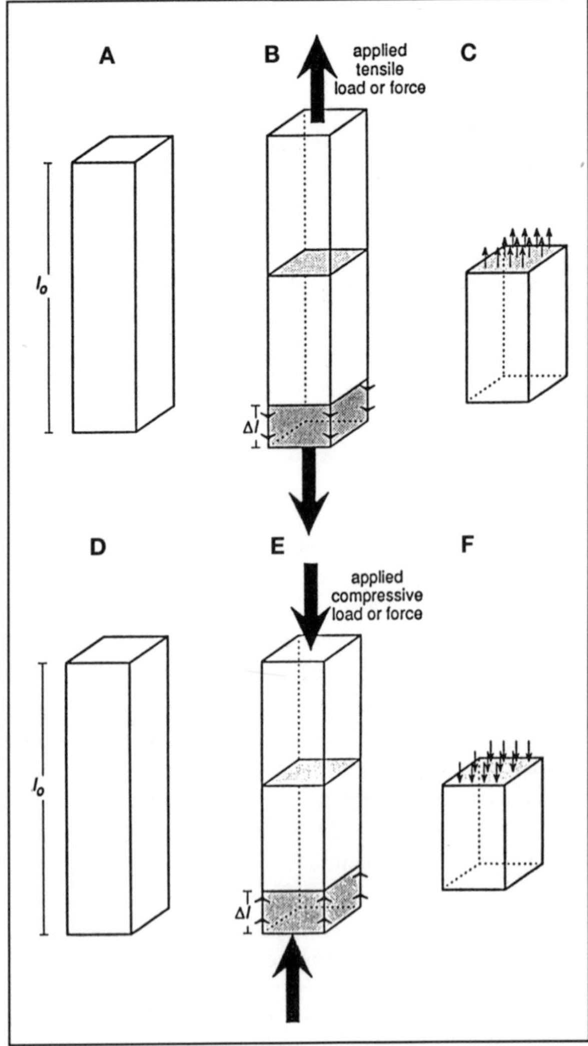

FIGURE 1.1. When a force is applied to a structure, it deforms, and its original length, l_O, changes by some amount Δl. If the applied force is tensile, the undeformed structure (**A**) will increase in length (**B**). The force applied to the ends of the structure acts everywhere within in it, such that if we look at any cross section of the structure (**C**), the tensile force would be acting everywhere in the cross section. The stress developed in the structure will be equal to the applied force divided by the cross-sectional area available to resist that force, as indicated by the shaded area. The situation for compressive loading is analogous (**D**, **E**, **F**).

amount of deformation simultaneously. When the musculature of primates exerts force on the skeleton, we immediately notice the resulting large-scale motions, but at the same time deformations of various elements are occurring as well. If the material from which an object is made is quite stiff, as is the case for bones and many of the structural materials in our everyday lives, the deformation may be microscopically small. Still, the length of the object in the direction of the force will change in direct proportion to the force applied. As the force increases, the object will deform more and more; for some materials, this relationship is linear, with deformation increasing in direct proportion to force (Fig. 1.2). Eventually the structure's molecular integrity will be pushed to its limits and it will fracture or fail in some other way.

The size of the objects under study can confound the relationship between applied force and resulting deformation. Imagine, for example, two columns composed of a single material, of equal height but of greatly differing thickness. When identical loads are applied to these structures, the slender one with the small diameter will shorten or elongate more than the other, and it will fail at a lower applied force. It is clearly desirable to be able to distinguish between behavior due to the nature of a given material and behavior due to the size and the shape of objects made of that material. To account for size-related effects, we can normalize force to some aspect of a structure's geometry that relates to how it will behave when loaded.

For structures in many loading circumstances, force is normalized by dividing by the cross-sectional area of the structure, a measure of the amount of material available to resist the applied load. This measure is defined as *stress,* typically abbreviated σ, and as a force divided by an area, is measured in units of Newtons/meter^2 (N/m^2), sometimes also known as Pascals (Pa). Stress can be thought of as a measure of force intensity. Similarly, we can think of deformations resulting from applied forces in terms of the change in length produced, normalized to the original length of the structure. This converts deformation to *strain* (ε) and because a length divided by a length is a dimensionless number, strain is typically expressed as a fraction, or percent length change. Forces that elongate structures will therefore result in positive strains; negative strains indicate length decreases.

If we observe the behavior of loaded structures in terms of applied stress and resulting strain, we often find that a large variety of structures made of a single material will produce a single, common stress-

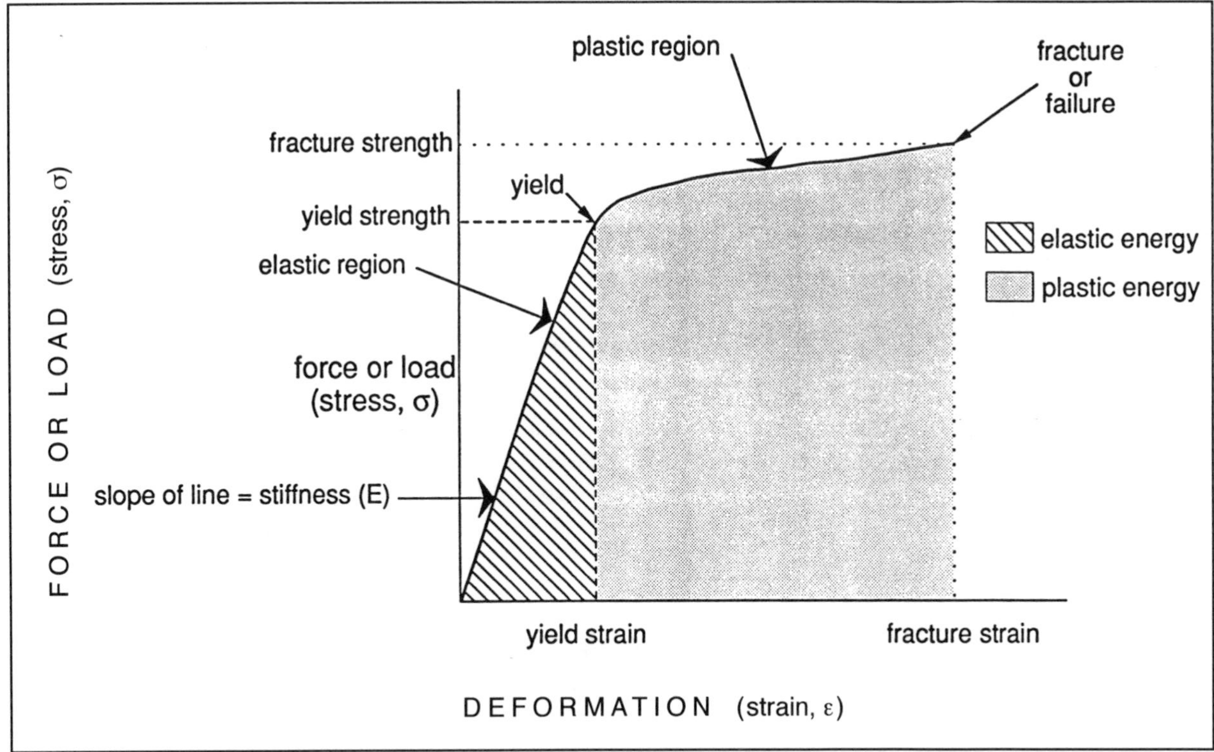

FIGURE 1.2. As increasing force is applied to an object, it deforms more and more, as illustrated in this plot of force versus deformation. For many materials, the deformation increases linearly with increasing force up until the yield point (elastic region) and then increases a great deal with very small increments in force as it enters the region of plastic behavior (yield and then plastic region). Ultimately, when it can no longer deform and still maintain its structural integrity, it fails or fractures. Plots of load versus deformation and of stress versus strain appear identical in shape, but they are reported in different units. The area under the stress-strain curve represents the energy imparted to the object (per unit volume) as it is stressed (elastic energy); the nature of the relationship between stress and strain (linearly elastic versus plastic, etc.) will therefore play a big role in determining how much energy can be absorbed by structures made of a given material. The area under the plastic region of the curve (plastic energy) is much larger than that in the elastic region, even though it represents a much smaller range of stresses.

strain plot, regardless of the rate of loading or the number of loading cycles applied (Fig. 1.2). This curve often has a characteristic shape, with a straight line relationship between stress and strain over a particular loading range. When the force applied to structures made of such materials is removed, the structure returns to its original dimensions; such behavior is termed linearly elastic, and materials of this kind are therefore called elastic materials. The slope of the linear portion describes a fundamental property of the material: its *elastic* or *Young's modulus* (*E*) or stiffness (Fig. 1.2). Elastic modulus is calculated as stress/strain and is expressed in units of N/m^2 or Pa.

Continuing past the elastic region, before fracture occurs, the relationship between load and deformation may change, with small increments of force producing ever-larger deformations. As damage to the molecular architecture of the structure accumulates, it reaches a point at which there is no longer sufficient integrity to remain whole, and fracture occurs. The point of demarcation between the elastic region of the curve and the region of large deformations (the *plastic* region) is known as the *yield point* of the material. After the yield point, structures begin to undergo unrecoverable damage; when the applied load is removed from the specimen, it will no longer return to its original dimensions. The stress (or strain) at which yield occurs is designated the *yield stress* (or *yield strain*), and the stress at failure the *ultimate, fracture,* or *failure stress* (or *strain*) (Fig. 1.2).

Loading Modes

The behavior of a material depends not only on its composition but also on the way it is loaded. The simplest loading modes are purely axial forces applied along the structure's main axis, with tension or tensile forces producing elongation, and compression or compressive forces producing shortening (Fig. 1.3B and C). In addition, *shear* forces are another important loading mode. Shearing forces act to distort a structure as though each plane of material is being slid past the adjacent layer, which would occur if we apply a pair of forces that are equal in magnitude, opposite in direction, and not lined up with one another to a structure (Fig. 1.3E). As J. E. Gordon has said, "If tension is about pulling and compression is about pushing, then shear is about sliding" (Gordon, 1978). We can define shear stress (τ) as shear force normalized to the area being sheared. When a shearing force is applied to an object, it changes shape rather than undergoing simple elongation; it follows that shear strain is the angle through which a material is distorted rather than a length change, and that shear modulus is shear stress/shear strain in analogy to tensile or compressive stress or strain. Twisting loads or *torsional* loads also produce shearing stresses, where the shearing is between concentric layers (Fig. 1.3F).

In most realistic biological situations, structures

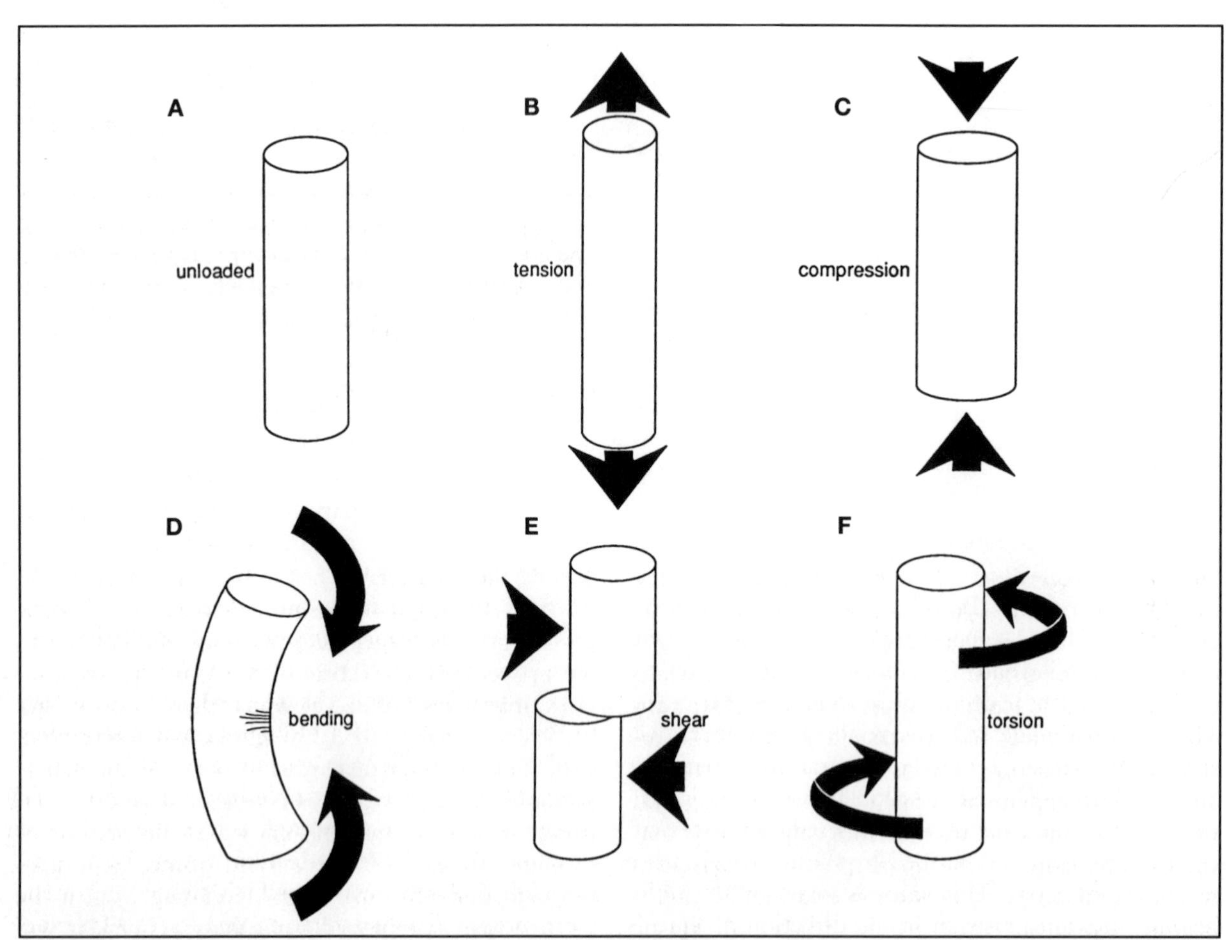

FIGURE 1.3. Schematic representation of a beam subjected to a number of different loading modes. Adapted from M. Nordin and V. H. Frankel: Basic Biomechanics of the Musculoskeletal System, 2nd ed. Philadelphia, Lea & Febiger, 1989. Used with permission.

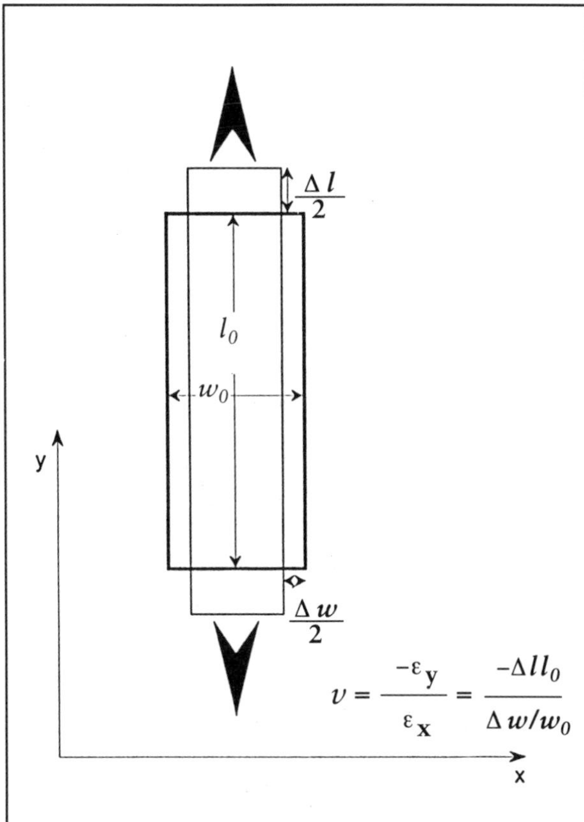

FIGURE 1.4. As a structure experiences an axial load in one direction (x axis), it will simultaneously deform in a direction at 90° to the loading axis (y axis). The degree to which a structure subjected to tension along its long axis will shorten along its transverse axis is determined by its Poisson's ratio *(ν)*, the ratio between the strain in the two directions.

are subjected to a combination of loading modes. In fact, even when a structure experiences a simple, purely axial load, a combination of stresses will be developed. This is apparent when we consider that if a structure is loaded in tension, it not only increases in length, it decreases in thickness (Fig. 1.4). That is, all tensile stresses generate compressive stresses at right angles to the orientation of tension, and vice versa. The relationship between the longitudinal stress and the transverse stress developed at right angles to it is described by the *Poisson's ratio* *(ν)* of the material. This value is given by the ratio between the tensile strain in the direction of uniaxial loading and the compressive strain at 90° to the primary loading direction multiplied by minus one (to make the ratio positive). According to elastic theory, Poisson's ratio can range from 0 to 0.5, although stiff materials such as bone typically have values between 0.3 and 0.4 (0.5 represents a constant volume material). Many soft biological materials have even higher Poisson's ratios, approaching or even exceeding 1.0.

Another important kind of complex loading mode is *bending,* probably the single most important loading mode for bones. When a beamlike structure is bent, one surface experiences tension and the opposite surface undergoes compression. In structures under axial loading and under bending, shearing stresses will also occur simultaneously within the structure along different planes than the planes undergoing tension or compression (Fig. 1.3D) (see Gordon [1978] for more on shear stresses developed in axial and bending loads).

Often the tensile, compressive, and shear strengths of a material will differ significantly, and an element's performance in its biological context may, therefore, depend on both the magnitude and the type of load experienced. If a structure is loading in a complex manner, it will fail whenever its lowest failure stress is reached; for example, a structure with high tensile and shear strength may fail at low bending stresses if its compressive strength is poor, since as soon as the portion of the structure experiencing compression fails, structural integrity is lost.

Anisotropy

In comparison to materials fabricated by engineers, one of the hallmarks of biological materials is that they are highly oriented or *anisotropic* (their mechanical properties are strongly dependent on the orientation at which the measurements are made), unlike steel or iron, which are *isotropic* (they behave in similar ways regardless of orientation). At an intuitive level, this characteristic relates readily to what we know of a biological material widely used by humans, wood, which typically has a characteristic grain. It is far simpler to split wood along the grain than to cut through across the grain, for example. Because of this feature, it is important to note carefully the way in which tissue specimens were prepared when relating results of materials tests to the biology of the organism; tissues may easily differ in properties in longitudinal and transverse directions by an order of magnitude. In some cases these differences may relate directly to how structures are oriented within the body, with differences in strength, for instance, reflecting differences in stress magnitudes in different directions (see sections on biological tissues to follow).

Energy and Its Storage and Release

When materials deform under force, energetic changes occur. It takes energy to do work, or, as physicists and engineers view work, to exert a force over a distance. As an object is deformed over a greater and greater distance (as strain increases), more and more external energy is used to distort its molecular structure. For linearly elastic materials, the energy that is put into deforming a structure can be returned to the environment; release the applied force, and the structure will return to its original length. This concept can be demonstrated by the behavior of a rubber band. To stretch a rubber band takes energy, and the greater the strain the rubber band experiences, the more energy is used. That energy, originating from the chemical energy used to power the muscles of the person stretching the rubber band, is stored in the stretched rubber band until the rubber band is released. Then, suddenly, the energy used to strain the rubber is returned to the environment. Rubber bands and springs are, however, simply extreme examples of an energy storage-and-release phenomenon that is common to many materials. All materials can act as if they are springs; the spring is cocked by applying a force, and by virtue of the energy stored in the spring because it has been deformed (its strain energy), the spring can do work.

For a given material, the amount of energy that is stored in the material when it is deformed is given by the area under the stress-strain curve up to the stress level applied (Fig. 1.3). The greater the applied stress and resulting strain, the more energy is put into the specimen. For linearly elastic materials, the area under the stress-strain plot is a triangle, so, using some basic geometry, 1/2 × stress (the height of the triangle) × strain (the length of the base of the triangle) will give the amount of energy per unit volume of material. Energy is typically measured in Joules (J), the amount of energy it takes to exert a force of one Newton over one meter; these units make good sense:

$$\sigma \times \varepsilon = \frac{\text{force}}{\text{area}} \times \frac{\text{length}}{\text{length}} = \frac{\text{work}}{\text{volume}} \qquad (1)$$

Just as elastic modulus and failure stress correspond to our everyday notions of stiffness and strength, energy-absorbing capacity has an intuitive reference as well: Materials that absorb a lot of energy before they fracture (have a high work of fracture) are said to be tough, while those that fracture at low energy levels are brittle. Both strength and stiffness will determine a material's toughness, but in general, materials that can deform a great deal before they fail have a high work of fracture because they can absorb a great deal of energy even before they reach high stress levels.

THE MECHANICAL BEHAVIOR OF BIOLOGICAL MATERIALS

Bone

Bone is a complex tissue that is incorporated into the skeleton in a variety of forms. On the grossest level, there is a basic distinction between cortical or *compact bone* and trabecular or *cancellous bone* (Fig. 1.5). Cortical tissue is solid, dense, and compact with no visible spaces in the bone material. It makes up the shafts of the long bones and the outermost surfaces of all bones, contributing 80 percent

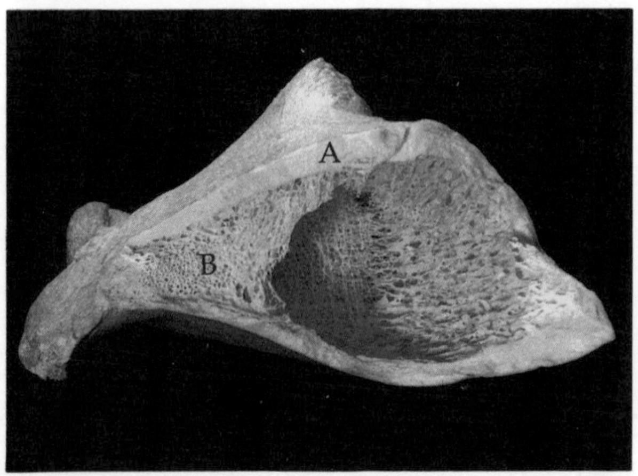

FIGURE 1.5. There are two basic kinds of bone tissue: A = compact cortical bone and B = trabecular or cancellous bone. The two kinds of tissues are clearly distinct in this proximal horse femur.

of the skeletal mass of humans (Gamble, 1987). Trabecular tissue, occurring in the epiphyses and metaphyses of long bones, the vertebral centra, and the interior of flat bones, consists of networks of bony plates and struts separated by large interconnecting spaces that are easily apparent to the naked eye. Because of its porosity, trabecular bone contributes most of the skeleton's volume despite its small fraction of the skeleton's mass. It is also characterized by a very high surface area:volume ratio, intimate proximity to bone marrow, and high vascularization.

These tissue types share fundamental similarities; they are formed by the same developmental processes, contain the same cell types, and have similar extracellular matrix protein and mineral content. They differ fundamentally, however, not only in their anatomical distribution but also in their morphology, and material behavior. Compressive strength of compact bone is about 250 MPa (1 MPa = 10^6N/m^2), and its compressive yield strain and ultimate strain are about 10,000 με and 18,000 με respectively (Gordon, 1978; Currey, 1984). The strength of bone tissue is thus far greater than concrete (≅ 4 MPa), wood (≅ 100 MPa), and glass (35–175 MPa), less than steel (400–1550 MPa), and about equal to cast iron (140–300 MPa), at much lower density and therefore a great weight savings to animals. Bone is not as strong in tension; strength is about 150 MPa (60 percent of the compressive strength), and tensile yield and ultimate strains are about 7,000 and 30,000 με (Currey, 1984).

The stiffness (elastic modulus) of cortical bone is generally reported to be between 17 and 25 GPa (1 GPa = 10^9N/m^2) (Currey, 1984). This makes bone about as stiff as wood (14 GPa), far stiffer than rubber (7×10^{-3} GPa), and far less stiff than most engineered materials (glass: 70 GPa; iron and steel: approximately 210 GPa) and some natural ones (diamond: 1200 GPa). Recent studies have demonstrated that the modulus of trabecular tissue is considerably less, probably between 3 and 5 GPa (Nordin and Frankel, 1989).

The architecture of cancellous bone is three-dimensionally complex and, for any particular site, is remarkably stereotyped, although the structural patterning differs from site to site within the body (Wolff, 1869; Lanyon, 1974; Treharne, 1981;

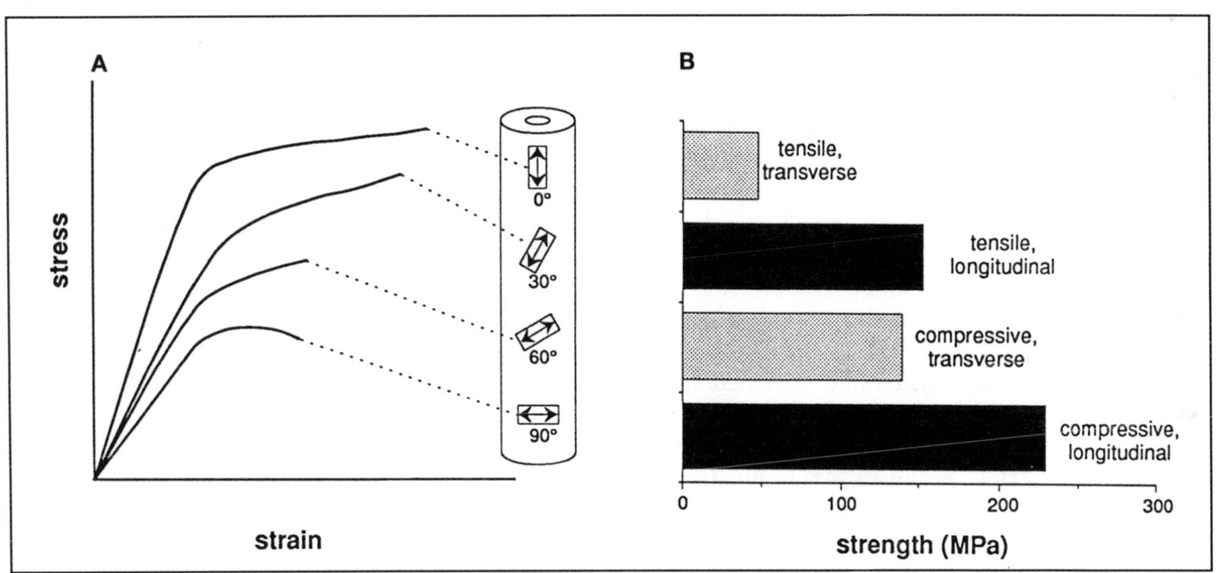

FIGURE 1.6. Bone is one example of an anisotropic material. A: Schematic stress-strain profiles of bone samples taken from a given bone but at different orientations to the long axis (adapted from Nordin, 1989). As the samples go from longitudinal (0°) to transverse (90°), the stiffness and strength decrease. B: Many materials show anisotropy in several loading modes; here, the difference in failure strength for bone at perpendicular orientations is significant for both tensile and compressive loading. Adapted from M. Nordin and V. H. Frankel: Basic Biomechanics of the Musculoskeletal System, 2nd ed. Philadelphia, Lea & Febiger, 1989. Used with permission.

Goldstein, 1987). Trabecular bone sampled from different anatomical sites differs in a variety of morphological characteristics, including porosity, predominant trabecular orientation, and the shape of individual trabeculae (Currey, 1984). Along with these structural differences, bone sampled from different sites may vary tremendously in strength and modulus. Because the trabecular bone from different parts of the body differs in density, it will, according to the results discussed previously, differ in material properties as well. Indeed, one of the most consistent and significant results of studies comparing material properties of bone from different regions has been the correlation between the bone properties and the function of the bony region tested (Goldstein, 1987). For example, the strength of trabecular bone samples taken from the non-weight-bearing upper-limb bones of humans (humerus and distal radius) is only from 0.03 to 6.3 MPa while the strength of samples from the weight-bearing proximal femur ranges from 0.2 to 16.2 MPa (Evans and King, 1961; Martens et al., 1983).

Both compact and cancellous bone are highly anisotropic, and virtually all material characteristics vary depending on orientation. This is most clearly established for the compact cortical bone of the long bones, from which most samples for materials testing have been taken. Stiffness, yield strength, and yield strain are all greatest in the longitudinal direction (along the bone's long axis) (Fig. 1.6). As the axis of test specimens to the longitudinal axis increases, stiffness decreases, reaching minimum values at 90° (Frankel and Burstein, 1970). Thus limb bones appear to have greatest strength and stiffness in the direction in which we would expect the greatest stresses. It is noteworthy that this generalization is based on samples of extremely limited taxonomic diversity and overall body size (primarily humans and cows, but see also Currey, 1990). If material properties do vary with direction of greatest stresses, there may be as yet unexplored variation in these characteristics among primates of differing patterns of mechanical usage of limbs.

Cartilage

Articular cartilage is an integral part of all the major joints of the postcranium. It plays a major role in the primary functions of the joints, allowing compressive forces to be transmitted over a large area at relatively low stress and permitting precise movements with a minimum of friction and wear. In primates, articular cartilage is typically between about 0.5 mm and 5 mm in thickness; it rarely reaches any greater thickness because it is an avascular tissue that must be nourished by diffusion of nutrients from the synovial fluid bathing the cartilage cells without direct nutrition from the blood supply. Water comprises from 60 to 80 percent of its tissue weight, and the remaining portion comprises collagen (60 percent), cartilage cells or chondrocytes (2 percent), and a gel made of large macromolecules known as proteoglycans (Mow et al., 1989) (Fig. 1.7). Collagen—the most ubiquitous of all structural proteins in the animal world—is found in the form of fibrils with great tensile strength but little ability to resist compression without simply buckling. The collagen molecules form a fibrous framework, some of which penetrates the bone surface to anchor the cartilage into its bony support. The large proteoglycan molecules draw large volumes of water into the tissue, by osmosis and by attraction of water molecules by the large number of negatively charged groups on the molecule's surface. These molecules also tend to repel one another. The result is a viscous gel with a tendency to expand because of the water that is continually attracted by the large charged molecules. This gel is prevented from fully expanding by the collagen framework; in fact, the swelling against the collagen fibers loads them in tension even when there is no external load on the cartilage. Cartilage might best be thought of, then, as a fiber-reinforced material that holds a large quantity of water and gel, where material behavior is determined both by the fibers and by the forced movements of water through a kind of macromolecule "jungle."

Because of this fine-scale organization, the mechanical behavior of cartilage is far more complicated than that of bone. If a compressive load is applied to cartilage, it will indeed deform, but if that load is left on the cartilage without changing in magnitude, the cartilage will continue to deform as water is exuded from the tissue under pressure. When tested in tension, the initial portion of the stress-strain curve has a very shallow slope, and then turns up sharply to get much steeper and quite linear. Because of this complex behavior, it is hard to define a single elastic modulus value for cartilage. Based on the more linear portions of the curve, val-

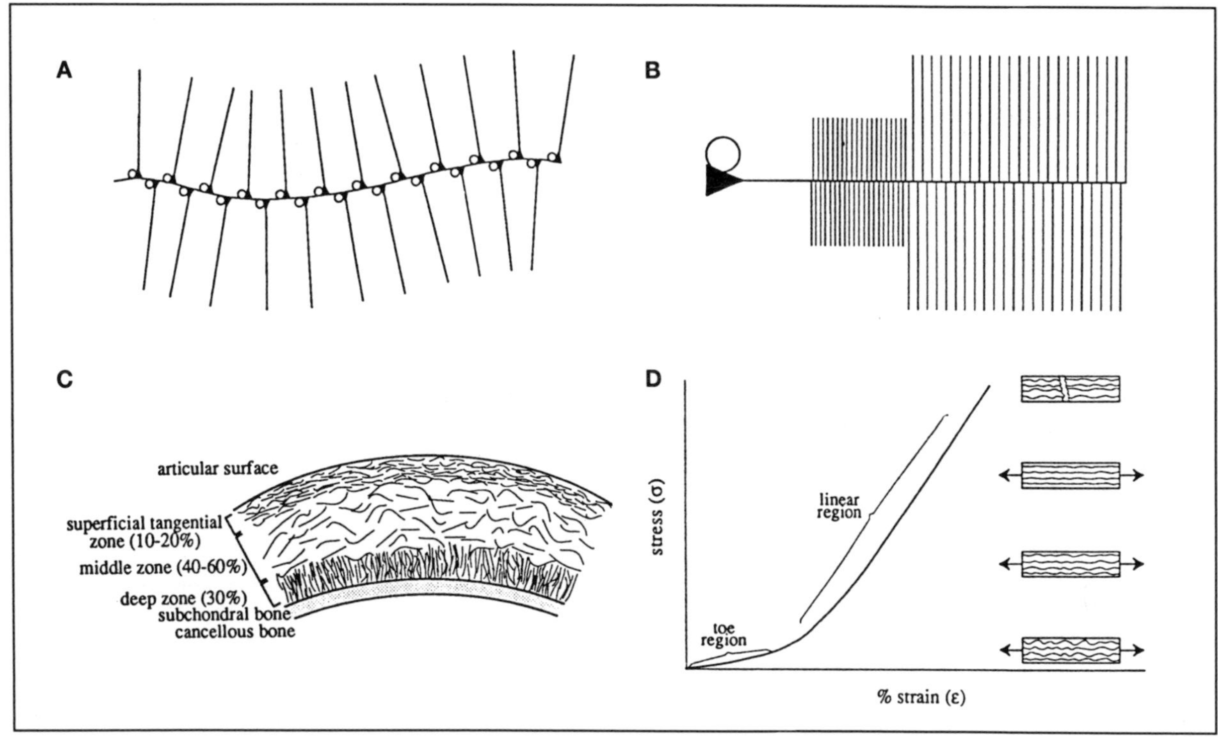

FIGURE 1.7. The main structural constituents of cartilage are large complex macromolecules. A: A typical large proteoglycan molecular aggregate is made of a long backbone of polymerized hyaluronic acid with a large number of complex protein and glycoprotein subunits extending out around it. B: Each of these subunits is itself made of smaller constituent macromolecules, primarily keratan sulfate and chondroitin sulfates, attached to their own backbone protein molecule, and linked to the major structural core by linking proteins designated here as triangles and circles. C: Along with proteoglycans, cartilage behavior is determined by the distribution and arrangement of collagen fibrils. These collagen molecules are oriented perpendicular to the bone surface at the interface between the cartilage and bone, penetrating into the bone itself to attach the joint surface to the underlying bony tissue. Superficial to this deep, dense collagenous cartilage is a large middle zone where fibers are sparse and oriented randomly. Here, the concentration of the non-collagenous proteoglycans and the water they attract is high, making this the most effective compressive load-bearing region. At the surface, the collagen increases again in density, and lies in tightly woven arrays parallel to the joint's bearing surface. D: Because of the complexity of its constituents, the mechanical behavior of cartilage differs from that of stiffer mineralized tissues. In the initial part of the stress-strain curve, the material deforms a great deal; this loading pulls on the loose, wavy collagen fibers embedded in matrix and begins to pull them straight and taut. As they straighten out, they become capable of bearing tensile stresses, and the curve takes on the characteristics of the collagen rather than the matrix. When the tensile strength of the collagen is exceeded, the cartilage ruptures. Adapted from M. Nordin and V. H. Frankel: Basic Biomechanics of the Musculoskeletal System, 2nd ed. Philadelphia, Lea & Febiger, 1989. Used with permission. (Reprinted with permission from Rosenberg et al., 1975.)

ues range from 10 to 60 MPa, from 300 to 2,000 times less stiff than bone. Similarly, cartilage fails at lower stresses than bone, with failure typically occurring at less than 100 MPa (Mow et al., 1989). The difference between the failure stress of these two tissues provides an explanation for one of the notable features of limb bones: The ends are typically expanded relative to the shafts. Hence, when a given force is applied to a bone with cartilage-covered ends, the stress (force/area) is less in the region where the cartilage sits than in the bone shaft, a cartilage-free area. In this way, bones can withstand forces that would otherwise be too great for the weak articular cartilage covering their ends.

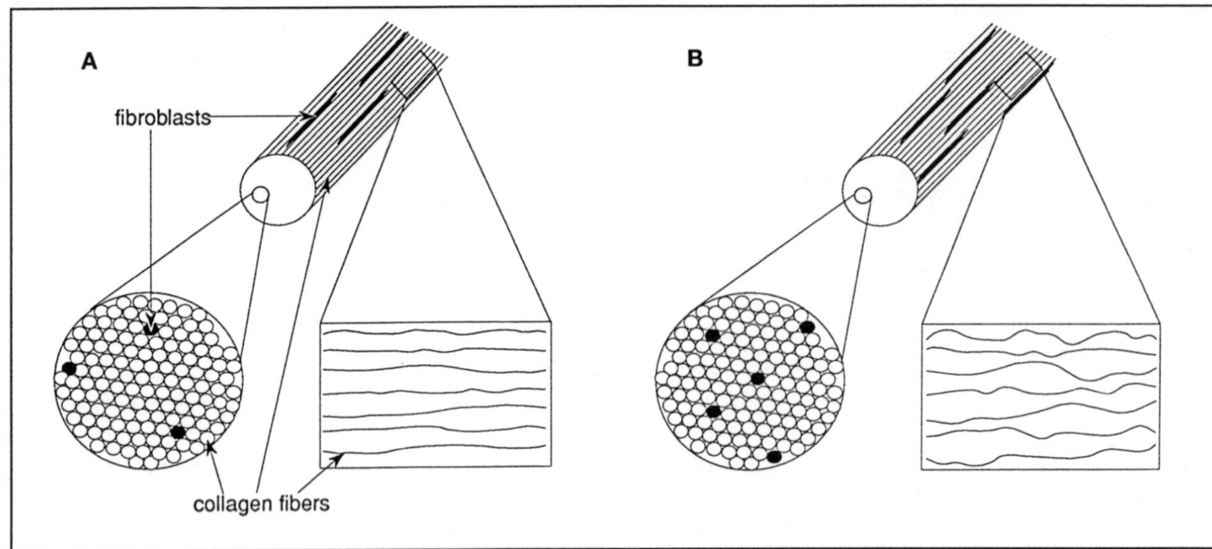

FIGURE 1.8. Tendons (A) and ligaments (B) are both complexes of connective tissue fibers (primarily collagen) and living cells (fibroblasts). Both have fibers arrayed in primarily longitudinal bundles, interspersed with cells and sometimes the elastic protein elastin. In tendons, the fibers stretch nearly their full length and are interspersed with few cells; hence they function in a manner similar to pure collagen bundles. Ligaments have a higher cell density and fibers that are wavier, often with some proportion of fibers that are oriented randomly with respect to the ligament's long axis rather than a purely longitudinal orientation. As a result, ligaments have a somewhat lower longitudinal tensile strength than tendons but are better at resisting forces applied in other directions. Tendons experience forces primarily via their own muscles; hence the predictability of loading orientation is greater. Adapted from M. Nordin and V. H. Frankel: Basic Biomechanics of the Musculoskeletal System, 2nd ed. Philadelphia, Lea & Febiger, 1989. Used with permission.

Tendons and Ligaments

Tendons, ligaments, and even skin are connective tissues made of the same basic constituent structural materials: collagen fibers, elastic fibers, and a gelatinous matrix (Fig. 1.8). The differences among these tissues in mechanical behavior stem from differences in the orientation of the fibers, differences in the properties of collagen and elastin (the elastic protein that contributes to the elastic fibers), and the relative proportion of collagen and elastin. Collagen is relatively inextensible and strong, with a failure stress of from 50 to 130 MPa, while elastin is extremely flexible and weak, failing at around 35 MPa (Carlstedt and Nordin, 1989).

In tendons, collagen is the predominant constituent and is typically oriented so that fibers are parallel to one another and to the line of attachment from muscle to bone. Tendons can, in fact, be thought of as large bundles of collagen fibers with little else of mechanical importance. This arrangement allows the collagen fibers to act efficiently in their primary task, transmitting large, unidirectional tensile loads from muscle to bone. The low compressive strength and stiffness of the collagen produces a flexible structure that can easily pass around curved and angled bone surfaces.

Ligaments, interconnecting adjacent or nearly adjacent bones, serve primarily to provide mechanical stability to joints, to help prevent dislocations, and to restrict joint motions. They have less consistent structural organization than tendons, with substantial variation from ligament to ligament, perhaps depending on the function of the ligament in question and the kind of loading the ligament is designed to resist, although few clear data on this subject have yet been collected. The orientation of collagen fibers in ligaments is approximately parallel but is not nearly as regular as the packing of collagen in tendons. Ligaments vary in their elastin contribution, with some ligaments containing relatively little elastin and some, especially ligaments associated with the vertebral column, containing twice as much elastin as collagen (Nachemson and Evans, 1968). Ligaments are deployed around joints in ways that subject them to tensile loads that may be predominantly in one direction, but also to loads in other orientations. This seems to parallel their less uniform collagen fiber organization. Ligaments, like tendons, are pliant, but have enough oriented collagen to offer resistance to distorting forces applied near joints.

For students of the primate postcranium, there are a number of particularly interesting aspects of tendon design. There are two important factors that influence the stress imposed on a tendon during locomotor activity: (1) the force of contraction of the muscle to which the tendon is attached and (2) the size of the tendon in relation to the size of the muscle. As a muscle contracts, the stress in the tendon will increase; anything else that stretches the tendon, like gravity causing a joint to rotate, will also increase the tendon stress. The power of contraction of the muscle, in turn, depends on the number of muscle fibers contracting; the muscle's cross-sectional area will therefore be a good estimate of its force-generating ability. Stress at failure is about twice as high in tendons as in muscle tissue, hence when the muscle reaches failure stress, even if its cross-sectional area is only two times that of its tendon, the muscle rather than the tendon will fail and the tendon will not be pushed to exceed its performance capacity.

Tendons can also play an important role in the overall energy budget of locomotion because of their material properties. In some cases, gravity or other external forces acting on the limbs or a part of a limb will cause a tendon to be stretched; for example, when you land on the ground after a jump or a step, your ankle dorsiflexes under the influence of gravity acting on body weight, and your Achilles' tendon at the back of the ankle is stretched. This movement has thereby cocked the tendinous spring, with gravity supplying the necessary energy that is

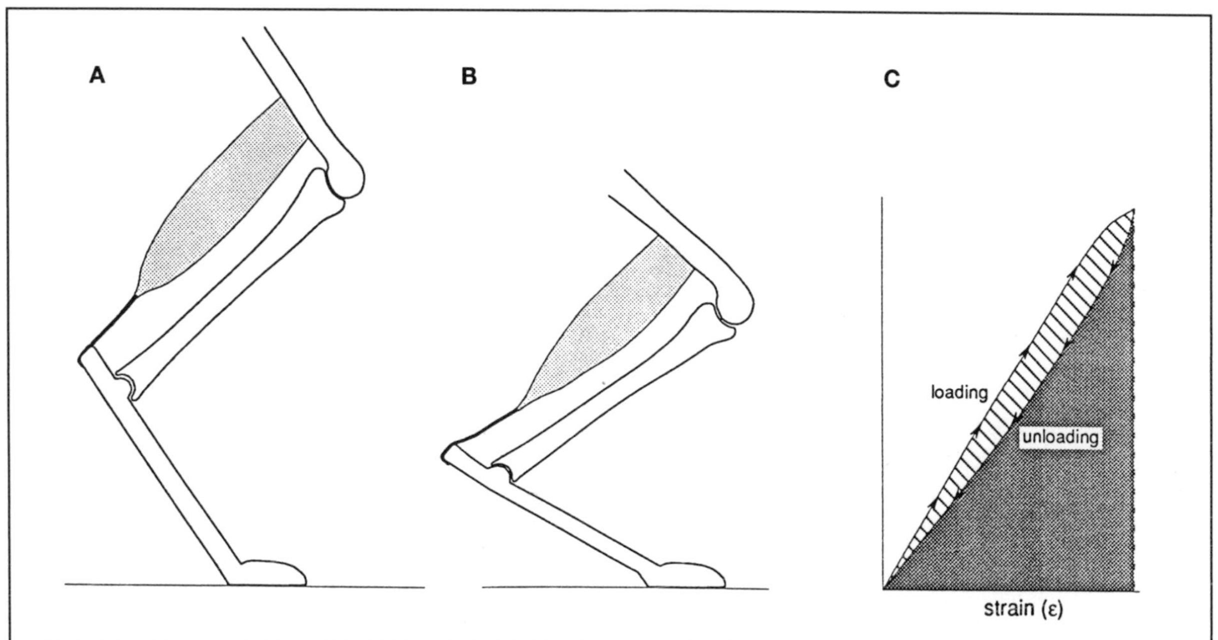

FIGURE 1.9. Elastic energy is stored in tendons during some locomotor movements. As a joint flexes under the force of gravity, tendons can go from a relaxed (**A**) to a stretched (**B**) configuration. This process of stretching the tendon stores energy within it, and the energy is then released to the animal as its joints return to their starting positions. If the tendon material displays a stress-strain curve, as in C, where the energy stored in the tendon during loading or deformation (shaded area plus hatched area) is identical to the energy released when the force is removed or unloading occurs (shaded area alone), then the tendon can be an efficient source of energy for locomotion.

now available to be returned to the working limb. This tendon recoil helps move the ankle joint in the opposite direction as your body moves into the next part of the locomotor cycle (Fig. 1.9). This kind of passive stretching of tendons without expending muscular energy, followed by elastic recoil, can make a major contribution to the metabolic cost of locomotion; in kangaroos it has been estimated that from 40 to 55 percent of the total energy cost for hopping is contributed by this mechanism (Alexander and Vernon, 1975). There is certainly a substantial contribution in other animals during walking and running as well (see Alexander, 1988, for discussions of the role of elastic mechanisms in the locomotion of a variety of animals). How important tendon strain energy may be depends in part on the thickness of the tendon; thick tendons experience low stresses for a given force magnitude (since σ = force/cross-sectional area), hence they deform little and store little energy. The importance of this effect has yet to be studied specifically within primates, but as new information is gathered, we

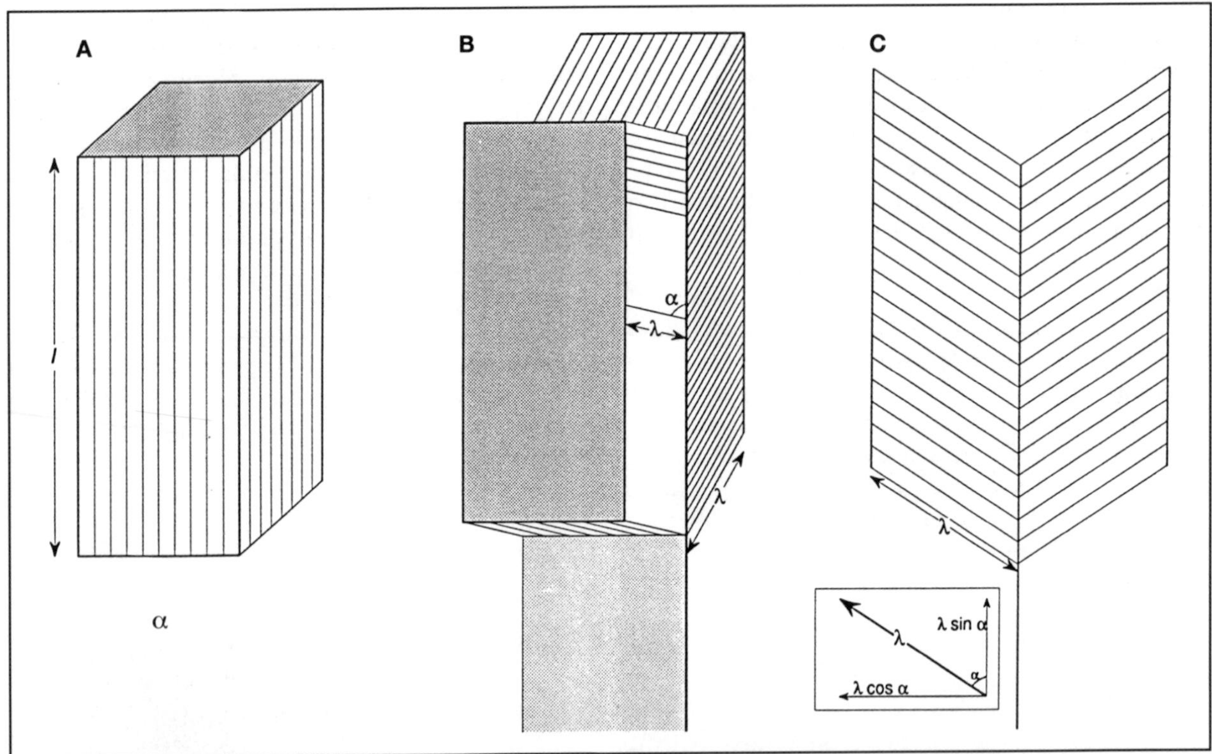

FIGURE 1.10. There are two basic types of muscle fiber architecture. Fibers may be arrayed in parallel (**A**) or may be oriented at an angle (designated here as α) to the line of action of the muscle (**B**). This schematic illustrates several key differences in these types of organization; fibers are typically much longer in parallel-fibered muscles (fiber length *l*), while those in pinnate-fibered muscles are short (fiber length λ). The direction of pull of each fiber in muscle **A** is similar and aligned with the line of action of the muscle. In **B**, fibers converge to a tendon at an angle and therefore generate force in a different direction than the line of action of the entire muscle. This is illustrated in two dimensions in **C**. Each fiber of length λ will generate a force along its axis at angle α to the tendon. This force (λ in the boxed inset) can be resolved into its components parallel and perpendicular to the line of action. The force component that will produce useful shortening is $\lambda \sin \alpha$, which is always less than λ. There is also a force component perpendicular to the line of action, $\lambda \cos \alpha$, which must be resisted; in bipinnate muscles like the one illustrated here, fibers on opposite sides of the central tendon will generate equal but opposite forces that will cancel one another out. Although the fibers in pinnate-fibered muscles are short, muscle force is a function of cross-sectional area of fibers and not length; hence the greatly increased number of fibers per unit volume of pinnate-fibered muscles allows them to generate large forces. Their total shortening distance is, however, much less than that of a parallel-fibered muscle.

will almost certainly find that tendon elastic mechanisms play an important role in many aspects of primate locomotion.

Muscle

Muscle is a fundamental component of the limb's mechanical operating system. By virtue of the key features of muscle cells, possession of force-generating contractile proteins and excitable membranes interconnected with the nervous system, they can, upon receiving a nerve signal, shorten up to about 30 percent, thereby generating force to produce and control movements. The biology of muscle is an enormous subject in and of itself; readers are encouraged to seek additional information (basic sources include McMahon, 1984; Goslow, 1985; Pitman and Peterson, 1989). Here we will treat the molecular contractile machinery of muscle as a black box, and consider larger-scale features of skeletal muscles that may have a large effect on the biomechanics of the limbs in locomotion.

Muscles are made up of a number of individual muscle cells or fibers. A single discrete muscle may have only a few fibers if it is a small muscle in a small animal, or it may be composed of many thousands of these fibers. The fibers are bundled together in a variety of geometries by the surrounding connective tissues linking the force-producing fibers into bone via tendons. The magnitude of the force a muscle can generate with these fibers is determined by both the number of fibers firing and the orientation of these fibers to the line of action of the muscle. The number of fibers acting together, or the muscle's *fiber cross-sectional area,* determines the total force-generating capacity. If all fibers are inline with the muscle's attachment points, all the force these fibers generate will go directly to producing rotation at the relevant joint (see discussion of lever mechanics). If the fibers are at an angle, as in pinnately fibered muscles, the useful force will be only a fraction of the net force generated, depending on the precise orientation (Fig. 1.10). It is important to recognize that the length of a muscle has no effect on the magnitude of force it can generate, although fiber length will determine the possible range of excursion of a bony attachment, given that muscles can only shorten to a certain percentage of their resting length. Similarly, because fiber cross-sectional area is the key to force-generating capacity, a muscle's volume or weight is not a good indicator of its strength. Because the capacity of a fiber to produce force that is useful for a particular movement depends strongly on its orientation relative to its bony attachments, the architecture of muscle fibers within muscles has a major effect on the biomechanics of limbs (see also Gans and Bock, 1965; Gans and Vree, 1987).

Different populations of muscle fibers vary in their biochemical properties, the magnitude of the force they can produce, the rate at which they shorten, and their capacity to resist fatigue. Often, three basic types of muscle fibers are distinguished, and they can be designated as fast-contracting fast-fatiguing type and fast-contracting fatigue-resistant type and a slow-contracting fatigue-resistant type. The maximum force generated by a fast-contracting (or fast-twitch) fast-fatiguing fiber is greater than that of a fast-twitch fatigue-resistant fiber, which is in turn greater than the greatest force exerted by a slow-twitch fiber. Along with the differences among fiber types in force production and contraction speed, these fiber types tend to be found in different muscles. As a simplified generalization, locomotory muscles can have a high proportion of fast-twitch fibers, while postural muscles have a preponderance of slow fibers. There may also be a layering of fiber types within a single muscle, with slow fibers on the muscle's deep surface and fast fibers more superficial. Many aspects of this functional and biochemical differentiation among fiber types can be quite complex, and this complexity may prove critical to our understanding of structural design of locomotor systems; good sources for further information include Stuart and Enoka, 1983; Goslow, 1985; Squire, 1986; and Pitman and Peterson, 1989.

Adaptive Variation

As biomechanical approaches increase in importance in the study of structural and functional design of limbs, data concerning differences among taxa in material characteristics are beginning to appear. This is most true for bone; little variation has yet been found in the mechanical properties of tendons from various mammals (Bennett et al., 1986). For bone, however, there appears to be substantial variation in mechanical properties among anatomical locations and taxa. Most notably, Currey has compared the bone making up deer antlers, whale

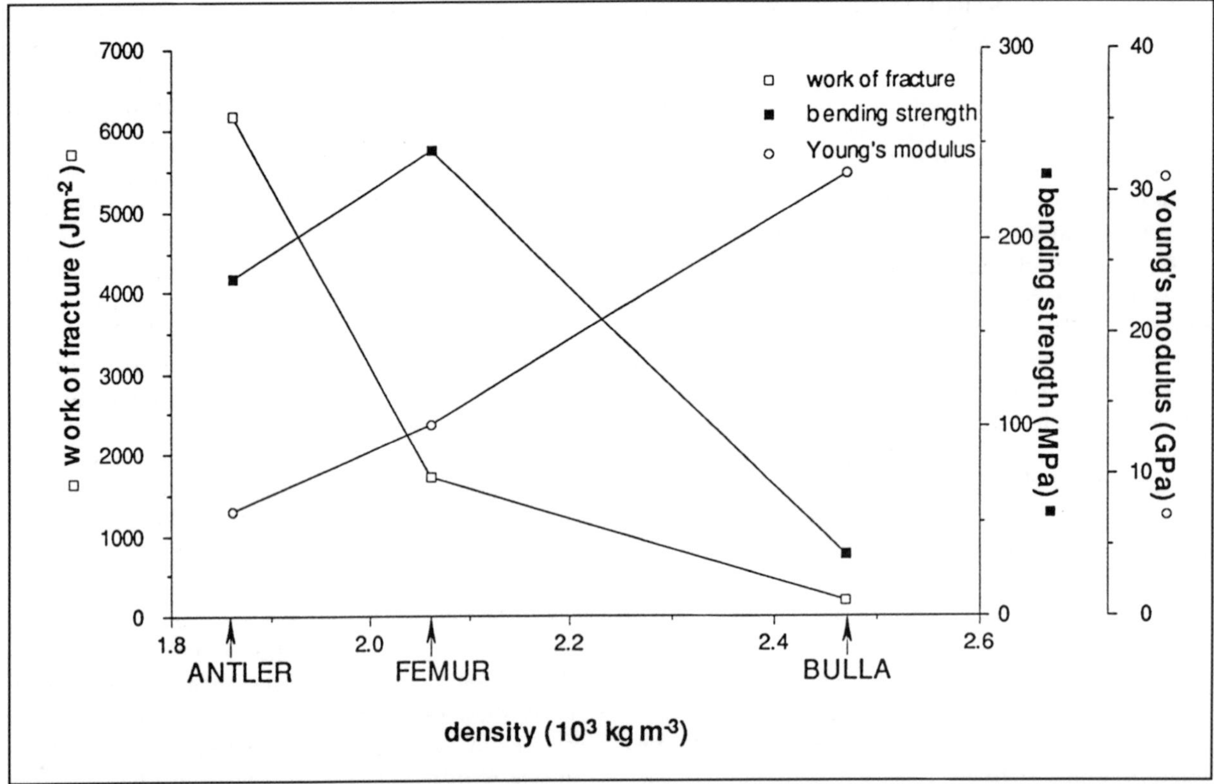

FIGURE 1.11. Mechanical characteristics of three distinct functional bone types. Bone mineral density is plotted along the x axis, with antler showing the lowest density, femur an intermediate density, and tympanic bulla the highest density of mineralization. Work of fracture (toughness, left axis) is indicated by open squares, maximum stress at failure in bending is indicated by closed squares (right axis), and Young's modulus is indicated by open circles (stiffness, far-right axis). Adapted from J. D. Currey (1984) with permission.

tympanic bullae, and cow femora. He found that the functional differentiation among these bones (absorbing impacts in male-male fights, reflecting sound vibrations from the external environment and thus isolating the auditory apparatus, and supporting locomotor forces) is mirrored in their material properties (Currey, 1984) (Fig. 1.11). He found that femora had the greatest strength, with reasonably high stiffness and toughness. Antler, on the other hand, does not support the body and limbs during posture and locomotion but must be able to withstand large impacts without fracturing; this tissue is less strong than limb bone but shows a substantially lower stiffness and hence a greatly enhanced capacity for absorbing energy upon impact. Bones of the ear region, like antler cores, need not withstand much stress but must have a high acoustic impedance to function well; in this case, bone samples showed very high mineral content and thus high stiffness and impedance, but low strength and toughness. Although the taxonomic diversity in primate locomotor and postural behavior must certainly impose a wide range of mechanical loading environments on the postcranium, adaptive variation in material characteristics of primate bone has yet to be assessed. Body size may well be a factor, with mechanical requirements for skeletal composition differing between very large and very small primates. Future studies could make a valuable contribution to our understanding of primate limbs by addressing some of these questions.

STRUCTURES

Size

The size of an organism has profound effects on all aspects of its biology. Both the behavioral repertoires of primates and the structural designs of primate limbs are influenced by body size in a variety of ways. From a biomechanical viewpoint, the forces that are imposed on portions of the body are often dictated primarily by body size, as is the ability of a structure to accommodate the forces it experiences. The relationship between size and shape, particularly of skeletal parts, has been a primary focus of morphology in general and of primatology in particular over the last ten years or so; further discussions of this subject can be found in Calder (1984), McMahon (1984), Schmidt-Nielsen (1984), Jungers (1985b), LaBarbera (1989), Swartz and Biewener (1992). Fleagle (1985) is an especially good introduction for students of primate biology.

For example, some aspects of a structure's ability to carry out mechanical tasks are direct functions of its cross-sectional area. Load support is one such task. To illustrate this, consider a cube-shaped element, analogous to a bone, that supports its own weight. The task, weight support, is a function of the element's weight or volume, or its length × width × height. Weight is therefore proportional to length cubed, while the ability to perform the task is a function of cross-sectional area (length × width) and is proportional to linear dimensions squared. The effectiveness of the structure can be characterized in a simplified way as having to do with the ratio of "ability" to "need," or cross-sectional area divided by volume, proportional to the ratio of length cubed to length squared. If the structure increases in total size by doubling each linear dimension, as might happen as young primates grow larger or when selection increases body size within a lineage over successive generations, cross-sectional area will increase by a factor of 2^2 or 4 while volume increases by a factor of 2^3 or 8 (Fig. 1.12). The structure's "ability-to-need" effectiveness has therefore decreased from one to one-half, a decrease of a factor of two, equivalent to the amount of size increase; if the structure had tripled in size, area would have gone up 9-fold, the volume 27-fold, and the ratio between the two would be one-third the original value.

It is clear from this example that we can express the changes in the size of some important aspect of organismal shape in terms of changes in other variables, especially overall size. In a size series of identically shaped structures, dimensions of different kinds will change with shape in a completely predictable way that can be specified a priori on the basis of geometry alone. Length will change in proportion to the cube root of volume or volume$^{0.33}$; cross-sectional area changes in proportion to the square of length or the two-thirds power of volume (volume$^{0.67}$). Similar rules can be constructed for other basic shape parameters and variables that can be expressed as combinations of simple lengths, times, and masses (see also Vogel, 1988).

These simple rules based on idealized shapes can form a series of baseline predictions for structural analysis. If length changes in proportion to volume$^{0.33}$ for similarly shaped structures, and mass changes in proportion to volume$^{1.00}$, then linear dimensions of primates should change in proportion to body mass$^{0.33}$ if the small and large animals have similar shapes. The match of this kind of hypothesis to the real world can be tested with linear regression analysis of logarithmically transformed data (Cock, 1966; Sokal and Rohlf, 1981; Sprugel, 1983; McArdle, 1988; LaBarbera, 1989). When organismal structure conforms to predictions based on maintaining the same shape over a range of sizes, the patterning of shape in relationship to size is said to be *isometric;* if a structure increases or decreases disproportionately with size it is said to show *positive* or *negative allometry* respectively (Fig. 1.13). Systematic differences from the null hypothesis of an isometric scaling relationship—either in terms of positive or negative allometry or when individual data points (representing particular taxa or individuals of a particular age or other class) deviate substantially from patterns observed in a wider group—provide clues that shape may be related to size in an interesting way.

Even when shape remains similar with respect to size (isometric scaling), animal limbs may function in different ways at different sizes. In the cube example, cross-sectional area decreases relative to volume when shape is constant, so maintaining similar function would entail disproportionate increases in cross-sectional area relative to volume; conversely, maintaining a constant cross-sectional area to volume ratio with increasing size requires a shape change where each solid becomes stockier with increasing size (Fig. 1.12). Setting up alternative hypotheses in scaling analysis therefore requires clear articulation of the predictions that form the basis for each hypothesis, whether they be constancy of shape or constancy of a particular function, and so on. Developing hypotheses regarding the relationship between shape and a particular function, in turn, requires a careful determination of which parameters of structural shape most directly determine the critical relevant aspects of mechanical function.

The most important aspects of organismal size for biomechanical analysis of the limbs concern the changing relationships between the forces exerted on parts of the body and the ability of structures to resist those forces (or, in the case of muscles, to generate those forces) with changes in body size. This

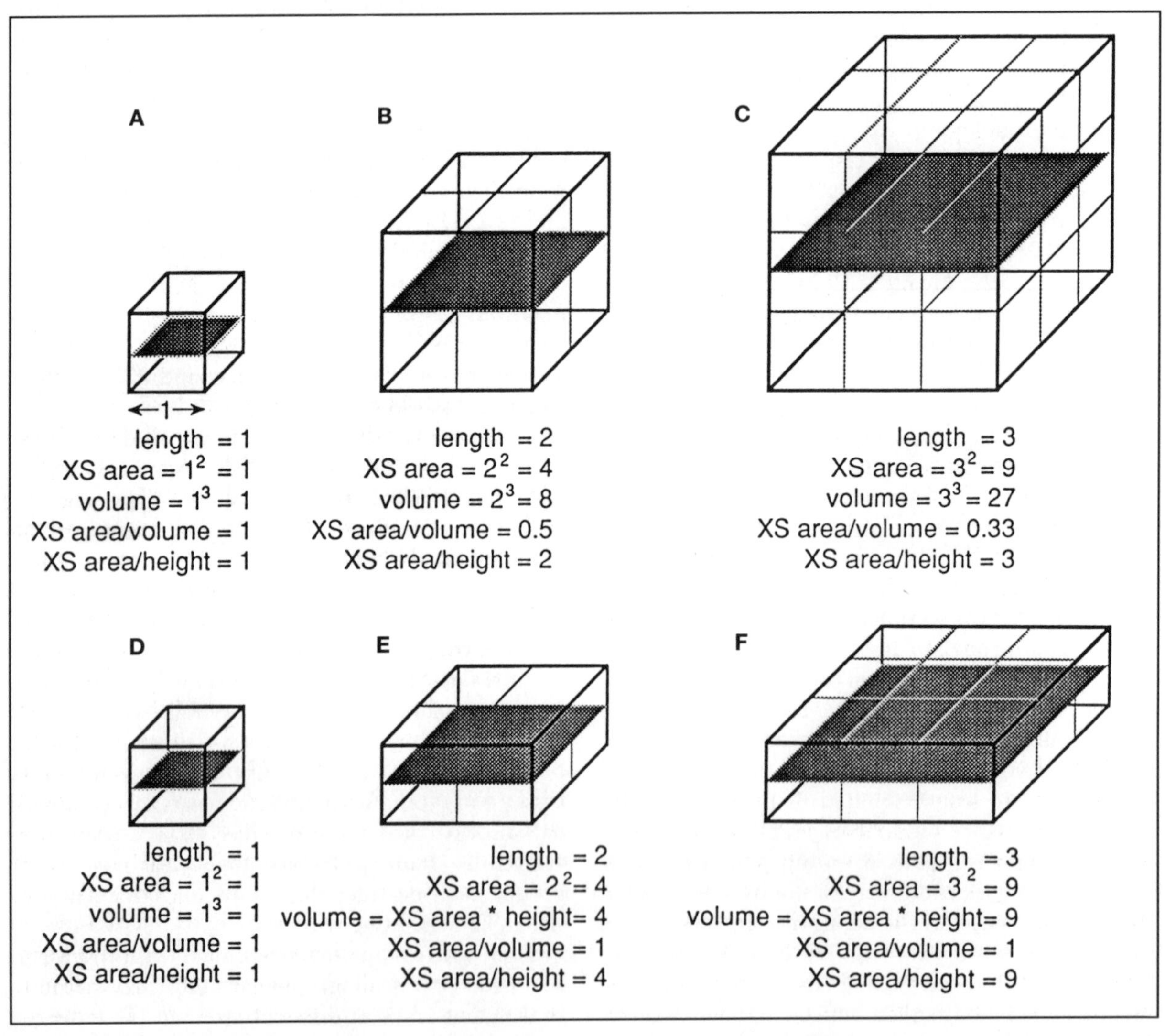

FIGURE 1.12. Schematic illustration of changing relationships among linear dimensions, surface areas, and volumes with size. **A–C**: As a cube triples in linear dimensions, area increases 9-fold and volume increases 27-fold, dramatically changing the ratio of area to volume and area to height. Functionally similar changes will occur in organisms of similar shape even when body shape is more complex than a simple cube. **D–F**: If the relationship between cross-sectional area (ability to support load) and volume (load) is kept constant as a solid increases 3-fold in one linear dimension (length), then overall shape changes dramatically in the size series. Adapted from A. A. Biewener, ed., Biomechanics: A Practical Approach. Vol. 2. Structures. Copyright 1992 by permission of the Oxford University Press.

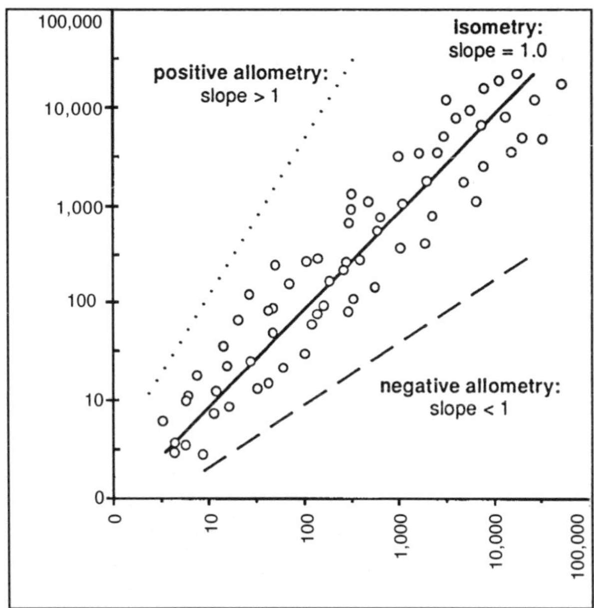

FIGURE 1.13. Log-log plot of data representing isometric scaling. If the two variables being compared are of the same dimensionality (both linear, areal, and so on), then the slope of the regression will be 1.00 when the structures are isometric. If the y variable increases more rapidly than the x variable, the slope is greater than 1 and y is said to display positive allometry with respect to x. Adapted from A. A. Biewener, ed., Biomechanics: A Practical Approach. Vol. 2. Structures. Copyright 1992 by permission of the Oxford University Press.

relationship is altered by virtue of simple geometric principles. Forces exerted on the postcranium typically arise from gravity acting on the mass of the body or some part of the body, or from their accelerations. Forces will therefore change in proportion to some function of body mass. The ability of a structure to resist a force is most often a function of the structure's cross-sectional area rather than its mass. Areas (products of two linear measurements, such as length and width) and masses or volumes (products of three linear measurements, that is, length, width, and height) will change in relative value with changes in size unless radical shape alterations occur. Throughout the next section, the relationship between structural shape and stress-bearing ability will be discussed within the context of the changes in these parameters with overall size.

Shape

A variety of aspects of the shape of biological entities may be important in biomechanical analysis; the performance of an anatomical structure will be determined by the interaction of material properties and the way the material is deployed in space. Which shape variables are most important depends on the nature of the behavior examined and of the question being asked; different aspects of limb shape will have various effects on performance, biomechanical or otherwise. Here, the focus of discussion will be structural features of greatest functional import for overall support and movement of the postcranium and we will examine the variables of broadest application in biomechanical analysis. More in-depth treatments can be found in a variety of sources, including Wainwright et al. (1976), Gordon (1978), Alexander (1981b), Vogel (1988), and Wainwright (1988).

Linear Measurements Simple linear measurements contain important information for some modes of analysis. Indeed, much of our understanding of the mechanics of movement and its anatomical substrate in primates comes from analyses of linear dimensions of limb segments or individual bones. Linear dimensions may convey important information about the relative proportions of an organism, and these proportions, in turn, may contain functionally relevant information; knowing that humans have exceedingly long hind limbs is a clue that these structures play a particularly important role in human locomotion. In addition, most allometric analysis of the primate postcranium (e.g., Shea, 1981; Aiello, 1984; Jungers, 1985a; Jungers, 1988; Demes and Günther, 1989) have also focused primarily on linear measurements. In the context of scaling analysis, we can look at the linear dimensions of skeletal elements or body segments with respect to body size and discover both the overall pattern of change with size and the patterning, if any, of deviations from the general pattern. In terms of scaling, if there are no shape changes with respect to size in a given sample, we can expect that linear dimensions will increase in proportion to the cube root of body size. Deviations from this prediction can tell us whether a particular element or part of an element increases or decreases in size disproportionately during growth or over evolutionary time, and comparisons of scaling patterns of lengths and diameters can depict changes in different aspects of an element's shape with size.

The functional meaning of many linear dimensions of the limbs can best be made clear and explicit in terms of lever mechanics and simplified engineering analysis; the apparatus primates use for locomotion basically comprises interacting lever systems. Placing the analysis of structural shape in the realm of lever mechanics is also helpful in clearly distinguishing mechanically important from less important dimensions, and in providing a framework for understanding the functional significance of particular dimensions.

The rigid structural supports of animal lever systems experience forces that cause them to rotate with respect to one another at joints. These forces may be exerted by muscles directly, may be exerted by gravity, or may arise from the accelerations of all or part of the animal. For illustration, consider a muscle exerting a force (F_1), sometimes termed the in-force (or F_i), on a bone near a proximal joint and thereby resulting in the exertion of a force (F_2), sometimes termed the load or out-force (F_o), at the distal end of the bone as it rotates at the joint or pivot (Fig. 1.14A). The effectiveness of force F_1 in producing rotation is measured by its *moment* or torque about the axis of rotation; torque is the force applied multiplied by the perpendicular distance between the pivot or center of rotation (the joint) and the line of action of the force (known as the *moment arm* or *lever arm*, here abbreviated l_1). At equilibrium, the moments of these two forces, M_1 (= $F_1 \times l_1$) is then equal to M_2 (= $F_2 \times l_2$), hence $F_1 l_1 = F_2 l_2$ (since $M_1 = M_2$), and thus F_2, the useful force output of the system, = $F_1 (l_1 / l_2)$. The moment arms of the respective forces, simple linear measures, determine the relative magnitude of the moments. This relationship holds regardless of the location of the pivot with respect to the points of application of the forces, that is, whether the anatomical system puts the joint between the muscle attachment and the out-force (a first-class lever system), the out-force between the joint and the muscle attachment (second-class lever), or the muscle in the middle (third-class lever) (Fig. 1.15). For velocities, the lever-arm relationship is reversed: $v_1 l_2 = v_2 l_1$, so v_2, the velocity of the moving lever, = $v_1 (l_2/l_1)$ (Fig. 1.14B).

If we wish to measure the effectiveness of a particular muscle in exerting a given force or eliciting a movement of a given velocity, we need to measure the lever arms of both the muscle force and the out-force (the force delivered by the skeletal element when the muscle contracts) as well as knowing something about the force and velocity of contraction of the muscle; the same applies to external forces as well. Similarly, if we wish to estimate how the force exerted at a particular anatomical location changes with body size, we might wish to examine the scaling of the lever arms as much as studying the changes in muscle size and architecture. Importantly, the length of a lever arm is often not a simple distance between two clear anatomical landmarks, and it is strongly dependent on joint angle. That is, lever mechanical relationships change dynamically throughout a motion, so understanding limb design might require analyzing lever mechanics in a variety of postures or points in the locomotor cycle. In turn, this implies that careful attention must be paid to the issue of when in the cycle the critical activities are occurring or when the largest forces are generated (Fig. 1.14C). When independent information suggests that forces or velocities are greatest at a particular position, measurements should be made for this lever configuration.

These principles of lever analysis can exemplify an important part of the selection of functionally relevant physical dimensions: in biomechanical analysis, we usually wish to know the length of an element or part of an element not just to help describe its shape but, more importantly, because of the significance of these measures as parts of functioning lever systems. Traditional morphometric analyses often describe skeletal shape using a somewhat arbitrary set of linear dimensions defined by convention. They often do not precisely match the critical lever arms, although they may be related to them in some cases; total bone length is highly correlated with the distance from joint to joint, and lengths of muscle attachment processes may be correlated with muscle lever arms.

Surface Areas, Cross-sectional Areas, Moments of Inertia, and Beam Analysis Although simple lever analysis often pays little attention to the pivot between the rigid skeletal levers, the size and shape of joint surfaces may also hold important biomechanical information. Joints must resist the forces transmitted through adjacent elements, while simultaneously permitting and controlling movement. The

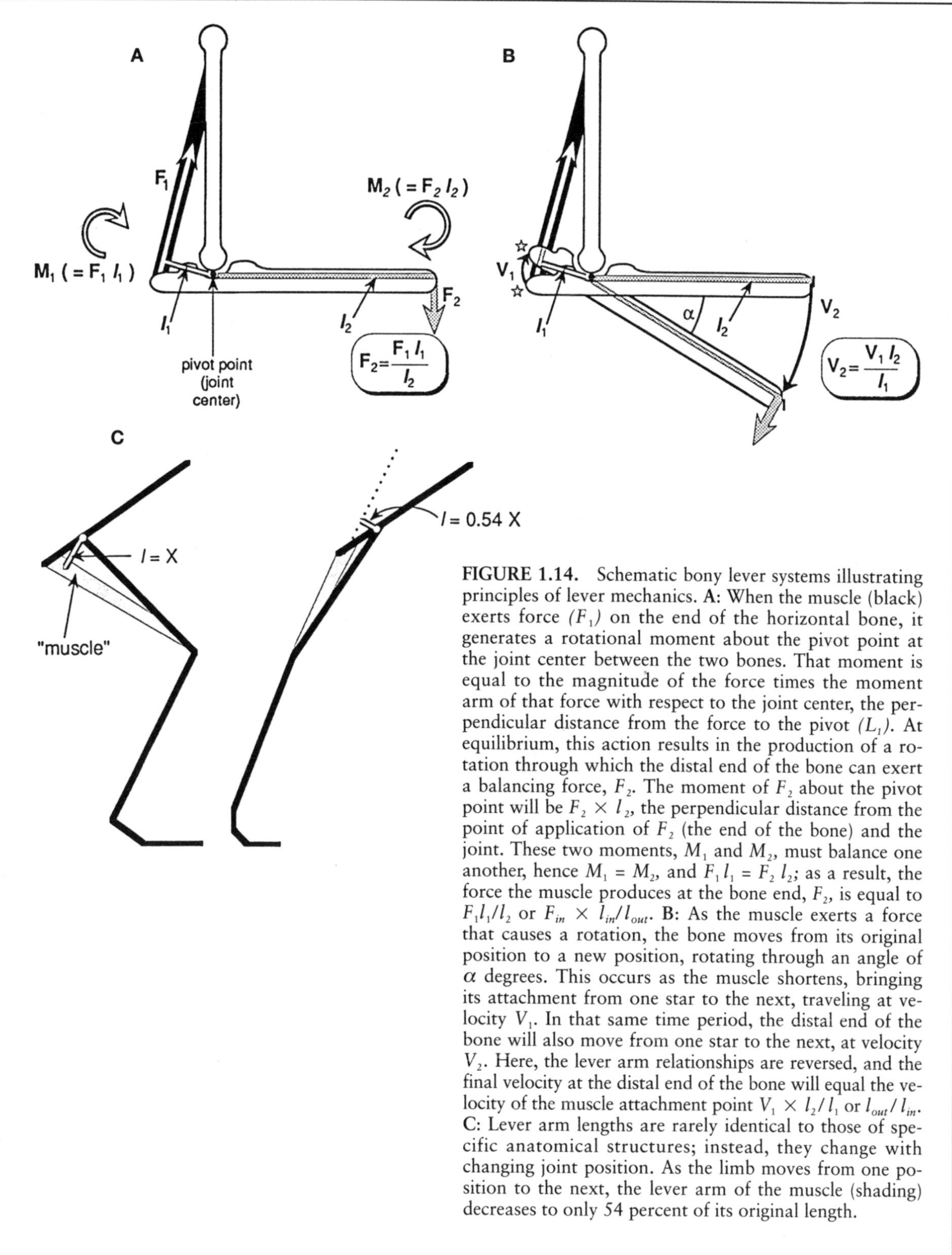

FIGURE 1.14. Schematic bony lever systems illustrating principles of lever mechanics. **A:** When the muscle (black) exerts force (F_1) on the end of the horizontal bone, it generates a rotational moment about the pivot point at the joint center between the two bones. That moment is equal to the magnitude of the force times the moment arm of that force with respect to the joint center, the perpendicular distance from the force to the pivot (L_1). At equilibrium, this action results in the production of a rotation through which the distal end of the bone can exert a balancing force, F_2. The moment of F_2 about the pivot point will be $F_2 \times l_2$, the perpendicular distance from the point of application of F_2 (the end of the bone) and the joint. These two moments, M_1 and M_2, must balance one another, hence $M_1 = M_2$, and $F_1 l_1 = F_2 l_2$; as a result, the force the muscle produces at the bone end, F_2, is equal to $F_1 l_1 / l_2$ or $F_{in} \times l_{in} / l_{out}$. **B:** As the muscle exerts a force that causes a rotation, the bone moves from its original position to a new position, rotating through an angle of α degrees. This occurs as the muscle shortens, bringing its attachment from one star to the next, traveling at velocity V_1. In that same time period, the distal end of the bone will also move from one star to the next, at velocity V_2. Here, the lever arm relationships are reversed, and the final velocity at the distal end of the bone will equal the velocity of the muscle attachment point $V_1 \times l_2 / l_1$ or l_{out} / l_{in}. **C:** Lever arm lengths are rarely identical to those of specific anatomical structures; instead, they change with changing joint position. As the limb moves from one position to the next, the lever arm of the muscle (shading) decreases to only 54 percent of its original length.

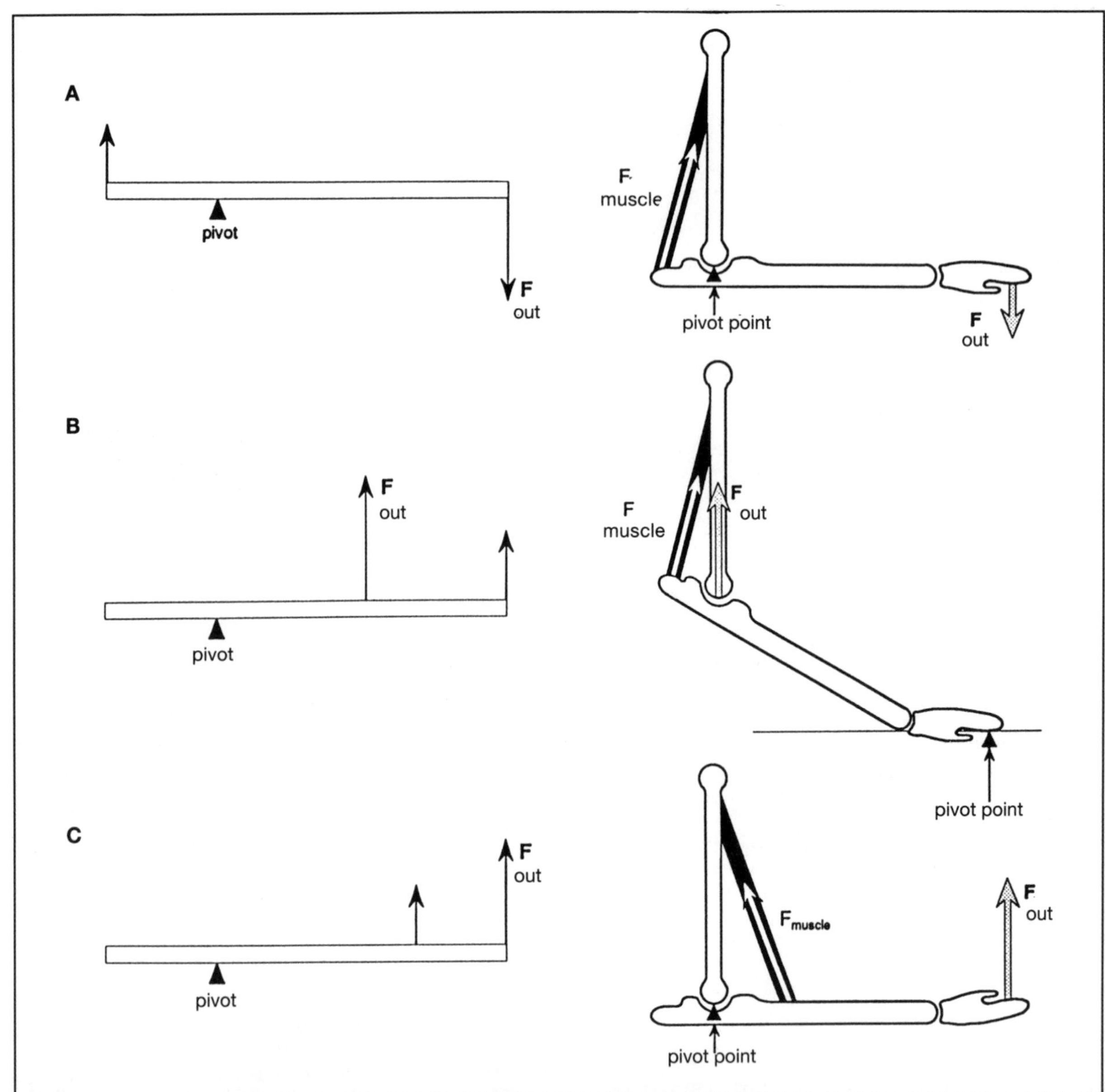

FIGURE 1.15. Lever mechanical analysis can be applied to bone/muscle systems regardless of the geometrical relationships between the force applied by muscles, the resulting force output, and the pivot or joint center. Here, the three possible relationships are illustrated for a schematic forelimb. Adapted from M. Hildebrand: Analysis of Vertebrate Structure, 2nd ed. John Wiley and Sons, 1982. Reprinted by permission of John Wiley & Sons, Inc. © 1982.

absolute surface area of a joint measures the material over which joint forces will be distributed, as well as providing an outside limit on the range of motion possible, a characteristic that is also critically determined by three-dimensional topology.

There are other biomechanically important structural shape characteristics that relate directly to their mechanical performance. The optimum shape for an engineering structure subjected to a simple loading mode (pure tension, compression, bending,

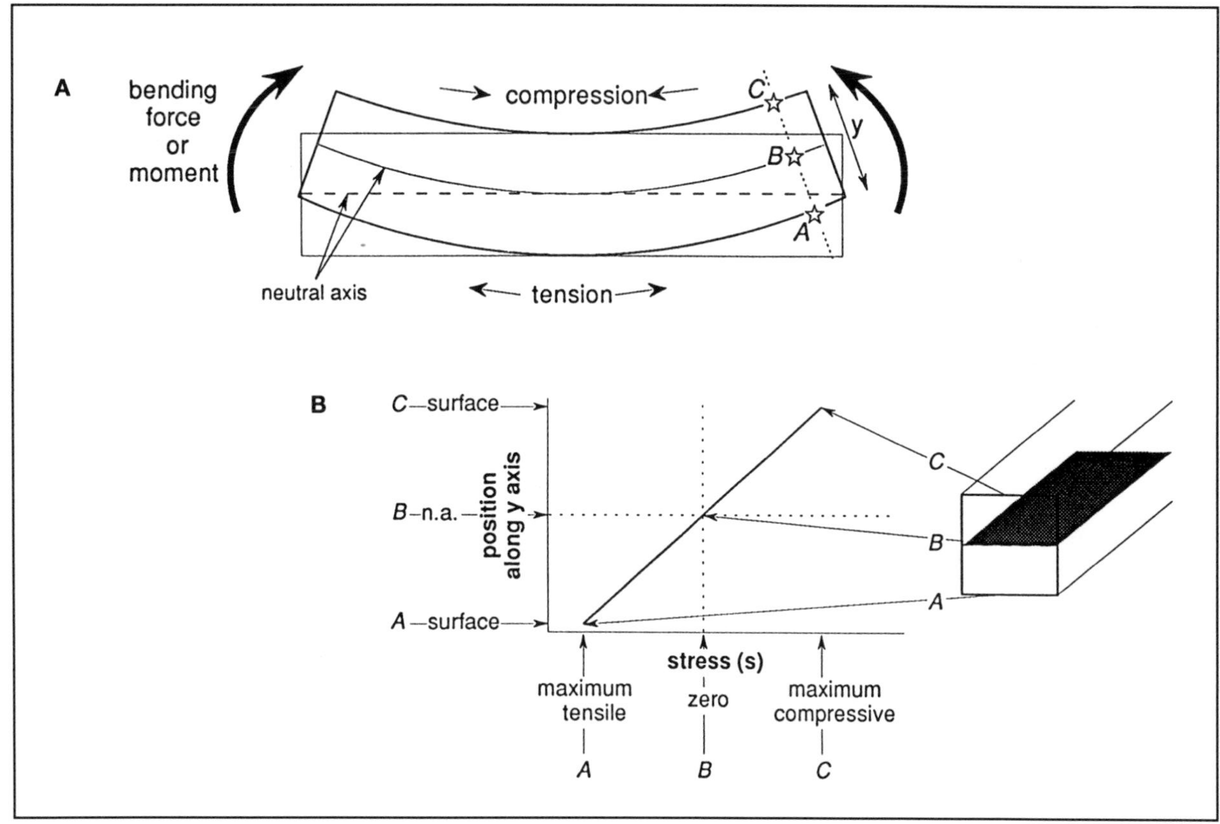

FIGURE 1.16. A: When an original prism-shaped beam is subjected to bending, the resulting deformation tends to curve the beam, placing one surface of the bone in compression and the opposite surface in tension. Between these two surfaces, there is a plane where the bending stresses are zero, the so-called neutral plane or neutral axis of bending. At three points along a y axis running from one surface of the beam to the other, points *A*, *B*, and C sit on the maximum-tension, neutral, and maximum-compression planes respectively. **B**: A plot of the changes in beam stress with changing position along the y axis. Stress in a beam-like structure varies continuously throughout the cross section, with highest stresses at the outermost surfaces of the structure (within the plane of bending).

or shear) can be derived from basic principles of structural design. Although biological structures typically experience complex, ever-changing loading regimes, simplified beam models can often provide a first approximation of the relationship between structural design and mechanical performance (see Wainwright et al., 1976; Vogel, 1988; Wainwright, 1988; and Biewener, 1992, for more general treatments). The ability of a structure to meet the requirements for performing important behaviors clearly depends strongly on its material composition, but also on its size and shape.

Adequate design to meet these requirements varies according to the loading regime to which an element is subjected. In the case of pure tension, both the stress developed in a beamlike structure and its deformation are dictated solely by cross-sectional area; the stress developed is simply force/cross-sectional area, and the deformation or strain is stress/stiffness (Young's modulus). Simple tension, however, is a relatively unusual form of loading in the biological world (although not unknown; see below and Swartz et al., 1989, for a primatological example), so in most structures, we must consider additional aspects of shape beyond cross-sectional area alone. Bending is by far the

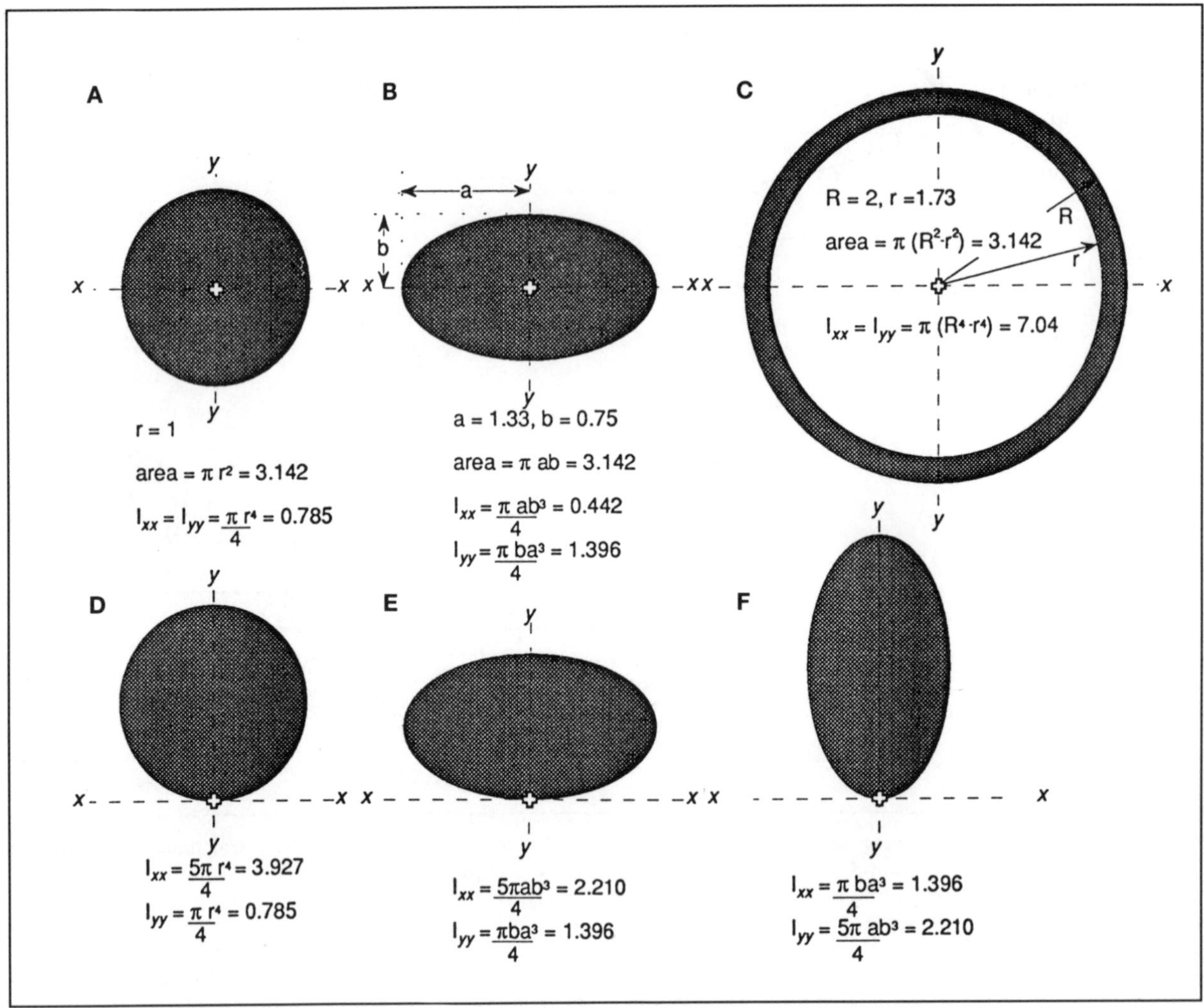

FIGURE 1.17. Second moment of area or moment of inertia depends on cross-sectional area, cross-sectional shape, and location of the axis about which the moment of area is calculated. For each shape, area is identical. When the second moments of area are calculated about the center of radially symmetrical shapes (A and C), I_{xx} and I_{yy} are identical. If the shape is not radially symmetrical (B, E, and F) or if the second moment of area is not calculated about axes passing through the center of the cross section (D, E, and F), then second moment of area will vary depending on whether it is calculated about a horizontal *(x-x)* or vertical *(y-y)* axis. Increments of area that are far from the intersection of the axes contribute disproportionately to the second moment of area, so shapes with area distributed at large distances have much larger second moment of area for the same cross-sectional area (C and E).

most common loading mode in limb bones and in limbs as a whole (Fig. 1.16). For structures made of stiff, linearly elastic materials, bending stress can be expressed as:

$$\sigma_b = \frac{M_y}{I} \tag{2}$$

where σ_b is the stress due to bending, M is the moment of the applied force about the structure's neutral axis, y is perpendicular distance from the neutral axis to the point at which we wish to consider bending stress (reaching its greatest value at the structure's greatest diameter), and I is the moment of inertia or second moment of area of the cross section (Fig. 1.17). I, determined as:

$$I = \int (y^2 dA), \tag{3}$$

is a measure of the placement of material in a given cross section with respect to a defined axis, generally the neutral axis of bending (Fig. 1.17). For a fixed area, I will increase as increments of area are placed at a greater distance from the axis. In fact, each area (dA) increment is multiplied by the square of its distance y, so I will rise rapidly as material is dispersed from the axis of interest. In terms of shape, bending stresses for a given loading situation (a particular load applied at a particular location, thus a particular value for M) will be smallest when y/I is maximized. The magnitude of M will equal the applied force multiplied by some fraction of the length of the structure. Therefore, accurate measurements of lengths, diameters, and moments of inertia are critical to analysis of shape with respect to bending.

It is important to note that I has both area (dA, see equation 3) and placement of area (y, see equation 3) incorporated within it. The relevant distance by which area increment if multiplied is measured from a specified location, usually an axis through the centroid of the shape (Fig. 1.17). Moments of inertia for a given shape, therefore, must always be defined with respect to a particular axis. Typically, I is calculated as I_{xx} and I_{yy} with respect to axes through the section centroid in the x or y direction (anatomically, the anteroposterior or mediolateral axes). However, these may not coincide with the directions of greatest I values (designated I_{max}) nor the directions in which bending typically occurs in the animal's locomotion. For this reason, it may be preferable to calculate I_{xx}, I_{yy}, and I_{max}, as well as θ, the angle between I_{max} and a predefined geometric or anatomical axis. This still begs the question of the precise direction of greatest bending; determination of this orientation requires direct measurement of in vivo strains.

The deflection of a loaded beam under bending or a bone experiencing locomotor forces depends on similar parameters. For the simplest case, a cantilever beam experiencing a concentrated load at its end, like a primate holding a heavy fruit in an outstretched arm,

$$\delta_{max} = \frac{FL^3}{3EI} \tag{4}$$

where θ_{max} is the deflection perpendicular to the beam's long axis, F is the force applied, L is beam length, and E is the Young's modulus or stiffness of the material of which the structure is made. If a beam is supported at both ends, or the force applied to it is distributed over its length, whether uniformly or in a varying fashion, the details of its pattern of deflection will differ. The maximum deflection, the most relevant concern for an animal, is always proportional to some function of length (ranging from length squared to length to the fourth power) and inversely proportional to I (see Beer and Johnston, 1981, and Muvdi and McNabb, 1991, for details of how deflection changes with variation in support and load distribution conditions).

Simple compression has certain similarities to simple tension; in general, the cross-sectional area of a structure will determine its behavior under a given force. In long, slender structures, however, axial compressive loading will cause the structure to begin to buckle, bowing it laterally in a mode of behavior known as *elastic* or *Euler buckling* (see also Wainwright et al., 1976) The magnitude of the compressive force that will produce critical buckling is given by the Euler buckling formula:

$$F_E = \frac{n\pi^2 EI}{L^2} \tag{5}$$

where F_E is the force to produce Euler buckling, L is the length of the structure, and n is a constant determined by the kinds of movement possible (technically

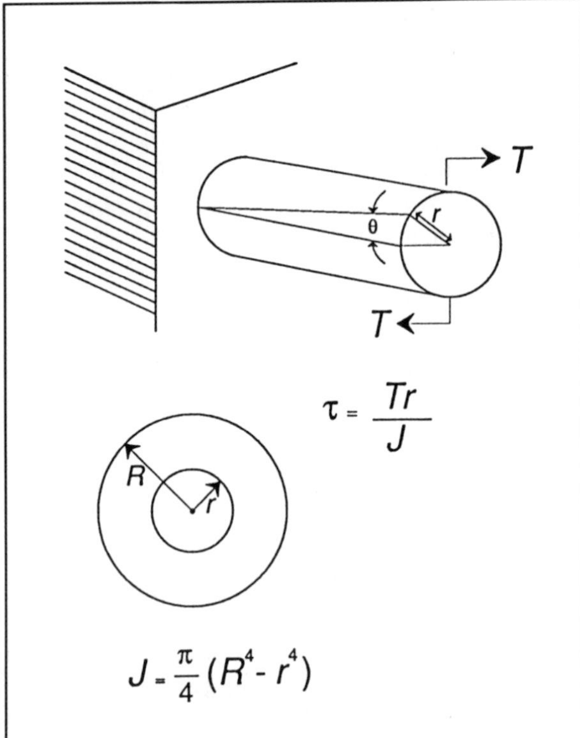

FIGURE 1.18. As a beam is subjected to torsion, the torsional moment *(T)* will produce a rotation at one end of the beam relative to the fixed end; this can be expressed as an angular deformation as the end rotated through angle θ. The stress developed from this torque is a function of the applied torque, the distance from the neutral axis of the beam *(r)*, and the polar moment of inertia *(J)*. As in beams in bending, the stresses will therefore increase linearly as one moves from the center of the section out to the outermost surface.

speaking, the degrees of freedom) at the ends of the column. The values of n can range from less than one (0.25 when one end is fixed and one end is free to rotate in its own plane) to a value of 4 when both ends are firmly fixed (rare in the dynamic world of organisms). The critical shape variables here are, once again, I and length. Because I is a function of distribution of material in the cross section, the relationship between I and length relates to the slenderness of the column. This idea is often made explicit by expressing I as Ar^2, where A is the cross-sectional area and r is the least radius of gyration, the distance at which one could place a ring of area equal to the entire cross section and achieve a moment of inertia equal to the moment of inertia of the original cross section.

When this expression is substituted into equation 5, it is apparent that F_E is proportional to $A/(L/r)^2$, and the resulting failure stress (FE/A) is inversely proportional to the square of L/r, known as the slenderness ratio. The critical stress for very slender columns will be quite low; for bone, the slenderness ratio at which bone is equally likely to fail by Euler buckling and by compressive yield is only $L/r = 32$ (Currey, 1984). Long, slender bones of living verte-

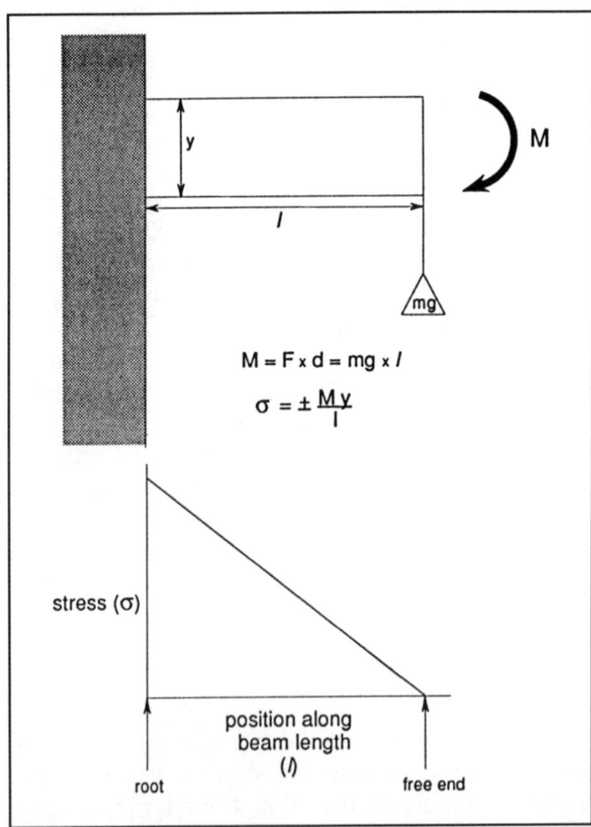

FIGURE 1.19. Stresses in a beam vary from tensile to compressive surfaces, but also with distance from the applied force. In this schematic, a weight (mass × gravity, *mg*) is applied to the end of a fixed beam. The bending stress resulting from the applied force then increases in direct proportion to the bending moment of the force, which is the force multiplied by the distance from the application point. This situation is analogous to many biological loading modes, such as the application of a muscle force to one end of a bone or the application of a ground reaction force at one end of a limb.

brates may come quite close to this value (e.g., bat and pterodactyl wing bones and, of interest to primatologists, gibbon forelimb bones). In cases of this kind, it is particularly important to assess the risk of Euler buckling, and thus to measure both moments of inertia and length of the structure.

A final possible mode of failure for columns under compressive loading is known as *local buckling*. Although considering Euler buckling might lead us to think that the force required to make a support element buckle will increase with increasing moment of inertia, as I goes up for a given cross-sectional area, the wall thickness of the cross section must decrease. Eventually, when the wall becomes very thin, it comes under risk of developing a local kink that becomes the initiation site for an adjacent catastrophic collapse. Local buckling generally occurs at a stress σ_L given by:

$$\sigma_L = \frac{k\,E\,t}{D} \qquad (6)$$

where k is an empirically determined constant that generally lies between 0.5 and 0.8, t is wall thickness, and D is the diameter of the cross section. Although failure in this mode does not depend directly on I, for a given cross-sectional area I cannot increase indefinitely before the ratio t/D becomes dangerously small.

The application of torque to a structure is really analogous to applying a bending moment, and torsional stress can be expressed as:

$$\tau = \frac{T\,r}{J} \qquad (7)$$

where τ is the shear stress due to torsion, T is the torque or torsional moment of the applied force about the structure's neutral axis, r is perpendicular distance from the neutral axis to the point at which we wish to consider shear stress (reaching its greatest value at the structure's greatest diameter), and J is the polar moment of inertia (polar second moment of area) of the cross section (Fig. 1.18). J is the sum of the moments of inertia about any two perpendicular axes through the centroid of a cross section, and just as I can be thought of as a shape measure encapsulating a structure's overall resistance to bending, J can be considered a shape summary of shearing resistance under torsion.

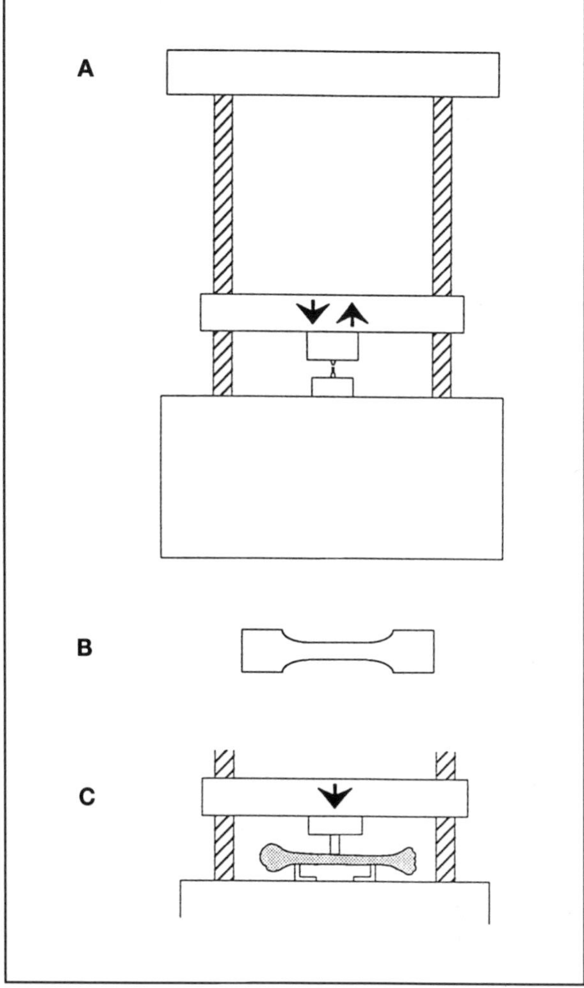

FIGURE 1.20. Illustration of basic materials testing apparatus. **A**: The test specimen is placed in grips between two fixating devices connected to sensitive devices for measuring applied loads and changes in height of the apparatus. The apparatus can then be moved up or down at controlled speed to stretch or compress the specimen during measurement of applied force and distance. **B**: Test specimens are typically machined to have relatively large ends so that the danger of the specimen failing at the grip point is minimized. **C**: This kind of apparatus can be modified to apply a three-point bending load to specimens when they are supported underneath by two spatially separated supports and the compressive load is applied midway between the supporting legs. Adapted from R. McNeill Alexander: Animal Mechanics. Packard Publishing Ltd., 1983. By permission of Packard Publishing Ltd.

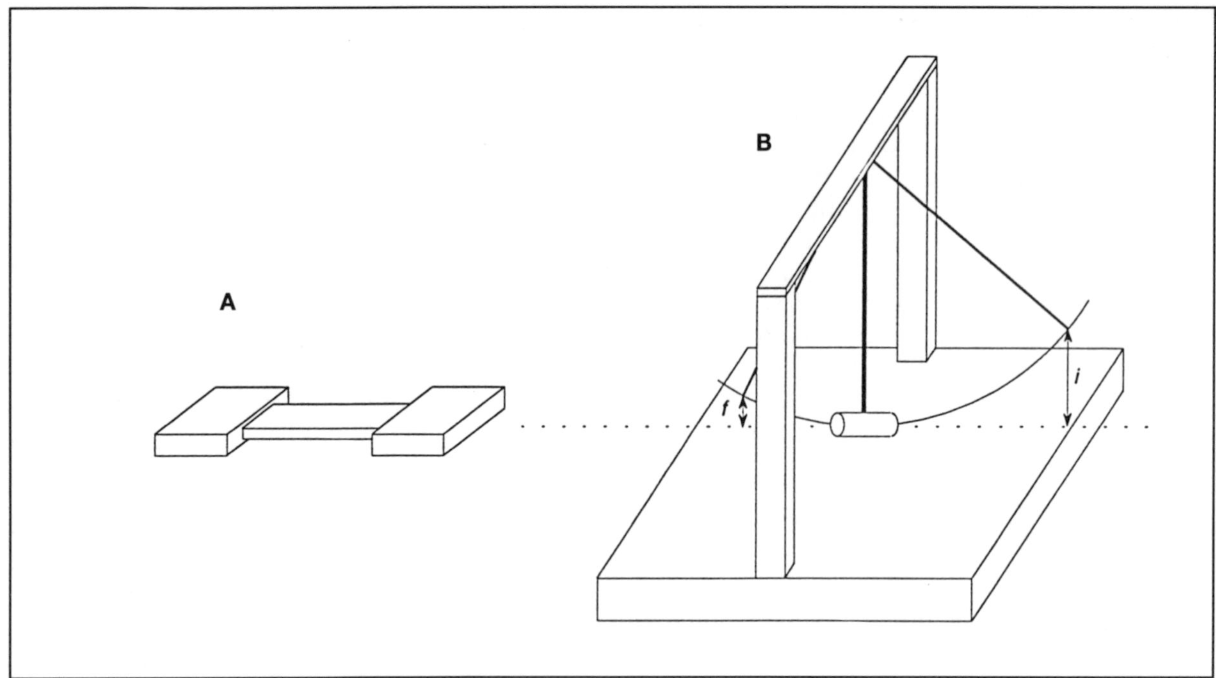

FIGURE 1.21. Pendular apparatus for testing energy-absorption capabilities of biological materials. **A**: Specimens are embedded in epoxy on either end for gripping into the testing apparatus. An adequate length of material is left between the grips so that a lead pendulum can swing through the specimen and break it. **B**: Once the specimen is prepared, it is placed in the path of the swinging pendulum bob. The energy lost in fracturing the specimen is equal to $[mg(i-f)]$, where m is the mass of the pendulum, g is the acceleration due to gravity (9.8 m/s^2), and i and f are the initial and final heights of the pendulum above the lowest point it reaches at the bottom of the swing. Additional corrections for losses due to friction may also be necessary. Adapted from R. McNeill Alexander: Animal Mechanics. Packard Publishing Ltd., 1983. By permission of Packard Publishing Ltd.

Many biological structures are not uniform in cross-sectional area and shape throughout their length, so values of I and J from one location may not accurately reflect shape elsewhere. The general distribution of forces in structure (estimated from models or determined empirically) will then help determine where cross-sectional shape should be assessed. In a beam loaded as a simple cantilever, for example, the bending moment, M, will increase linearly from the end to the root; similarly, the bending moment applied to a limb bone by a ground reaction force vector will increase linearly from distal to proximal (Fig. 1.19). If the ability to withstand a particular force is of interest, it is therefore important to measure moments of inertia at the anatomical locations where the effects of the force are likely to be greatest, or to assess cross-sectional geometry at a variety of locations within a structure.

EXPERIMENTAL TECHNIQUES FOR BIOMECHANICAL STUDY

A wide variety of techniques are used to understand biomechanical aspects of animal design. Some focus primarily on the material characteristics of particular tissues and are appropriate tools for assessing, for example, differences in mechanical capabilities of particular tissue type from different anatomical locations and from species that differ in behavioral capacity. Some focus more specifically on mechanically important aspects of structural geometry. Still others assess the behavior of mechanically important elements of limbs during normal or controlled locomotor activities.

Materials Testing

To ascertain the mechanical properties of a biological material such as bone, tendon, ligament, and so on, standardized samples of the tissue can be subjected to tests using equipment designed to monitor applied force and resulting deformation simultaneously (Jobbins, 1981). Typically, test specimens are prepared so that they have a uniform cross section in the testing region. For tests of mechanical behavior in tension, there is usually a larger gripping area on the ends (Fig 1.20); this gives the equipment an adequate grip on the sample while ensuring that the stress will be greater in the center so that the specimen will not fail at the grips. In tensile testing, the apparatus pulls the specimen's ends apart with a known amount of force while monitoring the displacement of the grips. From this, a graph of force versus displacement is produced and, when the original dimensions of the specimen are known, converted to a plot of stress versus strain. For compression tests, the specimens are subjected to crushing loads as they are compressed between two flat plates (platens). In this case, the force applied is mapped out against the decreasing distance between the two platens. For both of these tests, the raw results give only a record of force versus displacement; based on the size of the specimen (cross-sectional area and original length), these values can be simply converted to stresses and strains. Torsional testing machines can also be used to subject cylindrical specimens to twisting tests, measuring the shear strain produced as one end of the cylinder is rotated with respect to the other.

With a few minor modifications, compres-sion-testing equipment can also be put to use to assess the behavior of a material in bending. Instead of placing a sample between two flat plates, the specimen is rested horizontally on two knife-edge supports. A third edge, centered between the other two, sits on the facing platen. As it is brought down against the specimen, this edge will impose a bending moment (Fig. 1.20C).

Energy-absorbing capacity or work of fracture of a material can also be tested directly with impact tests. In these tests, a large, low-friction hammer or pendulum is allowed to swing downward and contact the test specimen at the bottom of its path, break the specimen, and swing through to the opposite side (Fig. 1.21). The amount of energy required to break the specimen will be reflected directly in the height the pendulum reaches after the specimen fracture. If breaking the specimen requires a great deal of the pendulum's energy, it will swing up only a short distance, and if the work to fracture is negligible the pendulum will reach a height close to that of its original position.

Several considerations are important when materials-testing procedures originally designed for study of man-made materials such as steel and concrete are applied to biological tissues. This kind of testing is most straightforward for stiff, linearly elastic materials that can easily be machined to ideal specimen shape. This poses some difficulties for biological tissues, particularly for very stretchy materials such as ligaments or skin. Special techniques or apparatus may be required for accurate measurement of tissues of very low stiffness. Also, it is critical that the specimens come from samples that accurately reflect the composition of the tissue in life; as tissues dehydrate, for example, their material properties may change enormously. Formalin-fixed or alcoholic tissues can rarely be used for these studies, and careful precautions must be taken if frozen samples are to maintain their normal characteristics following thawing. Perhaps most important, materials tests should reflect the normal biological use of the tissue. Ligaments and tendons probably never experience significant compressive loads; tensile testing is the appropriate analysis mode for these tissues. If one has reason to think that bending is the predominant loading mode for a tissue such as bone, then bending tests are most appropriate in understanding the mechanical characteristics of this tissue type. Many tissues, particularly bone, function in a variety of ways, depending on the anatomical region of interest and on the locomotor behavior of the particular taxa of interest, and it is therefore useful to use a combination of materials tests to examine these tissues.

Morphometrics

Statistical analysis of quantitative aspects of postcranial anatomy can be used to explore a wide range of issues in the functional biology, development, phylogenetic analysis, and biogeography of

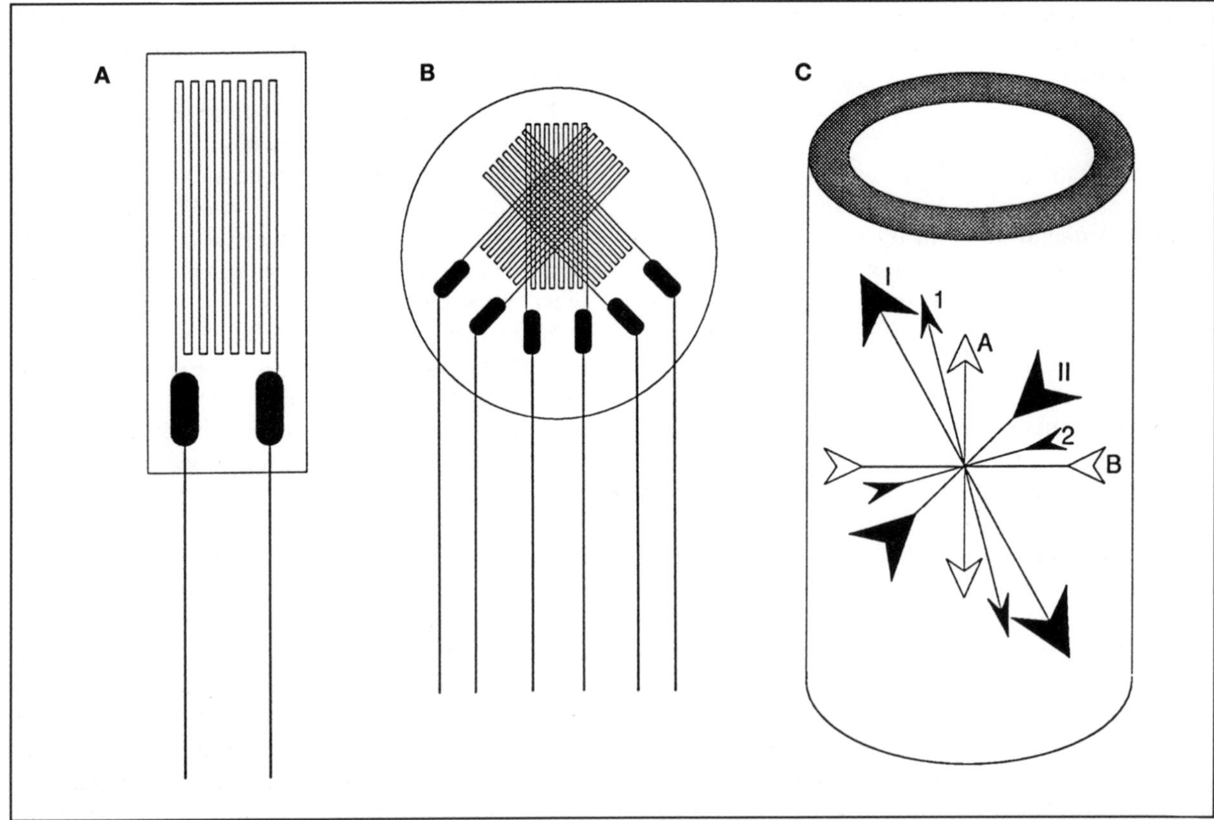

FIGURE 1.22. A: Single-element strain gauge, with wire grid and metal tabs onto which lead wires are soldered. Strain is measured along the gauge's long axis. B: Rosette strain gauge with three single-element gauges superimposed on one another at 45°. From these three independent measurements, the maximum and minimim principal strains can be calculated. C: On a bone with anatomical axes *A* and *B* (open arrowheads), rosette strain gauges can be used to calculate principal strains, which may or may not coincide in orientation with the anatomical or material axes; here, principal strain axes are 1 (tension) and 2 (compression) (narrow arrowheads). Because of the anisotropy of bone as a material, the axes of principal stresses (I and II, wide arrowheads) may not coincide with the principal strain axes. From S. M. Swartz (1991) Strain Analysis as a Tool for Functional Morphology. Amer. Zool. *31:655–669*. By permission of *American Zoologist.*

living and extinct primates; there have been many hundreds of studies using these techniques in the last ten years alone. In a specifically biomechanial context, morphometric analysis can be employed in a variety of ways. Analysis of quantitative measurements (linear, areal, or others) can uncover patterns of morphological similarity in taxa with shared functional characteristics with the goal of discovering structural features that may be mechanically important. In this kind of approach, morphometric analysis is used to generate hypotheses that can later be tested or explored more thoroughly with other analytical approaches. Alternatively, morphometric analysis can be employed to discover patterns of structural variation in features already known to be of mechanical importance, particularly in relation to variation in body size, locomotor mode, phylogenetic affinity, or evolutionary time (with fossils).

Using analysis based on lever mechanics, stress-resisting geometric characteristics, and so on, morphometric data can be analyzed in an allometric context as well. Subjects of these analyses can be adult members of different species (adult interspecific scaling, e.g., Jungers, 1984a, 1984b), different-sized adult members of a single species (adult intraspecific scaling), or ontogenetic series of one or several closely related species (ontogenetic scaling, e.g., Shea, 1985).

Strain Gauge Analysis

To understand how the shape of bones relates to the mechanical environment they experience during normal locomotor behavior, we usually want to know something about the forces that the bone actually encounters. Strain gauge analysis obtains measurements of local strains directly from the surface of a normally functioning bone, and can therefore provide information about normal bone loading patterns directly (see also Swartz, 1991). This method produces a direct record of the strains (and hence the stresses) at particular locations on bones during activities that the investigator believes are mechanically significant—at the point in an animal's stride when the stresses are greatest, for example, as a jumper leaves the ground, or as an arboreal animal hoists itself up in a tree. This technique involves the application of single-element or rosette strain gauges, small devices commonly used by engineers to monitor strains developed in various kinds of equipment, and so on, under conditions that duplicate their intended use. These gauges are basically precision resistors made of a long piece of wire folded up to lie in a small space (Fig. 1.22) and attached to a thin plastic film; when a small current is passed through the gauge via wire connections, one can measure a given voltage across the gauge in accordance with Ohm's Law ($V = IR$, where V = voltage, I = current, and R = resistance). They are applied directly to the surface of a relatively stiff object whose mechanical behavior is of interest. As the gauge is deformed, the length of the gauge wire will change; if the gauge experiences tension, the wire gets longer, and if it is compressed, the gauge wire gets shorter. As the gauge wire changes length, its electrical resistance changes slightly. This resistance change is reflected in a change in voltage that can be measured directly. Hence, strain gauges can make very sensitive measurements of length changes along their long axis (accuracy reaches approximately 0.0001 to 0.00001 percent). In order to use this technique to understand aspects of primate limb biomechanics, these gauges can be implanted surgically on the surface of a limb bone in a region that is free of muscle attachment, and then used to monitor the strains that particular locomotor behaviors engender in the bones.

Most biological structures experience somewhat complex forces, and primate limb bones are no exception. During some locomotor activities, a bone may be subjected to forces from several different muscles, each varying in size and force magnitude, each oriented in a different way, and each changing dynamically throughout the locomotor cycle. In addition, forces such as gravity and inertial forces due to accelerations of the body and limb segments will affect skeletal loading. This complexity makes it very hard to predict a priori how bone stresses will be oriented, and we usually need to determine the direction of primary bone loading (or principal strain axis) empirically. Using strain gauges put together by layering three individual or single-element gauges on top of one another, known as rosette gauges (Fig 1.22), we can use the strain information from three known orientations to calculate where the principal strain axis will lie and how large the strain along that axis will be. Building on the information about strain magnitude and orientation at a particular site, multiple strain gauges are often applied in a particular region to help ascertain a bone's overall loading mode. For example, if two or more gauges are applied to around the circumference of a limb bone and all experience compressive strains of comparable magnitude in the middle of a locomotor cycle, we could conclude that the bone experienced a primarily compressive load at that time. If, however, compressive strains were observed on one surface of the bone, tensile strains on the opposite surface, and close to zero strains in between, we could conclude that the bone experienced bending loads. Like many of the techniques summarized in what follows, strain data are acquired simultaneously with film or video recordings to coordinate the patterns of bone deformations observed with the particular movements producing the loads.

Force Plates, Cinematography, Cineradiography, and Electromyography

Although strain gauge analysis provides a clear and detailed picture of the loads actually experienced by a bone during normal movements, it requires surgical implantation of measuring devices and is most useful for studying fairly large animals.

Another important technique in biomechanical analysis of forces on limbs is force plate, or force platform, analysis. A force plate records the forces exerted by a limb on the ground as it steps on the plate; by recording the force the limb exerts, it also records the equal and opposite force experienced by the limb as a whole (Alexander, 1981a). Typically, force plates are instrumented to break this information down into the components of force exerted in the vertical, anteroposterior, and mediolateral directions. Hence, force platform records not only provide information about the overall magnitude of forces exerted on limbs, they also allow an assessment of the relative importance of these distinct force components (see Reynolds, 1985a, 1985b for primatological examples). Like strain gauge analysis, these recordings must be combined with videotape or film records to synchronize information about the changing mechanical environment of the limb with the pattern of limb movements.

Force plate data depict the net force exerted on a limb as a whole as that limb contacts the plate, and although analysis of the net force exerted on the limb combined with a knowledge of the position of limb segments at a given moment can be used to calculate net bending moments exerted about each limb segment (see Biewener, 1983; Biewener et al., 1983, for examples of the application of this method), they provide only indirect information about the loads experienced by each bone. Nonetheless, this information can provide great insight into the biomechanics of limb use. For example, force plate data allow us to compare how forces experienced by fore and hindlimbs may differ in a particular locomotor mode. From this base, it is possible to compare relative limb loading (perhaps expressing load as a multiple of body weight) of a single taxon in a variety of locomotor behaviors, or limb loading of a number of related but divergently sized taxa in similar locomotor behaviors.

Cineradiography films (conventional or high-speed X-ray movies) refine this procedure so that we can compare ground forces (or something else like strain, tendon force, or EMG) directly to instantaneous location of bones and joints, rather than estimate those things as we derive them from regular films. This technique combines X-ray capacity with the ability of movies to show locomotor movements in detail. More important, a great deal of additional biomechanically useful information can be gained from cineradiographs. Because the position of bones themselves can be seen accurately, it is possible to precisely monitor movements of skeletal elements rather than the general outline of limb or body segments, and hence gain a much more detailed appreciation of the skeletal substrate for particular motions. Better visualization of bony anatomy—and in particular joint location—also allows more precise calculation of skeletal and joint stresses engendered during locomotor movements, with or without complementary force plate data.

Although anatomists have been able to deduce "the action" of a muscle based on the geometry of its origin and insertion and the topology of the joints it crosses, this purely anatomical information is rarely adequate to assess the mechanical behavior of muscles. Electromyography (EMG) is a technique by which we can directly measure the activity in muscles as those muscles contract. The technique is based on the principle that muscle contraction produces (or is produced by, however you care to think of it) electrical activity, and that by using electrodes to monitor electrical activity in a muscle we can learn things about the muscle's activity, including what movements activate the muscle and, to a lesser extent, what proportion of its total force capacity the muscle is producing. As in the case of loads experienced by bones, it is the complexity of the structure and function of limbs that makes this kind of sophisticated measurement technique necessary to understand muscle biology. We think of the function of muscles in the context of jointed lever systems and how muscles produce movements at joints by shortening. If we are interested in knowing when the biceps brachii muscle crossing the hingelike elbow joint is active, it seems evident that shortening will flex the elbow. Looked at more closely, however, there are several sources of indeterminacy in this situation. More than one muscle will cross almost any joint we wish to study. Some of them will produce a particular action such as flexion, and others will produce an opposing or antagonistic action, such as extension. Some of these muscles will cross a single joint, and others will cross two or more joints; some may be parallel-fibered and some pinnate, some may cross near the joint and others have larger lever arms.

EMG takes advantage of the fact that muscles operate as electrical systems. When a nerve impulse commanding muscle contraction arrives at its at-

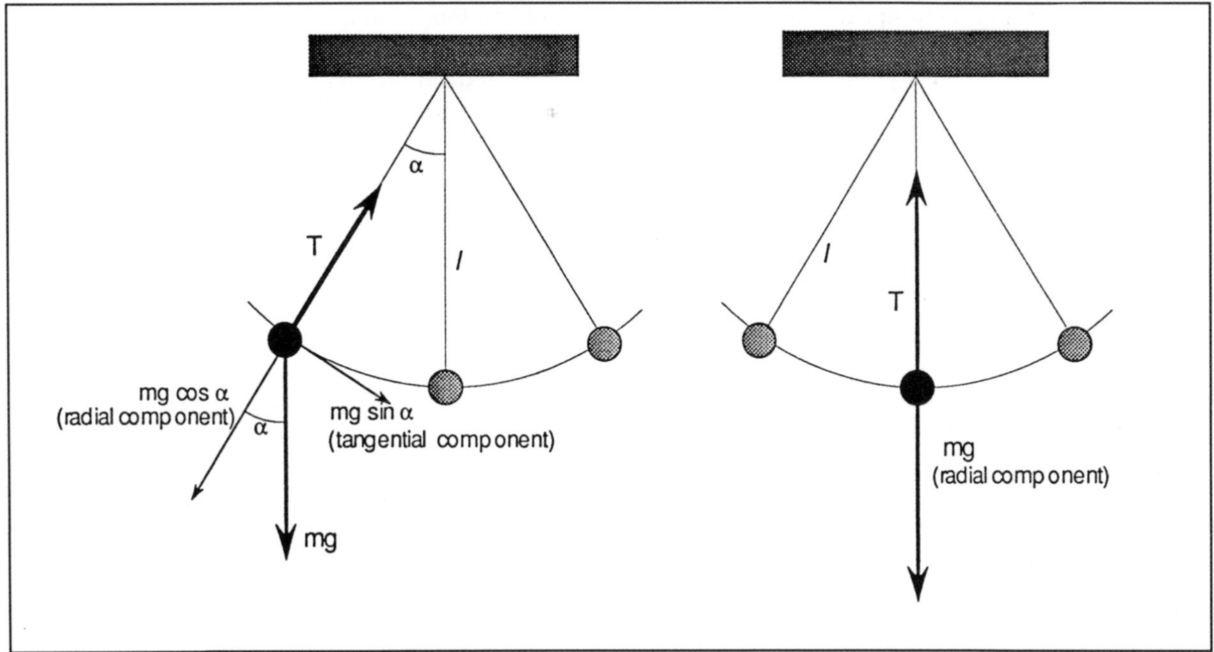

FIGURE 1.23. As a brachiating primate swings suspended by a single handhold, the animal mimics pure pendular motion. In a simple pendulum, at the beginning of a swing, two forces are exerted on the pendular bob: gravity acting downward on the bob mass (*mg*) and cord tension acting along the line connecting the center of mass to the center of rotation at the handhold. The gravitational force makes an angle to the cord tension force (α). The gravitational force can be thought of in terms of its vector components in line with the tensile force and perpendicular to it (in line with the forward velocity of the pendulum). These components are given by appropriate trigonometric functions of the weight vector. At the middle of the swing, the two vectors are in line with one another. This kind of model can be applied to understanding forces generated in limb structures of brachiating primates. Adapted from Swartz (1989). By permission of Plenum Publishing Corp.

tachment to the muscle (the neuromuscular junction), the electrical disturbance in the nerve ending results in the release of a neurotransmitter substance. Through a sequence of chemical interactions, the membrane surrounding the muscle cell undergoes an electrical depolarization, and this electrochemical reaction spreads along the length of the entire muscle fiber. After depolarization, the muscle's membrane returns to its inital state of polarization, and this polarization spreads along the membrane in the same wavelike fashion as the depolarization spread. This entire chain of events (for more detailed information see McMahon, 1984; Loeb et al., 1986) occurs in from 2 to 5 milliseconds; even muscle fibers that are several centimeters long will experience the entire action potential in few milliseconds.

By monitoring these electrical potentials, EMG recordings can therefore produce detailed records of the onset, duration, and relative intensity of a muscle or muscle partition's activity pattern with respect to particular movements. This is an important element of how particular movements are produced, and hence how plausible scenarios for evolution of particular configurations of muscular anatomy are worked out (for examples concerning primatological applications, see Tuttle and Basmajian, 1974; Tuttle et al., 1975; Stern et al.; Jungers and Stern, 1984; Larson and Stern, 1987). This technique also contributes significantly to other biomechanical investigations; for example, information concerning which muscles are active during particular activities and the relative intensity of their activity, coupled to anatomical data on muscle size, position, and architecture, can help provide a more complete picture of the mechanical environment experienced by the skeleton.

GIBBONS AND BRACHIATION: A CASE STUDY

Amid the broad spectrum of locomotor repertoires found within primates, the gibbons (Hylobatidae) and spider monkeys (*Ateles*) stand out for their spectacular and highly specialized arm-swinging behavior, known as brachiation. Along with this peculiar pattern of locomotor behavior, it has long been recognized that brachiators differ systematically from nonbrachiating primates in trunk and limb morphology. Indeed, the unique arm-swinging locomotor behavior and unusual morphology of brachiating primates have been subjects of research and discussion among biologists for decades for a number of reasons (among many others, see Keith, 1903, 1923; Gregory, 1928; Schultz, 1936, 1953, 1973; Avis, 1962; Tuttle, 1972; Fleagle, 1974; Jungers and Stern, 1980, 1984; Fleagle et al., 1981; Preuschoft and Demes, 1984; Swartz, 1989).

Brachiating locomotion is particularly well suited to biomechanical analytical approaches for several reasons. It is in some respects simpler than terrestrial locomotion, because only one pair of limbs is used. Furthermore, biomechanical analyses can often take advantage of similarities between organismal design and performance and the design and function of human-made mechanical systems. Here, the similarities are obvious between the locomotor movements of brachiators and the kinematics of pendular motion, and the mechanics and energetics of pendular systems have been thoroughly understood for a long time, thus providing an appropriate starting point for biomechanical analysis (Fig. 1.23). By virtue of functioning much like a zoological pendulum, at least superficially, the mechanical environment of brachiator forelimbs should be substantially different from those of other primates, and relationships between mechanical performance and structural design should be particularly robust. This offers the advantage that even if form-function relationships in limbs are not strictly one-to-one, it may be reasonable to attempt to decipher important aspects of structural design, even if each anatomical configuration can carry out many functions and a variety of anatomies are at least adequate for carrying out particular behaviors. Specialized jumping behaviors would constitute another primate example of locomotor adaptation that is well suited to a biomechanical approach.

Our understanding of brachiator anatomy and behavior has benefited from many of the approaches discussed earlier. Fleagle (1974) was the first to adopt biomechanical approaches to brachiation. He used conventional and high-speed films (24 and 64 frames per second) to demonstrate that siamangs not only appear grossly to mimic pendular motions but can be compared to pendular motion in a detailed way. By analyzing the repositioning of body segments throughout the swing cycle, he was able to suggest that (1) by extending their legs and free arm downward, siamangs can increase the kinetic energy gained during the downswing; and (2) by flexing the legs on the upstroke, siamangs decrease their moment of inertia and increase their angular velocity.

More recent work has probed the movements of brachiators more deeply by looking at the patterns of movements of the forelimb skeleton with cineradiography (Jenkins et al., 1978; Jenkins, 1981). A number of important insights into the functional anatomy of postcranial elements were gained in these studies. Jenkins was able to document a series of bone-movement patterns that could not have been predicted a priori and to use this information to interpret some important skeletal features of brachiators. He demonstrated that, during the portion of the locomotor cycle in which the animal is supporting its body weight by the arm in question (the "support phase"), the shoulder joint moves in a complex way. The joint moves simultaneously in a caudal, dorsal, and medial direction, while at the same time rotating on the dorsum of the thorax. Jenkins linked this large and multidimensional motion to the specialized anatomy of the brachiator clavicle; the clavicle is elongated and sigmoidally shaped, and its proximal end is twisted in relation to its distal end. By virtue of this anatomy, the clavicle is able to maintain its strong interconnection to the scapula in its specialized position on the animal's dorsal thorax (scapulae of quadrupedal monkeys and most other mammals are located on the lateral aspect of the trunk). Furthermore, he pointed out that while hanging, brachiator clavicles are not oriented laterally, as in quadrupeds, but at an upward and dorsal orientation. In this position, if the clavicle were straight and/or relatively short it would impinge on important structures that leave the thorax for the limb. Finally, he used his cineradiographic evidence to clarify the role of shoulder musculature in brachiation; previous authors had suggested that scapular rotation in the plane of the scapular blade was a critical characteristic driving the musculoskeletal anatomy of suspensory species. Jenkins documented that most of the scapular rotation occurs during the nonpropulsive part of the locomotor cycle, and that rotation is only a small part of the movements the scapula must experience for effective propulsion. With this information, it was then possible to more accurately evaluate the architecture of brachiator shoulder musculature with respect to the functional demands of arm-swinging movements.

Allometric analyses have also clarified mechanically relevant aspects of the skeletal anatomy of brachiators (e.g., Aiello, 1981, 1984; Jungers, 1984a, 1984b). These studies confirm the clear visual impression that the forelimbs of suspensory taxa are greatly elongated, and Jungers' careful analysis shows clearly that this is not likely to have resulted from broad, all-catarrhine trend toward forelimb elongation, but rather to specialization in relation to the unique requirements of suspensory locomotion.

Basic biomechanical principles of motion and stress analysis have also been used to model brachiator locomotion (Preuschoft and Demes, 1984, 1985; Swartz, 1989). A mechanical model of forces exerted at the hand of a brachiator on a branch shows that the largest animal that could utilize arm-swinging locomotion and still maintain the integrity of a handhold grip far exceeds the body size of all living primates, thereby rejecting the hypothesis that siamangs are the largest living brachiators because of fundamental mechanical constraints (Swartz, 1989). This model also generated several hypotheses that could be tested in future biomechanical studies, suggesting that forces tending to extend the fingers and loosen the grip will be strongly dependent on both branch size and velocity. With the development of a branchlike force "platform" analog, it will be possible to test these hypotheses more fully, and to determine if the limits on locomotor speed and handhold size predicted by theoretical models operate during the naturalistic locomotor behavior of suspensory animals.

EMG data have also been used to help interpret the functional significance of the distinctive anatomy of brachiating primates (Stern et al., 1977; Stern et al., 1980b; Jungers et al., 1980, 1984; Fleagle et al., 1981). In one particularly innovative and noteworthy set of experiments, Jungers and Stern used EMG to test a specific hypothesis developed from anatomical observation. They called attention to the fact that gibbon forelimbs show a striking specialization in their muscular anatomy: Several muscles have moved or extended their insertion areas to attach at least partly into one another rather than directly to the bone alone. On the ventral side of the forelimb, the short head of the biceps brachii takes origin in part from the pectoralis major and, in turn, gives rise to part of the flexor digitorum superficialis, while on the dorsum, the dorsoepitrochlearis also originates in part from the short head of the biceps brachii. It had long been suggested that this anatomical modification was directly linked to conducting force from the large proximal muscles (pectoralis major, latissimus dorsi) across more distal joints, particularly the metacarpophalangeal and interphalangeal joints crossed by the flexor digitorum superficialis. In this way, it was suggested, activity or passive tension in any of the muscles of the "muscle chains" could produce automatic finger flexion and hence gripping.

Jungers and Stern proposed that EMG could directly test this hypothesis. By monitoring the activity levels in each muscle of the chains, they could determine whether activation of multiple "links" in the chain always fired simultaneously, and whether activity in proximal elements effected movement in the fingers. They showed conclusively, at both low and high speeds, that the anatomically based hypotheses were incorrect, and that these muscle groups do not function as single "supermuscles." Instead, they demonstrated that the biceps brachii is the key element in this anatomical arrangement, and that in the course of the evolution of increased force-production capacity in this muscle, it has increased both in size and in lever arm and has thereby moved from its primitive anatomical location.

Finally, we have used in vivo strain gauge analysis to determine what kinds of loads are experienced by the forelimb bones of brachiating gibbons (Swartz et al., 1989). We predicted that the adoption of arm-swinging behavior should modify overall limb loading in brachiators from a regime dominated by compressive and bending stresses to one

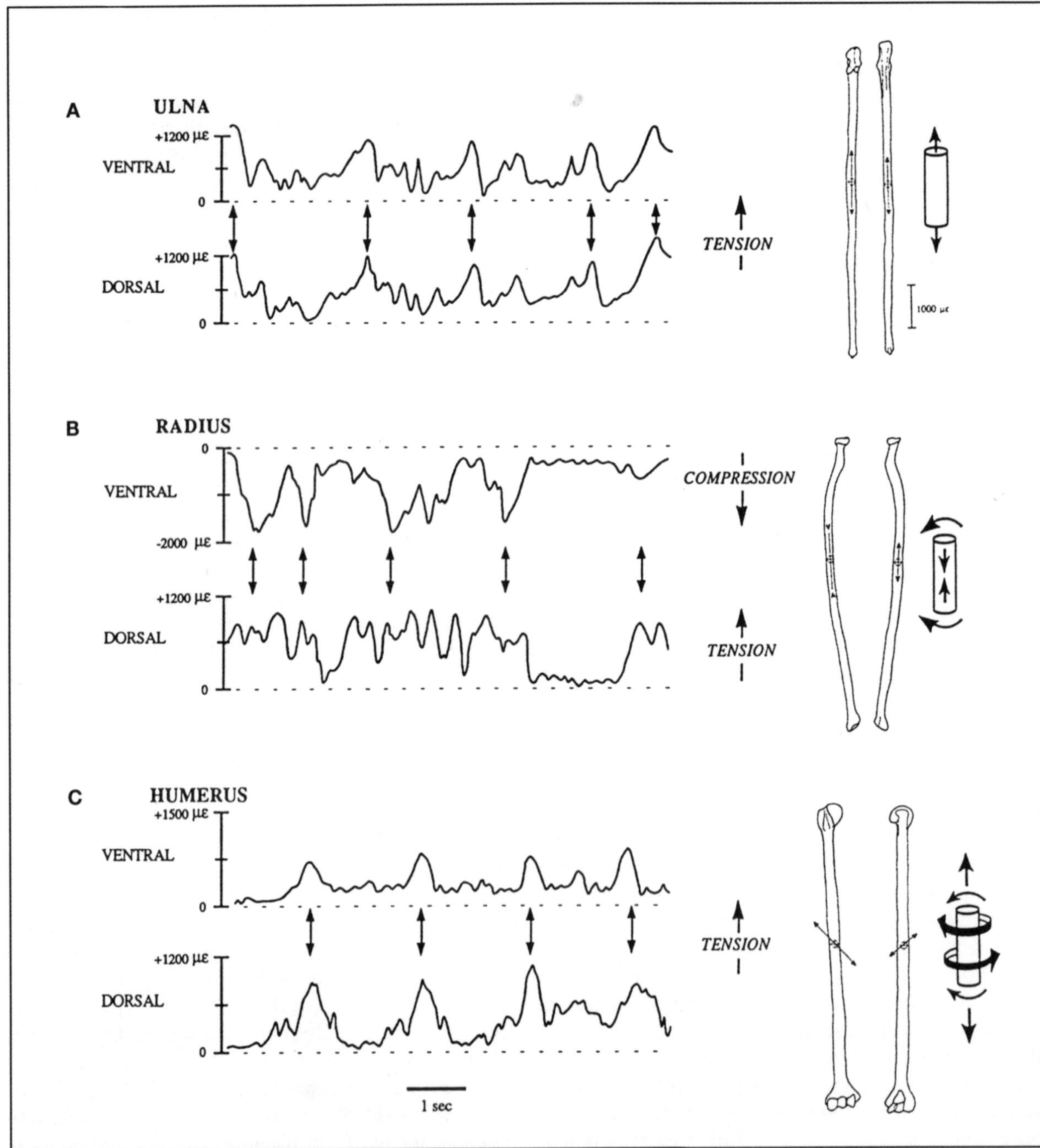

FIGURE 1.24. Each illustration (A, B, and C) gives, on the left, representative strain recordings from the dorsal and ventral midshaft surfaces of gibbon limb bones. Time is across the horizontal axis; each tracing represents about 8 seconds. Strain magnitudes in all bones reach their maximum values at the midpoint of the downswing, indicated by the small double-pointed arrows. The magnitude and orientation of the peak strains differ, however, among bones. On the right, each bone is depicted in dorsal and ventral view with vectors depicting the magnitude and orientation of peak principal strains superimposed on the gauge attachment sites. On the far right is a simplified view of the predominant loading pattern for each bone, showing axial tension for the ulna, bending and compression for the radius, and tension, torsion, and bending for the humerus. Adapted from Swartz et al. (1989).

where tensile stresses predominate. According to basic pendular models of brachiation, a gibbon forelimb functions like a string suspending the animal's body weight and would therefore experience purely tensile stress; the magnitude of that stress and the resulting strain can be predicted from the animal's mass, velocity, and bone cross-sectional area and the stiffness of the bone material. Our records from the dorsal and ventral midshaft of each bone show that, much as one would expect for a pendulum, strains peak in the middle of the swing, where the animals' velocity is greatest. The forces a pendulum experiences would, however, be expected to produce pure axial tension; in contrast, each of the three bones of the limb is loaded in a distinctly different fashion (Fig. 1.24). The ulna does experience axial tension, but the positive (elongating) strains on the dorsal surface and the compressive strains on the ventral surface of the radius show that this bone experiences substantial bending. Strains on the humerus are even more complex, evidencing tension combined with both bending and torsion; the torsion is clearly indicated by the fact that the principal strains are not lined up with the bone's axis, but deviate from it at near the 45° value that would be expected in pure torsional loading.

The studies described above can be used to help us interpret this complex pattern. Cineradiography shows that in brachiating spider monkeys, who presumably share many characteristic movement patterns with brachiating gibbons, the ulna remains in line with the centripetal acceleration predicted by pendular models (Preuschoft and Demes, 1984, 1985; Swartz, 1989) and shows little or no rotation along its long axis (Jenkins, 1981). In fact, even the magnitude of the tension recorded from the ulna was similar (within 11 percent) to values expected for a bone functioning as a passive suspending cord of a pendulum.

Why then is the radius loaded in bending and compression? The radius and ulna obviously must differ in the ways they transmit muscular force, the shape of the diaphysis, and/or the way they support weight. Muscle force may be particularly important, because it is typically the greatest constituent of axial force exerted on long bones (generally two to three times the magnitude of the reaction force of the body acting against the ground). EMG studies show that the forearm flexor muscles are especially active in brachiating gibbons (Stern et al., 1980; Jungers and Stern, 1981, 1984). These muscles cross the bone's ventral surface so that their pull will bend the bone in a ventrally concave fashion. During the support phase, the forearm and carpus are pronated, and by this movement the flexor muscles change their position relative to the radius and ulna. At the midsupport position, the flexor muscles have much greater leverage about the radius than the ulna, causing the radius to bend. In addition, weight is transferred in tension directly from the humerus to the ulna via the interlocking humero-ulnar joint, while the direct connections from humerus to radius are made exclusively of soft connective tissues that will themselves deform rather than transfer stress to the radius.

The torsion of the humerus arises because of the rotation of the gibbon's body about its supporting hand (Fleagle, 1974). As this rotation occurs, Jenkins' cineradiographic analyses show that the shoulder joint is stabilized but that the radius and proximal carpus rotate freely (Fleagle, 1974; Jenkins et al., 1978; Jenkins, 1981). This radius\-proximal carpus rotation can then provide propulsive motion, while at the same time freeing the forearm bones from torsion. Stabilization at the shoulder combined with the hingelike morphology of the humero-ulnar joint can then result in torsional loading of the humerus.

Together, these biomechanical techniques allow a new kind of in–depth view of the functional and ultimately the biological significance of anatomy. In the case of brachiating gibbons, these approaches may change our ideas about how limb elongation may evolve. In general, any animal is better off with long limbs; limb elongation increases stride length, locomotor velocity, and reach. In biology as in everyday life, however, there is no free lunch: as limb length increases, stresses due to bending may easily climb precipitously, in direct relation to bone length. For this reason, limb length is probably constrained in most terrestrial mammals. In vivo strain data show that the way gibbons use their unique anatomy reduces the amount of bending experienced, so that selection for increased bone length could be favored with minimal risk of increased strain and bone failure. This is an excellent example of how the application of biomechanics to the biology of the primate postcranium can enlighten our understanding of the evolution of structural and functional design.

LITERATURE CITED

Aiello, LC (1981) The allometry of primate body proportions. Symp. Zool. Soc. Lond. *48:*334–358.

——— (1984) Applications of allometry: The postcrania of higher primates. In H Preuschoft, DH Chivers, WY Brockelman, and N Creel (eds.): The Lesser Apes: Evolutionary and Behavioral Biology. Edinburgh: Edinburgh University Press, pp. 170–179.

Alexander, RM (1981a) Analysis of force platform data to obtain joint forces. In D Dowson and V Wright (eds.): An Introduction to the Biomechanics of Joints and Joint Replacement. London: Mechanical Engineering Publications Ltd., pp. 30–35.

——— (1981b) Animal Mechanics. Oxford: Blackwell Scientific Publications.

——— (1988) Elastic Mechanisms in Animal Movement. Cambridge: Cambridge University Press.

Alexander RM, and Vernon A (1975) The mechanics of hopping in kangaroos (Macropodidae). J. Zool. Lond. *177:*265–303.

Avis V (1962) Brachiation: The crucial issue for man's ancestry. S.W. J. Anthrop. *18:*119–148.

Beer FP, and Johnston ER, Jr. (1981) Mechanics of Materials. New York: McGraw Hill.

Bennett MB, Ker RF, Dimery NJ, and Alexander RM (1986) Mechanical properties of various mammalian tendons. J. Zool. Lond. *209:*537–548.

Biewener AA (1983) Locomotory stresses in the limb bones of two small mammals: The ground squirrel and chipmunk. J. Exp. Biol. *103:*131–154.

——— (1992) Biomechanics: A Practical Approach. Vol. 2, Structures. Oxford: Oxford University Press, pp. 1–20.

Biewener AA, Thomason J, Goodship AE, and Lanyon LE (1983) Bone stress in the horse forelimb during locomotion at different gaits: A comparison of two experimental techniques. J. Biomech. *16:*565–576.

Calder WA III (1984) Size, Function and Life History. Cambridge: Harvard University Press.

Carlstedt CA, and Nordin M (1989) Biomechanics of tendons and ligaments. In M Nordin and VH Frankel (eds.): Basic Biomechanics of the Musculoskeletal System. Philadelphia and London: Lea & Febiger, pp. 59–74.

Cock AG (1966) Genetical aspects of metrical growth and form in animals. Quart. Rev. Biol. *41:*131–190.

Currey JD (1984) The Mechanical Adaptations of Bones. Princeton: Princeton University Press.

——— (1990) Physical characteristics affecting the tensile failure properties of compact bone. J. Biomech. *23:*837–844.

Demes B, and Günther MM (1989) Biomechanics and allometric scaling in primate locomotion and morphology. Folia Primatol. *53:*125–141.

Evans FG, and King AL (1961) Regional differences in some physical properties of human spongy bone. In FG Evans (ed.): Biomechanical Studies of the Musculoskeletal System. Springfield, Ill.: C. C. Thomas, pp. 49–67.

Fleagle JG (1974) Dynamics of a brachiating siamang (*Hylobates [Symphalangus] syndactylus*). Nature *248:*259–260.

——— (1985) Size and adaptation in primates. In WL Jungers (ed.): Size and Scaling in Primate Biology. New York: Plenum Press, pp. 1–20.

Fleagle JG, Stern JT, Jr., Jungers WL, Susman RL, Vangor AK, and Wells JP (1981) Climbing: A biomechanical link with brachiation and with bipedalism. Symp. Zool. Soc. Lond. *48:*359–375.

Frankel VH, and Burstein AH (1970) Orthopaedic biomechanics. Philadelphia: Lea & Febiger.

Gamble JG (1987) The Musculoskeletal System. New York: Raven Press.

Gans C, and Bock WJ (1965) The functional significance of muscle architecture: A theoretical analysis. Ergeb. Anat. Entwicklgesch. *38:*115–142.

Gans C, and d. Vree F (1987) Functional bases of fiber length and angulation in muscle. J. Morph. *192:*63–85.

Goldstein SA (1987) The mechanical properties of trabecular bone: Dependence on anatomic location and function. J. Biomech. *20:*1055–1061.

Gordon JE (1978) Structures. New York: Plenum Press.

Goslow GE (1985) Neural control of locomotion. In M Hildebrand, DM Bramble, KF Liem, and DB Wake (eds.): Functional Vertebrate Morphology. Cambridge: Harvard University Press, pp. 338–365.

Gregory WK (1928) Were the ancestors of man primitive brachiators? Proc. Am. Philos. Soc. *67:*129–150.

Jenkins FA, Jr. (1981) Wrist rotation in primates: A

critical adaptation for brachiators. Symp. Zool. Soc. Lond. *48*:429–451.

Jenkins FA, Jr., Dombroski PJ, and Gordon EP (1978) Analysis of the shoulder in brachiating spider monkeys. Am. J. Phys. Anthrop. *48*:65–76.

Jobbins B (1981) Experimental stress analysis and material testing. In D Dawson and V Wright (eds.): An Introduction to the Biomechanics of Joints and Joint Replacement. London: Mechanical Engineering Publications, Ltd., pp. 42–48.

Jungers WL (1984a) Aspects of size and scaling in primate biology with special reference to the locomotor skeleton. Yrbk. Phys. Anthrop. *27*:73–97.

——— (1984b) Scaling of the hominoid locomotor skeleton with special reference to lesser apes. In H Preuschoft, DJ Chivers, WY Brockelman, and N Creel (eds.): The Lesser Apes: Evolutionary and Behavioral Biology. Edinburgh: Edinburgh University Press, pp. 146–169.

——— (1985a) Body size and scaling of limb proportions in primates. In WL Jungers (ed.): Size and Scaling in Primate Biology. New York: Plenum Press, pp. 345–382.

——— (1985b) Size and Scaling in Primate Biology. New York: Plenum Press.

——— (1988) Relative joint size and hominoid locomotor adaptations with implications for the evolution of hominid bipedalism. J. Hum. Evol. *17*:247–265.

Jungers WL, and Stern JT, Jr. (1980) Telemetered electromyography of forelimb muscle chains in gibbons (*Hylobates lar*). Science *208*:617–619.

——— (1981) Preliminary electromyographic analysis of brachiation in gibbon and spider monkey. Int. J. Primatol. *2*:18–33.

——— (1984) Kinesiological aspects of brachiation in lar gibbons. In H Preuschoft, DJ Chivers, WV Brockelman, and N Creel (eds.): The Lesser Apes: Evolutionary and Behavioral Biology. Edinburgh: Edinburgh University Press, pp. 119–134.

Keith A (1903) On the degree to which the posterior segments of the body have been transmuted and suppressed in the evolution of man and allied primates. J. Anat. *37*:18–40.

——— (1923) Man's posture: Its evolution and disorders. Brit. Med. J. *1*:451–454, 499–502, 545–548, 587–591, 624–626, 669–672.

LaBarbera M (1989) Analyzing body size as factor in ecology and evolution. Ann. Rev. Ecol. Syst. *20*:97–118.

Lanyon LE (1974) Experimental support for the trajectorial theory of bone structure. J. Bone Joint Surg. *56B*:160–166.

Larson SG, and Stern JT, Jr. (1987) EMG of shoulder muscles during knuckle-walking: problems of terrestrial locomotion in a suspensory adapted primate. J. Zool. Lond. *212*:629–655.

Loeb GE, and Gans C (1986) Electromyography for Experimentalists. Chicago: University of Chicago Press.

McArdle BH (1988) The structural relationship: Regression in biology. Can. J. Zool. *66*:2329–2339.

McMahon TA (1984) Muscles, Reflexes and Locomotion. Princeton: Princeton University Press.

Martens M, VanAudekercke R, Delport P, DeMeester P, and Muelier JC (1983) The mechanical characteristics of cancellous bone at the upper femoral region. J. Biomech. 16:971–983.

Mow VC, Proctor CW, and Kelly MA (1989) Biomechanics of articular cartilage. In M Nordin and VH Frankel (eds.): Basic Biomechanics of the Musculoskeletal System. Philadelphia and London: Lea & Febiger, pp. 31–58.

Muvdi BB, and McNabb JW (1991) Engineering Mechanics of Materials. New York: Springer–Verlag.

Nachemson AL, and Evans JH (1968) Some mechanical properties of the third human lumbar interlaminar ligament (*ligamentum flavum*). J. Biomech. *1*:211–220.

Nordin M, and Frankel VH (1989) Biomechanics of bone. In M Nordin and VH Frankel (eds.): Basic Biomechanics of the Musculoskeletal System. Philadelphia and London: Lea & Febiger, pp. 3–29.

Pitman MI, and Peterson L (1989) Biomechanics of skeletal muscle. In M Nordin and VH Frankel (eds.): Basic Biomechanics of the Musculoskeletal System. Philadelphia and London: Lea & Febiger, pp. 89–111.

Preuschoft H, and Demes B (1984) Biomechanics of brachiation. In H Preuschoft, DJ Chivers, WV Brockelman, and N Creel (eds.): The Lesser Apes: Evolutionary and Behavioral Biology. Edinburgh: Edinburgh University Press, pp. 96–118.

——— (1985) Influence of size and proportions on the biomechanics of brachiation. In WL Jungers (ed.): Size and Scaling in Primate Biology. New York and London: Plenum Press, pp. 383–399.

Reynolds TR (1985a) Mechanics of increased support of weight by the hindlimbs in primates. Am. J. Phys. Anthrop. *67:*335–349.

——— (1985b) Stresses on the limbs of quadrupedal primates. Am. J. Phys. Anthrop. 67:351–362.

Schmidt-Nielsen K (1984) Scaling: Why Is Animal Size So Important? Cambridge: Cambridge University Press.

Schultz AH (1936) Characters common to higher primates and characters specific for man. Quar. Rev. Biol. *11:*259–283, 425–455.

——— (1953) The relative thickness of the long bones and the vertebrae in primates. Am. J. Phys. Anthrop. *11:*277–311.

——— (1973) The skeleton of the Hylobatidae and other observations on their morphology. In D Rumbaugh (ed.): Gibbon and Siamang. Basel: Karger, pp. 1–54.

Shea BT (1981) Relative growth of the limbs and trunk in the African apes. Am. J. Phys. Anthrop. *56:*179–201.

——— (1985) Ontogenetic allometry and scaling: A discussion based on the growth and form of the skull in African Apes. In WL Jungers (ed.): Size and Scaling in Primate Biology. New York: Plenum Press, pp. 175–207.

Sokal RR, and Rohlf FJ (1981) Biometry: The Principles and Practice of Statistics in Biological Research. San Francisco: W. H. Freeman and Co.

Sprugel DG (1983) Correcting for bias in log–transformed allometric equations. Ecology *64:*209–210.

Squire JM (1986) Muscle: Design, Diversity and Disease. Menlo Park, Calif.: Benjamin/Cummins.

Stern JT, Jr., Wells JP, Jungers WL, and Vangor AK (1980b) An electromyographic study of serratus anterior in atelines and *Alouatta.* Am. J. Phys. Anthrop. *52:*323–334.

Stern JT, Jr., Wells JP, Jungers WL, Vangor AK, and Fleagle FG (1980a) An electromyographic study of the pectoralis major in atelines and *Hylobates* with special reference to the evolution of a pars clavicularis. Am. J. Phys. Anthrop. *52:*13–25.

Stern JT, Jr., Wells JP, Vangor AK, and Fleagle JG (1977) Electromyography of some muscles of the upper limb in *Ateles* and *Lagothrix.* Yrbk. Phys. Anthrop. *20:*498–507.

Stuart DG, and Enoka RM (1983) Motoneurons, motor units and the size principle. In RG Grossman and WD Willis (eds.): The Clinical Neurosciences. Section 5, Neurobiology. New York: Churchill Livingston, pp. 338–365.

Swartz SM (1989) Pendular mechanics and the kinematics and energetics of brachiating locomotion. Int. J. Primatol. *10:*387–418.

——— (1991) Strain analysis as a tool for functional morphology. Amer. Zool. *31:*655–669.

Swartz SM, Bertram JEA, and Biewener AA (1989) Telemetered in vivo strain analysis of locomotor mechanics of brachiating gibbons. Nature *342:*270–272.

Swartz SM, and Biewener AA (1992) Shape and scaling. In AA Biewener (ed.): Biomechanics: A Practical Approach. Vol. 2, Structures. Oxford: Oxford University Press, pp. 21–43.

Treharne RW (1981) Review of Wolff's Law and its proposed means of operation. Orthop. Rev. *10:*35–47.

Tuttle RH (1972) Functional and evolutionary biology of hylobatid hands and feet. In DM Rumbaugh (ed.): Gibbon and Siamang. Basel: Karger, pp. 136–206.

Tuttle RH, and Basmajian JV (1974) Electromyography of *Pan gorilla:* An experimental approach to the problem of hominization. Symp. 5th Cong. Intl. Primat. Soc. pp. 313–314.

Tuttle RH, Basmajian JV, and Ishida H (1975) Electromyography of the gluteus maximus muscle in *Gorilla* and the evolution of hominid bipedalism. In RH Tuttle (ed.): Primate Functional Morphology and Evolution. The Hague: Mouton, pp. 253–269.

Vogel S (1988) Life's Devices: The Physical World of Animals and Plants. Princeton: Princeton University Press.

Wainwright SA (1988) Axis and Circumference: The Cylindrical Shape of Plants and Animals. Cambridge: Harvard University Press.

Wainwright SA, Biggs WD, Currey JD, and Gosline JM (1976) Mechanical Design in Organisms. London: Edward Arnold.

Wolff JD (1869) Über die Bedentung der Architectur der spongiosun Substanz. Centralbl. die Med. Wissensch. *54:*849–851.

Part II The Primate Body

CHAPTER TWO

Functional Morphology of the Shoulder in Primates

Susan G. Larson

The primate shoulder is a region of considerable complexity for the functional morphologist. In large part this complexity stems from the anatomy itself. The shoulder region includes three bony elements—the clavicle, scapula, and humerus—and four articulations—the glenohumeral joint or shoulder proper, the sternoclavicular joint, the acromioclavicular joint, and the so-called scapulothoracic "joint." Most of these articulations are capable of undergoing motions in several planes. Possible motions at the glenohumeral joint include virtually all combinations of abduction-adduction, flexion-extension, and axial rotation. The scapula can rotate on the rib cage in a glenoid-up glenoid-down fashion, undergo cranial and caudal translation, and can be protracted (move forward) or retracted (move toward the vertebral column). The sternoclavicular joint provides the only bony anchor for the shoulder girdle, yet this joint also permits elevation and depression, protraction and retraction, and rotation of the clavicle.

Each of these articulations is an independent entity, capable of independent movement, yet all contribute to overall shoulder motion in a complex coordination of simultaneous movements. For example, the simple act of arm-raising in humans involves abduction and lateral rotation at the glenohumeral joint; elevation, rotation, and retraction of the clavicle; plus rotation and cranial translation of the scapula.

With this complex interplay of motions, it is often necessary to assess the contribution of each joint to the overall movement in order to understand how some particular motion is brought about. Unfortunately, because of the relatively free movement of the scapula under the skin, this information is not easily attained without some form of radiographic analysis. Several studies have used radiographs to estimate the ratio of glenohumeral to scapulothoracic motion during arm-raising in humans, that is, to document what Codman (1934) called "the scapulohumeral rhythm," (Flecker, 1929; Inman et al., 1944, Yamshon and Bierman, 1948; Saha, 1961; Freedman and Munro, 1966; Doody et al., 1970; Poppen and Walker, 1976; Bagg and Forrest, 1988). Yet even the analysis of this relatively simple movement has produced conflicting results. Some workers report a constant, although not always the same, ratio of glenohumeral to scapulothoracic motion (Inman et al., 1944, Yamshon and Bierman, 1948; Poppen and Walker, 1976), whereas others indicate that the ratio changes as elevation proceeds (Freedman and Munro, 1966; Doody et al., 1970; Bagg and Forrest, 1988). Only recently have researchers begun to examine any movements besides simple abduction of the arm, and again, only in humans (e.g., Howell et al., 1988). With the significant exception of the cineradiographic study by Jenkins et al. (1978) on the spider monkey shoulder during

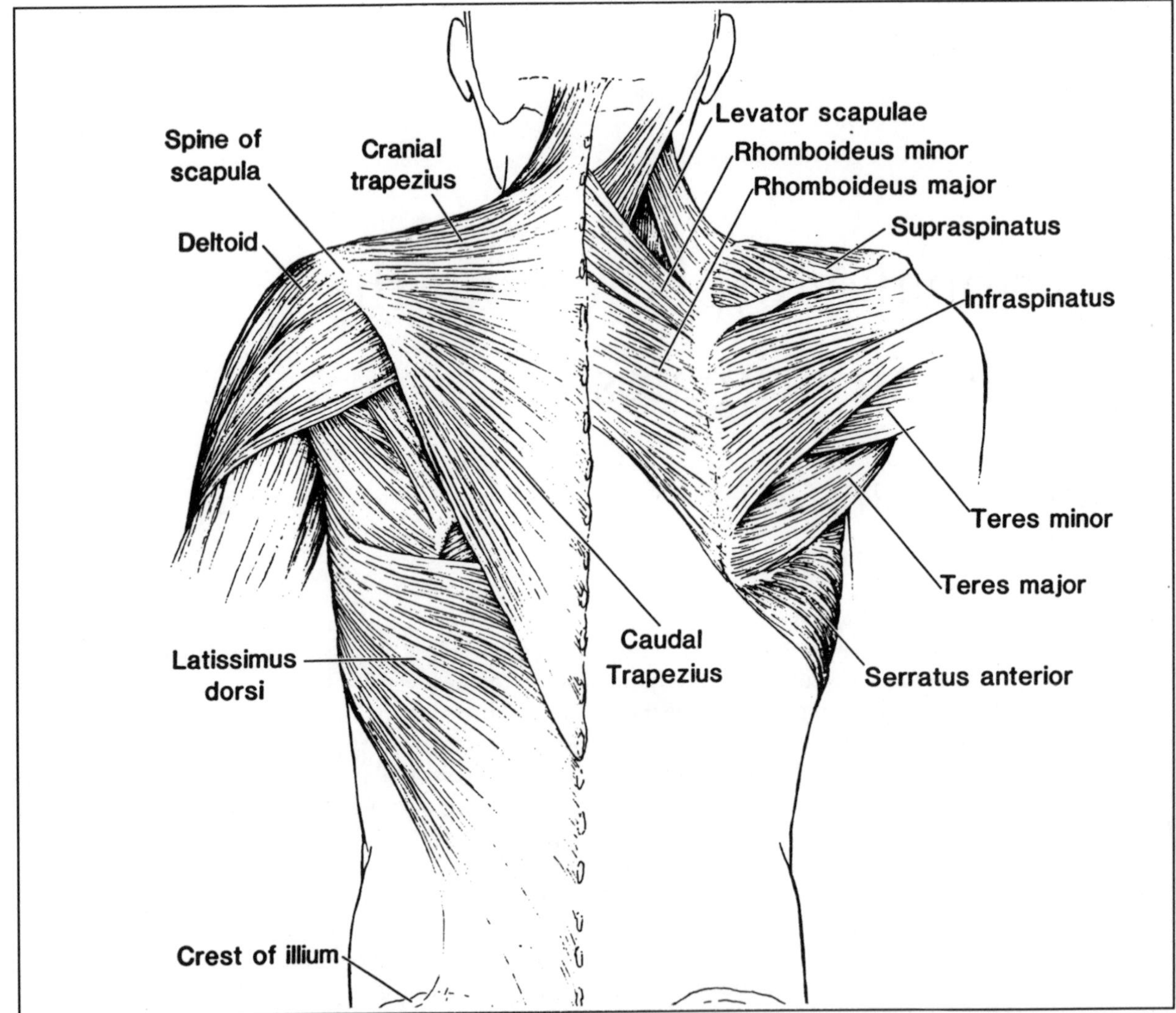

FIGURE 2.1. Dorsal musculature of the human shoulder region. Redrawn from Hollinshead and Jenkins (1981).

arm-swinging, no comparable data on shoulder motion in nonhuman primates exist.

Another aspect of the morphological complexity of the shoulder region is the large number of muscles involved in producing shoulder motion (Figs. 2.1, 2.2, and 2.3). Anatomically, these muscles can be sorted into three groups (Inman et al., 1944): (1) the axioscapular muscles, that is, those arising from the thorax and inserting onto the scapula, including the trapezius, rhomboids, serratus anterior, and levator scapulae; (2) the axiohumeral muscles, that is, those arising from the thorax and inserting onto the humerus, including the latissimus dorsi and the pectoral muscles;[1] and (3) the scapulohumeral muscles, that is, those arising from the scapula and inserting onto the humerus, including the teres major, the deltoid, and the muscles collectively known as the rotator cuff. Consideration of function produces a different sorting of muscles such as that used by Ashton and Oxnard (1963): (1) muscles responsible for propulsion including the latissimus dorsi, the pectoralis major, and the teres major; (2) muscles responsible for arm-raising including the deltoid, the trapezius, and the serratus anterior; (3) muscles responsible for glenohumeral stabilization including the rotator cuff members; and (4) muscles responsible for shoulder girdle stabilization including the rhomboids, the levator scapulae, and the subclavius. Both systems have their value, but neither is ideal. The anatomical classification is important in indicating which joints particular muscles can act upon, but does not reveal any information about what they do. The functional classification groups together muscles involved in a common motion, but does not describe any specific muscle's contribution to that motion. It also oversimplifies muscle function by not indicating the variety of motions that a single muscle can be involved in and the diversity of roles it can play in the production of motion. But as heuristic devices, these classification schemes have gone a long way in reducing the overall complexity in muscular anatomy of the shoulder for the functional morphologist.

Beyond the basic intricacy of the anatomy, functional interpretation of the morphology of this region can be made difficult because of the shoulder's complicated adaptive role. Among the evolutionary trends that Le Gros Clark (1959) lists as characterizing the order Primates are specializations of the extremities for grasping related to survival in an arboreal habitat. The role of the shoulder in this context has been to provide the mobility necessary to reach the irregular supports of a discontinuous arboreal substrate. Among higher primates, the grasping hands have evolved into manipulatory organs, again emphasizing the need for mobility at the shoulder. This emphasis on mobility is reflected in many aspects of the morphology of the primate shoulder such as a large, globular humeral head and relatively small glenoid fossa. Such adaptations for

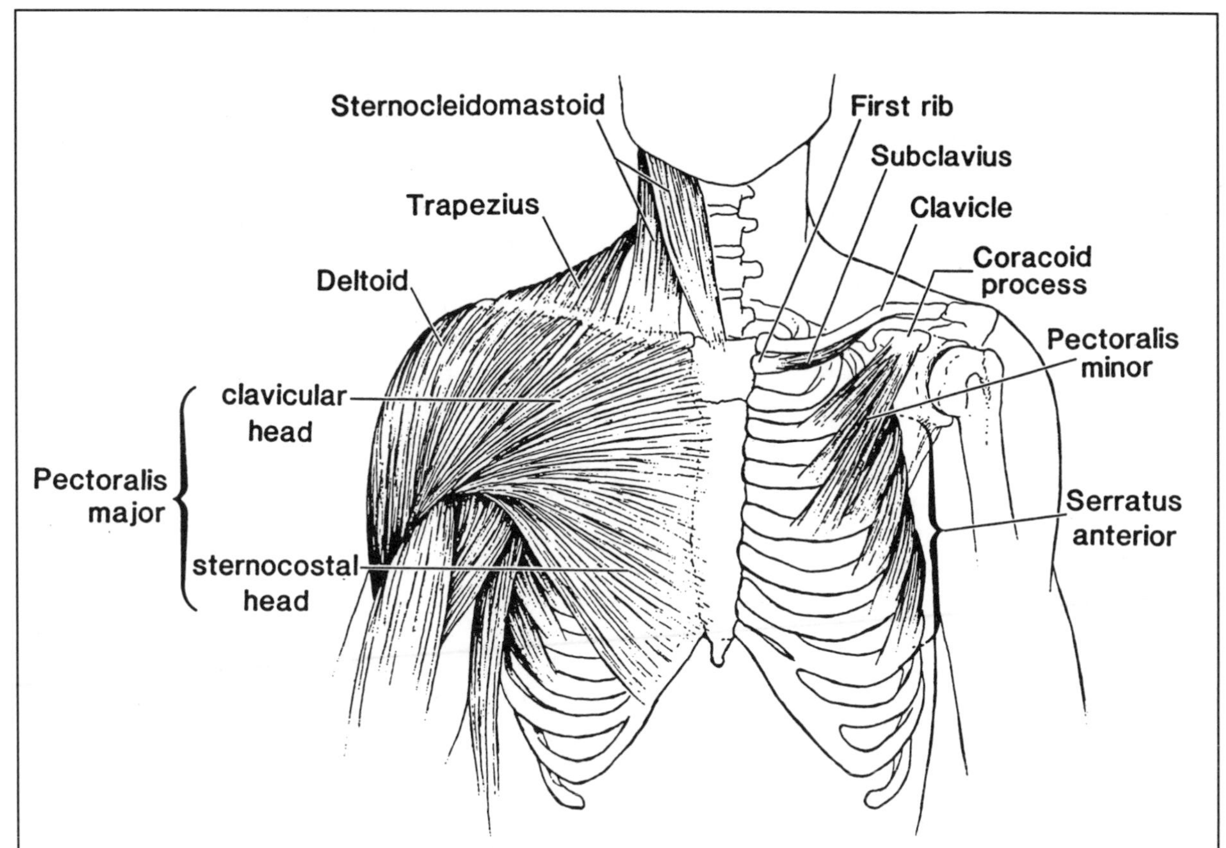

FIGURE 2.2. Ventral musculature of the human shoulder region. Redrawn from Hollinshead and Jenkins (1981).

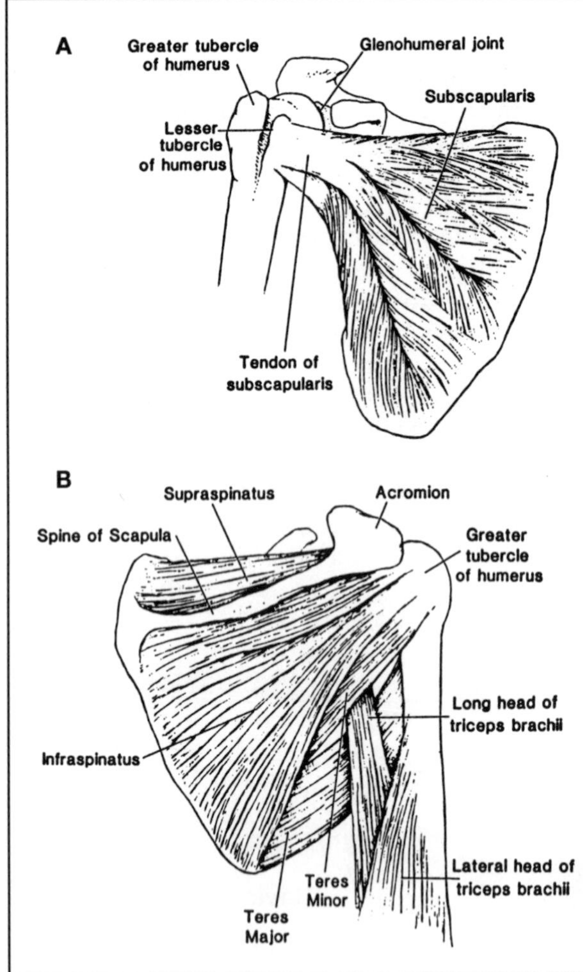

FIGURE 2.3. Deep scapulohumeral muscles of the human shoulder. **A**: The ventral rotator cuff muscle the subscapularis. **B**: Dorsal members of the rotator cuff. Redrawn from Hall-Craggs (1990).

enhanced mobility generally result in reduced stability of the joint, yet in no primate except humans has the upper limb been completely freed from support and locomotor roles. Thus the morphology of the shoulder region is always a compromise between the conflicting functional demands for free mobility in order to reach and grasp arboreal supports and to allow effective use of the hand as a manipulatory organ, versus the need for sufficient stability to support the weight of the body and to withstand the disruptive forces generated during locomotion. Therefore, in attempting to understand the functional significance of some particular character, one must realize that the morphology of that trait may not be ideal for a single purpose; the functional morphologist must attempt to understand how the balance of conflicting demands has been achieved in the species under study.[2]

It is this complexity, however, that makes the shoulder such an intriguing and challenging region to study. It is also a region that displays a significant amount of morphological variation across different species, suggesting the existence of a variety of evolutionary modifications in accommodation to different habits and habitats. The shoulder thus presents a great deal of potential for the investigation of the relationship between form and function. In this regard, primates represent a nearly ideal group to study due to the great diversity in postures and modes of locomotion displayed across the order.

The shoulder is also a region of particular interest to physical anthropologists because of the potential insights it might provide in understanding one of the most significant events in human evolution, namely, the origins of erect posture and bipedalism. Many of the theories about the evolution of bipedalism have emphasized the importance of an arboreal stage in human evolution as a preadaptation to upright posture. For example, for many years the general consensus held that arm-swinging was the major factor preadapting a protohominid species to the development of orthograde posture and bipedalism (Keith, 1902, 1923; Gregory, 1927, 1928a, 1928b; Morton, 1926, 1927; see Tuttle, 1974, 1975 for reviews). Others have argued that the progenitors of humans were at most "prebrachiators" possessing a combination of primitive and derived characters relating to upper limb mobility, but without the extreme specializations of advanced brachiators such as the extant apes (Straus, 1949; Schultz, 1950, 1968; Le Gros Clark, 1959; Napier, 1963; Napier and Napier, 1967). More recently, several theorists have suggested that climbing can explain the preadaptation of the hind limb to bipedalism (Stern, 1971, 1976; Tuttle, 1975; Prost, 1980, 1985; Fleagle et al., 1981; Stern and Susman, 1981). Being able to understand and interpret the significance of features of the upper limb shared between humans and certain nonhuman primates, and to apply this understanding to the interpretation of fossils, could help us choose between these alternatives. Although understanding the morphology and evolution of the hominid upper limb might not reveal any direct information about the development of bipedalism, it should give us insight into the nature of the locomotor stage preceding or accompanying the assumption of upright posture.

COMPARATIVE MORPHOLOGY OF THE PRIMATE SHOULDER

Although the variation in osteological and muscular anatomy of the shoulder among nonhuman primates forms a fairly continuous spectrum, consistent distinctions can be made between the species at either end of that spectrum. This pair-wise contrast between so-called suspensory primates, on the one hand, and dedicated terrestrial quadrupedal primates, on the other, creates a useful device for summarizing the range of morphological features observed within the order. It is readily acknowledged, however, that such a dichotomy is artificial and largely ignores the many interesting and unique aspects of the forms intermediate between these two extremes. With this caveat in mind, in the following paragraphs I will attempt to give an overview of the contrasting features that have been described for anthropoid primates.[3]

The element of the shoulder that has received by far the most attention is the scapula. Numerous studies have focused on the scapula's shape and configuration (Erikson, 1963; Ashton and Oxnard, 1964; Ashton, Healy, Oxnard, and Spence, 1965; Ashton, Oxnard, and Spence, 1965; Ashton et al., 1971, 1976; Corruccini and Ciochon, 1976; Oxnard, 1963, 1967; Roberts, 1974; Manaster, 1979; Kimes et al., 1979; Kimes et al., 1981; Doyle et al., 1980; Schultz, 1986; Shea, 1986; Takahashi, 1990). In one group, primate scapulae are longer (measured from the vertebral border to the glenoid fossa) than they are wide (measured from the superior to inferior angle) (Fig. 2.4A). Because the blade of the scapula is not very wide, the supra- and infraspinous fossae are rather narrow. The glenoid fossa in these species is fairly tall and narrow and is directed ventrally, and the acromion does not project very far beyond the glenoid. At the opposite end of the spectrum are scapulae that are very broad as measured along the vertebral border, and the supra- and infraspinous fossae are, therefore, wide (Fig. 2.4B). The distance from the vertebral border to the glenoid fossa is shorter than is the case in the first group, and the scapular spine itself is more oblique in its orientation. The glenoid fossa in these primates is wider and faces more cranially than in the scapulae of the first group, and the acromion tends to project significantly beyond the glenoid. There are many primate scapulae that are intermediate in form between these two extremes.

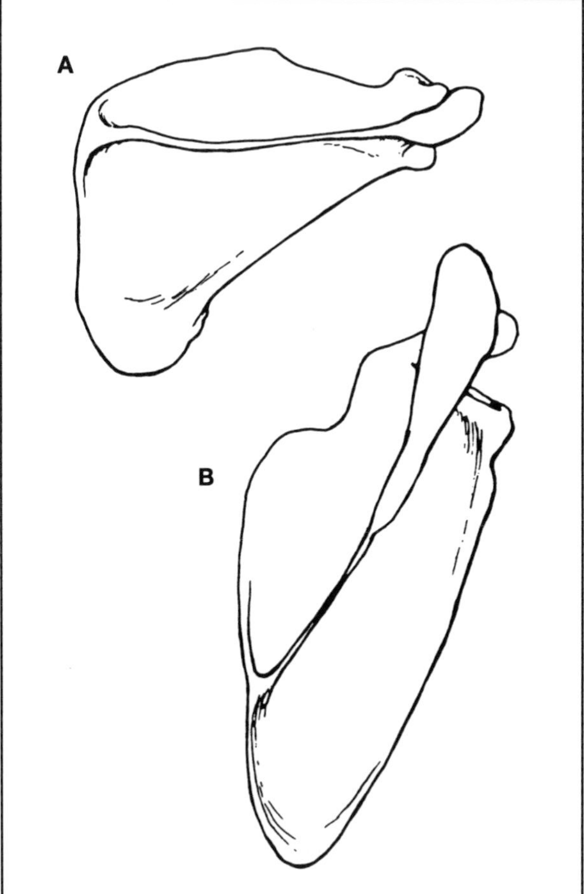

FIGURE 2.4. Comparison of the scapulae from a quadrupedal and a suspensory primate. **A** is a dorsal view of a vervet (*Cercopithecus aethiops*) scapula, and **B** is a similar view of a siamang (*Hylobates syndactylus*) scapula. The vervet scapula is long (measured along the scapular spine) and narrow, the dorsal fossa are relatively narrow, and the acromion does not project very far beyond the glenoid. The siamang scapula is wider along the vertebral border, the dorsal fossae are relatively large, the scapular spine is obliquely placed, and the acromion projects significantly beyond the glenoid.

The primate species whose scapulae fall into the

first group are primarily quadrupedal, either on the ground or in the trees, such as macaques, baboons, several cercopithecines, and many nonateline New World monkeys. The second group includes primates that engage in some amount of suspensory behavior as well as climbing, bridging, and other arboreal activities involving use of the upper limb in overhead postures. They include the hominoids as well as the atelines, although some authors would place the latter into an intermediate group. This intermediate group consists of quadrupedal monkeys that are more versatile in the trees than the first group, and can be described as being highly agile engaging in a wide variety of scrambling, bridging, and leaping behaviors as well as branch running and walking. Of course, as with morphological groupings, such generalizations about locomotor behaviors are clearly artificial and ignore many unique aspects of the locomotor repertoires of individual species. They do, however, suggest an initial direction to take in the investigation of the functional significance of the shared traits.

The proximal humerus has received far less attention than the scapula in morphological studies of the primate shoulder. Some features that have been investigated are the size and shape of the humeral head, its orientation relative to the transverse axis of the distal articular surface, the degree to which the greater tubercle projects proximally beyond the level of the humeral head, and the position of the deltoid tuberosity on the shaft of the humerus. Roughly the same groupings of primates can be achieved on the basis of contrasts in any one of these features. The hominoid primates and atelines have large, rounded humeral heads that are directed medially (to a greater or lesser degree depending on the taxon) relative to the transverse axis of the elbow (Inman et al., 1944; Evans and Krahl, 1945; Erikson, 1963). The superior surface of the head is usually at least equal to the level of the greater tubercle, if not higher than it, and the deltoid tuberosity is halfway or more down the shaft of the humerus (Oxnard, 1963; Napier and Davis, 1959; Le Gros Clark, 1959). The most strictly quadrupedal primates have more oval-shaped humeral heads that are narrower medio-laterally and somewhat flattened proximally (Gebo et al., 1988; Rose, 1989; Harrison, 1989). The articular surface of the humeral head faces almost directly posteriorly relative to the transverse axis of the elbow. The greater tubercle often projects proximally beyond the level of the humeral head, especially in terrestrial quadrupedal monkeys (Jolly, 1967; Birchette, 1982; Ciochon, 1986), and the deltoid insertion is only one-quarter to one-third the way down the humeral shaft. Again, the more agile arboreal quadrupedal monkeys tend to have features intermediate between these two groups.

The clavicle has been largely ignored in most studies of comparative primate anatomy. It is generally acknowledged, however, that most hominoids and atelines have relatively long clavicles that are obliquely oriented so that their lateral end is higher than the medial end, giving these species a "hunched-shoulder" appearance (Erikson, 1963; Schultz, 1930, 1973; Napier and Napier, 1967). In addition, the apes have marked clavicular torsion (Oxnard, 1968). Quadrupedal primates tend to have fairly short, flat clavicles with a relatively small amount of torsion.

Muscular differences also distinguish these groups of primates. Among the axioscapular muscles, the trapezius and the caudal portion of serratus anterior display contrasts correlated with locomotor differences. In quadrupedal monkeys, the upper and lower parts of the trapezius are approximately equal in size, while in the hominoids the cranial portion is broader and thicker than the caudal portion (Miller, 1932; Ashton and Oxnard, 1963; Oxnard, 1963). The individual digitations of the caudal part of serratus anterior are longer and thicker in hominoid primates than in quadrupedal monkeys, and they arise from more ribs, so that the most inferior digitations are oriented along an almost direct craniocaudal axis (Miller, 1932; Ashton and Oxnard, 1963; Howell and Straus, 1971; Stern et al., 1980). The two major axiohumeral muscles—the latissimus dorsi and sternocostal pectoralis major—comprise a larger mass in the apes than in quadrupedal monkeys (Oxnard, 1963, 1967). In addition, the lateral margin of the latissimus dorsi in the apes has been expanded to include fibers arising from the iliac crest and traveling directly cranially to the humerus (Miller, 1932; Ashton and Oxnard, 1963). Finally, among the scapulohumeral muscles, the deltoid is hypertrophied in "suspensory" primates (Oxnard, 1967) and the members of the rotator cuff are estimated to be relatively larger on the basis of the size of the scapular fossae (Roberts, 1974).

Besides these individual features, there is a major structural contrast that distinguishes the hominoids from quadrupedal monkeys. This difference involves the shape of the rib cage and the position of the scapula upon it. Like most other mammals, the thorax in quadrupedal monkeys is deep dorso-ventrally, and more narrow medio-laterally (Schultz, 1930). The scapula is placed lateral to the rib cage and is directly in line vertically with the humerus, radius, and ulna in a quadrupedal posture (Keith, 1923; Miller, 1932; Roberts, 1974). In the apes, the thorax is broader than it is deep, and the scapula is positioned dorsally so that the glenoid faces laterally rather than ventrally (Keith, 1923; Le Gros Clark, 1959; Schultz, 1956, 1973). In a quadrupedal posture, therefore, the scapula and humerus are no longer aligned.[4]

FUNCTIONAL INTERPRETATIONS OF THE LOCOMOTOR CONTRASTS IN MORPHOLOGY

The recognition of morphological similarities between humans and apes and the implications raised by a presumedly shared locomotor history have been major stimuli for investigation of the primate shoulder. Although arm-swinging has generally been assumed to be the major factor influencing the evolutionary modification of the hominoid upper limb, the functional analysis of the hominoid shoulder has actually focused mainly on features associated with enhancement of the arm-raising mechanism rather than with arm-swinging per se (e.g., Erikson, 1963; Oxnard, 1963, 1967; Ashton and Oxnard, 1964; Ashton, Healy, Oxnard, and Spence, 1965; Ashton, Oxnard, and Spence, 1965; Ashton et al., 1971, 1976; Ciochon and Corruccini, 1977; Corruccini and Ciochon, 1976). Relatively less attention has been paid to how suspending the body below the supporting hand may have influenced shoulder anatomy. With a few notable exceptions, (e.g., Jolly, 1967; Manaster, 1979; Kimes et al., 1979; Doyle et al., 1980; Schultz, 1986), the shoulders of quadrupedal monkeys have been studied mainly as contrasts to the morphology observed among hominoids and atelines.

The Scapula

Many of the contrasts in scapular shape between hominoids and atelines versus more strictly quadrupedal monkeys have been related to scapular rotation. During arm-raising, the scapula rotates in a "glenoid-up" fashion to provide a mobile platform for the humerus at the glenohumeral joint and thereby increases the total range of movement for the upper limb. In a now-classic electromyographic study (see chapter 1 for a description of the technique of electromyography), Inman et al. (1944) described a

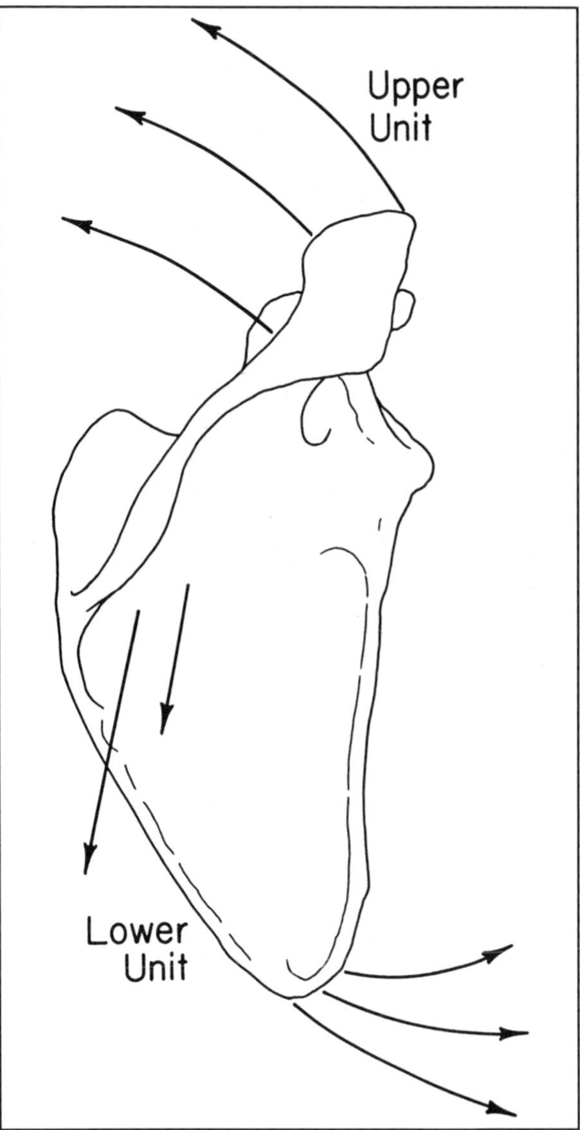

FIGURE 2.5. Scapulothoracic force couple as described by Inman et al. (1944). According to this model, scapular rotation is brought about by the action of the cranial trapezius as the upper unit of the couple and the caudal serratus anterior and caudal trapezius as the lower unit of the couple. Although the cranial portion of trapezius appears to participate in human scapular rotation, it is inactive during arm elevation in chimpanzees.

muscular force couple—made up principally of the cranial portion of trapezius and the caudal serratus anterior—that brought about this scapular rotation in humans (Fig. 2.5). The scapulothoracic force couple model thus suggested an obvious interpretation for some of the muscular differences observed in hominoids and atelines versus quadrupedal monkeys. That is, the attachments and increased size of the upper trapezius and the caudal serratus anterior in the brachiators as compared to quadrupedal monkeys were said to reflect an enhancement of the arm-raising mechanism (Ashton and Oxnard, 1963; Oxnard, 1963, 1967).

Drawing upon this interpretation of muscular function, Ashton and Oxnard went on to analyze the possible influences that changes in the leverage of these muscles might have on scapular shape (Ashton and Oxnard, 1964; Oxnard, 1963, 1967). They suggested that an increase in the angulation of the scapular spine would change the orientation of the fibers of the upper trapezius so that they could act more as rotators and less as cranial translators. In addition, they reasoned that increasing the distance of the inferior angle from the axis of rotation (i.e., lengthening the lever arm of the caudal serratus anterior) would enhance the ability of the caudal serratus anterior to produce scapular rotation, and would increase the efficiency of the force couple by maximizing the length of the couple arm. Much of the expansion of the cranio-caudal width of the scapula could therefore be related to this combination of factors acting to improve the scapular rotation mechanism. They also suggested that a cranial orientation of the glenoid fossa would be associated with a habitual raised-arm position.

Based on these biomechanical deductions, Ashton, Oxnard, and their colleagues devised a series of measurements on the scapula designed to document a number of features including the manner of attachment and the lever-arm lengths of the upper trapezius and the caudal serratus anterior among different primate species (Ashton and Oxnard, 1964; Ashton, Healy, Oxnard, and Spence, 1965; Ashton, Oxnard, and Spence, 1965; Ashton et al., 1971, 1976; Oxnard, 1963, 1967). First with univariate, then later with multivariate techniques, these authors were able to demonstrate that primates could be accurately sorted into locomotor groups in large part on the basis of these scapular characters.

The scapular metrics devised by Ashton, Oxnard, and their colleagues were successful in discriminating different locomotor groups within primates, which seemed largely to validate their biomechanical interpretations. But the scapulothoracic force couple model had only been documented by electromyography in humans. Tuttle and Basmajian (1977) were the first to use electromyography to attempt to test hypotheses of muscle function in nonhuman primates. In their study of "scapular rotators" in the great apes, they observed that contrary to expectations the upper trapezius and the caudal serratus anterior were only minimally active during arm-raising (Tuttle and Basmajian, 1977). To account for these surprising results, Tuttle and Basmajian suggested that the cranial orientation of the glenoid in the apes had reduced the need for scapular rotation during arm-raising.

Subsequent electromyographic studies on caudal serratus anterior activity in gibbons and spider monkeys (Jungers and Stern, 1984; Stern et al., 1980) have revealed that the recruitment of this muscle is more complex than earlier analyses might have indicated. These authors reported that in gibbons and spider monkeys there is internal differentiation within the fan-shaped array of muscular digitations of the caudal serratus anterior. Although most of the digitations did participate in scapular rotation during arm-raising, the most extreme caudal digitations were inactive during such behaviors. These cranio-caudally oriented muscular bundles were active instead to help transmit the weight of the trunk to the limb during suspensory postures and locomotion.

In a reanalysis of caudal serratus anterior function in chimpanzees, Larson et al. (1991) observed a similar pattern of recruitment to that reported by Stern et al. (1980) and Jungers and Stern (1984). The more dorso-ventral bundles were active in all arm-raising behaviors, whereas the cranio-caudal digitations were active during pendant suspension and the support phase of arm-swinging.

The EMG results on gibbons, spider monkeys, and chimpanzees, therefore, support the association that has been made between the caudal elongation of the inferior angle of the scapula, as a reflection of caudal serratus anterior leverage for scapular rotation, and the importance of arm-raising in the positional repertoire of a primate species (Ashton and Oxnard, 1964; Ashton, Healy, Oxnard, and Spence,

1965; Oxnard, 1963, 1967). They refute the view that the unique caudal expansion of the muscle to include cranio-caudally oriented bundles, as found in the apes, is also a reflection of an enhancement of the arm-raising mechanism. Instead, these extreme caudal bundles appear to be more directly related to suspensory postures per se.

Although Tuttle and Basmajian (1977) observed little or no activity in the cranial trapezius during arm-raising in their great ape subjects, Jungers and Stern (1984) found the cranial trapezius to be more active in the gibbon, including during the arm elevation that occurs in the second half of the swing phase of brachiation. They suggested that the recruitment of the cranial trapezius could often be linked to head-turning rather than scapular movement. This has been investigated in some detail by Larson et al. (1991) in their reexamination of cranial trapezius recruitment in the chimpanzee. They found that in nearly all cases, activity in the cranial trapezius was directly linked to turning the head and not to scapular movement. They concluded that if head position was taken into account, then the cranial trapezius was essentially inactive during all arm-raising behaviors in the chimpanzee.

The negligible participation of the cranial trapezius in arm-raising is a surprising result. It sets chimpanzees (and perhaps other apes) apart from humans who apparently do use the cranial trapezius to help rotate the scapula during arm elevation (Inman et al., 1944; Yamshon and Bierman, 1948; Wiedenbauer and Mortensen, 1952; Bearn, 1961; Bull et al., 1984; Bagg and Forrest, 1986; pers. obs.). This nonparticipation in arm elevation raises the question of why the cranial portion of trapezius has been enlarged in the great apes. Larson et al. (1991) have proposed that perhaps its size can be related to the muscular demands for head control in these large primates arising from some combination of relative head size, the posterior position of the atlanto-occipital joint, and the high degree of facial prognathism. The merits of this explanation aside, they concluded that the nonparticipation of the cranial trapezius in scapular rotation calls into question the association that has been made between the orientation of the scapular spine and the importance of arm-raising in "brachiating" primates (Ashton and Oxnard, 1964; Ashton, Healy, Oxnard, and Spence, 1965; Oxnard, 1963, 1967). They suggested that the angulation of the scapular spine may instead be related to the dorsal members of the rotator cuff and the roles they play during arm-swinging and suspension (Larson and Stern, 1986; Larson et al., 1991).

The muscles known collectively as the rotator cuff (supraspinatus, infraspinatus, teres minor, and subscapularis—see Fig. 2.3) are, in fact, the other muscular factors thought to have a significant influence on scapular shape. Because they arise from the dorsal and ventral scapular fossae, enlarging or reducing their relative sizes is presumed to have a major impact on scapular dimensions. Wolffson (1950) tested this relationship empirically by removing one or more of these muscles in newborn rats and observing the resulting changes in scapular form at the end of postnatal growth. Removal of members of the rotator cuff resulted in severe reductions in the sizes of the scapular fossae.

Research on relative fossa size has concentrated on the supra- and infraspinous fossae. Several workers have investigated differences in the relative sizes of these dorsal scapular fossae across different primate species, including Frey (1923), Schultz (1930), Inman et al. (1944), Roberts (1974), Corruccini and Ciochon (1976), Manaster (1979), and Shea (1986). Generally speaking, the more arboreal acrobatic species tend to have larger supra- and infraspinous fossae than the more strictly terrestrial primates, and the hominoids and atelines have the largest fossae of all. The increase in size is brought about mainly through a widening of the fossae, and therefore is thought to contribute to the greater cranio-caudal breadth of the scapula observed in hominoids and atelines.[5] Roberts (1974) reported that although the supra- and infraspinous fossae both vary in size across primates, the infraspinous fossa is invariably larger than the supraspinous fossa. In addition, among the hominoids, the infraspinous fossa displays a wide range of relative sizes whereas the relative size of the supraspinous fossa is fairly constant. One outstanding exception to this general trend is the gorilla, with a supraspinous fossa that is completely outside the range of other primates. Because of these variations within locomotor categories, Ashton, Oxnard, and Spence (1965) included measurements on the dorsal scapular fossae among their list of "residual dimensions" of the shoulder thought to reflect taxonomic rather than functional differences among primates. However, these residual dimensions were only marginally successful in making taxonomic distinctions and the absence of a functional "signal" was never clearly demonstrated.

The generally accepted functional explanation for the differences in scapular fossa size among primate species again focuses mainly on the arm-raising mechanism and the work of Inman et al. (1944). In their EMG study on human shoulder muscles, Inman et al. (1944) described a second force couple, this time operating at the glenohumeral joint (Fig. 2.6). The upper unit of the couple included the deltoid and the supraspinatus, and the lower unit consisted of the infraspinatus, the teres minor, and the subscapularis plus the pressure and friction of the humeral head against the glenoid. According to this model, the deltoid, assisted by the supraspinatus, provided the power for arm elevation. The remaining members of the rotator cuff created a downward pull on the humeral head to resist the tendency for the deltoid to produce superior displacement of the humerus rather than elevation. The important roles played by the rotator cuff muscles during arm elevation, as proposed by the scapulohumeral force couple model, thus suggested an explanation for their differential development (as reflected in scapular fossa size) in acrobatic arboreal primates.

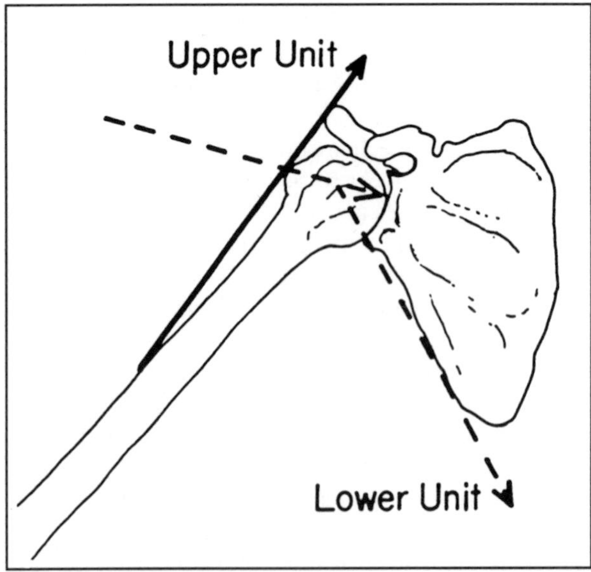

FIGURE 2.6. Scapulohumeral force couple as described by Inman et al. (1944). In this model, the power for glenohumeral elevation is provided by the deltoid (the main component of the upper unit), while the tendency for the deltoid to produce superior humeral displacement is resisted by the lower unit, made up of the infraspinatus, the teres minor, and the subscapularis. The third vector included by Inman et al. (1944) was said to represent the pressure and friction of the head of the humerus at the glenoid.

Roberts (1974) also emphasized the importance of the cuff muscles in maintaining glenohumeral joint stability. According to Roberts, in order for primates to be able to use their forelimbs in a more versatile fashion than most nonprimate mammals, they have had to sacrifice some intrinsic joint stability in favor of increased joint mobility. Primates therefore must rely more heavily on muscular effort to provide stabilization, mainly through the action of the members of the rotator cuff. This should be especially true for primates engaging in suspensory postures and locomotion where presumably the glenohumeral joint would be under transarticular tensile stress, and for the knuckle-walking hominoids where the repositioning of the scapula onto the dorsum of the thorax has introduced high shearing stress at the glenohumeral joint during quadrupedal postures.

Roberts (1974) proposed that the twofold function of the rotator cuff musculature—that is, its roles in glenohumeral movement and stabilization—could be used to account for many of the observed differences in relative sizes of the scapular fossae. This interpretation of rotator cuff function was based only on the EMG data published for humans by Inman et al. (1944). According to this model, the cuff muscles should act in concert, either in their capacities as participants in the force couple during behaviors involving arm elevation, or as joint stabilizers during suspensory and/or quadrupedal postures and locomotion. It therefore cannot account for the differential development of either the supra- or infraspinous fossa in particular primate species as documented by Roberts (1974).

Tuttle and Basmajian (1978) examined the activity patterns of the rotator cuff muscles and deltoid in the great apes during a variety of behaviors involving arm elevation, hanging and hoisting. Although their data set was not complete for each animal, they concluded that the patterns of recruitment they observed largely agreed with those reported for human subjects with the notable exception of the teres minor in the orangutan. This muscle was surprisingly inactive during brachial elevation. They also reported that none of the cuff muscles were active during pendant bimanual or unimanual hanging and concluded that joint integrity was maintained solely by osseoligamentous structures, contrary to predictions by Roberts (1974).

Larson and Stern (1986, 1987) conducted detailed studies of scapulohumeral muscle recruitment in the chimpanzee during reaching and active locomotion. Their results indicate that the members of the rotator cuff behave more as individuals than previous conceptions of their functional roles would suggest. An additional discovery was that the subscapularis does not act as a single unit, but rather shows internal differentiation and can be roughly divided into three parts: upper, middle, and low (Fig. 2.7).

Larson and Stern (1986) found that the early phase of arm elevation is brought about by the combined activity of only the deltoid and supraspinatus. They concluded that the force couple model for glenohumeral elevation suggested by Inman et al. (1944) was inapplicable to chimpanzees, and perhaps incorrect in general. Instead they proposed that the tendency for the deltoid to displace the humeral head superiorly during the initial phase of elevation was prevented by those fibers of the rotator cuff that pass superior to the joint (Fig. 2.8). Thus the supraspinatus with its attachment to the top of the greater tubercle bears the main responsibility for preventing humeral displacement in the early phase of any elevation. Depending on the orientation of the humerus and the direction of movement, this responsibility may be shared by, or shifted to the infraspinatus or the upper subscapularis as the movement proceeds.

Larson and Stern (1986) emphasized the distinct role of each member of the rotator cuff in free arm movements. The supraspinatus is a more or less

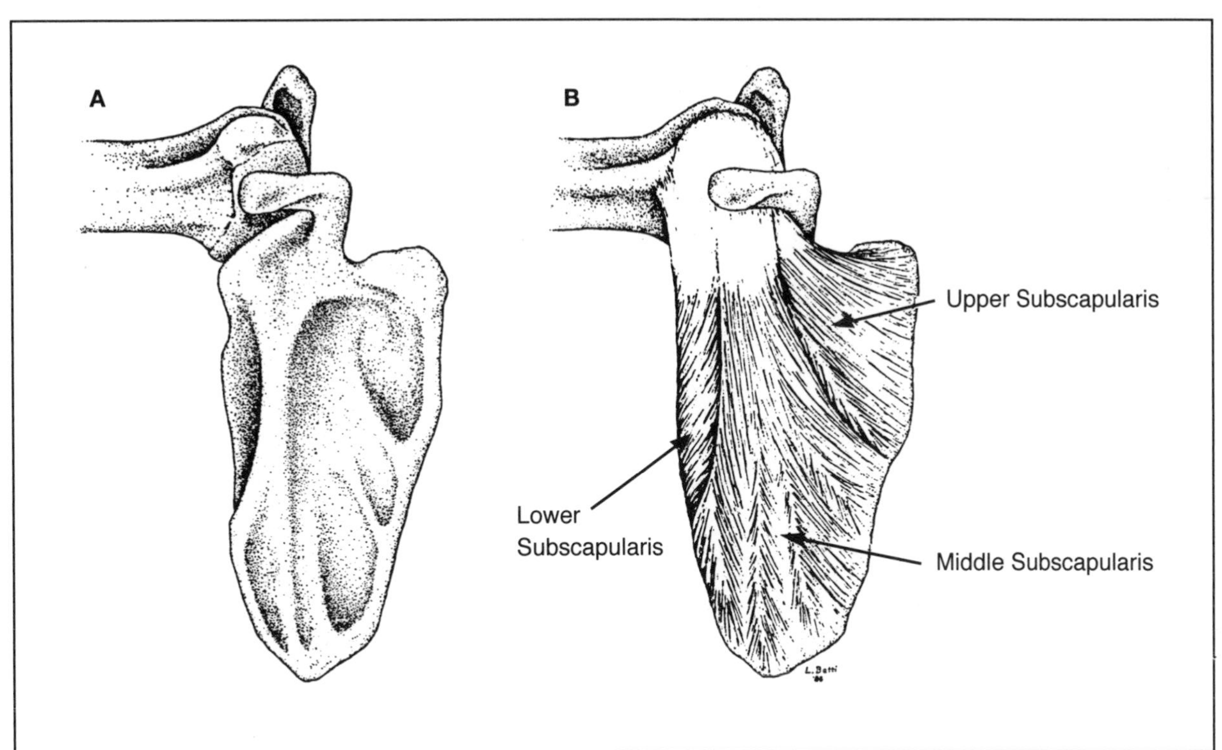

FIGURE 2.7. Chimpanzee subscapular fossa (**A**) and subscapularis muscle (**B**). The muscle appears to be divided into three contiguous parts: An upper portion near the superior angle, a middle portion, and a lower portion arising from the ventral sulcus near the axillary border. Although there is a gradient of EMG activity from the uppermost to the lowermost parts, in general terms the upper portion of subscapularis acts as an abductor-medial rotator, the middle portion as a pure medial rotator, and the lower portion as an adductor-medial rotator. See Larson and Stern (1986) for additional information.

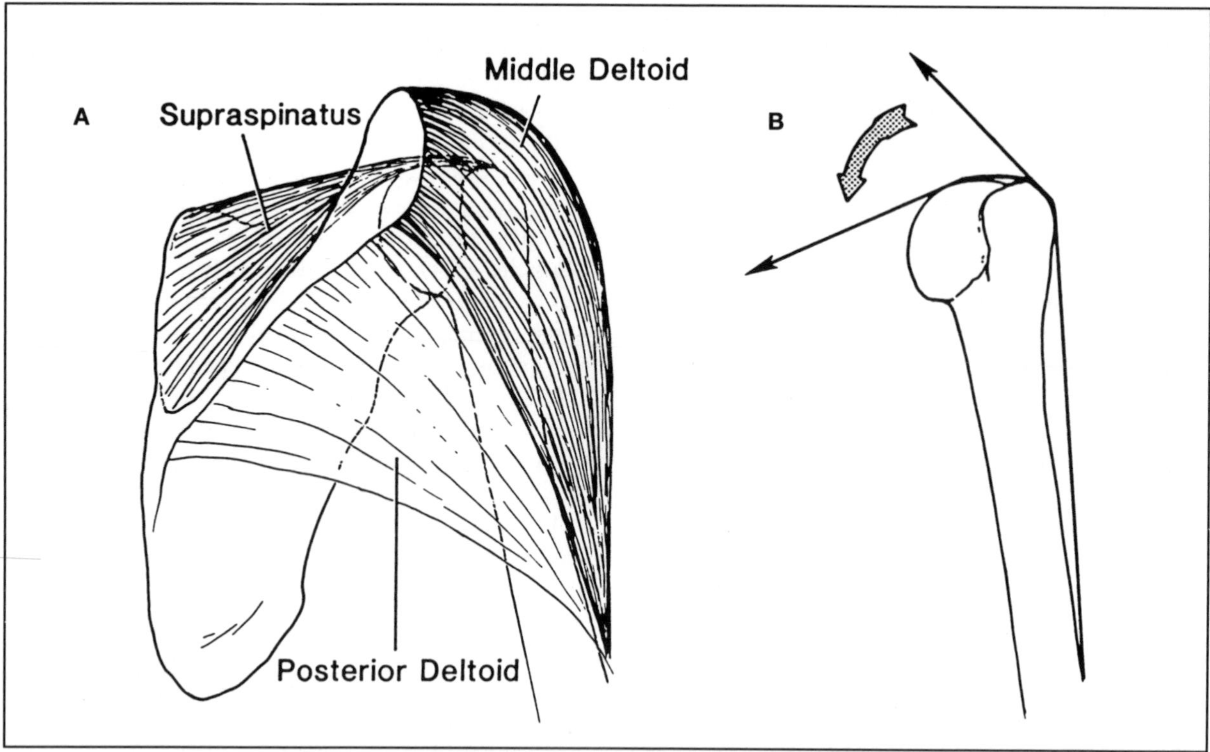

FIGURE 2.8. Mechanism for glenohumeral elevation as proposed by Larson and Stern (1986). **A**: schematic view of the principal muscles involved in the model. **B**: isolated view of forces acting to elevate arm. As in the scapulohumeral force couple model described by Inman et al. (1944), the power for arm elevation is provided by the deltoid. However, the tendency for the deltoid to produce superior displacement of the humerus is resisted, not by the lower members of the rotator cuff but by the fibers of the cuff that pass superior to the joint, mainly the supraspinatus. From its attachment on the greater tubercle the supraspinatus produces a rotatory torque to the humeral head causing it to spin downward in an abductory motion (hatched arrow). It is possible that the supraspinatus also acts as a literal barrier to superior humeral displacement acting in a sense as a dynamic ligament. In primates with highly mobile shoulders where the glenoid is very shallow and the humeral head is higher than the tubercles, the supraspinatus may provide additional stabilization to the joint by the downward force component produced by the passage of the muscle over the superior surface of the humeral head.

pure abductor playing the dual role of resisting humeral displacement and assisting the deltoid in providing abductory power. The infraspinatus is an abductor/lateral rotator adding additional abductory power through the middle of an elevation to assist the deltoid and to laterally rotate the arm. The upper subscapularis acts as an abductor/medial rotator becoming active when lateral rotation is to counteracted, or if medial rotation is needed. The middle and lower portions of subscapularis and the teres minor did not contribute to arm-raising in any significant way, but on independent evidence, Larson and Stern (1986) concluded that the middle portion of the subscapularis is a pure medial rotator; the lower subscapularis is an adductor/medial rotator; and the teres minor is an adductor/lateral rotator.

During locomotor behaviors, Larson and Stern (1986, 1987) observed additional evidence for functional differentiation between individual members of the rotator cuff. The subscapularis was unique among the cuff muscles in playing an important role during the support or "pull-up" phase of climbing up a vertical tree trunk. Contrary to the conclusion of Tuttle and Basmajian (1978) that joint integrity during pendant suspension was maintained solely by osseoligamentous structures, Larson and Stern (1986) found that the infraspinatus was consistently active during unimanual or bimanual hanging and during the support phase of arm-swinging, and they concluded that this muscle plays a special role as a joint stabilizer resisting transarticular stress during suspensory behaviors. During knuckle-walking, a different type of joint stabilization is required, in this case, resistance to shear stress produced by the lateral orientation of the glenoid and the resulting tendency for the humerus to displace dorsally (Roberts, 1974). The supraspinatus and the infraspinatus were observed to perform this important stabilizing role alone, without the assistance of the other cuff muscles (Larson and Stern, 1987). Finally, Larson and Stern (1986, 1987) have emphasized the distinct nature of teres minor recruitment compared to other members of the rotator cuff. It plays no important role during arm-raising behaviors; during arm-swinging its recruitment is more similar to the posterior deltoid and the teres major than to any of the other members of the cuff; and it plays a completely distinct role from the other cuff muscles during knuckle-walking.

The individualization of the rotator cuff muscles described by Larson and Stern for chimpanzees provides a framework for interpretation of the variations in scapular fossa size that have been documented for different primate species (as previously discussed). Additional research is required to verify the general applicability of these interpretations of muscular function to other species.

Two additional features of scapular form that have received some attention in the literature are the degree of projection of the acromion laterally beyond the glenoid, and the shape of the glenoid fossa. Inman et al. (1944) pointed out that the acromion of hominoid primates and *Ateles* projected noticeably beyond the glenoid, and linked this greater development directly to the importance of the deltoid in glenohumeral elevation. This feature has been quantified in primates by Corruccini and Ciochon (1976) who reported that acromion length was one of the most significant features distinguishing hominoids from quadrupedal monkeys. Ciochon and Corruccini (1977) analyzed acromion length in conjunction with coracoid length as a coraco-acromial projection index. They associated a high value for this index with the existence of a coraco-acromial ligament, a feature that appears to be unique to hominoid primates. Ciochon and Corruccini (1977) suggested that this ligament—and by association, a projecting acromion and coracoid—form a protective cuff that limits the superior movement of the humerus. This explanation may have been derived from the human clinical literature where the coraco-acromial ligament is often associated with a fairly common disability known as impingement syndrome, which involves painful elevation of the arm due to contact between the proximal humerus and the ligament. A more biomechanically motivated explanation for the existence of this ligament has been presented by Putz et al. (1988). They attached strain gauges (see chapter 1) to the coracoid process and acromion of cadavers and applied simulated muscular tension to the two bony processes through the tendons of the attached muscles. Putz et al. (1988) concluded that the ligament acts as a dynamic brace shunting distorting forces from one bony process to the other. They also reported that after division of the ligament, significantly more distortion was registered in the acromion than in the coracoid process implying that the "stay" effect was more important for the acromion than for the coracoid. This would suggest that the development of the coraco-acromial ligament in the hominoids was probably an adaptation to stabilize a laterally projecting acromion. It seems most likely that the elongation of the acromion itself was related to improvement of the leverage for the deltoid as originally suggested by Inman et al. (1944).

Roberts (1974) has discussed the functional significance of differences in glenoid fossa shape. He suggested that the cranio-caudally elongate form of the glenoid in quadrupedal monkeys permits a wider range of motion in a parasagittal plane. The elongation of the cranial margin to form a liplike structure acts to prevent dislocation of the glenohumeral joint when the humerus is highly retracted. An ovate glenoid, most extremely developed in the hominoids and atelines, creates a more nearly true ball-and-socket joint that permits a wide range of motion in many directions. It is therefore an accommodation to the high degree of mobility necessary at the shoulder in primates engaged in acrobatic arboreal activities.

The Humerus

Rose (1989) has recently published an elegant morphological and functional analysis of the shape of the humeral articular surface in nonhuman primates. Although it is well recognized that the expansion of articular surface area in hominoids and atelines is associated with a greater range of motion

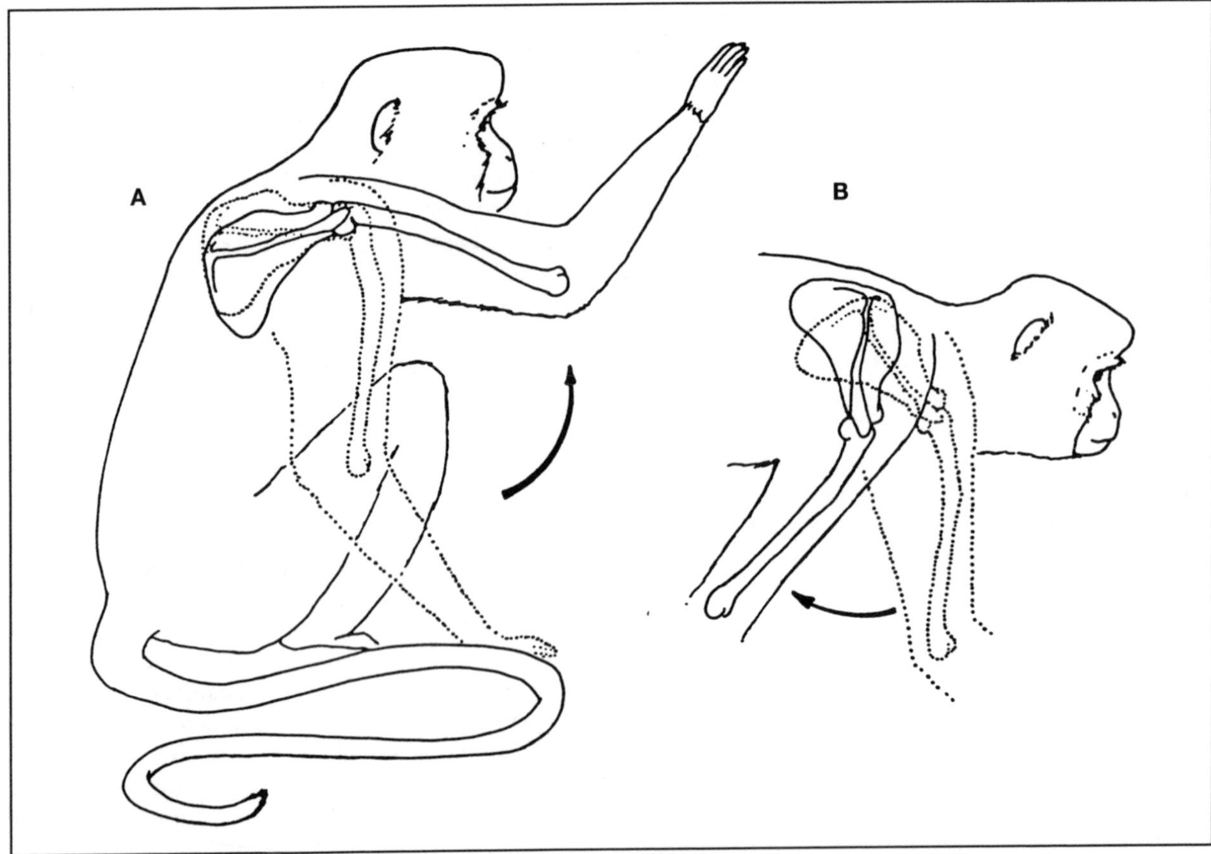

FIGURE 2.9. Functional configurations of the glenohumeral joint of a quadrupedal monkey. When the animal is in a sitting posture, as in **A**, the joint is extended (retracted) and the articular surface of the humeral head is nearly spherical in outline. Thus the glenohumeral articulation approximates a ball-and-socket joint permitting relatively free mobility for feeding and manipulative activities. During quadrupedal locomotion, as in **B**, the joint is fully flexed (protracted) and the glenoid is articulating with the flatter, more narrow region of the superior humeral articular surface. In this configuration, movement is severely limited and the joint is highly stabile and able to resist the disruptive forces generated during locomotion. Adapted from Rose (1989).

in these forms, Rose (1989) demonstrates how articular surface area must be considered in conjunction with the contour of the humeral head. In quadrupedal monkeys the humeral head is flattened proximally, whereas the distal region is more spherical in outline (see also Harrison, 1989). This creates two functional regions within the glenohumeral joint: one in which the joint is fully flexed (protracted), and one in which the joint is extended (retracted) (Fig. 2.9) (Rose, 1989). In the latter configuration, the glenohumeral joint approximates a ball-and-socket joint permitting abduction/adduction and axial rotation as well as flexion/extension. This is the position the joint would assume when the monkey was sitting and thus permits a wide range of motion for manipulative activities including food gathering. When the joint is fully flexed, such as during the weight-bearing phase of quadrupedal posture or locomotion, the glenoid articulates with the flatter, more narrow proximal articular surface of the humeral head. Movement is therefore severely limited and the joint is in a highly stable configuration resistant to the disruptive forces generated during locomotion (Rose, 1989).

The humeral head of the apes and the atelines is inflated (more spherical) overall compared to quadrupedal monkeys, but the most important distinction occurs in the proximal region of the articular surface. Here the articular surface area has expanded due to the migration of the tubercles (Fleagle and Simons, 1982; Rose, 1989), and the contour is uniformly rounded (Rose, 1989; Harrison, 1989). This results in an extensive range of motion at the glenohumeral joint even when it is in a fully flexed position, and therefore gives the hominoids and the atelines the necessary shoulder mobility for overhead postural, locomotor, and manipulative activities (Rose, 1989).

In addition to the shape of the articular surface of the humeral head, several workers have commented on the height of the superior surface of the head relative to the height of the greater tubercle (Fig. 2.10) (Jolly, 1967; Walker, 1974; Ziemer, 1978; Birchette, 1982; Fleagle and Simons, 1982; Ciochon, 1986; Gebo et al., 1988; Harrison, 1989). Jolly (1967) related the size of the greater tubercle to the action of the supraspinatus because the tubercle represents its lever arm at the glenohumeral joint. He suggested that a prolonged greater tubercle, as seen in terrestrial quadrupedal monkeys, gives the supraspinatus improved leverage for resisting passive extension of the glenohumeral joint during quadrupedal postures and the weight-bearing phase of locomotion. In addition, he proposed that the nonprojecting tubercle of more arboreal primates was associated with a short lever arm for the supraspinatus favoring fast (though weak) protraction of the humerus such as during the recovery (swing) phase of arboreal quadrupedalism (see chapter 1 for a discussion of the significance of different lever arm lengths).

Subsequent workers have tended to emphasize the presumed role of supraspinatus as a humeral protractor (flexor) in their interpretations of differences in greater tubercle height. A high greater tubercle was said to provide a more powerful recovery of the forelimb during terrestrial quadrupedal walking, whereas a low tubercle favored quick recovery in arboreal quadrupedalism (Walker, 1974; Birchette, 1982; Ciochon, 1986). Others have suggested that a high greater tubercle is associated with rapid humeral protraction during fast quadrupedal gaits (Fleagle and Simons, 1982; Gebo et al., 1988). Fleagle and Simons (1982) linked a low greater tubercle to glenohumeral stability suggesting that the positioning of the tubercles along the side of the humeral head in arboreal monkeys gave the rotator cuff muscles more extensive insertions around the head and thus improved muscular stabilization of the glenohumeral joint.

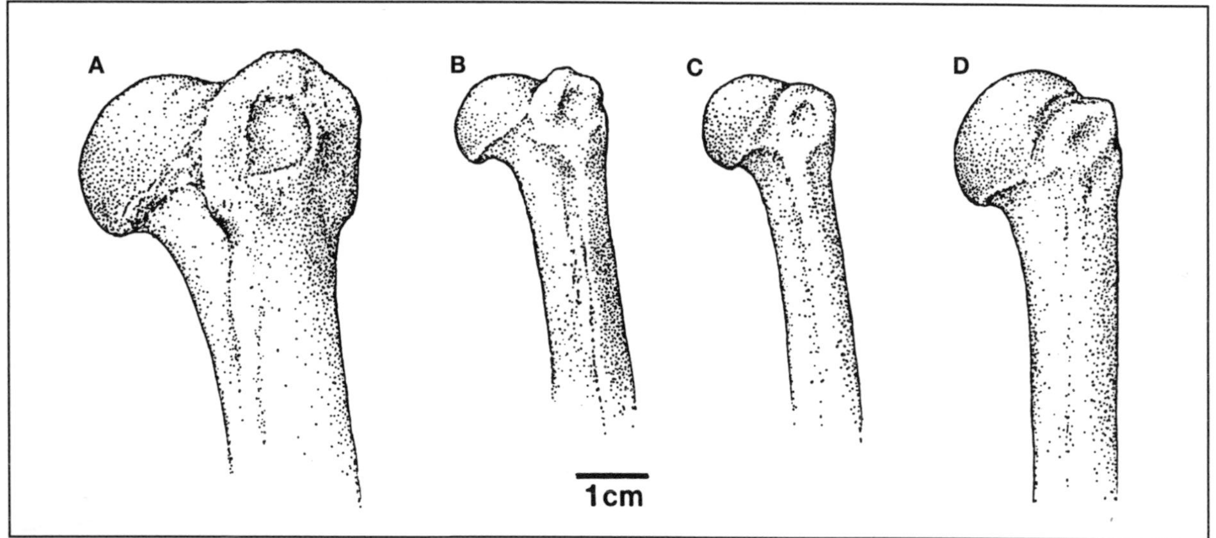

FIGURE 2.10. Lateral view of primate proximal humeri. A: *Papio anubis;* B: *Cercopithecus aethiops;* C: *Presbytis melalophos;* and D: *Ateles* sp. In A and B the greater tubercle extends proximally beyond the level of the head. In C the tubercle and head are approximately equal in height, whereas in D the head rises above the level of the tubercle. Reprinted from Larson and Stern (1989).

The common assumption upon which many of these interpretations are based, namely, that the supraspinatus acts as a humeral protractor (flexor) during the swing phase of quadrupedalism, has not, however, been borne out by electromyographic evidence. An initial indication that this was the case is found in a study on shoulder muscle recruitment during knuckle-walking in chimpanzees by Larson and Stern (1987). They reported that the supraspinatus was not consistently active during the swing phase of knuckle-walking. When it was recruited during swing phase, its activity was most directly related to the abductory state of the arm. Larson and Stern (1987) found that the most important role played by the supraspinatus in knuckle-walking came during support phase when it, with the infraspinatus, acted to stabilize the glenohumeral joint against dorsal displacement of the humerus.

Because chimpanzees are in many ways very atypical quadrupeds, the noninvolvement of the supraspinatus in humeral protraction during the swing phase of knuckle-walking could be attributed to their uniquely derived morphology. In a more direct test of the significance of differences in greater tubercle height in quadrupedal monkeys, Larson and Stern (1989) examined supraspinatus recruitment in the vervet monkey, a more cursorial primate species. They observed a nearly complete absence of activity during the swing phase of quadrupedal walking and galloping. Recently, they have reported a similar pattern of recruitment for the baboon and macaque (Larson and Stern, 1992). As with the chimpanzee, the most significant recruitment of the supraspinatus during quadrupedal locomotion in each of these species was in support phase again to help stabilize the joint. Also like the chimpanzee, when the arm was elevated against gravity, such as during a reach for food or during the swing phase of climbing, then the supraspinatus was active to help protract (flex) the arm. Larson and Stern (1989, 1992) concluded that differences in the height of the greater tubercle among primate species were clearly not related to "speed or power" of humeral protraction during the swing phase of quadrupedalism. Instead, they proposed that the explanation for differences in the leverage of the supraspinatus lay in its role in elevation of the arm against gravity.

Larson and Stern (1989, 1992) suggested that the so-called lowering of the greater tubercle in arboreal primate species, usually associated with a reduction in the lever arm for supraspinatus, is better viewed as a raising or angling of the humeral head upward to increase the overall mobility of the joint. Similar conclusions regarding the significance of an elevated humeral articular surface have been offered by Grand (1968), Walker (1974), Ziemer (1978), Rose (1989), and Harrison (1989). Larson and Stern (1989, 1992) indicated, however, that the resulting short lever arm produced by the elevation of the humeral head poses a disadvantage for an arboreal primate, especially because the frequency and adaptive importance of arm-raising is increased in an arboreal habitat. This difficulty is compounded by the fact that the greater mobility of the glenohumeral joint in arboreal primates probably also makes it inherently less stable, and therefore, places an even greater burden on the supraspinatus for joint stabilization (see Fleagle and Simons, 1982). Larson and Stern (1989, 1992) concluded that the only way for an animal with a low greater tubercle to deal with these heavy demands on the supraspinatus for brachial elevation and joint stabilization is to increase the overall size of the supraspinatus itself. As described previously, it is indeed the primate species with the greatest degree of glenohumeral mobility, namely the hominoids and the atelines, that have the largest supraspinous fossae.

According to Larson and Stern (1989, 1992), reduced glenohumeral mobility is the central factor in understanding the elongation of the greater tubercle in terrestrial quadrupedal primates. The supraspinatus in these species performs exactly the same roles in brachial elevation and glenohumeral stabilization as it does in more arboreal species. Because movement at the glenohumeral joint in terrestrial quadrupeds is mainly confined to a parasagittal plane, however, a projecting greater tubercle is no longer an obstacle to joint mobility. Therefore, terrestrial primates are able to utilize the mechanical advantage of a long lever arm to reduce the amount of force needed from the supraspinatus during arm elevation. They can thus take advantage of the energetic efficiency offered by a relatively smaller supraspinatus muscle (as reflected by the small size of their supraspinous fossae) while still maintaining its functional role.

The last feature of the primate shoulder that I wish to discuss is humeral torsion. In most quadrupedal mammals and monkeys, the humeral head faces posteriorly and superiorly to articulate with a ventral-facing glenoid fossa of a scapula that is positioned laterally on a dorso-ventrally deep thorax. In the hominoids, the scapula has been repositioned onto the dorsum of a medio-laterally widened rib cage, and the glenoid, therefore, faces laterally. Concomitantly, the humeral articular surface in hominoids has been rotated medially to maintain the glenohumeral articulation (Le Gros Clark, 1959). A number of researchers have analyzed the degree of humeral torsion in primate humeri including Martin (1933), Inman et al. (1944), Evans and Krahl (1945), Le Gros Clark and Thomas (1951), Napier and Davis (1959), Zapfe (1960), Preuschoft (1973), Fleagle and Simons (1982), Sarmiento (1985), Larson (1988), and Rose (1989).

Most early workers believed that humeral torsion was the result of real twisting of the humeral head (e.g., Martin, 1933; Inman et al., 1944; Evans and Krahl, 1945). According to Fleagle and Simons (1982), however, the appearance of torsion is largely due to a lateral migration and reduction in size of the lesser tubercle. This produced an expansion of the articular surface on the medial side of the head, thus giving the illusion of a shift in the central axis of the whole head. According to Fleagle and Simons (1982), the position of the greater tubercle has remained virtually unchanged. The implication of this view is that "torsion" is in fact mainly an expansion of the articular surface rather than a migration of the head.

Rose (1989) concurs with Fleagle and Simons (1982) that torsion is not produced by twisting of the proximal end of the humerus. He maintains, however, that the extensiveness of the articular surface is a distinct feature from torsion although an expanded articular surface and marked torsion may go together. Rose concludes that expansion of the articular surface can be brought about by migration of either or both tubercles. If it is mainly the result of lateral migration of the lesser tubercle, then an appreciable amount of torsion will also be produced. An even higher level of torsion results from the concomitant posterior migration of the greater tubercle. Rose (1989) asserts that the former condition describes *Pongo* and *Ateles,* whereas the latter is seen in the African apes. On the other hand, if the lateral migration of the lesser tubercle is accompanied by an anterior migration of the greater tubercle, then a highly inflated articular surface will be produced, but with a very low degree of torsion. According to Rose, this is the condition seen in the lesser apes.

Larson (1988) has offered a functional explanation for the differences in humeral torsion between gibbons and chimpanzees based primarily upon elbow positioning requirements. Although humeral torsion is often viewed simply as a component of the modifications in the hominoid shoulder for increased mobility (Miller, 1932; Le Gros Clark, 1959), humeral torsion is only necessary if it is important for the flexion/extension movements at the elbow joint to occur in a roughly parasagittal plane (Inman et al., 1944). Very likely through the mechanisms described by Rose (1989) summarized above, the gibbon humerus has a very large articular surface but with a relatively small amount of torsion. This results in a glenohumeral joint with a high level of mobility, but an elbow joint that has a lateral orientation when the arm is at rest (Fig. 2.11). Although this lateral set to the elbow must be overcome by extra muscular activity during free arm movements, Larson asserts that the gibbon tolerates this energetic inefficiency because of the advantages that a lateral orientation to the elbow joint offers during arm-swinging.

During slow pendular arm-swinging, the trunk undergoes approximately 180° of rotation under the supporting hand (Avis, 1962; Fleagle, 1974, 1977; Jungers and Stern, 1984; Preuschoft and Demes, 1984; Larson and Stern, 1986). At the termination of support phase, the upper limb is in an extreme state of lateral rotation. Larson (1988) noticed that this position was achieved in chimpanzees through hypersupination of the forearm and wrist, and lateral rotation of the shoulder so that the cubital fossa of the elbow faces up and inward (Fig. 2.12A). In gibbons, on the other hand, there seemed to be less pronounced supination of the forearm and wrist and more extreme rotation at the shoulder, judging by the orientation of the cubital fossa, which is actually visible from behind the animal (Fig. 2.12B). According to Larson, the extreme position of the elbow in gibbons is actually due to the lateral set of the elbow stemming from their low degree of humeral torsion, and there probably is no more rotation at the shoulder than occurs in the chimpanzee. Larson argued that the ability of gibbons to

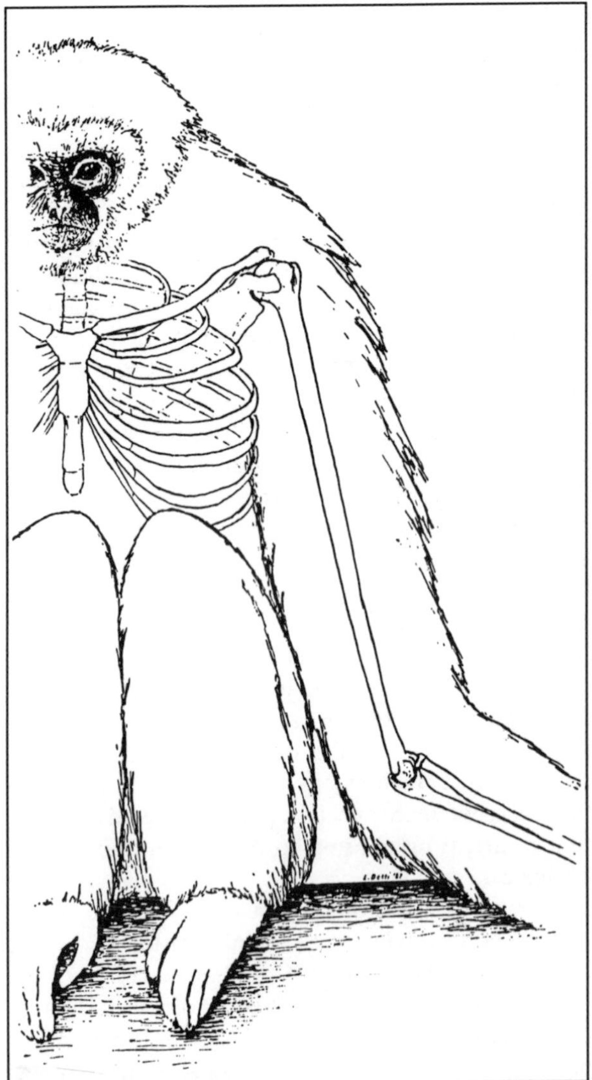

FIGURE 2.11. Gibbon in a typical resting posture. The arm is in a relaxed position, splayed out to the side and lightly resting on the supporting tree trunk. In this posture, the humerus is positioned such that the long axis of the head is roughly perpendicular to the glenoid fossa of the scapula. Because of the low degree of humeral torsion, the elbow is in a position that appears as if the arm is laterally rotated. Thus the low degree of humeral torsion gives the gibbon elbow a "lateral set." Reprinted from Larson (1988).

achieve such an extreme elbow position becomes critical during the more rapid form of arm-swinging known at ricochetal brachiation (Tuttle, 1969). In this mode of suspensory locomotion, the trunk undergoes only approximately 90° of rotation. At the end of support phase, the elbow undergoes sharp flexion to help raise the body's center of mass for a brief aerial phase before the next hand grasp. Because of the limited trunk rotation, to perform this motion the arm must be placed in an extreme position with the humerus extended and adducted behind the trunk and the transverse axis of the elbow horizontal. This position of the elbow would seem possible only with low humeral torsion and the resulting lateral set to the elbow joint. According to this proposal, therefore, a low degree of humeral torsion is associated with arm-swinging habits rather than marked torsion being associated with arm-swinging, as has been commonly believed (e.g., Miller, 1932; Le Gros Clark, 1959; Campbell, 1966).

If low humeral torsion is associated with suspensory locomotion, then what accounts for the marked level of torsion observed in the great apes, especially the African apes? Larson (1988) has suggested that the high level of humeral torsion in the African apes represents an accommodation to quadrupedal postures and locomotion by animals with dorsally placed scapulae. With a laterally facing glenoid, the only way to maintain an elbow joint that operates in a parasagittal plane is by having a medially directed humeral head. In support of this explanation, Larson noted that the most terrestrial of the great apes, the gorilla, has the most marked degree of humeral torsion, and the least terrestrial, the orangutan, has almost as little humeral torsion as the gibbon. Preuschoft (1973) has presented a brief explanation of humeral torsion among the apes along similar lines.

Sarmiento (1985) has also examined the functional implications of humeral torsion among the apes and, like Larson, linked the degree of torsion with the requirements for elbow joint positioning. In his analysis, the low amount of torsion in gibbons is seen as an advantage in arm-swinging to give the correct orientation to the elbow for flexion when the arm is abducted, and the high degree of torsion in the African apes is associated with the necessity for a parasagittally functioning elbow joint during quadrupedal postures and locomotion, conclusions similar to those later drawn by Larson (1988). In addition, however, Sarmiento (1985) proposed that vertical climbing is a major factor in the relative degree of humeral torsion among the apes. In order to grasp an overhead vertical support, the ape must be able to orient its cubital fossa medially to some degree. According to Sarmiento, the extent

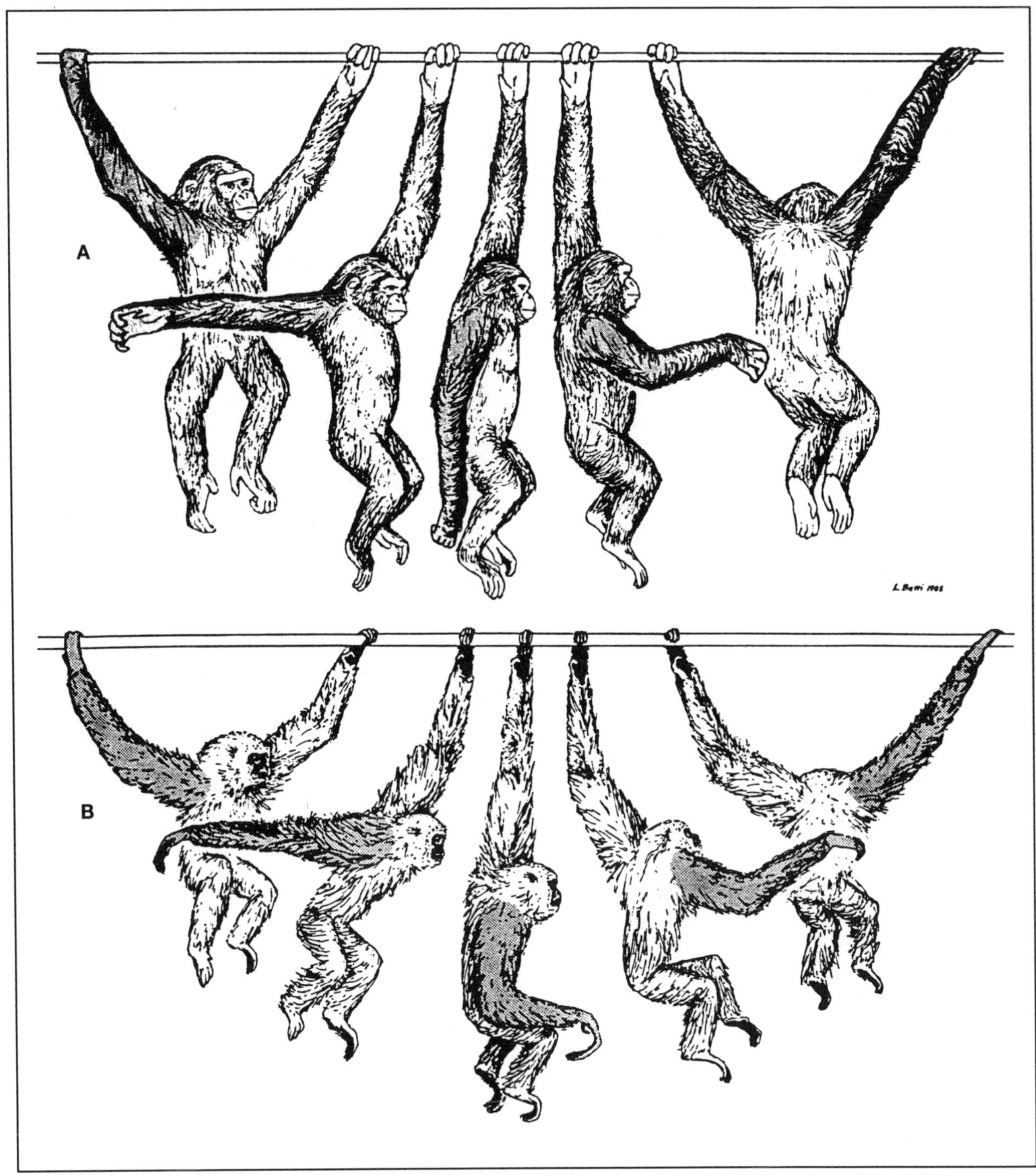

FIGURE 2.12. Chimpanzee (**A**) and gibbon (**B**) engaged in pendular arm-swinging. The body of either the gibbon or the chimpanzee swings like a pendulum beneath the supporting (unshaded) left arm as the trunk rotates around the long axis of the upper limb. The supporting limb begins in a state of neutral rotation, but by the end of support phase, the rotation of the trunk has brought the limb into a state of extreme rotation. For the chimpanzee in **A**, the position of the supporting arm at terminal support phase appears to be the result of a combination of lateral rotation of the arm and hypersupination of the forearm. In **B**, the cubital fossa is visible from behind the animal at terminal support phase, which suggests that the gibbon's arm is more extremely rotated than was the chimpanzee's. This is largely due to the "effective lateral rotation" of the distal end of the humerus resulting from the low degree of humeral torsion in gibbons. Reprinted from Larson (1988).

of the demands for medial rotation depend on the diameter of the vertical support being climbed, and how high the animal reaches. Gibbons—with low torsion and therefore limited ability to orient the cubital fossa medially—tend to reach very high and thus do not require a very extreme elbow position. African apes, on the other hand, often climb large diameter supports that require a greater degree of medial rotation of the arm, made possible by their high degree of humeral torsion. It is unclear, however, if the choice of supports has dictated the degree of humeral torsion in these species, or if they tend to favor particular kinds of supports because of the limitations to medial rotation brought about by different degrees of humeral torsion. Evidence seems to suggest that the marked torsion in the African apes, at least, is primarily driven by the requirements presented by quadrupedalism rather than climbing. In Sarmiento's comparison of captive versus wild orangutans, he observed much higher torsion measurements in the captive sample, a group that had spent much more time walking quadrupedally than orangutans do in the wild.

CONCLUSIONS

In this chapter I have tried to give a rough overview of some of the issues pertaining to the functional morphology of the primate shoulder, and to review the research that has been performed to address them. To some degree I have attempted to present this material in a historical perspective to not only document what is known, but also to suggest something about the process by which this information has grown and developed over time. We are all indebted to the landmark studies of Inman et al. (1944), Ashton and Oxnard (1963, 1964), and Roberts (1974), among others, for the great strides they made in organizing a wealth of material. More important than whether the interpretations were right or wrong, these studies have given focus to many subsequent workers and stimulated much research.

Because I have attempted to give a sense of the range of morphological variation across anthropoids by contrasting the ends of the spectrum, I have necessarily glossed over many morphological differences within broadly defined groups. It is the task of future research to fill in some of the details within these roughly painted outlines. Several recent studies have in fact begun to look for more subtle differences in morphology within locomotor groups, especially in light of more accurate and quantitative data on behavior in the wild (Manaster, 1979; Schultz, 1986; Takahashi, 1990).

One of the major goals of functional morphological research is the eventual application of our understanding of the relationships between form and function to the interpretation of fossil material. While fossil scapulae and humeri are fairly rare, in recent years the number of specimens known has been growing. The sample presently includes humeri of *Aegyptopithecus zeuxis* (Fleagle and Simons, 1982); a scapular fragment of *Apidium phiomense* (Anapol, 1983); two scapular fragments, one complete humerus, and parts of three others of *Pliopithecus vindobonensis* (Zapfe, 1960); associated scapular and proximal humeral fragments of *Proconsul africanus* (Napier and Davis, 1959; Walker and Pickford, 1983; Rose, 1983); a humerus missing only the head from *Dendropithecus macinnesi* (Le Gros Clark and Thomas, 1951), as well as an isolated proximal humeral fragment of either that species or *P. africanus* (Gebo et al., 1988); a proximal humerus of *Victoriapithecus* (Harrison, 1989); associated right and left humeri and scapulae of *Paracolobus chemeroni* (Birchette, 1982); two humeri (missing only the heads) of *Sivapithecus* (Pilbeam et al., 1990); a Miocene proximal humeral fragment of uncertain attribution (Rose, 1989); humeri of *Mesopithecus pentelici* (Szalay and Delson, 1979; Ciochon, 1986); a humerus missing only the head of *Dolichopithecus ruscinensis* (Szalay and Delson, 1979; Ciochon, 1986); and several humeral and scapular fragments of *Theropithecus (Simopithecus) oswaldi* (Jolly, 1972) and *T. brumpti* (Ciochon, 1986). In addition, there are three scapular fragments and eight proximal humeri among the collection of early hominid fossils, including associated scapular and humeral fragments from both *Australopithecus africanus* and *A. afarensis* (Broom et al., 1950; Howell and Coppens, 1976; Howell, 1978; Day et al., 1976; Johanson et al., 1982; Lovejoy et al., 1982; Pickford et al., 1983).[6] Although considerable effort has been expended in the interpretation of this fossil material, at this point in time it is not possible to draw firm conclusions concerning the evolutionary history of the primate shoulder. Questions still remain about what constitutes the primitive condition in humeral morphology for catarrhines (see Fleagle and Simons, 1982; Rose, 1989), about the origins of suspensory locomotion and the role it has played in the evolution of the hominoids, and to what degree understanding the evolution of the upper limb can illuminate the origins of the hominid lineage. In many cases these questions can have taxonomic as well as functional significance (see Pilbeam et al., 1990). As more fossil material is discovered and our ability to understand the significance of subtle differences in morphology increases, we will be able to answer these and many related questions about the function and evolution of the primate shoulder.

NOTES

Acknowledgments: I would like to thank Jack Stern for helpful comments on this manuscript, and Luci Betti for preparation of the figures. This material is based upon work supported by the National Science Foundation under grants nos. BNS 8819621 and 8905476.

1. In most nonhuman primates the pectoral muscles all insert onto the humerus. Among the hominoids, however, the pectoralis minor inserts onto the coracoid process of the scapula.

2. It should go without saying that other factors besides function have a significant influence on morphology. Among the most important is phylogeny, which establishes the basic configuration of the morphology of the region under study. Nonetheless, within the limits of this historical context, natural selection can bring about modifications, for instance, increasing the glenohumeral joint's capacity to resist compressive forces or enhance its range of motion, and so on.

3. In this chapter I shall only discuss monkeys and apes. Although prosimians present an interesting and diverse array of forms, the shoulders of anthropoids have been more extensively studied and comparative data are therefore more available.

4. The thorax of *Ateles* is broader than the typical quadrupedal monkey, but not as broad as a hominoid. Jenkins et al. (1978) have shown that the scapula is dorsally placed during arm-swinging in spider monkeys, as it is in apes. It is unclear, however, whether the scapula remains dorsally placed during quadrupedal locomotion in spider monkeys.

5. It should be noted that there is no clear way of distinguishing a change in width of the infraspinous fossa associated with the size of the infraspinatus muscle from a change in width of the fossa associated with prolongation of the inferior angle and an increase in the lever arm of the caudal serratus anterior.

6. For lists of fossil scapulae and humeri of more recent hominids, see Mann and Trinkaus (1973), Smith (1976), Trinkaus (1983), and Radovcic et al. (1988).

LITERATURE CITED

Anapol F (1983) Scapula of *Apidium phiomense:* A small anthropoid from the Oligocene of Egypt. Folia Primatol. *40:*11–31.

Ashton EH, Flinn RM, Oxnard CE, and Spence TF (1971) The functional and classificatory significance of combined metrical features of the primate shoulder girdle. J. Zool. Lond. *163:*319–350.

——— (1976) The adaptive and classificatory significance of certain quantitative features of the forelimb of primates. J. Zool. Lond. *179:*515–556.

Ashton EH, Healy MJR, Oxnard CE, and Spence TF (1965) The combinaton of locomotor features of the primate shoulder girdle by canonical analysis. J. Zool. Lond. *147:*406–429.

Ashton EH, and Oxnard CE (1963) The musculature of the primate shoulder. Trans. Zool. Soc. Lond. *29:*553–650.

——— (1964) Functional adaptations of the primate shoulder girdle. Proc. Zool. Soc. Lond. *142:*49–66.

Ashton EH, Oxnard CE, and Spence TF (1965) Scapular shape and primate classification. Proc. Zool. Soc. Lond. *145:*125–142.

Avis V (1962) Brachiation: The crucial issue for man's ancestry. S.W. J. Anthrop. *18:*119–148.

Bagg SD, and Forrest WJ (1986) Electromyographic study of the scapular rotators during arm abduction in the scapular plane. Am. J. Phys. Med. *65:*111–124.

——— (1988) A biomechanical analysis of scapular rotation during arm abduction in the scapular plane. Am. J. Phys. Med. Rehab. *67:*238–245.

Bearn JG (1961) An electromyographic study of the trapezius, deltoid, pectoralis major, biceps and triceps muscles, during static loading of the upper limb. Anat. Rec. *140:*103–108.

Birchette M (1982) The Postcranial Skeleton of *Paracolobus chemeroni.* Ph.D. dissertation, Harvard University, Cambridge.

Broom R, Robinson JT, and Schepers GWH (195) Sterkfontein Ape-man, *Plesianthropus.* Transvaal Museum Memoir No. 4. Pretoria, South Africa.

Bull ML, DeFreitas V, and Vitti M (1984) Electromyographic study of the trapezius (pars superior) and levator scapulae muscles in the movements of the head. Electromyogr. Clin. Neurophysiol. *24:*217–223.

Campbell BG (1966) Human Evolution. London: Heinemann Ed. Books.

Ciochon RL (1986) The Cercopithecoid Forelimb: Anatomical Implications for the Evolution of African Plio-Pleistocene Species. Ph.D. dissertation, University of California, Berkeley.

Ciochon RL, and Corruccini RS (1977) The coracoacromial ligament and projection index in man and other anthropoid primates. J. Anat. *124:*627–632.

Codman EA (1934) The Shoulder. Brooklyn: G. Miller.

Corruccini RS, and Ciochon RL (1976) Morphometric affinities of the human shoulder. Am. J. Phys. Anthrop. *45:*19–38.

Day MH, Leakey REF, Walker AC, and Wood BA (1976) New hominids from East Turkana, Kenya. Am. J. Phys. Anthrop. *45:*369–436.

Doody SG, Freedman L, and Waterland JC (1970) Shoulder movements during abduction in the scapular plane. Arch. Phys. Med. Rehab. *51:*595–604.

Doyle WJ, Siegel MI, and Kimes KR (1980) Scapular correlates of muscle morphology in *Macaca mulatta.* Acta Anat. *106:*493–501.

Erikson GE (1963) Brachiation in the New World monkeys. In J Napier and NA Barnicot (eds.): The Primates. Symp. Zool. Soc. Lond. No. 10, pp. 135–165.

Evans FG, and Krahl VE (1945) The torsion of the humerus. A phylogenetic survey from fish to man. Am. J. Anat. *76:*303–337.

Fleagle JG (1974) Dynamics of brachiating siamang *(Hylobates symphalangus syndactylus).* Nature *248:*259–260.

——— (1977) Brachiation and biomechanics: The siamang as example. Malay. Nat. J. *30:*45–51.

Fleagle JG, and Simons EL (1982) The humerus of *Aegyptopithecus zeuxis:* A primitive anthropoid. Am. J. Phys. Anthrop. *59:*175–193.

Fleagle JG, Stern JT, Jr., Jungers WL, Susman RL, Vangor AK, and Wells JP (1981) Climbing: A biomechanical link with brachiation and bipedalism. In MH Day (ed): Vertebrate Locomotion. Symp. Zool. Soc. Lond. No. 48, pp. 359–375.

Flecker H (1929) Roentgenographic study of movements of abduction at normal shoulder joint. Med. J. Aust. *2:*122–128.

Freedman L, and Munro RR (1966) Abduction of the arm in the scapular plane: Scapular and glenohumeral movements. J. Bone Joint Surg. *48A:*1503–1510.

Frey H (1923) Untersuchungen über die Scapula, speciell über ihre äussere Form und deren Abhängigkeit von der Funktion. Zeitschr. f. Anat. u. Entwicklungsgesch. *68:*277–324.

Gebo DL, Beard KC, Walker A, Larson SG, Jungers WL, and Fleagle JG (1988) A hominoid proximal humerus from the Early Miocene of Runsinga Island, Kenya. J. Hum. Evol. *17:*393–401.

Grand T (1968) Functional anatomy of the upper limb. In MR Malinow (ed.): Biology of the Howler Monkey *(Alouatta caraya).* Bibl. Primat. No. 7. Basel: Karger, pp. 104–125.

Gregory WK (1927) The origin of man from the anthropoid stem—when and where? Proc. Am. Phil. Soc. *66:*439–463.

——— (1928a) The upright posture of man: A review of its origin and evolution. Proc. Am. Phil. Soc. *67:*339–378.

——— (1928b) Were the ancestors of man primitive brachiators? Proc. Am. Phil. Soc. *67:*129–150.

Hall-Craggs ECB (1990) Anatomy as a Basis for Clinical Medicine, 2d ed. Baltimore: Urban & Schwarzenberg.

Harrison T (1989) New postcranial remains of *Victoriapithecus* from the middle Miocene of Kenya. J. Hum. Evol. *18:*3–54.

Hollinshead W, and Jenkins DB (1981) Functional Anatomy of the Limbs and Back, 5th ed. Philadelphia: W.B. Saunders.

Howell AB, and Straus WL (1971) The muscular system. In CG Hartman and WL Straus, Jr. (eds.): The Anatomy of the Rhesus Monkey. New York: Hafner, pp. 89–175.

Howell FC (1978) Hominidae. In VJ Maglio and HBS Cooke (eds.): Evolution of African

Mammals. Cambridge: Harvard University Press, pp. 154–248.

Howell FC, and Coppens Y (1976) An overview of Hominidae from the Omo succession, Ethiopia. In Y Coppens, FC Howell, GL Isaac, and REF Leakey (eds.): Earliest Man and Environments in the Lake Rudolf Basin. Chicago: University of Chicago Press, pp. 522–532.

Howell SM, Galinot BJ, Renzi AJ, and Marone PJ (1988) Normal and abnormal mechanics of the glenohumeral joint in the horizontal plane. J. Bone Joint Surg. *70A*:227–232.

Inman VT, Saunders JB deCM, and Abbott LC (1944) Observations on the function of the shoulder joint. J. Bone Joint Surg. *26*:1–30.

Jenkins FA, Dombrowski PJ, and Gordon EP (1978) Analysis of the shoulder in brachiating spider monkeys. Am. J. Phys. Anthrop. *48*:65–76.

Johanson DC, Lovejoy CO, Kimbel WH, White TD, Ward SC, Bush ME, Latimer BM, and Coppens Y (1982) Morphology of the Pliocene partial hominid skeleton (A.L. 288–1) from the Hadar Formation, Ethiopia. Am. J. Phys. Anthrop. *57*:403–451.

Jolly CJ (1967) The evolution of the baboons. In H Vagborg (ed.): The Baboon in Medical Research, Vol 2. Austin: University of Texas Press, pp. 23–50.

——— (1972) The classification and natural history of *Theropithecus (Simopithecus)* (Andrews, 1916), baboons of the African Plio-Pleistocene. Bull. Brit. Mus. (Nat. Hist.), Geol. *22*:1–122.

Jungers WL, and Stern JT, Jr. (1984) Kinesiological aspects of brachiation in lar gibbons. In H Preuschoft, DJ Chivers, WY Brockelman, and N Creel (eds.): The Lesser Apes. Edinburgh: Edinburgh University Press, pp. 119–134.

Keith A (1902) The extent to which the posterior segments of the body have been transmuted and suppressed in the evolution of man and allied primates. J. Anat. Lond. *37*:18–40.

——— (1923) Man's posture: Its evolution and disorders. Brit. Med. J. 1:451–454, 499–502, 545–548, 587–590, 624–626, and 669–672.

Kimes KR, Doyle WJ, and Siegel MI (1979) Scapular correlates of muscle morphology in *Papio cynocephalus*. Acta Anat. *104*:414–420.

Kimes KR, Siegel MI, and Sadler DH (1981) Musculoskeletal scapular correlates of planitgrade and acrobatic positional activities in *Papio cynocephalus anubis* and *Macaca fascicularis*. Am. J. Phys. Anthrop. *55*:463–472.

Larson SG (1988) Subscapularis function in gibbons and chimpanzees: Implications for interpretation of humeral head torsion in hominoids. Am. J. Phys. Anthrop. *76*:449–462.

Larson SG, and Stern JT, Jr. (1986) EMG of scapulohumeral muscles in the chimpanzee during reaching and "arboreal" locomotion. Am. J. Anat. *176*:171–190.

——— (1987) EMG of chimpanzee shoulder muscles during knuckle-walking: Problems of terrestrial locomotion in a suspensory adapted primate. J. Zool. Lond. *212*:629–655.

——— (1989) The role of supraspinatus in the quadrupedal locomotion of vervets *(Cercopithecus aethiops)*: Implications for interpretation of humeral morphology. Am. J. Phys. Anthrop. *79*:369–377.

——— (1992) Further evidence for the role of supraspinatus in quadrupedal monkeys. Am. J. Phys. Anthrop. *87*:359–363.

Larson SG, Stern JT, Jr., and Jungers WL (1991) EMG of serratus anterior and trapezius in the chimpanzee: Scapular rotators revisited. Am. J. Phys. Anthrop. *85*:71–84.

Le Gros Clark WE (1959) The Antecedents of Man. Edinburgh: Edinburgh University Press.

Le Gros Clark WE, and Thomas DP (1951) Associated jaws and limb bones of *Limnopithecus macinnesi*. Fossil Mammals of Africa, No. 3. Brit. Mus. (Nat. Hist.).

Lovejoy CO, Johanson DC, and Coppens Y (1982) Hominid upper limb bones recovered from the Hadar Formation: 1974–1977 collections. Am. J. Phys. Anthropol. *57*:637–649.

Manaster BJ (1979) Locomotor adaptations within the *Cercopithecus* genus: A multivariate approach. Am. J. Phys. Anthrop. *50*:169–182.

Mann A, and Trinkaus E (1973) Neandertal and Neandertal-like fossils from the Upper Pleistocene. Yrbk. Phys. Anthrop. *17*:169–193.

Martin CP (1933) Cause of torsion of the humerus and the notch on the anterior edge of the glenoid cavity of the scapula. J. Anat. *67*:573–582.

Miller RA (1932) Evolution of the pectoral girdle and forelimb in the primates. Am. J. Phys. Anthrop. *17*:1–56.

Morton DJ (1926) Evolution of man's erect posture. Preliminary report. J. Morphol. Physiol. *43:*147–149.

——— (1927) Humans origins, correlation of previous studies of primate feet and posture with other morphologic evidence. Am. J. Phys. Anthrop. *10:*173–203.

Napier JR (1963) Brachiation and brachiators. In J Napier and NA Barnicot (eds.): The Primates. Symp. Zool. Soc. Lond. No. 10, pp. 183–194.

Napier JR, and Davis PR (1959) The fore-limb skeleton and associated remains of *Proconsul africanus*. Fossil Mammals of Africa, No. 16. Brit. Mus. (Nat. Hist.).

Napier JR, and Napier PH (1967) A Handbook of Living Primates. New York: Academic Press.

Oxnard CE (1963) Locomotor adaptations of the primate forelimbs. In J Napier and NA Barnicot (eds.): The Primates. Symp. Zool. Soc. Lond. No. 10, pp. 165–182.

——— (1967) The functional morphology of the primate shoulder as revealed by comparative anatomical, osteometric, and discriminant function techniques. Am. J. Phys. Anthrop. *26:*219–240.

——— (1968) The architecture of the shoulder in some mammals. J. Morph. *126:*249–290.

Pickford M, Johanson DC, Lovejoy CO, White TD, and Aronson JL (1983) A hominoid humeral fragment from the Pliocene of Kenya. Am. J. Phys. Anthrop. *60:*337–346.

Pilbeam D, Rose MD, Barry JC, and Ibrahim Shah SM (1990) New *Sivapithecus* humeri from Pakistan and the relationship of *Sivapithecus* and *Pongo*. Nature *348:*237–239.

Poppen NK, and Walker PS (1976) Normal and abnormal motion of the shoulder. J. Bone Joint Surg. *58A:*195–201.

Preuschoft H (1973) Functional anatomy of the upper extremity. In GH Bourne (ed.): The Chimpanzee, Vol. 6. Basel: Karger, pp. 34–120.

Preuschoft H, and Demes B (1984) Biomechanics of brachiation. In H Preuschoft, DJ Chivers, WY Brockelman, and N Creel (eds.): The Lesser Apes. Edinburgh: Edinburgh University Press, pp. 96–118.

Prost JH (1980) Origin of bipedalism. Am. J. Phys. Anthrop. *52:*175–190.

——— (1985) Chimpanzee behavior and models of hominization. In S Kondo (ed.): Morphophysiology, Locomotor Analyses and Human Bipedalism. Tokyo: University Tokyo Press, pp. 289–303.

Putz R, Liebermann J, and Reichelt A (1988) Funktion des Ligamentum coracoa-cromiale (Function of the coracoacromial ligament). Acta Anat. *131:*140–145.

Radovcic J, Smith FH, and Trinkaus E (1988) The Krapina Hominids: An Illustrated Catalog of Skeletal Collection. Zagreb: Maladost Publ. House.

Roberts D (1974) Structure and function of the primate scapula. In FA Jenkins (ed.): Primate Locomotion. New York: Academic Press, pp. 171–200.

Rose MD (1983) Miocene hominoid postcranial morphology: Monkey-like, ape-like, neither, or both? In RL Ciochon and RS Corruccini (eds.): New Interpretations of Ape and Human Ancestry. New York: Plenum Press, pp. 405–417.

——— (1989) New postcranial specimens of catarrhines from the Middle Miocene Chinji Formation, Pakistan: Descriptions and a discussion of proximal humeral functional morphology in anthropoids. J. Hum. Evol. *18:*131–162.

Saha AK (1961) Theory of Shoulder Mechanism: Descriptive and Applied. Springfield: Charles C. Thomas.

Sarmiento EE (1985) Functional Differences in the Skeleton of Wild and Captive Orangutans and Their Adaptive Significance. Ph.D. dissertation, New York University, New York.

Schultz AH (1930) The skeleton of the trunk and limbs of higher primates. Hum. Biol. *2:*303–438.

——— (1950) The specializations of man and his place among catarrhine primates. Cold Spr. Harb. Symp. Quant. Biol. *15:*37–52.

——— (1956) Postembryonic age changes. Primatologia *1:*887–964.

——— (1968) The recent hominoid primates. In SL Washburn and PC Jay (eds.): Perspectives on Human Evolution. New York: Holt, Rinehart, and Winston, pp. 122–195.

——— (1973) The skeleton of the Hylobatidae and other observations on their morphology. In DM Rumbaugh (ed.): Gibbon and Siamang,

Vol. 2. Basel: Karger, pp. 1–54.

——— (1986) The forelimb of the Colobinae. In DR Swindler and J Erwin (eds.): Comparative Primate Biology. Vol. 1, Systematics, Evolution, and Anatomy. New York: Alan R. Liss, pp. 559–669.

Shea BT (1986) Scapular form and locomotion in chimpanzee evolution. Am. J. Phys. Anthrop. *70*:475–488.

Smith FH (1976) The Neandertal remains from Krapina, Northern Yugoslavia: An inventory of the upper limb remains. Zeitschrift für Morphologie und Anthropology, *67*:275–290.

Stern JT, Jr. (1971) Functional myology of the hip and thigh of cebid monkeys and its implications for the evolution of erect posture. Bibliotheca Primatologica, No. 14. Basel: Karger.

——— (1976) Before bipedality. Yrbk. Phys. Anthropol. *19*:59–68.

Stern JT, Jr., and Susman RL (1981) Electromyography of the gluteal muscles in *Hylobates, Pongo,* and *Pan:* Implications for the evolution of hominid bipedality. Am. J. Phys. Anthrop. *55*:153–166.

Stern JT, Jr., Wells JP, Jungers WL, and Vangor AK (1980) An electromyographic study of serratus anterior in Atelines and *Alouatta:* Implications for hominoid evolution. Am. J. Phys. Anthrop. *52*:323–334.

Straus WL (1949) The riddle of man's ancestry. Quart. Rev. Biol. *24*:200–223.

Szalay FS, and Delson E (1979) Evolutionary History of the Primates. New York: Academic Press.

Takahashi LK (1990) Morphological basis of armswinging: Multivariate analysis of the forelimbs of *Hylobates* and *Ateles*. Folia Primatol. *54*:70–85.

Trinkaus E (1983) The Shanidar Neandertals. New York: Academic Press.

Tuttle RH (1969) Quantitative and functional studies on the hands of the Anthropoidea. 1. The Hominoidea. J. Morphol. *128*:309–364.

——— (1974) Darwin's apes, dental apes, and the descent of man: Normal science in evolutionary anthropology. Curr. Anthropol. *15*:389–398.

——— (1975) Parallelism, brachiation, and hominoid phylogeny. In WP Luckett and FS Szalay (eds.): Phylogeny of the Primates. New York: Plenum Press, pp. 447–480.

Tuttle RH, and Basmajian JV (1977) Electromyography of pongid shoulder muscles and hominoid evolution. Part I, Retractors of the humerus and "rotators" of the scapula. Yrbk. Phys. Anthrop. *20*:491–497.

——— (1978) Electromyography of pongid shoulder muscles. Part II, Deltoid, rhomboid, and "rotator cuff." Am. J. Phys. Anthrop. *49*:47–56.

Walker AC (1974) Locomotor adaptations in prosimian primates. In FA Jenkins (ed.): Primate Locomotion. New York: Academic Press, pp. 349–381.

Walker AC, and Pickford M (1983) New postcranial fossils of *Proconsul africanus* and *Proconsul nyanzae*. In RL Ciochon and RS Corruccini (eds.): New Interpretations of Ape and Human Ancestry. New York: Plenum Press, pp. 325–351.

Wiedenbauer MM, and Mortensen OA (1952) An electromyographic study of the trapezius muscle. Am. J. Phys. Med. *31*:363–372.

Wolffson DM (1950) Scapula shape and muscle function, with special reference to the vertebral border. Am. J. Phys. Anthrop. *8*:331–341.

Yamshon LJ, and Bierman W (1948) Kinesiological electromyography. Part II, The trapezius. Arch. Phys. Med. *29*:647–651.

Zapfe H (1960) Die Primatenfunde aus der miozänen Spaltenfüllung von Neudorf an der March (Devinská Nová Ves) Tschechoslowakei. Schweiz. Palaeontol. Abhanlungen *78*:1–293.

Ziemer LK (1978) Function and morphology of forelimb joints in the woolly monkey *Lagothrix lagotricha*. Contrib. Primatol. *14*:1–130.

CHAPTER THREE

Functional Anatomy of the Elbow and Forearm in Primates

Michael D. Rose

The elbow region incorporates a number of joints that share a common joint cavity. This articular complex includes the humeroulnar, humeroradial, and proximal radioulnar joints, each of which has further functional subdivisions. The humeroulnar and humeroradial joints are involved in flexion-extension (Fig. 3.1). The humeroradial and proximal radioulnar joints, in combination with the distal radioulnar joint that is situated just proximal to the wrist, are involved in the axial rotatory movement of pronation-supination (Fig. 3.2). These four joints and the movements they make possible play important roles in the functioning of the forelimb as a grasping, manipulative, supporting, and propulsive mechanism.

During the recovery phase of various locomotor activities, and during hand placement for nonlocomotor grasping and manipulative activities, the hand does not contact the substrate. In these situations, the joints of the elbow and forearm provide the capability for flexion-extension and axial rotation of the forearm on the arm. This is important for bringing the hand into position for its ultimate contact with the substrate. After the hand has contacted a substrate or superstrate, especially during the propulsive phase of locomotor activities, flexion-extension mostly involves movement of the arm on the forearm, and forearm rotation may accompany movements of the forearm on the hand at the wrist. At the same time, the joints concerned must effectively transmit loading forces, and withstand any tendency of the bones involved to separate under the influence of these forces.

It must be stressed that the elbow and forearm function as part of the integrated link system of the forelimb as a whole. This is particularly evident in activities such as suspensory locomotion, where movement is three dimensional and where, for example, rotatory movement involves the coordinated action of all the joints of the limb from the scapulothoracic to the carpometacarpals. This integration will not be emphasized in the following account, but is implicit in most of the functional interpretations discussed. The following account is divided into two parts. In the first part, I consider variation among primates in each of the functional complexes of the elbow and forearm. In the second part, I discuss the unique specializations of various primate groups. These discussions are exclusively concerned with extant primates. Fossil taxa may exhibit suites of functional features that differ from those of their living relatives. Although most of the data presented below are discussed in qualitative terms, some of the discussion is based on quantitative data presented in Rose (1988a) and other sources that are cited where relevant. The primate taxonomy followed is that of Fleagle (1988).

THE FUNCTIONAL COMPLEXES OF THE ELBOW AND FOREARM

Variation in the form and function of the elbow and forearm regions among primates accompanies variation in the internal and external proportions of

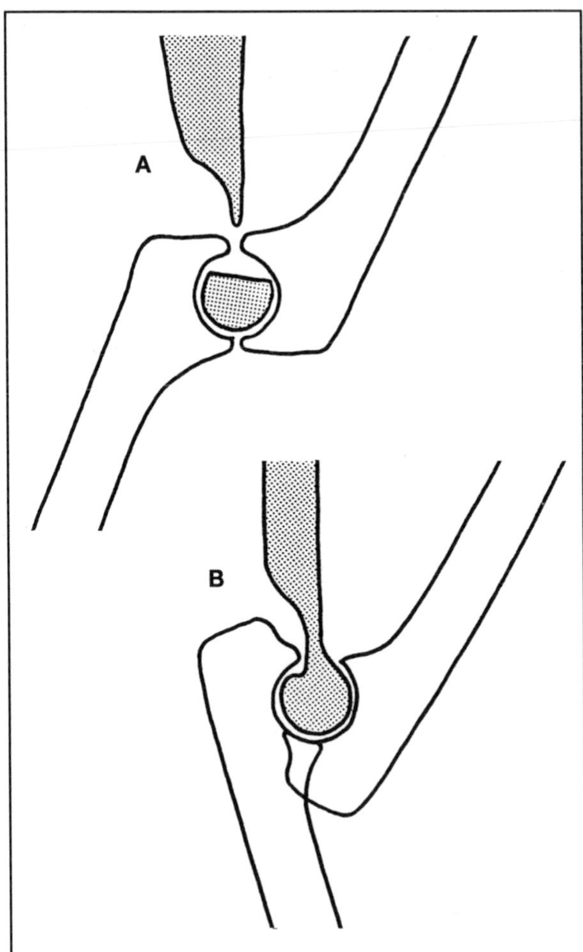

FIGURE 3.1. Flexion-extension at the humeroulnar joint. Sections through the midline of the ulnar trochlear notch and the equivalent part of the humeral trochlea and shaft, drawn to the same anteroposterior diameter of the trochlea. A: *Gorilla:* total excursion is 175°. There is a supratrochlear foramen in this humerus. B: *Nasalis:* total excursion is 130°. Morphological features underlying the difference in excursion are discussed in the text.

the forelimb, and all of these features are linked to absolute size. Unless otherwise indicated, the following discussion of proportions and body size is taken from Jungers (1978, 1979, 1984, 1985; see also Jouffroy and Lessertisseur, 1979). The overall trend in primates is for forelimb length to scale positively (i.e., to become relatively longer) with increasing body size. However, prosimians (with the exception of lorisids), callitrichid platyrrhines, and humans have shorter forelimbs than would be predicted from their body size. Lorisids, atelines, *Pongo,* and hylobatids have longer forelimbs than would be predicted. As hindlimb length scales negatively the intermembral index generally scales positively with increasing body size. These features may provide the biomechanical basis for a greater reliance on forelimb-dominated locomotion, climbing in larger primates, for example, compared with their smaller relatives. Hylobatids and *Pongo* have higher intermembral indices than would be predicted from their body size. In hylobatids the hindlimbs are relatively long, but the forelimbs are extremely long. In *Pongo* the high index results from quite short hindlimbs and quite long forelimbs. Indriids and humans have small intermembral indices for animals of their size. In both cases this results from having relatively long hindlimbs and shortened forelimbs.

As with the intermembral index, the brachial index scales positively with body size. A higher-than-predicted brachial index for the long forelimb of hylobatids is the result of a particular elongation of the forearm. The relatively high brachial index in the springing or leaping cheirogaleids, galagids, indriids (all with relatively short forelimbs), and *Tarsius* (normal forelimb length) is generally the result of an especially short arm. In African apes a low brachial index results from a relatively long arm.

A fairly consistent pattern is found with the positive scaling of joint surface area with body size (Jungers, 1988; Swartz, 1989). Nevertheless, within hominoids, great apes have relatively large humeroulnar joints and hylobatids have relatively large humeroradial joints, while humans have relatively small elbow joints. Long bone shaft curvatures in anthropoid primates have been studied by Swartz (1990). These curvatures scale positively with body size. Exceptions to the general trend are found in the humeri of atelids and hylobatids. These are relatively straight, so as to withstand the torque and off-axis loading associated with suspensory activities. Hylobatid radii have a relatively high mediolateral curvature, especially in the region of insertion of the major forearm rotating muscles. Primate long bones are generally less curved than those of other mammals (see also Biewener, 1983a), probably because they are relatively longer, and are subject to greater bending stresses.

Using data from Tuttle (e.g., 1972), Swartz (1990) shows that there is a regular scaling of the masses of major forearm-moving muscle groups

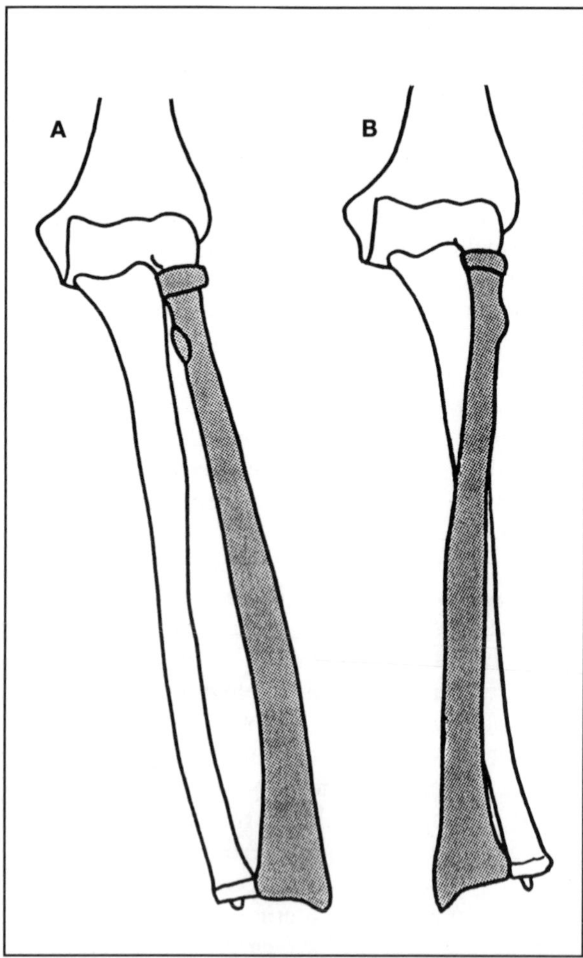

FIGURE 3.2. Pronation-supination in the left forearm of *Homo*. A: supination, B: pronation. In A there is an angulation of the ulna and radius with respect to the long axis of the humerus (arm angle). This is made up of 5° due to the angulation of the distal humeral articular surface (cubital angle) and 5° due to the angulation of the ulnar trochlear notch with respect to the long axis of the ulna (ulnar notch angle). These are close to the numerical values given by Sarmiento (1985). In B the radius has changed its orientation by 10° and is now in line with the long axis of the humerus. There has also been a slight lateral (abduction) movement of the ulna during pronation. The significance of these kinematic features is discussed in the text. See also Figure 3.15.

with body size. The single exception is the relatively large mass of the supinating musculature in hylobatids. This muscle group has an obvious importance in initiating and controlling body rotations below the supporting limb during brachiation. The major musculature of the elbow and forearm is illustrated in Figure 3.3. The principal flexing muscle of the elbow is m. brachialis. M. biceps brachii is a combined flexor and supinator, and m. brachioradialis is an accessory flexor. In all primates m. brachialis arises mostly from the anterior surface of the humeral shaft and inserts proximally on the ulna (Fig. 3.3C). This insertion shows little variation, except that in hominoids the area is in the form of a relatively broad trough. In *Daubentonia* the laterally situated brachialis flange is developed to a degree only matched (among extant taxa) by nonprimate mammals such as anteaters (e.g., *Tamandua)* and koalas *(Phascolarctos)* that are diggers and/or climbers (see also Godfrey, 1988). In most primates M. biceps brachii arises from two origins on the shoulder girdle, and inserts into the bicipital tuberosity of the radius (Fig. 3.3B). A second insertion, into the deep fascia of the forearm, occurs in humans. The length of the radial neck, the lever arm for m. biceps as a flexor, scales slightly negatively with body size, so that larger animals have relatively shorter radial necks (Fig. 3.4D and C). Exceptions to this trend are small bodied lorisoids (e.g., Fig. 3.4A; *Galago senegalensis, Loris, Arctocebus),* cheirogaleids, and *Tarsius* (Fig. 3.4B), which have relatively short radial necks. These are scurrying and small-bodied climbing or leaping primates. Great apes and humans have a relatively long lever arm for biceps (Fig. 3.4E). In most primates the bicipital tuberosity faces more or less anteriorly. However, in hominoids it faces more laterally. This is associated with a general torsion of the proximal radius, and will be discussed later. M. brachioradialis, an accessory flexor, arises from the lateral supracondylar crest of the humerus (Fig. 3.3C). In most primates it inserts on the distolateral part of the radial shaft, but in nonhuman hominoids it inserts more proximally.

M. triceps is the extending muscle of the elbow (Fig. 3.3A). It arises from the scapula and the posterior shaft of the humerus, and inserts into the tip of the olecranon process of the ulna. This process, which makes up most of the length of the lever arm for triceps, scales slightly positively with body size (Fig. 3.5D and E). It is relatively long in the scurrying cheirogaleids and callitrichids (Fig. 3.5F and G), possibly to provide a favorable lever arm for m. triceps to withstand the relatively high turning moments acting at the elbow during locomotion utiliz-

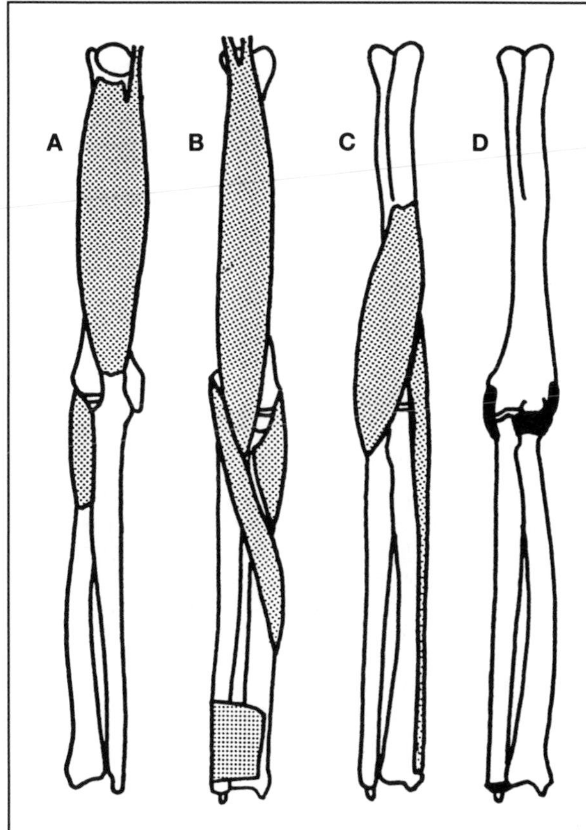

FIGURE 3.3. The major musculature and ligaments associated with the joints of the left elbow and forearm of *Papio*. **A**: posterior view, showing m. triceps (above) and the origin of m. supinator from the humerus and ulna (below). **B**: anterior view showing m. biceps (above), m. pronator teres (center left), the insertion of m. supinator into the radius (right center), and m. pronator quadratus (below). **C**: anterior view showing m. brachialis (above) and m. brachioradialis (below). M. brachialis lies deep to m. biceps. **D**: anterior view showing the medial and lateral collateral ligaments of the elbow, the annular ligament surrounding the radial head, and the triangular ligament connecting the distal ends of the radius and ulna.

ing fairly flexed limbs. The process is relatively short in the climbing lorisids and large hominoids, and in the suspensory *Ateles* and hylobatids (Fig. 3.5B, C, and A). This may be related to the rapid extension of the elbow required for reaching during the swing phase of the upper limb during climbing and suspensory activities. In most primates the olecranon process projects directly proximally, but in terrestrial cercopithecines it is directed proximoposteriorly (Fig. 3.6E), possibly to provide a favorable lever arm for m. triceps when the elbow is in a position close to full extension, as occurs toward the end of the propulsive stance phase of quadrupedal locomotion (Jolly, 1967, 1972). The muscle masses that arise from the medial and lateral humeral epicondyles are mostly responsible for movements of the hand and fingers. Although they cross the elbow joint and may have an action on it, their lines of action all pass quite close to the axis of the joint, so any major action at the elbow is unlikely.

The forearm pronating muscles are m. pronator teres, which arises from the medial humeral epicondyle and inserts into the radial shaft, and m. pronator quadratus, which passes between the ulna and the radius in the distal forearm (Fig. 3.3B). A second ulnar head of m. pronator teres is present in

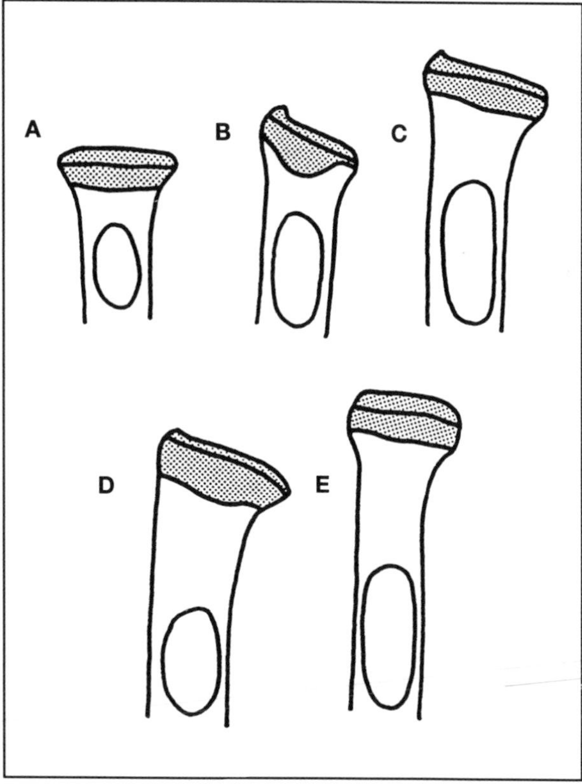

FIGURE 3.4. Lever length for m. biceps. Anterior views of left radial heads and necks, drawn to the same total proximodistal radial length. **A**: *Arctocebus;* **B**: *Tarsius;* **C**: *Papio;* **D**: *Varecia;* **E**: *Pan*. In **A** and **B** the neck is short for the size of the animals, and in **E** the neck is long for the size of the animal.

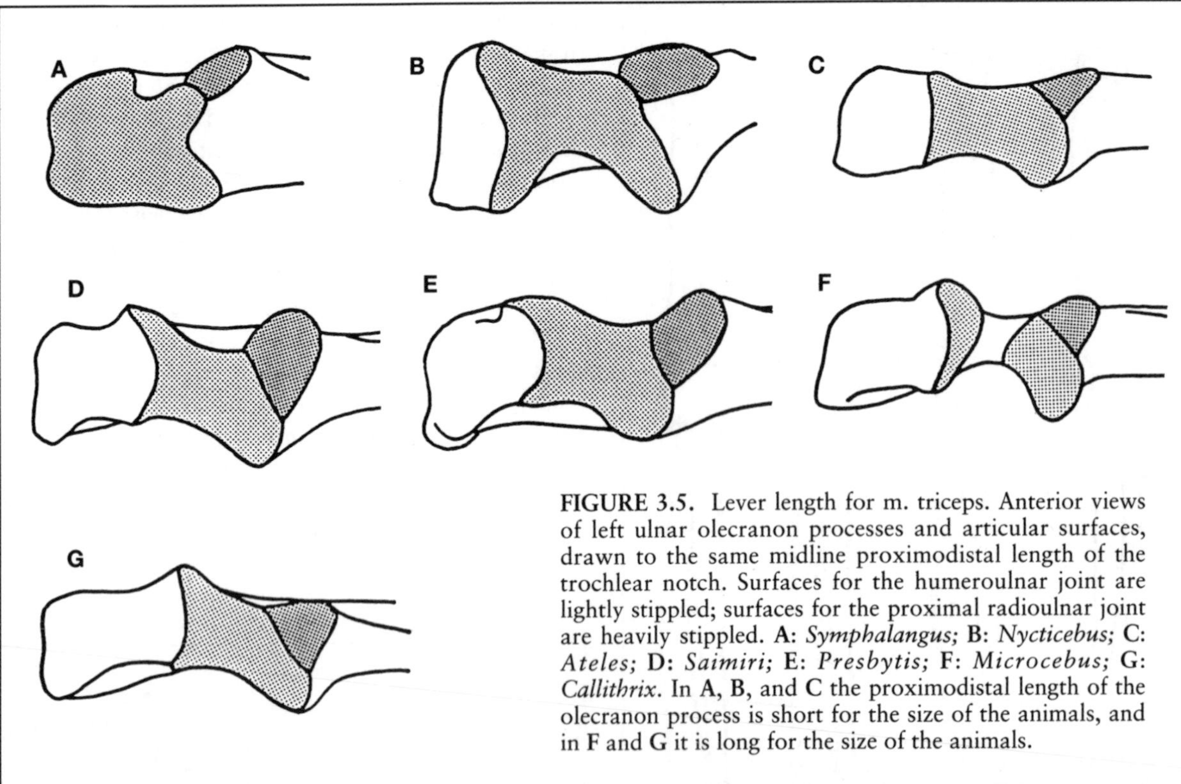

FIGURE 3.5. Lever length for m. triceps. Anterior views of left ulnar olecranon processes and articular surfaces, drawn to the same midline proximodistal length of the trochlear notch. Surfaces for the humeroulnar joint are lightly stippled; surfaces for the proximal radioulnar joint are heavily stippled. **A**: *Symphalangus;* **B**: *Nycticebus;* **C**: *Ateles;* **D**: *Saimiri;* **E**: *Presbytis;* **F**: *Microcebus;* **G**: *Callithrix.* In **A**, **B**, and **C** the proximodistal length of the olecranon process is short for the size of the animals, and in **F** and **G** it is long for the size of the animals.

humans. The major supinating muscle is m. supinator, assisted by mm. biceps and brachioradialis. M. supinator arises from the lateral humeral epicondyle and the proximal ulnar shaft and inserts into the radial shaft (Fig. 3.3A and B). The ulnar head is absent or weakly developed in nonhominoids. These axially rotating muscles are particularly important for the brachiating hylobatids. Thus, m. supinator is particularly well developed and m. pronator teres inserts into the anterior rather than the lateral aspect of the radius.

Two small muscles in the elbow region may serve to prevent any tendency for the ulna to rotate about its long axis during elbow and forearm movements. These are m. anconeus, which passes posteriorly between the lateral epicondyle and the posterior aspect of the proximal ulna, and m. epitrochleo-anconeus (absent in humans), which passes posteriorly between the medial epicondyle and the posterior aspect of the olecranon process (see also Sarmiento, 1985).

The humeroulnar joint, between the humeral trochlea and the trochlear notch of the ulna, is a uniaxial joint that only allows flexion-extension movements (Fig. 3.1). As is usual in uniaxial joints the major ligaments of the joint are in the form of medial and lateral collateral ligaments (Fig. 3.3D). These are somewhat eccentrically placed with respect to the axis of the joint, so that they come under tension in extension and help stabilize the joint in this position. As might be expected, the major variations in the morphology of the joint can be linked either to the amount of movement allowed at the joint or to the stabilization of the joint. Flexion-extension excursion is generally somewhat less than 180° in primates (Fig. 3.1B), but is 180° or more in hominoids (Fig. 3.1A), due to a greater extension range (Harrison, 1982).

If the forearm continues the line of the arm, then flexion-extension movements are in a parasagittal plane. The forearm frequently diverges laterally so that there is an angle between the arm and forearm. The following discussion of this angulation is largely taken from the more extensive account of Sarmiento (1985). In most primates this angle is small but in hominoids and *Ateles* it is larger, espe-

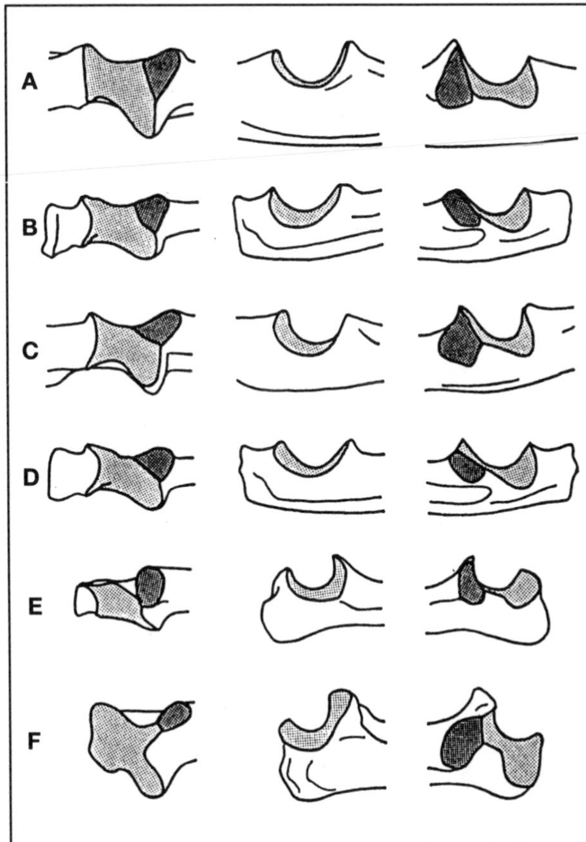

FIGURE 3.6. Anterior (left), medial (center), and lateral (right) views of left proximal ulnae, drawn to the same proximodistal length of the trochlear notch. The trochlear (light stipple) and radial (dark stipple) notches are indicated. **A**: *Lemur;* **B**: *Arctocebus;* **C**: *Tarsier;* **D**: *Aotus;* **E**: *Papio;* **F**: *Pan.* Only a part of the olecranon process is shown in **A** and **C**.

cially in humans, where it is about 10° (Fig. 3.2). In hominoids the angle results from an angulation of the distal humeral articular surface, the cubital angle, which is particularly large in Asian apes (Fig. 3.7G) and an angulation between the trochlear notch of the ulna and the long axis of the ulna, the ulnar notch angle, which is particularly large in African apes and humans (Fig. 3.2). To the extent that the cubital angle is high, parasagittal movement will be accompanied by adduction (with flexion) and abduction (with extension). A high ulnar notch angle results in an additional supination (with flexion) and pronation (with extension) movement. More importantly, the angulation can result in a biomechanical carrying angle. This will be discussed in more detail in what follows.

The humeroulnar joint has two parts with different functional morphologies. This partitioning is sometimes clearly evident on the trochlear notch of the ulna, which may have a nonarticular region separating proximal and distal articular surfaces (Figs. 3.5F, 3.8B, and 3.9A, F, and G). A similar boundary, in the form of a low ridge, may be present between the anterior and posterior parts of the humeral trochlea. The size of the anterior part of the trochlea—relative to the area for articulation with the radius—is not obviously linked to absolute size. Lorisids (Fig. 3.7A), indriids, and callitrichids have relatively small anterior trochleas, while cheirogaleids, lemurids, and hominoids (Fig. 3.7F, G, and H) have relatively large anterior trochleas.

There are three regular convex surfaces that permit uniaxial movement in combination with reciprocally curved concave surfaces: cylindrical, conical, and trochleiform (pulley-shaped) surfaces. In primates the anterior part of the humeral trochlea may have any of these shapes, or some intermediate shape. It is cylindrical in most prosimians (Fig. 3.8D), but is the shape of a truncated cone in lorisids (except *Perodicticus*), galagids (Fig. 3.8A and B), and *Tarsius*. The shape in platyrrhines is that of a cylinder that slants distomedially to proximolaterally, but with proximodistally oriented margins (Fig. 3.8F). This shape may appear almost conical in some specimens (Fig. 3.8E; see also Ford, 1988). In *Saimiri, Cebus,* and cercopithecids, the anterior part of the trochlea is best described as having a relatively cylindrical lateral part and a pronounced medial flange or keel that is best developed distally (Fig. 3.8G). There is also a minimally developed lateral keel adjacent to the area for articulation with the radius in large arboreal colobines such as *Nasalis* (Fig. 3.9C). In hominoids, medial and lateral keels are present all the way round the trochlea margins. The trochlear surface is concave between these keels, so that the overall shape of the anterior trochlea is truly trochleiform (Figs. 3.7G and H, 3.8C, and 3.9F and H). The distal part of the trochlear notch of the ulna largely mirrors the size and form of the anterior trochlea. Thus it is transverse when the trochlea is cylindrical (Figs. 3.8D and 3.6A), has a strong mediodistal slant when the trochlea is conical (Fig. 3.8A), and has mediodistal and laterodistal surfaces separated by a ridge when the trochlea is trochleiform (Fig. 3.8C). In cercopithecids, the

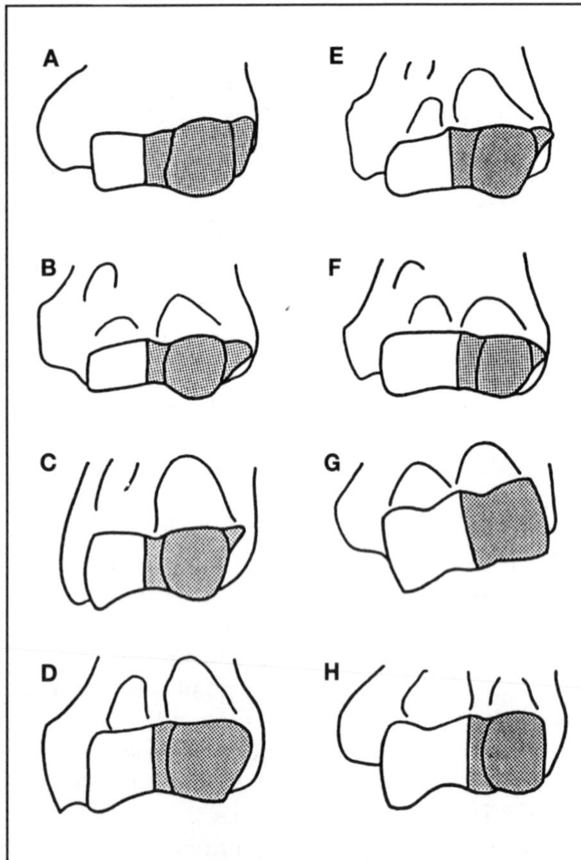

FIGURE 3.7. Anterior views of left distal humeri, drawn to the same total mediolateral width of the articular surfaces. Surfaces for articulation with the radius are stippled. The relative mediolateral width of the trochlea increases from **A** through **H**. **A**: *Perodicticus;* **B**: *Lepilemur;* **C**: *Saimiri;* **D**: *Callicebus;* **E**: *Pithecia;* **F**: *Varecia;* **G**: *Symphalangus;* **H**: *Homo.*

strongly developed medial trochlear keel is mirrored by a particularly strong inclination of the trochlear notch (Fig. 3.8G). When the anterior trochlea is conical, trochleiform, or a cylinder plus keel, the corresponding inclination of the mediodistal part of the ulnar trochlear notch is visible as an appreciable flange in medial view (Fig. 3.6B, E, and F).

Medially, the posterior trochlea is a continuation of the anterior trochlea as it passes distally and posteriorly to terminate as the distal border of the olecranon fossa (Figs. 3.9 and 3.10). Laterally the trochlea expands to occupy the area posterior to the surface for articulation with the radius. As might be expected, the degree of this lateral expansion is greater in taxa with relatively broad areas for articulation with the radius (Fig. 3.9E) than in those with relatively narrow areas for the radius (Fig. 3.9G and H). However, clinging and leaping strepsirhines generally have a relatively small lateral expansion of the posterior trochlea (Fig. 3.9A), while lorisids have a relatively large expansion (Fig. 3.9B). In cercopithecids there is only a modest expansion (Fig. 3.9C and D), while in hominoids, where the anterior trochlea is relatively wide and the area for the radius extends far posteriorly, the modest expansion of the posterior trochlea is situated more on the posterior, rather than the distal aspect of the humerus (Fig. 3.9F and G). In distal view the posterior trochlea surface is evenly concave. In larger

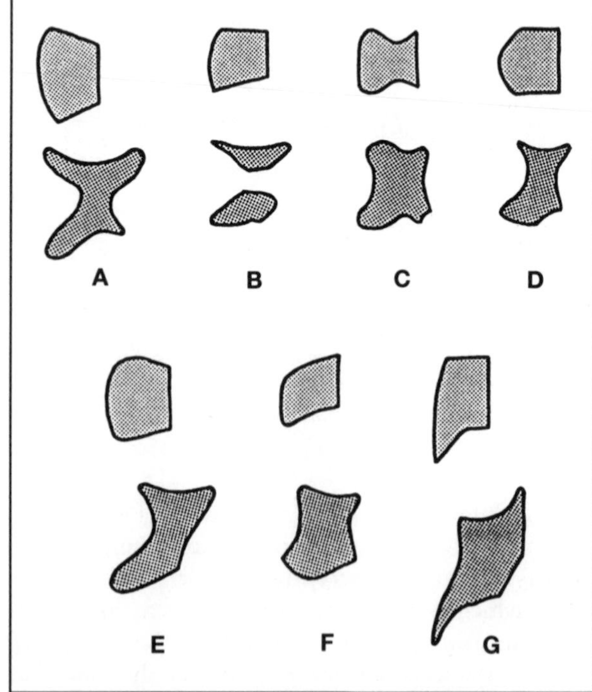

FIGURE 3.8. Silhouettes of anterior views of left humeral trochleas (light stipple) and ulnar trochlear notches (heavy stipple), drawn to the same mediolateral trochlear width. **A**: *Loris;* **B**: *Galago;* **C**: *Gorilla;* **D**: *Indri;* **E**: *Callithrix;* **F**: *Alouatta;* **G**: *Papio.* The trochlea is in the shape of a truncated cone in **A**, is trochleiform in **C**, and is cylindrical in **D**. The other examples are variations on the conical and cylindrical forms. The orientation of the distal part of the ulnar trochlear notch corresponds to the trochlear shape.

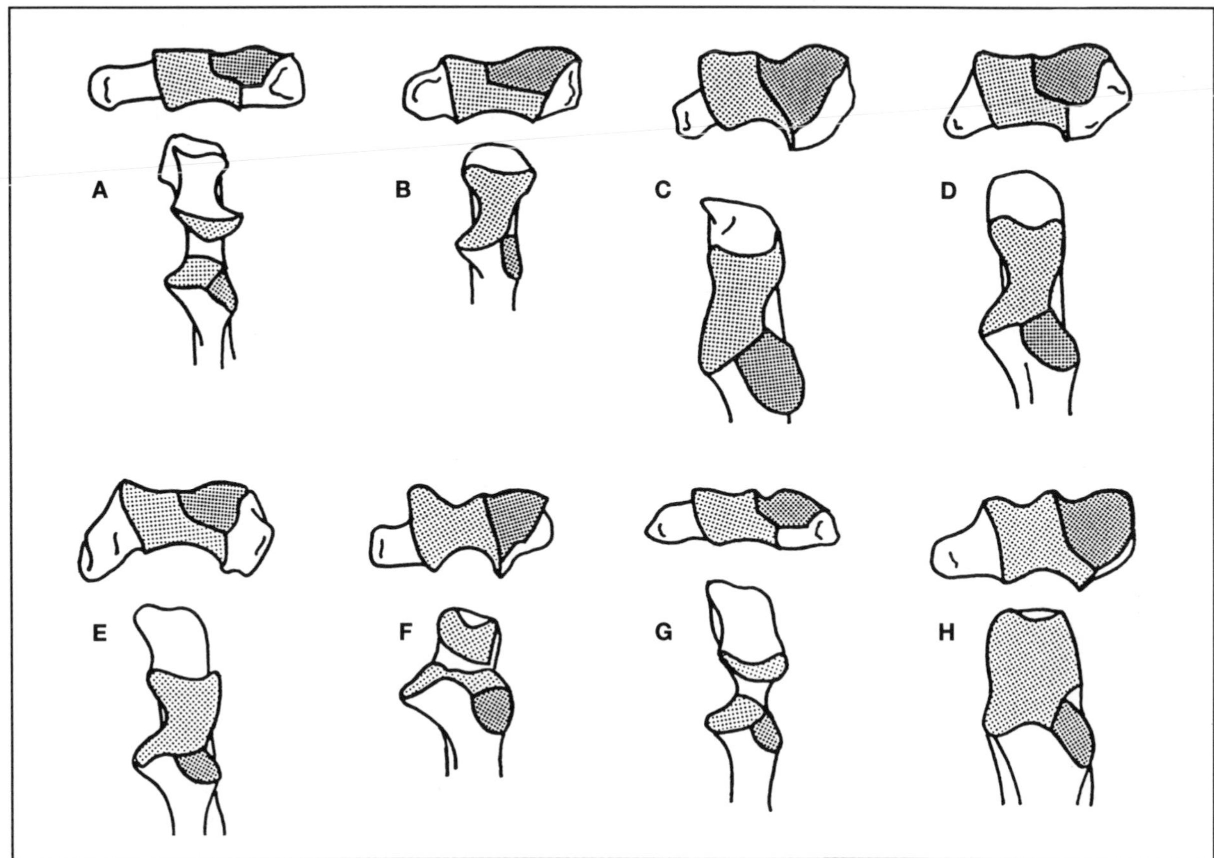

FIGURE 3.9. Distal views of left humeri and anterior views of left proximal ulnae, drawn to the same mediolateral width of the posterior trochlea. Humeroulnar surfaces (light stipple) and humeral and ulnar surfaces for the radius (heavy stipple) are indicated. The posterior part of the humeral trochlea extends laterally, posterior to the surface for the radius, to different degrees in the various specimens illustrated. The proximal part of the ulnar trochlear notch mirrors the morphology of the posterior trochlea. A: *Galago;* B: *Nycticebus;* C: *Nasalis;* D: *Colobus;* E: *Cebus;* F: *Pongo;* G: *Microcebus;* H: *Symphalangus.*

platyrrhines and especially in catarrhines, however, the lateral edge of the posterior trochlea projects to form an anteroposteriorly aligned keel (Fig. 3.9C, D, and F). In catarrhines this surface extends proximally as a vertical surface in the lateral wall of the olecranon fossa (Fig. 3.10G and H), whereas in noncatarrhines it stops short at the distal margin of the olecranon fossa (Fig. 3.10A–F). In hominoids the keel is a continuation of the lateral keel of the anterior trochlea (Fig. 3.9F and H). It is best developed in the Asian apes (Sarmiento, 1985). Because of the presence of well-developed medial and lateral keels in hominoids, the posterior trochlear surface is markedly concave and is a direct continuation of the equally concave anterior trochlea (Fig. 3.9F). In noncatarrhines, where the lateral expansion of the posterior trochlea tends to be more marked, the trochlear surface as a whole spirals around the distal humerus as a screw surface. This adds a medial translatory movement to flexion and a lateral translatory movement to extension, which may help to stabilize the joint during these movements.

Features of the proximal trochlear notch of the ulna match those of the posterior humeral trochlea (Figs. 3.6, 3.8, and 3.9). Thus, this surface is convex, is relatively broad in taxa with a greatly expanded posterior humeral trochlea (Figs. 3.8A and 3.9B), and is relatively narrow in great apes (Figs. 3.8C, 3.9F, and 3.6F). In catarrhines especially, the lateral part of the proximal trochlear notch forms a

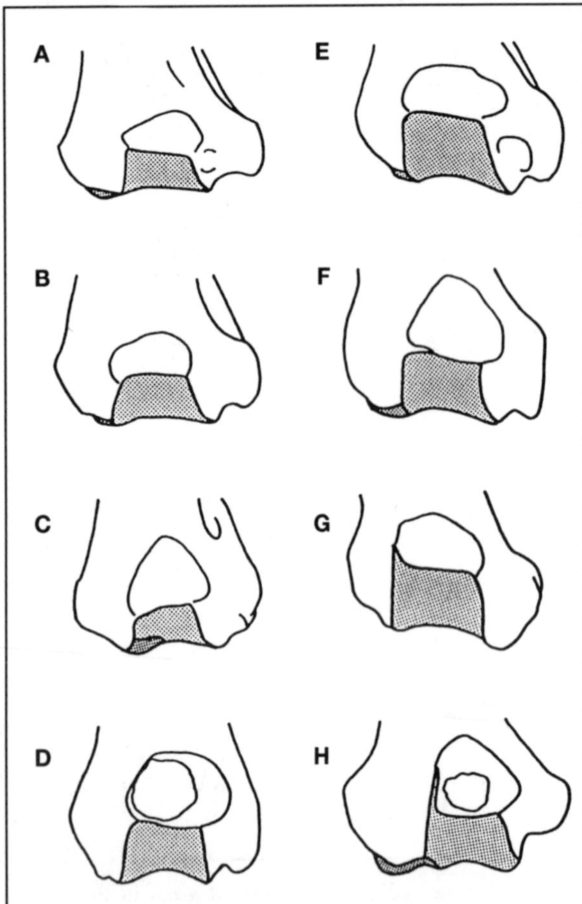

FIGURE 3.10. Posterior views of left distal humeri, drawn to the same total mediolateral width of the distal articular surface. Trochlear (light stipple) and capitular (dark stipple) surfaces are indicated. A: *Hapalemur;* B: *Cheirogaleus;* C: *Propithecus;* D: *Loris;* E: *Cacajao;* F: *Ateles;* G: *Cercopithecus;* H: *Gorilla.* The trochlear surface extends into the lateral wall of the olecranon fossa in G and H.

flat, laterally facing surface for articulation with the lateral flange of the posterior humeral trochlea (Fig. 3.6E and F).

The features of the humeroulnar joint previously discussed are associated with translatory movements and with the range of flexion-extension movements. The remaining features of the joint are most likely to be associated with load transfer and the stabilization of the joint against shear forces and axially rotating forces. These features are most marked in hominoids: the deeply socketed articular surfaces, the medial and lateral keels present all the way around the joint margins, and the presence of humeroulnar contact in the lateral part of the olecranon fossa in extension are all features that provide maximum stability throughout the flexion-extension range. The situation in cercopithecids and some platyrrhines is slightly different. The humeral and ulnar surfaces are less curved and the medial humeral trochlear keel is best developed anteriorly and distally, while the lateral keel is only developed posteriorly, suggesting that stabilization is maximal in the midpart of the movement range, as occurs during the weight-bearing phase of quadrupedal progression. Jenkins (1973) categorizes this functional complex as the *advanced therian* pattern and notes that it has evolved in a number of quadrupedal mammalian taxa. Stabilizing features are less obvious in other primates, where a generally small body size may make definitive stabilizing features less necessary. However, the concave curvature of the posterior trochlear together with the reciprocally curved proximal ulnar trochlear notch must stabilize the humeroulnar joint toward the extension end of the movement range. Similarly, a conical, or tilted, cylinder-shaped anterior trochlea may resist medially directed shear forces acting on a more flexed limb. The cylindrical anterior trochlea of many prosimians, while protecting against axial rotation, would seem to provide minimal protection against shear forces. The relatively large size of the humeroulnar articulation in, for example, lemurids might provide stability via a broad area of contact between the humerus and the ulna.

At the humeroradial joint, the head of the radius moves on the humeral surface as it accompanies the ulna during flexion-extension movement, and spins on the humeral surface during pronation and supination at the radioulnar joints. The radial head is held in place by the annular ligament that surrounds the side of the head and is attached to the ulna anterior and posterior to its radial notch (Fig. 3.3D). It is difficult to use standard anatomical terminology consistently with respect to the radial head and, in some species, with respect to the distal radius. If the (generally) flattened anterior surface of the radial shaft is taken as a reference plane, then with respect to landmarks found on the head in all primates, there is a torsion of the head and neck in the pronation (internal rotation) sense. This is least evident in leaping and slow-climbing strepsirhines and most evident in hominoids. Thus a landmark such as the "lateral" lip, discussed later, may in fact be located lateroanteriorly or even, in hominoids, anterolaterally. In hominoids, where there is a general symmetry to the radial head, the most important effect of the torsion is on the positioning of the bicipital tuberosity, as mentioned above. In indriids especially, there is an additional supination of the distal radius with respect to the shaft. For ease of comparative explanation, these torsions are not taken into account in the discussion of most of the radial features given later.

On the distal humerus there are three articular regions, which articulate with different parts of the radial head (Fig. 3.11). The central region is the globular capitulum, articulating with a central depression on the proximal surface of the radial head. The zona conoidea lies between the capitulum and the lateral margin of the trochlea, and articulates with the peripheral part of the proximal surface of the radial head and, to a variable extent, with the articular surface on the side of the head. In nonhominoids a third region, the capitular tail, is an extension from the proximolateral part of the capitulum. It also articulates with the peripheral rim of the proximal surface of the radial head.

Typically the capitulum is globular in shape, extends partly onto the lateral surface to the region of the lateral epicondyle, and terminates toward the posterior part of the distal humerus (Fig. 3.9). In indriids, the capitulum is markedly inflated so that, in anterior view, it extends distally to the trochlea (Fig. 3.11E). The central depression on the radial head is correspondingly deep. This feature may help to stabilize the radial head as it moves on the capitulum. The capitulum is also quite inflated in leaping and in climbing strepsirrhines (Fig. 3.7A and B), and in callitrichids. In hominoids, especially great apes, the capitulum extends farther posteriorly than in other primates, allowing the radius to move with the ulna into a hyperextended position (Fig. 3.9F and H). In cercopithecids the distal part of the capitulum is somewhat flattened (Fig. 3.11F), possibly providing a larger area of contact with the radius during load bearing in more extended forearm positions.

The zona conoidea is particularly broad in indriids, callitrichids, cercopithecids, and hylobatids (Fig. 3.11D, E, F, and G), and particularly narrow in lorisids (Fig. 3.11A; see also Gebo, 1989) and *Tarsius*. In most quadrupedal primates and in lorisids the zona is in the form of a relatively flat plane between the trochlea and the capitulum that terminates distally (Fig. 3.11A, C, and F). In leaping prosimians (Fig. 3.11E), cheirogaleids, callitrichids (Fig. 3.11D), some large colobines (Fig. 3.9C), and hominoids (Figs. 3.9H, 3.10H, and 3.11B), the zona is in the form of a recessed gutter. The gutter flattens distally and is not markedly evident in anterior view, except in indriids and hominoids, where the gutter extends onto the distal surface (Fig. 3.11B and E). As with the capitulum, the gutter continues posteriorly in hominoids. The overall width of the peripheral rim of the proximal articular surface of the radial head reflects the width of the humeral zona. In most primates, however, this rim is broader laterally than elsewhere. This results in a lateral lip to the surface and an oval outline to the radial head as a whole. Although its degree of development is quite variable, the lateral lip is generally more emphasized in the mostly quadrupedal species with a flat, relatively broad humeral zona (e.g., Fig 3.12A). The lateral lip articulates with the zona in the fully pronated position of the forearm and provides a relatively large area of contact during the load-bearing phase of quadrupedal locomotion. In lorisids there is a posterior rather than a lateral lip (Fig. 3.12C). As would be expected, the peripheral rim is more symmetrical in species with a gutterlike humeral zona (Fig 3.12B and D). This is particularly the case in large hominoids (Fig. 3.12F). Similarly, the peripheral rim is

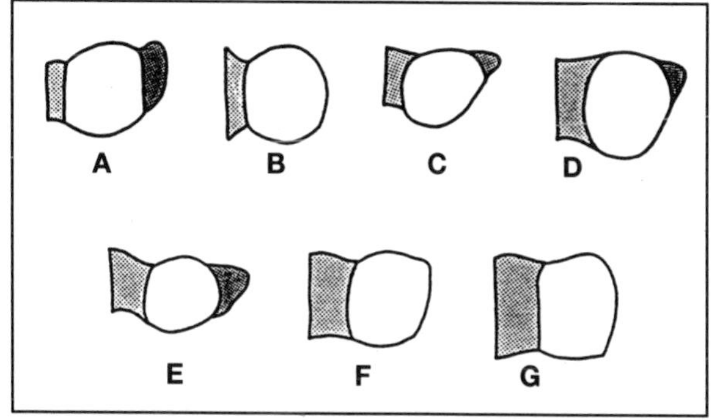

FIGURE 3.11. Anterior views of the left humeral surfaces for articulation with the radius, drawn to the same mediolateral width of the capitulum. The zona conoidea and the capitular tail are indicated with light and dark stipple respectively. The relative mediolateral width of the zona conoidea increases from A through G. A: *Perodicticus;* B: *Pan;* C: *Cacajao;* D: *Callithrix;* E: *Propithecus;* F: *Presbytis;* G: *Symphalangus.*

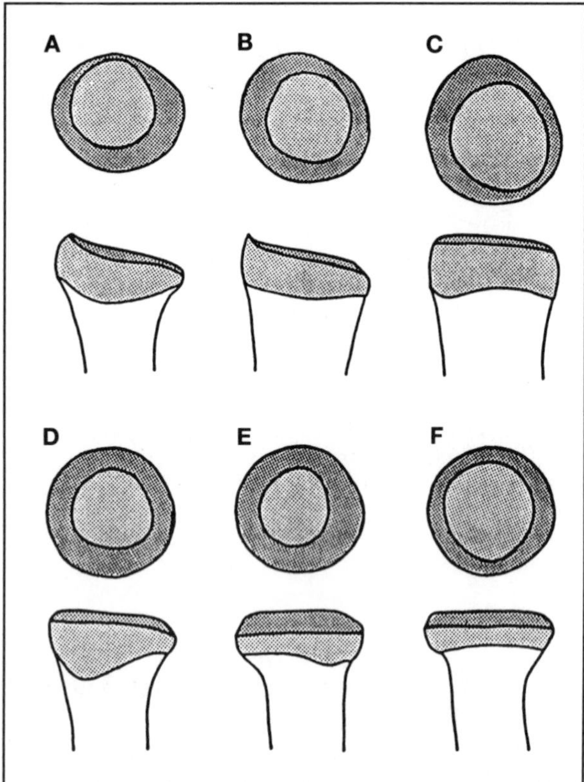

FIGURE 3.12. Proximal and anterior views of left proximal radii, drawn to the same maximum diameter of the head. Surfaces for articulation with the capitulum (light stipple on proximal views), the zona conoidea and capitular tail (dark stipple), and radial notch and annular ligament (medium stipple on anterior views) are indicated. **A**: *Daubentonia;* **B**: *Galago;* **C**: *Loris;* **D**: *Ateles;* **E**: *Symphalangus;* **F**: *Pan.*

relatively flat and proximally facing when the zona is flat (Fig. 3.12A, C, and D, and Fig. 3.13E) but extends more onto the side of the head (Fig. 3.13G), producing a bevelled appearance in side view, when the zona is gutterlike (Fig. 3.12E, F). Even in the former group, however, the peripheral rim tends to be somewhat bevelled in the region of the lateral lip (Figs. 3.12A and B, and 3.13A–D). This feature will be considered in the discussion of the radioulnar joints.

The capitular tail is well developed in leaping strepsirhines (Figs. 3.7B and 3.11E) and in lorisids (Figs. 3.7A and 3.11A; see also Szalay and Dagosto, 1980; Dagosto, 1983). It is small in cercopithecids and absent in hominoids (Fig. 3.11F and G). The tail faces anteriorly except in lorisids, where its lateral margin is turned anteriorly to form a flange. As well as articulating with the zona, the peripheral rim of the radial head also articulates with the capitular tail. Depending on the degree of development of the tail, this articulation occurs in positions between 90° of flexion and the fully flexed position. Although the peripheral rim is generally narrowest medially, this part of the radial head is typically tilted proximally. The central depression for articulation with the capitulum proper is also deepest adjacent to this region. The net effect is to produce an "upturned" medial lip opposite the "downturned" lateral lip on the radial head (Figs. 3.12A and B, and 3.13A and B). This medial lip contacts the capitular tail while the lateral lip contacts the zona, in positions close to full pronation of the forearm, and is probably a load-transferring and stabilizing mechanism for the medial side of the humeroradial joint (Fig. 3.14A–C). The medial lip is best developed in leaping strepsirhines rather than in the quadrupedal primates, in which the lateral lip mechanism is most evident. The medial lip is minimally developed in catarrhines, where the capitular tail is small or absent (Figs. 3.12E and F, and 3.14E and F). It is also minimally developed in lorisids, where the abovementioned lateral flange to the large capitular tail, together with the posterior lip on the radial head, provide an alternative stabilizing mechanism (Figs. 3.11A, 3.12C, and 3.13E).

Scattered reports in the literature indicate that pronation-supination excursions in primates range between approximately 90° and 180° (Darcus and Salter, 1953; Yalden, 1972; Ziemer, 1978; O'Connor and Rarey, 1979; Sarmiento, 1985). Excursions are maximal in hominoids. Thus for most primates adopting a quadrupedal posture, full pronation of the forearm does not bring the palm of the hand into a position to rest flat on the substrate. Additional movements, at the shoulder and within the carpus, are required to bring about full hand contact with the substrate (Jenkins, 1981; Rose, 1983, 1984; Sarmiento, 1985). During pronation-supination the radial head spins in place, while the distal radius rotates around the distal ulna at the distal radioulnar joint. In humans at least, the ulna moves slightly as the radius rotates (Fig. 3.2). Ulnar adduction accompanies supination, while abduction accompanies pronation (e.g., Palmer et al., 1982; Pirela-Cruz et al., 1991). This implies a lack of complete congruency at the humeroulnar joint.

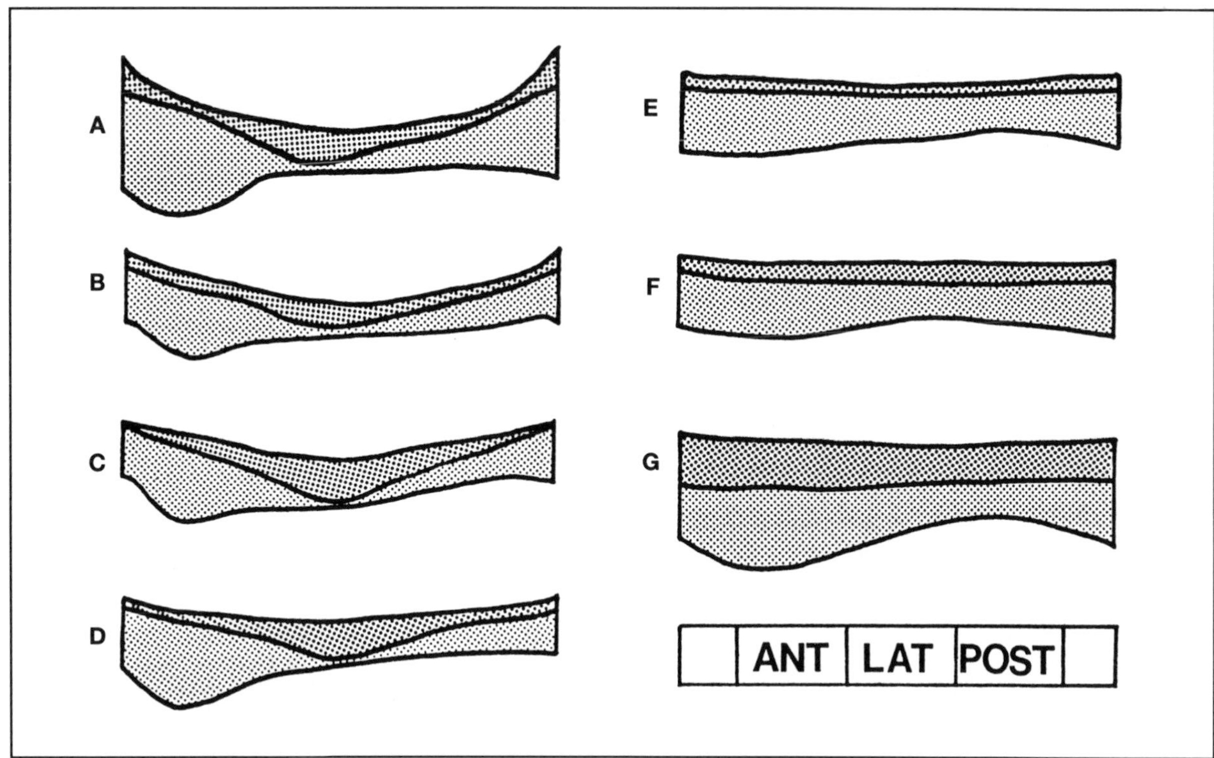

FIGURE 3.13. Views of the articular surface of the sides of left radial heads as seen if removed and laid flat, drawn to the same head circumference. Surfaces for the humerus (dark stipple) and radial notch and annular ligament (medium stipple) are indicated. **A**: *Tarsius;* **B**: *Indri;* **C**: *Callithrix;* **D**: *Cercopithecus;* **E**: *Perodicticus;* **F**: *Pongo;* **G**: *Gorilla.* Tilt of the head is indicated by the slope toward the center of the strip (especially evident in **A**, **B**, and **C**). An upturned medial lip is indicated by the peaks at each end of the strip in **A** and **B**. An angulated lateral lip is represented in the center of the strip in **A–D**.

The proximal radioulnar joint is between the radial notch of the ulna (Figs. 3.5, 3.6, and 3.9) and the articular surface on the side of the radial head (Figs. 3.4, 3.12, and 3.13). Typically the radial notch has a curvature that matches the curvature of the region of the radial head that articulates with it in full pronation (Fig. 3.14). In most strepsirhines and cebids, the radial notch faces anterolaterally (Fig 3.14A, B, and D). In cercopithecids, the notch faces more anteriorly, so that the radial head itself lies anterior to the ulna (Fig. 3.14E). This may be an additional mechanism for the stabilization of the radial head. In mammals with the radius and ulna fused together in the fully pronated position, the radius invariably lies directly anterior to the ulna at the elbow joint. In lorisids and hominoids, the radial notch faces more laterally than in other primates (Fig. 3.14C and F).

The radial surface for the proximal radioulnar joint is continuous with the peripheral rim of the proximal surface of the radial head. It is proximodistally deepest on the anterior and medial aspects of the head (Figs. 3.12 and 3.13). This is the region of the head that always articulates with the radial notch of the ulna during pronation-supination. In species with a restricted pronation-supination excursion, the rest of the surface, especially the posterior part, only contacts the annular ligament. This part of the surface is proximodistally shallow and may be absent (Fig. 3.13A–D). In proximal view, the outline of the radial head is usually flattened in this region (Fig. 3.12A–C). In species with a markedly oval shaped radial head there is maximum contact between the ulnar and radial surfaces

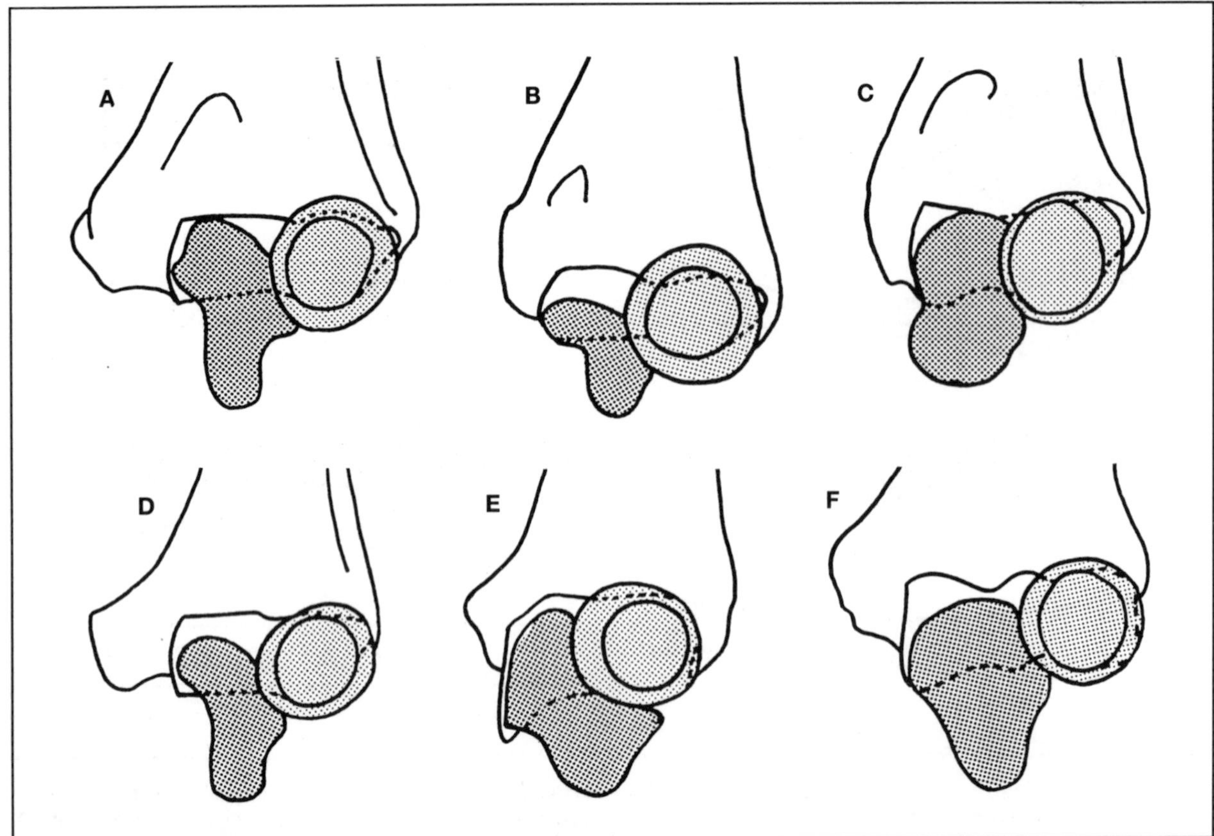

FIGURE 3.14. Anterior views of left distal humeri, with superimposed proximal ulnar sections and outlines of radial heads, drawn to the same maximum mediolateral width of the radial head, and in a position of 90° of flexion. The radial heads are shown in the position of full pronation, which is the most stable position of both the humeroradial and proximal radioulnar joints. A: *Galago;* B: *Propithecus;* C: *Loris;* D: *Callithrix;* E: *Papio;* F: *Pan.*

in full pronation (Fig 3.14E). All of these features of the radial head are more symmetrical in hominoids, due to the greater pronation-supination excursion (Figs. 3.12E and F, 3.13F and G, and 3.14F). As a result, the proximal radioulnar joint is equally stable in all positions.

In most primates, the radial head is tilted with respect to the long axis of the bone, so that the anatomically proximal surface of the head faces proximolaterally rather than proximally (Figs. 3.12A, B, and D, and 3.13A–C). Tilt is greatest, up to about 15°, in leaping strepsirhines (Fig. 3.12B) and in callitrichids (Fig. 3.13C). The above-mentioned local angulations produced by the medial and especially the lateral lips of the peripheral rim are superimposed on head tilt. To a certain extent, head tilt may serve to bring the head into a more transverse plane when there is a carrying angle at the elbow (Fig. 3.15B, supination). Except in hominoids, however, the head-tilt angle is considerably greater than the carrying angle. A more crucial function for head tilt is to maintain humeroradial contact as the radius changes its orientation with respect to the ulna during pronation-supination (Fig. 3.15). In supination, the long axes of the two bones are roughly parallel (Fig. 3.2). As the radius pronates, however, its distal end moves medially as it rotates around the ulna. In the absence of head tilt, this would have the effect of opening up the humeroradial joint on the lateral side. In primates with a flat zona and a well-developed lateral lip, the angulation of the lateral lip serves with head tilt to

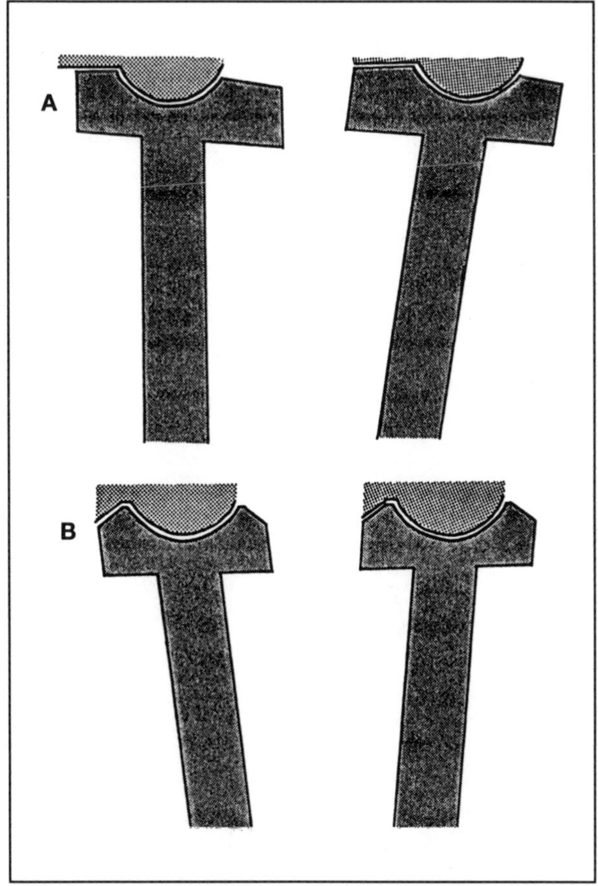

FIGURE 3.15. Diagrammatic sectional representation of left humeroradial congruency in supination (left) and pronation (right). **A**: primate with a flat zona conoidea and a radial head with a pronounced lateral lip. There is a head tilt of 5° and a lateral lip angle of 5°. These features allow congruency to be maintained as the long axis of the radius changes its orientation by 10° during pronation (see also Fig. 3.2). Additional stability in pronation is provided by the large area of contact between the lateral lip and the zona conoidea. **B**: hominoid, with a recessed zona conoidea and a beveled radial head. There is a head tilt of 5°. In the supinated position this allows the radius to lie parallel to the ulna with a 5° arm angle. Congruency at the humeroradial joint is maintained throughout the movement to full pronation, which again involves a change in orientation of 10° of the long axis of the radius. In pronation the radius is closer to the vertical than in **A**. These diagrams represent idealized simplifications.

maintain humeroradial contact in pronation (Fig. 3.15A).

The distal radioulnar joint is between a generally quite flat and relatively small surface on the medial side of the distal radius and a convex surface on the anterior surface of the distal ulna. Except in hominoids and lorisoids (Fig. 3.16D, E, and C) the joint surfaces are quite small for a joint with such a large excursion (Fig. 3.16A). Indeed, in many noncatarrhines, an articular surface on the radius may be lacking and the distal radioulnar joint is little more than a bursa-like space (Lewis, 1965). In most primates, soft-tissue structures play a crucial role in distal radioulnar joint function (Lewis, 1969, 1974; Lewis et al., 1970; Cartmill and Milton, 1977; Sarmiento, 1985, 1988; King et al., 1986). The most important of these structures in guiding movement is a ligament (in nonhominoids) or triangular fibrocartilaginous disc (in hominoids) that extends from its apical origin from the lateral side of the base of the ulnar styloid process to a linear insertion on the distomedial radius, between the surfaces for the distal radioulnar and radiocarpal joints (Fig. 3.3D). The disk or ligament's ulnar insertion is close to the center of rotation for the radioulnar joint (Kapandji, 1981) and—together with the interosseous membrane of the forearm, m. pronator quadratus, and other soft-tissue structures—serves to prevent separation of the two bones.

The ulnar facet for the distal radioulnar joint is generally quite small and faces laterodistally (Fig. 3.16A and B). The corresponding radial facet, when present, faces medioproximally. In lorisids, however, the ulnar facet is relatively large and evenly curved (Fig. 3.16C). The ulnar and radial facets face more distally and proximally (respectively) than in other nonhominoid primates. In hominoids, most of the very large ulnar facet (Corruccini et al, 1975) faces distally and, typically, articulates with the triangular articular disk (Fig. 3.16D). A laterally facing, convex peripheral rim articulates with a laterally facing, large and concave, striplike radial facet. In hominoids, the elaboration of the joint is related to a greater pronation-supination excursion.

SPECIALIZATIONS OF THE ELBOW AND FOREARM IN PRIMATES

In the following summary, primates are considered in groups that are convenient for discussion, rather than in groups based on consistent taxonomic or other criteria. Considerable differences in positional behavior, size, and morphology occur among the members of some of these groups. A

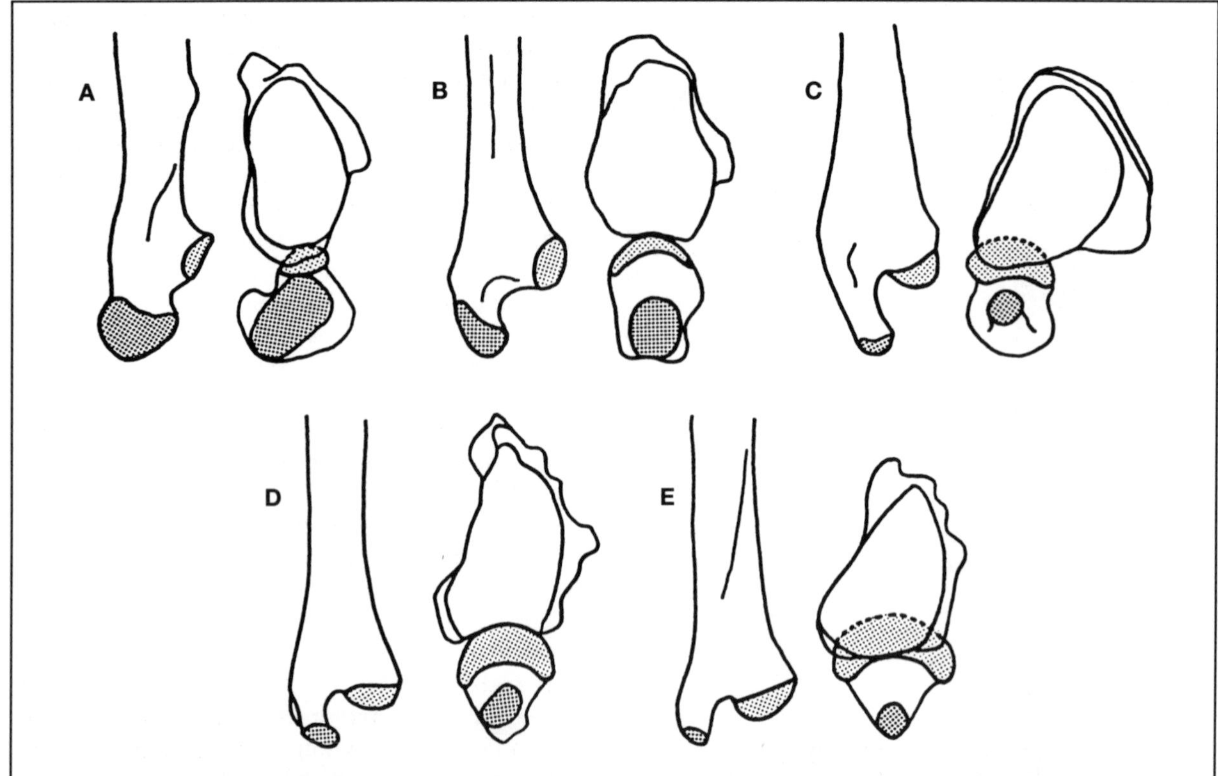

FIGURE 3.16. Anterior views (left) of distal left ulnae and distal views (right) of left radii and ulnae, drawn to the same total mediolateral width of the distal radius and ulna. Surfaces for the distal radioulnar joint (light stipple) and for contact between the ulnar styloid and soft or bony structures in the wrist (dark stipple) are indicated. **A**: *Varecia;* **B**: *Cercopithecus;* **C**: *Arctocebus;* **D**: *Pan;* **E**: *Pongo*. The radial surface overlaps the ulnar surface distally in **A**, **C**, and **E**.

complete examination of these differences is beyond the scope of this discussion. Some consideration is made of the well-documented differences among hominoids, however, and note is made of some major differences among the members of other groups.

Cheirogaleids include the smallest of the primates (Table 3.1). Locomotor behavior is dominated by a combination of scurrying, bent-limbed quadrupedalism and springing (e.g., Walker, 1979; Gebo, 1987). In this respect their locomotor biomechanics resembles that of some nonprimate small mammals (e.g., Biewener, 1983a, 1983b, 1989; Kram and Taylor, 1990; Alexander and Ker, 1990). Other cheirogaleid activities include clinging postures adopted during feeding, and rapid reaching and grasping associated with the capture of insect prey (e.g., Martin, 1972; Fleagle, 1988). The center of mass is carried close to the substrate (Gebo, 1987). This stabilization during locomotion is a result of the combination of relatively short forelimbs and a bent-limbed gait. The relatively large anterior humeral trochlea and proximal ulnar trochlear notch, and the anteriorly recessed zona conoidea results in particular elbow stability in flexion (Figs. 3.5F and 3.9G). The relatively long olecranon process provides a favorable lever arm for m. triceps to prevent overflexion during level-surface locomotion, inverted clinging positions that are sometimes adopted during feeding, and at the end of leaps (Figs. 3.5F and 3.9G). The short radial neck may be associated with m. biceps brachii producing rapid flexion, which is of use in the hand-to-mouth retrieval of captured insects. To the extent that they utilize clinging postures, cheirogaleids resemble clinging and leaping primates. This is reflected in

TABLE 3.1
Size Range in Primates

GENUS	FAMILY	WEIGHT (g)
Microcebus	Cheirogaleidae	60
Tarsius	Tarsiidae	123
Galago	Galagidae	215
Cheirogaleus	Cheirogaleidae	220
Arctocebus	Lorisidae	265
Euoticus	Galagidae	274
Loris	Lorisidae	275
Callithrix	Callitrichidae	310
Callimico	Callitrichidae	630
Lepilemur	Lepilemuridae	750
Saimiri	Cebidae	875
Nycticebus	Lorisidae	920
Perodicticus	Lorisidae	1,150
Hapalemur	Lemuridae	1,200
Aotus	Cebidae	1,220
Callicebus	Cebidae	1,400
Lemur	Lemuridae	2,400
Pithecia	Atelidae	2,500
Daubentonia	Daubentoniidae	2,800
Cacajao	Atelidae	3,000
Cebus	Cebidae	3,000
Varecia	Lemuridae	3,800
Cercopithecus	Cercopithecidae	4,200
Propithecus	Indriidae	6,500
Alouatta	Atelidae	6,850
Presbytis	Cercopithecidae	7,600
Ateles	Atelidae	9,000
Colobus	Cercopithecidae	9,000
Indri	Indriidae	10,000
Symphalangus	Hylobatidae	10,750
Nasalis	Cercopithecidae	15,095
Papio	Cercopithecidae	19,600
Pan	Pongidae	43,600
Pongo	Pongidae	59,000
Homo	Hominidae	61,500
Gorilla	Pongidae	125,500

Familial taxonomy and body weights are from Fleagle (1988). Body weights are averages of male and female values. For multispecific genera values are from the middle of the range of values.

their possession of a short forelimb, a short radial neck, and an anteriorly recessed zona conoidea.

In most indriids and galagids, and in *Tarsius*, positional behavior is dominated by clinging postures and leaping locomotion, although quadrupedalism may also be important, especially in the larger galagids (e.g., Napier and Walker, 1967; Charles-Dominique, 1977; Pollock, 1977; Crompton, 1983, 1984; Niemitz, 1983, 1985; Crompton and Andau, 1986). Clingers and leapers are characterized by having relatively short forelimbs, especially the arm segment. Within the humeroulnar joint, the posterior trochlea does not expand markedly posterior to the area for the radius (Fig. 3.9A). The humeroradial joint is characterized by an inflated capitulum and a relatively large capitular tail (Figs. 3.10C and 3.11E) that is associated with a pronounced medial lip on the markedly tilted radial head (Figs. 3.4B, 3.12B, and 3.13A and B). In galagids and *Tarsius*, the trochlea is typically cone-shaped (Fig. 3.8B), the zona conoidea (particularly narrow in *Tarsius)* is recessed anteriorly (Fig. 3.9A), and the radial neck is short (Fig. 3.4B). In indriids, the small anterior trochlea (Szalay and Dagosto, 1980; Godfrey, 1988) is cylindrical (Fig. 3.8D). Proximally it blends in with the humeral shaft instead of terminating at an excavated coronoid fossa (Szalay and Dagosto, 1980). All three components of the humeroradial joint are emphasized in indriids. The highly inflated capitulum is matched by a correspondingly deep central depression on the radial head (Fig. 3.10C). The broad, incised zona conoidea extends from the anterior onto the distal surface of the humerus and the capitular tail is particularly broad (Fig. 3.11E). The radial head is correspondingly strongly bevelled and has a strongly developed medial lip (Fig. 3.13B).

Szalay and Dagosto (1980) point out that forelimb features are more likely to be associated with the clinging rather than with the leaping components of the positional repertoire. They suggest that many of the features of the elbow region in clinging strepsirhines indicate the habitual use of the forelimb in flexion, during grasping associated with vertical clinging. Many features of the elbow region reflect a greater importance of the humeroradial over the humeroulnar joint. Thus the posterior trochlea is relatively narrow, and the anterior trochlea is either small or cone-shaped. The emphasized capituloradial articulation suggests that this is the site of most of the loading of the arm onto the forearm, and that tracking of the radius across the capitulum may be as important as ulnohumeral contact in determining flexion-extension movement. In addition, well-developed zona/bevel and capitular tail/medial lip mechanisms ensure stability, especially in the flexed position (Fig. 3.14A and B). The lack of a marked lateral expansion of the posterior trochlea points to the lesser importance of elbow stability in extension. The continuation distally of the zona in the relatively large-bodied indriids suggests, however, that more extended forelimb postures might also be of importance in this group. As in cheirogaleids, the short radial neck in *Tarsius* and smaller galagids implies rapid elbow flexion, possibly associated with insectivory.

Although the morphological similarities among clingers is an obvious reflection of their common positional behavior, the differences in their morphology are related to differences in absolute size or to different phylogenetic histories (Stern and Oxnard, 1973; McArdle, 1981; Oxnard et al., 1981; Oxnard, 1984). This partial overlap of morphological and/or behavioral features is also evident in members of other groups that have convergently acquired clinging capabilities, such as some cheirogaleids (Stern and Oxnard, 1973), the lemurids *Hapalemur* (Walker, 1979), *Lepilemur* (Oxnard, 1973; Jouffroy and Gasc, 1974), and *Lemur rubriventer* (Overdorff, 1988), and the platyrrhine *Cebuella* (Kinzey et al., 1975). Oxnard et al. (1990) provide a general overview of many of these similarities and differences.

Lorisids exhibit a number of distinct features associated with their deliberate quadrupedal and climbing locomotion, and suspensory postures (e.g., Walker, 1969, 1974, 1979; Charles-Dominique, 1977; Dykyj, 1980; Jouffroy et al., 1983; Glassman and Wells, 1984). Nevertheless, like cheirogaleids, they resemble clinging strepsirhines in some respects of the elbow region. Thus the capitular tail is large (Figs. 3.7A, 3.11A, and 3.14C). There is a small (as in indriids) conical (as in galagids) anterior trochlea (Figs. 3.7A and 3.8A). The zona conoidea is narrow as in *Tarsius* (Figs. 3.7A and 3.11A). This is not surprising: Gebo (1987) points out that all strepsirhines cling to a certain extent. Lorisids differ from other strepsirhines in having long forelimbs for animals of their size. Also, although the limbs are not as highly stressed as in primates with more acrobatic locomotor behavior, the humeral shaft is equally resistant to forces acting in the frontal and parasagittal planes (Demes and Jungers, 1989). The olecranon process of the ulna (Gebo, 1989) and the radial neck are both short (Figs. 3.4A and 3.5B), so that both m. triceps and m. biceps brachii can act to bring about rapid changes in limb orientation. The trochlea is small and cone-shaped anteriorly, but broadens considerably posteriorly (Figs. 3.8A and 3.9B), indicating a greater stability of the medial side of the elbow in extension. The humeroradial joint as a whole is relatively broad, despite the fact that the zona conoidea is narrow and flat (Figs. 3.7A and 3.14C). The capitulum is inflated and the capitular tail is large and bears a flange laterally (Fig. 3.7A). The radial head shows little torsion and there is a posterior rather than a lateral lip (Fig. 3.12C). These features suggest that the humeroradial joint is particularly stable in more flexed and semipronated positions and especially resists forces tending to displace the radial head laterally. This may occur during eccentric loading of the flexed forelimb during cautious quadrupedalism and climbing. An evenly curved, laterally facing radial notch of the ulna (Fig. 3.14C) and a large, evenly curved, and distally facing ulnar surface for the distal radioulnar joint (Fig. 3.16C), together with the equivalent radial features, resemble similar features in hominoids. Hominoid-like features have been noted in other regions of the lorisid postcranium (see Cartmill and Milton, 1977 and references therein). Their significance has been the subject of some debate (e.g., Cartmill and Milton, 1977; Corruccini, 1977; Lewis, 1985). Sarmiento (1985) notes that pronation-supination in lorisids is not as extensive as in hominoids: hominoid-like features of the forearm and wrist are most likely related to loading between the forearm and an adducted hand, as occurs during climbing.

Lemurids combine quadrupedalism with climbing and leaping (e.g., Walker, 1979; Gebo, 1987). The only outstanding feature of the elbow region is the large, cylindrical anterior trochlea that results in extensive humeroulnar contact (Figs. 3.6A and 3.7F). The humeroradial joint is stabilized in pronation by contact between the flat zona conoidea and the distinct lateral lip on the radial head (Fig. 3.12A). Other aspects of lemurid morphology are illustrated in Figs. 3.4D, 3.10A, and 3.16A. *Lepilemur* shares features of both its positional behavior (Charles-

Dominique, 1979) and its elbow morphology with indriids (Oxnard et al., 1990). Thus it has a small anterior trochlea that blends with the humeral shaft proximally, a recessed zona conoidea that extends quite far distally, an inflated capitulum, and a large capitular tail (Fig. 3.7B). Apart from its distinctive humeral shaft in the region of origin of m. brachialis, *Daubentonia* is quite similar to lemurids in its elbow and forearm regions (Fig. 3.12A).

The callitrichid platyrrhines are relatively small-sized primates with clawed digits. Locomotor behavior is dominated by quadrupedal scurrying and springing (Kinzey et al., 1975; Fleagle and Mittermeier, 1980; Garber, 1980, 1984). As might be expected, they share a number of elbow and forearm features with cheirogaleids and clinging and leaping strepsirhines in particular. Thus, as in cheirogaleids, the forelimbs are relatively short and the olecranon process of the ulna is relatively long (Fig. 3.5G). Both of these features are associated with a bent-limbed gait that maintains the center of mass close to the substrate. As in clinging and leaping strepsirhines, particularly indriids, callitrichids have a small, almost cylindrical anterior trochlea (Figs. 3.8E and 3.14D), an inflated capitulum, a broad zona conoidea that is recessed anteriorly (Fig. 3.11D), and a strongly tilted radial head (Fig. 3.13C).

Although there is a large range of body size and locomotor habits among other platyrrhines, they are all largely quadrupedal (Kinzey, 1976; Mendel, 1976; Mittermeier, 1978; Fleagle and Mittermeier, 1980; Schön Ybarra, 1984; Cant, 1986; Fleagle and Meldrum, 1988; Fontaine, 1990). Other positional activities used by some platyrrhines, such as hindlimb suspension or leaping, are not strongly linked to forelimb morphology. There is thus a considerable amount of uniformity in the morphology of the elbow and forearm. The anterior trochlea is typically in the form of a slanting cylinder (Figs. 3.6D, 3.7C, D, and E, and 3.8F) and the posterior trochlea bears a small lateral keel (Fig. 3.9E). The zona conoidea is flat (Figs. 3.7C, D, and E, and 3.11C) and there is a lateral lip on the radial head (Figs. 3.12 and 3.13). These are all features ensuring relative stability in the elbow region throughout the flexion-extension range, and with the forearm in a fully pronated position, as occurs during quadrupedal locomotion (see also Fleagle and Simons, 1982; Schön Ybarra, 1988). In the cebines, *Saimiri* and *Cebus,* additional stability is provided by a moderately developed medial keel on the anterior trochlea, similar to that found in cercopithecids (Fig. 3.7C). In pitheciines, *Chiropotes,* which is mostly quadrupedal, has a distal humerus similar to those of the cebines, while *Pithecia,* a leaper (Fleagle and Meldrum, 1988), resembles indriids in having a cylindrical anterior trochlea, a narrow posterior trochlea, and an inflated capitulum (Fig. 3.7E). Atelids, particularly *Ateles,* share many features of their positional behavior and postcranial morphology with, especially, suspensory hominoids (Erikson, 1963; Lewis, 1971; Mittermeier, 1978; Jenkins, 1981; Sarmiento, 1985; Cant, 1986). Thus the forelimb is relatively long, the humerus is straight shafted, there is a high carrying angle at the elbow, the olecranon process is short (Fig. 3.5C), the radial head is more circular than oval in outline (Fig. 3.12D), and there is a strong ligament associated with the distal radioulnar joint. These are all features related especially to the enhanced forearm rotation that accompanies suspensory activities. In other features of both the elbow and forearm, however, atelids resemble other platyrrhines.

Quadrupedalism predominates in the locomotor behavior of cercopithecids (Ripley, 1967; Mittermeier and Fleagle, 1976; Morbeck, 1977; Rose, 1977; Fleagle, 1977a, 1977b, 1980). The elbow and forearm regions show numerous specialized features related to an emphasis on the parasagittal limb movements that characterize cercopithecid quadrupedalism. Cercopithecids are among the more dextrous primates and, although this is mostly reflected in the morphology of the hand, it is also associated with moderately well-developed rotatory capabilities in the forearm. The humeroulnar joint is stabilized, especially in a partly flexed position, by contact between the articular lips of the ulnar trochlear notch and the anterodistal keel on the medial side of the anterior trochlea and the keel on the lateral side of the posterior trochlea (Figs. 3.5E, 3.6E, 3.8G, 3.9C and D, and 3.14E). The lateral keel continues as an articular surface in the lateral wall of the olecranon fossa, ensuring humeroulnar contact in more extended elbow positions (Figs. 3.9C and D, and 3.10G). Humeroradial contact is maximal in semiflexion and full pronation, due to the zona conoidea/lateral lip mechanism (Fig. 3.15A) and contact between the flattened anterodistal capitulum and the central depression of the radial head. Stability at the radioulnar joints results from the anterior positioning of the radius on the radial notch of the ulna and by maximal contact between these surfaces in full pronation (Fig. 3.14E). The joint surfaces of the distal radioulnar joint are slightly more extensive than in noncatarrhines (Fig. 3.16B).

A retroflexed ulnar olecranon process in terrestrial cercopithecines (Fig. 3.6E) allows m. triceps to act with a relatively long lever arm when the elbow is close to full extension (Jolly, 1967, 1972). A recessed zona conoidea in the large, relatively acrobatic, arboreal colobine *Nasalis* promotes stability at the humeroradial joint (Fig. 3.9C). Fleagle (1977a) notes that a relatively short olecranon process and a posteriorly extensive capitulum in the colobine *Presbytis melalophos* are associated with greater extension at the elbow, and more suspensory habits than in the closely related *P. obscura*.

Hominoids show a number of features associated with large flexion-extension and pronation-supination ranges and with stability of the joints of the elbow and forearm throughout their excursions. These features are all associated with the extensive use of the forelimb, in both tension and compression, in eclectic climbing and suspensory behaviors. These are used, at least to a degree, by all nonhuman hominoids. In addition, great apes share a number of features that are related to quadrupedalism. (The following discussion of the general and particular specializations of hominoids relies heavily on information from the detailed studies of Sarmiento [1985, 1988], but also includes some original observations and interpretations. Other sources are cited where relevant.)

The hyperextension capability of the hominoid elbow (Fig. 3.1A) results from a number of morphological features. In all primates the ulnar trochlear notch is roughly semicircular in midline proximodistal section, and in most primates the notch faces anteriorly, with a more or less equally developed coronoid process and olecranon beak. In hominoids, the trochlear notch faces anteroproximally, however, the coronoid process is pronounced, and the olecranon beak is reduced (Figs. 3.1A and 3.6F). This—together with the fact that the olecranon process does not extend proximal to the trochlear notch (Figs. 3.5A, 3.6F, and 3.9F)—delays full engagement of the proximal part of the trochlear notch with the olecranon fossa of the humerus until the hyperextended position is reached. The potential limitation to flexion that results from a projecting coronoid process is partly obviated by the presence of a deep humeral coronoid fossa (Fig 3.1A) and a notched proximal border to the trochlea (Figs. 3.7G and H, and 3.8C), which allows the coronoid process to engage quite deeply in the coronoid fossa. This feature results in a very thin plate of bone intervening between the olecranon and coronoid fossas, which may break down so that a supratrochlear foramen results (Figs. 3.1A and 3.10H). The posterior extension of the capitulum in hominoids allows the radius, moving with the ulna, also to attain the hyperextended position (Fig. 3.9F and H). A relatively long lever arm for m. biceps, resulting from the relatively long radial neck (Fig. 3.4E), and a powerful m. brachialis, with an extensive insertion area on the proximal ulna, reflect the importance of elbow flexion in the pulling and hoisting movements associated with climbing and suspensory activities. A midshaft insertion of m. brachioradialis on the radius in nonhuman hominoids suggests that it is maximally effective in more flexed forearm positions, toward the end of pulling movements.

A large pronation-supination excursion results from the circular shape of the radial head (Fig. 3.12E and F), and from the generous ulnar surface for the distal radioulnar joint (Fig. 3.16D and E). The additional presence of the strategically attached triangular articular disc at the distal radioulnar joint allows overriding of the radial surface on the ulnar surface at the extremes of movement, at least in humans (Fig. 2 in King et al., 1986). M. pronator teres and m. supinator are the main muscles producing these movements. Both of them are relatively massive (Oxnard, 1963) and may have well-developed ulnar as well as humeral origins in hominoids. However, m. biceps, acting as a supinator, also has an important role. Because of the placement of the bicipital tuberosity that results from proximal radial torsion, the tuberosity faces almost fully laterally in the fully pronated forearm. This allows m. biceps to act throughout the 180° or more of rotation associated with supination from the fully pronated position (Fig. 3.2).

Many morphological features of the hominoid elbow and forearm enable the ulna to lock against the humerus when the two bones rotate together below the fixed hand in suspensory activities, or they prevent the ulna from rotating about its long axis at the extremes of the pronation-supination movements of the radius that are associated with limb movement during climbing and quadrupedal activities. As expressed on the humerus, these features include the trochleiform shape of the mediolaterally expanded anterior humeral trochlea (Figs. 3.7G and H, and 3.8C) and the similarly expanded, deeply concave posterior trochlea, with a well-developed keel on its lateral side that continues as articular surface in the lateral wall of the olecranon fossa (Figs. 3.9F and H, and 3.10H). These features are all matched by reciprocal features of the ulnar trochlear notch (Figs. 3.5A, 3.8C, 3.9F and H, and 3.14F). By the same token, these features also resist any forces tending to abduct or adduct the ulna at the elbow, or to produce mediolateral or anteroposterior shearing. Because, in combination, these features are present around the whole extent of the humeroulnar joint, it is equally stable in all positions (Fig. 3.14F). The socketing of the radial head into the humeral zona conoidea resists medial migration of the radial head, particularly in full pronation (Fig. 3.15B). Soft-tissue structures—particularly the annular ligament, the interosseous membrane, the triangular articular disc, and musculature inserting into the radius, wrist, and hand—serve, in various combinations, three important functions. They help to brake movement at the extremes of forearm rotation; they are implicated in loading, and resist the tendency of the radius and ulna to separate during the compressive loading that occurs in quadrupedal activities; they resist the tendency of the ulna to displace proximally during tensile loading in suspension from a fixed-hand position. (For humans, see also Palmer and Werner, 1984; Linscheid, 1986.)

In addition to these general features, the various hominoids show additional features related to particular positional specializations. In hylobatids, brachiating locomotion—involving forceful rotations of the body and forelimb below a fixed handhold—and unimanual suspensory postures dominate the positional repertoire (Fleagle, 1976, 1980). These activities are associated with a number of unique morphological features. The long forelimb and particularly long forearm provide long lever arms for pendular movement. The straight humerus resists torque, while a mediolaterally curved radius, acting as a crank handle (Oxnard, 1963) provides good lever arms for the pronating and the particularly massive supinating musculature. An anterior rather than lateral insertion of m. supinator provides a similar advantage to that provided by the bicipital tuberosity for m. biceps' supinating action. A high cubital angle (Fig. 3.7G) provides a carrying angle that places the elbow in line between the hand and the center of mass of the body, and reduces bending moments at the elbow. The posterior trochlea is broad and its lateral keel is particularly well developed (Fig. 3.9H). This provides maximum stability at the humeroulnar joint in full extension. A mediolaterally broad (for hominoids) humeroradial joint (Fig. 3.7G) that includes an incised and broad humeral zona conoidea (Figs. 3.9H and 3.11G) provides high stability for the radial head. Indeed, in anterior view, the distal humeral articular surface has the appearance of a double trochlea (Fig. 3.7G). In addition, the radial notch of the ulna faces proximolaterally rather than laterally (Figs. 3.5A and 3.9H). This feature, together with a particularly strong annular ligament, prevents distal dislocation of the radial head during suspensory activities. The collateral ligaments of the elbow are also well developed. At the distal radioulnar joint, stability without the sacrifice of mobility is ensured by the triangular articular disc that is strongly anchored to the ulna, and by a relatively large area of contact between the two bones.

In *Pongo,* positional behavior includes quadrumanous climbing and clambering, together with suspensory activities (Sugardjito, 1982; Cant 1987a, 1987b; Sugardjito and van Hooff, 1989). The long forelimbs provide an ample reach and are extensively used in all positional activities (see also Rose, 1988b). A relatively high degree of humeral torsion results in the anatomically anterior aspect of the elbow and forearm facing medially in the neutral position of the forelimb. This is advantageous for the clasping associated with climbing activities. External rotation at the shoulder brings the elbow into its true anatomical position. Depending on the exact orientation of the forelimb, the high cubital angle and the hyperextensibility of the elbow provide carrying angles that serve the same function as in hylobatids. Of all the hominoids, *Pongo* has the most exaggeratedly trochleiform anterior trochlea and the best-developed lateral keel of the posterior trochlea (Fig. 3.9F). Although the trochlear margins are parallel and more or less anteroposteriorly aligned, the anterior trochlea is markedly asymmetrical in that the medial part is much larger than the lateral part. In addition, the medial trochlear lip is markedly convex, and extends onto the medial aspect of the distal humerus. This is accompanied by a marked concavity and lipping of the distomedial part of the ulnar trochlear notch (Fig. 3.9F). During suspensory activities these features promote humeroulnar contact as the two bones rotate together with the body beneath a fixed hand grip (see illustrations in Jenkins, 1981; and Larson, 1988, for other suspensory anthropoids). Mediolateral stability at the elbow is also promoted by strong collateral ligaments. A wide flexion-extension excursion is indicated on the humerus by particularly deep coronoid and radial fossas and by a high incidence of a supratrochlear foramen.

Although the humeral zona conoidea is incised and, together with the capitulum, extends far posteriorly, it is not as incised as in other apes, and the peripheral rim of the radial head is correspondingly less bevelled (Fig. 3.13F). This suggests that radial head stability is promoted more at the proximal radioulnar joint than at the humeroradial joint in *Pongo* (Fig. 3.16E). There is also a lateral extension of the distal part of the ulnar trochlear notch, proximal to the radial notch, that articulates with the zona (Fig 3.9F). This feature combines with the above-mentioned humeroulnar joint features to promote ulnar stability. The radial surface for the radiocarpal joint extends medially on a shelf of bone, so that the surface for the distal radioulnar joint, on the proximal surface of the shelf, faces proximomedially rather than medially (Fig. 3.16E). This results in a joint orientation that is orthogonal to loading forces between the hand and the ulna, via the radial shelf, when the hand grasps in an adducted position during climbing (illustrated in Sarmiento, 1985).

The locomotor behavior of African apes includes knuckle-walking quadrupedalism, climbing, and some suspensory behavior (Susman et al., 1980; Susman, 1984; Tuttle and Watts, 1985; Doran, 1989; Hunt, 1992). Although African apes share many features of the elbow and forearm regions with *Pongo,* they also exhibit a number of differences. The medial part of the anterior humeral trochlea lacks the marked curvature found in *Pongo* (Figs. 3.8C and 3.10H). In *Gorilla,* the medial trochlea is relatively small, so that the anterior trochlea is less trochleiform than in other apes (Fig. 3.8C). The African ape humeroantebrachial joint as a whole is large and faces medially in the neutral position due to torsion of the humerus. The zona conoidea is strongly incised and, unlike *Pongo,* there is no contact between the ulna and the zona (Figs. 3.6F and 3.14F). The forearm carrying angle (greater in *Gorilla* than in *Pan)* serves to bring the radius in line with the humerus in the fully pronated load-bearing position. The distal radioulnar joint lacks those features that in *Pongo* promote loading between the radial side of the hand and the ulna (Fig. 3.16E). Instead, especially in *Gorilla,* the triangular articular disc is strong and, together with features of the wrist joint, promotes direct loading between the hand and the ulna. African apes undoubtedly share with other apes an enhanced rotational capability of the radius, and a concomitant stabilization of the ulna. The previously mentioned features also indicate a well-developed capability to transfer compressive load between the arm and the hand during the frequently used knuckle-walking quadrupedalism.

In humans, the relatively short forelimb is relieved of most of its locomotor functions and is used more for manipulative, carrying, and throwing activities. But humans lack the hyperextension capability of apes, due to a less extensive posterior trochlear surface. Humans do have a full range of pronation-supination (Fig. 3.2) that is utilized in throwing and in the twisting movements associated with manipulation. Features related to stability in apes are expressed in muted form in humans. Thus the elbow joint as a whole is relatively small, the anterior trochlea is minimally trochleiform, the lateral keel of the posterior trochlea is weakly developed, and the zona conoidea is barely incised (Fig. 3.7H). High humeral torsion and a high carrying angle (Fig. 3.2) are most likely to be associated with hand placement during manipulatory activities (see also Larson, 1988).

DISCUSSION

The primate elbow and forearm provides an excellent example of the variety of different morphological patterns that can occur in the same anatomical region among different taxa. This variety is possible because there are a number of different joints in the elbow and forearm, each of which has functionally distinct components, the morphology of which can vary considerably. These different functional complexes do not reflect any fundamentally different types of movement taking place in the elbow and forearm. Rather, they reflect differences in excursion ranges—and, especially, patterns of loading—that are associated with forearm use during different types of locomotor, postural, and manipulative activities by the various taxa concerned. A part of this variety is related to absolute body size, both in the form of regular scaling effects in morphology or proportions and in the form of functionally significant departures from scaling trends.

Many of the analyses presented above represent "common sense" functional interpretations based on the visual inspection of skeletal or dissected material. This approach has obvious limits, exposed in the analyses of Figures 3.14 and 3.15, for example. These and many other aspects of elbow and forearm function will lack a complete explanation in the absence of more sophisticated analytical approaches. Studies on humans apart, there have been very few experimentally directed studies on the functioning of the primate elbow and forearm. Particular features of interest, the functional significance of which might be more fully elucidated by studies of these types, include the magnitude and direction of loading forces and articular contact areas, especially at the humeroulnar joint, in different positions of flexion-extension and during different types of activity; the exact changes in contact areas and relationships at the humeroradial and radioulnar joints during pronation-supination, particularly with respect to the humeral zona conoidea/radial head articulation, radial head tilt, and carrying angles; the precise role of the major muscles of the elbow and forearm, as prime movers, antagonists, and fixators; the precise role in elbow and forearm functioning of the musculature (mainly concerned with wrist and hand movements) arising from the humeral medial and lateral epicondyles.

This chapter has been exclusively concerned with the functional implications of variation in elbow and forearm morphology in extant primates. The fact that different primate taxa possess different functional complexes, together with the fact that the elbow region is relatively well represented in the fossil record, also makes this region an attractive one for phylogenetic analyses. Although much is already known about both its function and phylogeny, there is still much to be learned about this intriguing region of the primate postcranium.

Acknowledgments: I would like to thank DL Gebo for inviting me to prepare this chapter. The research reported here was supported by NSF grant BNS-9004502.

LITERATURE CITED

Alexander RMcN, and Ker RF (1990) Running is priced by the step. Nature *346:*220–221.

Biewener AA (1983a) Allometry of quadrupedal locomotion: The scaling of duty factor, bone curvature and limb orientation to body size. J. Exp. Biol. *105:*147–171.

——— (1983b) Locomotory stresses in the limb bones of two small mammals: The ground squirrel and chipmunk. J. Exp. Biol. *103:*131–154.

——— (1989) Scaling body support in mammals: Limb posture and muscle mechanics. Science *245:*45–48.

Cant JGH (1986) Locomotion and feeding postures of spider and howling monkeys: Field study and evolutionary interpretation. Folia Primatol. *46:*1–14.

——— (1987a) Effects of sexual dimorphism in body size on feeding postural behavior of Sumatran orangutans (*Pongo pygmaeus).* Am. J. Phys. Anthrop. *74:*143–148.

——— (1987b) Positional behavior of female Bornean orangutans (*Pongo pygmaeus).* Am. J. Primatol. *12:*71–90.

Cartmill M, and Milton K (1977) The lorisiform wrist joint and the evolution of brachiating adaptations in the Hominoidea. Am. J. Phys. Anthrop. *47:*249–272.

Charles-Dominique P (1977) Ecology and Behavior of Nocturnal Primates. New York: Columbia University Press.

Corruccini RS (1977) Features of the prosimian wrist joint in relation to hominoid specializations. Acta. Anat. *99:*440–444.

Corruccini RS, Ciochon RL, and McHenry HM (1975) Osteometric shape relationships in the wrist joint of some anthropoids. Folia Primatol. *24:*250–274.

Crompton RH (1983) Age differences in locomotion of two subtropical Galaginae. Primates *24:*241–259.

——— (1984) Foraging, habitat structure and locomotion in two species of *Galago*. In PS Rodman and JGH Cant (eds.): Adaptations for Foraging in Non-Human Primates. New York: Columbia University Press, pp. 73–111.

Crompton RH, and Andau PM (1986) Locomotion and habitat utilization in free-ranging *Tarsius bancanus:* A preliminary report. Primates *27:*337–355.

Dagosto M (1983) Postcranium of *Adapis parisiensis* and *Leptadapis magnus* (Adapiformes, Primates). Folia Primatol. *41:*49–101.

Darcus HD, and Salter N (1953) The amplitude of pronation and supination with the elbow flexed to a right angle. J. Anat. Lond. *85:*55–67.

Demes B, and Jungers WL (1989) Functional differentiation of long bones in lorises. Folia Primatol. *52:*58–69.

Doran DM (1989) Chimpanzee and Pygmy Chimpanzee Positional Behavior: The Influence of Environment, Body Size, Morphology, and Ontogeny on Locomotion and Posture. Ph.D. dissertation, State University of New York, Stony Brook.

Dykyj D (1980) Locomotion of the slow loris in a designed substrate context. Am. J. Phys. Anthrop. *52:*577–586.

Erikson GE (1963) Brachiation in the New World monkeys. Symp. Zool. Soc. Lond. *10:*135–164.

Fleagle JG (1976) Locomotion and posture of the Malayan siamang and implications for hominoid evolution. Folia Primatol. *26:*245–269.

——— (1977a) Locomotor behavior and skeletal anatomy of sympatric leaf-monkeys in West Malaysia. Yearb. Phys. Anthrop. *20:*440–453.

——— (1977b) Locomotor behavior and muscular anatomy of sympatric Malaysian leaf-monkeys (*Presbytis obscura* and *Presbytis melalophos).* Am. J. Phys. Anthrop. *46:*297–308.

——— (1980) Locomotion and posture. In DJ Chivers (ed.): Malayan Forest Primates: Ten Years' Study in the Tropical Forest. New York: Plenum Press, pp. 191–207.

——— (1988) Primate Adaptation and Evolution. New York: Academic Press.

Fleagle JG, and Meldrum DJ (1988) Locomotor behavior and skeletal morphology of two sympatric pitheciine monkeys, *Pithecia pithecia* and *Chiropotes satanas*. Am. J. Primatol. *16:*227–249.

Fleagle JG, and Mittermeier RA (1980) Locomotor behavior, body size and comparative ecology of seven Surinam monkeys. Am. J. Phys. Anthrop. *52:*301–314.

Fleagle JG, and Simons EL (1982) The humerus of *Aegyptopithecus zeuxis:* A primitive anthropoid. Am. J. Phys. Anthrop. *59:*175–193.

Fontaine R (1990) Positional behavior in *Saimiri boliviensis* and *Ateles geoffroyi.* Am. J. Phys. Anthrop. *82:*485–508.

Ford SM (1988) Postcranial adaptations of the earliest platyrrhine. J. Hum. Evol. *17:*155–192.

Garber PA (1980) Locomotor behavior and feeding ecology of the Panamanian tamarin (*Saguinus oedipus geoffroyi,* Callitrichidae, Primates). Int. J. Primatol. *1:*185–201.

——— (1984) Use of habitat and positional behavior in a neotropical primate, *Saguinus oedipus.* In PS Rodman and JGH Cant (eds.): Adaptations for Foraging in Non-Human Primates. New York: Columbia University Press, pp. 112–133.

Gebo DL (1987) Locomotor diversity in prosimian primates. Am. J. Primatol. *13:*271–281.

——— (1989) Postcranial adaptation and evolution in Lorisidae. Primates *30:*347–367.

Glassman DM, and Wells JP (1984) Positional and activity behavior in a captive slow loris: A quantitative assessment. Am. J. Primatol. *7:*121–132.

Godfrey LR (1988) Adaptive diversification of Malagasy strepsirhines. J. Hum. Evol. *17:*93–134.

Harrison T (1982) Small Bodied Apes from the Miocene of East Africa. Ph.D. dissertation, University of London, London.

Hunt KD (1992) Positional behavior of *Pan troglodytes* in the Mahale Mountains and Gombe Stream National Parks, Tanzania. Am. J. Phys. Anthrop. *87:*83–105.

Jenkins FA, Jr. (1973) The functional anatomy and evolution of the mammalian humeroulnar articulation. Am. J. Anat. *137:*281–298.

——— (1981) Wrist rotation in primates: A critical adaptation for brachiators. Symp. Zool. Soc. Lond. *48:*429–451.

Jolly CJ (1967) Evolution of the baboons. In H Vagtborg (ed.): The Baboon in Medical Research. Vol. 2. Austin: University of Texas

Press, pp. 323–338.

——— (1972) The classification and natural history of *Theropithecus (Simopithecus)* (Andrews, 1916), baboons of the African Plio-Pleistocene. Bull. Brit. Mus. (Nat. Hist.) Geol. Ser. *22*:1–123.

Jouffroy FK, and Gasc JP (1974) A cineradiological analysis of leaping in an African prosimian *(Galago alleni)*. In FA Jenkins, Jr. (ed.): Primate Locomotion. New York: Academic Press, pp. 117–142.

Jouffroy FK, and Lessertisseur J (1979) Relationships between limb morphology and locomotor adaptations among prosimians: An osteometric study. In ME Morbeck, H Preuschoft, and N Gomberg (eds.): Environment, Behavior, and Morphology: Dynamic Interactions in Primates. New York: Fischer, pp. 143–181.

Jouffroy FK, Renous S, and Gasc JP (1983) Etude cinéradiographique des déplacements du membre antérieur du Potto de Bosman (*Perodicticus potto,* P.L.S. Müller, 1766) au cours de la marche quadrupède sur une branche horizontale. Am. Sci. Nat., Zool., Paris *13*(*5*):75–87.

Jungers WL (1978) The functional significance of skeletal allometry in *Megaladapis* in comparison to living prosimians. Am. J. Phys. Anthrop. *49*:303–314.

——— (1979) Locomotion, limb proportions and skeletal allometry in lemurs and lorises. Folia Primatol. *32*:8–28.

——— (1984) Scaling of the hominoid locomotor skeleton with special reference to lesser apes. In H Preuschoft, DL Chivers, WY Brockelman, and N Creel (eds.): The Lesser Apes: Evolutionary and Behavioural Biology. Edinburgh: Edinburgh University Press, pp. 146–169.

——— (1985) Size and scaling of limb proportions in primates. In WL Jungers (ed.): Size and Scaling in Primate Biology. New York: Plenum, pp. 345–381.

——— (1988) Relative joint size and hominoid locomotor adaptations with implications for the evolution of hominoid bipedalism. J. Hum. Evol. *17*:247–265.

Kapandji IA (1981) The inferior radioulnar joint and pronosupination. In R Tbiana (ed.): The Hand. Philadelphia: Saunders, pp. 121–129.

King GJ, McMurtry RY, Rubenstein JD, and Ogston NG (1986) Computerized tomography of the distal radioulnar joint: Correlation with ligamentous pathology in a cadaveric model. J. Hand Surg. *11A*:711–717.

Kinzey WG (1976) Positional behavior and ecology in *Callicebus torquatus*. Yearb. Phys. Anthrop. *20*:468–480.

Kinzey WG, Rosenberger AL, and Ramirez M (1975) Vertical clinging and leaping in a neotropical anthropoid. Nature *255*:327–328.

Kram R, and Taylor CR (1990) Energetics of running: A new perspective. Nature *346*:265–267.

Larson SG (1988) Subscapularis function in gibbons and chimpanzees: Implications for interpretation of humeral head torsion in hominoids. Am. J. Phys. Anthrop. *76*:449–462.

Lewis OJ (1965) Evolutionary change in the primate wrist and inferior radio-ulnar joints. Anat. Rec. *151*:275–285.

——— (1969) The hominoid wrist joint. Am. J. Phys. Anthrop. *30*:251–267.

——— (1971) The contrasting morphology found in the wrist joints of semi-brachiating monkeys and brachiating apes. Folia Primatol. *16*:248–256.

——— (1974) The wrist articulations of the Anthropoidea. In FA Jenkins, Jr. (ed.): Primate Locomotion. New York: Academic Press, pp. 143–169.

——— (1985) Derived morphology of the wrist articulations, and theories of hominoid evolution, Part I, The lorisine joints. J. Anat. *140*:447–460.

Lewis OJ, Hamshere RJ, and Bucknill TM (1970) The anatomy of the wrist joint. J. Anat. (Lond.) *106*:539–552.

Linscheid RL (1986) Kinematic considerations of the wrist. Clin. Orthopaed. *202*:27–39.

McArdle JE (1981) Functional anatomy of the hip and thigh of the Lorisiformes. Contrib. Primatol. *17*:1–132.

Martin RD (1972) Adaptive radiation and behaviour of the Malagasy lemurs. Phil. Trans. Roy. Soc. B *264*:295–352.

Mendel F (1976) Postural and locomotor behavior of *Alouatta palliata* on various substrates. Folia Primatol. *26*:36–53.

Mittermeier RA (1978) Locomotion and posture in

Ateles geoffroyi and *Ateles paniscus*. Folia Primatol. *30:*161–193.

Mittermeier RA, and Fleagle JG (1976) The locomotor and postural repertoires of *Ateles geoffroyi* and *Colobus guereza,* and a re-evaluation of the locomotor category of semibrachiation. Am. J. Phys. Anthrop. *45:*235–256.

Morbeck ME (1977) Positional behavior, selective use of substrate and associated non-positional behavior in free ranging *Colobus guereza* (Ruppel 1835). Primates *18:*35–58.

Napier JR, and Walker A (1967) Vertical clinging and leaping: A newly recognized category of locomotor behavior of primates. Folia Primatol. *6:*204–219.

Niemitz C (1983) New results on the locomotion of *Tarsius bancanus* Horsfield, 1821. Ann. Sci. Nat. Zool. Paris *13:*89–100.

——— (1985) Leaping locomotion and the anatomy of the tarsier. In S Kondo (ed.): Primate Morphophysiology, Locomotor Analyses and Human Bipedalism. Tokyo: University of Tokyo Press, pp. 235–250.

O'Connor BL, and Rarey KE (1979) Normal amplitudes of pronation and supination in several genera of anthropoid primates. Am. J. Phys. Anthrop. *51:*39–44.

Overdorff D (1988) Preliminary report on the activity cycle and diet of the red-bellied lemur (*Lemur rubriventer)* in Madagascar. Am. J. Primatol. *16:*143–153.

Oxnard CE (1963) Locomotor adaptations of the primate forelimb. Symp. Zool. Soc. Lond. *10:*165–182.

——— (1973) Some locomotor adaptations among lower primates. Symp. Zool. Soc. Lond. *33:*255–299.

——— (1984) The place of *Tarsius* as revealed by multivariate statistical morphometrics. In C Niemitz (ed.): Biology of Tarsiers. Stuttgart: Fischer Verlag, pp. 17–32.

Oxnard CE, Crompton RH, and Lieberman SS (1990) Animal Lifestyles and Anatomies: The Case of the Prosimian Primates. Seattle: University of Washington Press.

Oxnard CE, German R, Jouffroy FK, and Lessertisseur J (1981) A morphometric study of limb proportions in leaping prosimians. Am. J. Phys. Anthrop *54:*421–430.

Palmer AK, Glisson RR, and Werner FW (1982) Ulnar variance determination. J. Hand Surg. *7:*376–379.

Palmer AK, and Werner FW (1984) Biomechanics of the distal radioulnar joint. Clin. Orthop. *187:*26–35.

Pirela-Cruz MA, Goll SR, Klug M, and Windler D (1991) Stress computed tomography analysis of the distal radioulnar joint: A diagnostic tool for determining translational motion. J. Hand Surg. *16A:*75–82.

Pollock JI (1977) The ecology and sociology of feeding in *Indri indri*. In T Clutton-Brock (ed.): Primate Ecology. London: Academic Press, pp. 37–69.

Ripley S (1967) The leaping of langurs: A problem in the study of locomotor behavior. Am. J. Phys. Anthrop. *26:*149–170.

Rose MD (1977) Positional behaviour of olive baboons (*Papio anubis)* and its relationship to maintenance and social activities. Primates *18:*59–116.

——— (1983) Miocene hominoid postcranial morphology: Monkey-like, ape-like, neither, or both? In RL Ciochon and RS Corruccini (eds.): New Interpretations of Ape and Human Ancestry. New York: Plenum, pp. 405–417.

——— (1984) Hominoid postcranial specimens from the Middle Miocene Chinji Formation, Pakistan. J. Hum. Evol. *13:*503–516.

——— (1988a) Another look at the anthropoid elbow. J. Hum. Evol. *17:*193–224.

——— (1988b) Functional anatomy of the cheiridia. In JH Schwartz (ed.): Orang-Utan Biology. New York: Oxford University Press, pp. 299–310.

Sarmiento EE (1985) Functional Differences in the Skeleton of Wild and Captive Orang-Utans and their Adaptive Significance. Ph.D. dissertation, New York University, New York.

——— (1988) Anatomy of the hominoid wrist joint: Its evolutionary and functional implications. Int. J. Primatol. *9:*281–345.

Schön Ybarra MD (1984) Locomotion and postures of red howlers in a deciduous forest-savanna interface. Am. J. Phys. Anthrop. *63:*65–76.

——— (1988) Positional behavior and limb bone adaptations in red howling monkeys (*Alouatta seniculus)*. Folia Primatol. *49:*70–89.

Stern JT, Jr., and Oxnard CE (1973) Primate locomotion: Some links with evolution and morphology. Primatologia *4*(11):1–93.

Sugardjito J (1982) Locomotor behaviour of the Sumatran orang utan (*Pongo pygmaeus abelii)* at Ketambe, Gunung Leuser National Park. Malay. Nat. J. *35:*57–64.

Sugardjito J, and van Hooff J (1986) Sex-age class differences in positional behavior of Sumatran orangutan (*Pongo pygmaeus abelii)* in the Gunung Leuser National Park, Indonesia. Folia Primatol. *47:*14–25.

Susman RL (1984) The locomotor behavior of *Pan paniscus* in the Lomako Forest. In RL Susman (ed.): The Pygmy Chimpanzee: Evolutionary Biology and Behavior. New York: Plenum Press, pp. 369–393.

Susman RL, Badrian NL, and Badrian AJ (1980) Locomotor behavior of *Pan paniscus* in Zaire. Am. J. Phys. Anthrop. *53:*69–80.

Swartz SM (1989) The functional morphology of weight bearing: Limb joint surface area allometry in anthropoid primates. J. Zool. Lond. *218:*441–460.

——— (1990) Curvatures of the forelimb bones of anthropoid primates: Overall allometric patterns and specializations in suspensory primates. Am. J. Phys. Anthrop. *83:*477–498.

Szalay FS, and Dagosto M (1980) Locomotor adaptations as reflected on the humerus of Paleogene primates. Folia Primatol. *34:*1–45.

Tuttle RH (1972) Relative mass of cheiridial muscles in catarrhine primates. In RH Tuttle (ed.): Functional and Evolutionary Biology of Primates. Chicago: Aldine-Atherton, pp. 262–291.

Tuttle RH, and Watts DP (1985) The positional behavior and adaptive complexes of *Pan gorilla.* In S Kondo (ed.): Primate Morphophysiology, Locomotor Analyses and Human Bipedalism. Tokyo: University of Tokyo Press, pp. 261–288.

Walker AC (1969) The locomotion of lorises, with special reference to the potto. E. Afr. Wildlife J. *8:*1–5.

——— (1974) Locomotor adaptations in past and present prosimian primates. In FA Jenkins, Jr. (ed): Primate Locomotion. New York: Academic Press, pp. 349–381.

——— (1979) Prosimian locomotor behavior. In GA Doyle and RD Martin (eds.): The Study of Primate Behavior. New York: Academic Press, pp. 543–566.

Yalden DW (1972) The form and function of the carpal bones in some arboreally adapted primates. Acta Anat. *82:*383–406.

Ziemer LK (1978) Functional morphology of the forelimb joints in the woolly monkey *Lagothrix lagotricha.* Contrib. Primatol. *14:*1–130.

C H A P T E R   F O U R

Aspects of the Anthropoid Wrist and Hand

Paul F. Whitehead

Primate hands are intricate organs that accomplish a variety of tasks. We are familiar with the complex sets of movements that can be accomplished during activities such as playing the piano and with the delicate precision of picking up a pin. For the majority of primates, however, hands function both for manipulation and for locomotion. Each of the major modes of primate locomotion, with the exception of human bipedalism, requires constant use of the hand as an organ of support. Therefore, the various demands on primate hands are complex. Hands must be flexible to accomplish the sorts of behaviors that allow primates to manipulate their environment but must also absorb the stresses generated during movement. Thus, the multiple functional demands inherent in hand use in primates make this organ extremely fascinating for study. This chapter will discuss the functional anatomy of anthropoid hands, particularly of catarrhine primates. Observations on hand use and function will be integrated with anatomical descriptions as well as with experimental studies, especially cineradiographic and electromyographic techniques.

HAND POSTURES DURING LOCOMOTION

In Kowalski's (1976) treatise on mammals, the Order Primates is described as plantigrade (i.e., the method of placing a hand or foot onto a support). Such a characterization is too simplistic and, in fact, is incorrect considering the variety of locomotor modes found within the Order Primates (Napier and Napier, 1967). Likewise, the term *plantigrade* probably should not be applied to both the hand and foot. A wide variety of research has discussed the kinds of hand postures used during quadrupedal locomotion in primates, and most of this work has been directed toward hominoid locomotion. Unfortunately, cercopithecoids (Old World monkeys) have not received as much attention, and most publications on hands (e.g., Hill, 1966; Napier and Napier, 1976) mention hand postures only in the context of locomotor classifications or species descriptions. For example, Rose (1973) divides cercopithecoid hand postures into two main types—palmigrade/plantigrade and digitigrade.

The term *plantigrade* has been used to denote a support phase for the foot in which heelstrike occurs first and is followed by contact of the sole (Brooks, 1986; Hildebrand, 1988). The typical example is that of adult-human striding (Grillner, 1981). Hildebrand (1988) uses *digitigrade* to describe animals that support their weight on the "ball" (metatarsal heads) of the foot. The important difference is that in plantigrady the proximal part of the foot (i.e., the heel) strikes the substrate first and body weight is subsequently transferred to more distal elements. By implication, palmigrade hand postures (Fig. 4.1) should follow the same sequence of weight transmission as is found in the plantigrade

FIGURE 4.1. Palmigrade hand posture during quadrupedal walking in *C. aethiops* at Lake Naivasha, Kenya. Palmigrady helps to maintain stability during arboreal locomotion.

foot. In contrast, digitigrady has no sequence of "roll-over" (i.e., the distal elements support the weight at the beginning of the support phase). In cercopithecoid digitigrady, weight is borne on the volar surfaces of the phalanges without subsequent transfer of weight onto the palm (Fig. 4.2).

Field observations in Kenya on palmigrade vervet monkeys (*Cercopithecus aethiops)* demonstrated that there is no "roll-over" during handstrike (Whitehead, 1991). The proximal hypothenar pad does not contact the substrate before the remainder of the hand. Instead, touchdown includes the digits, as well as the interdigital, thenar, and hypothenar pads. The pollex contacts the substrate during support, and the palm bears weight until the animal begins to initiate swing. Cinematography and cineradiography on captive vervet monkeys clearly confirm this observation during both slow and fast walking sequences (Fig. 4.3). Both techniques showed that, contrary to the plantigrade model, the phalanges are actually inclined downward toward the substrate prior to actual touchdown. This gives the hand movements greater similarity to digitigrade touchdown than had previously been expected.

The significance of palmigrady probably lies in a need for stability during arboreal locomotion (Fig. 4.1). Not only does a palmigrade hand posture provide greater surface area of contact with the substrate, but it also lowers the animal's center of gravity. Vervets behaviorally lower their centers of gravity during locomotion on inclined surfaces (Fig. 4.4). Full-time palmigrady would aid in preventing loss of balance, especially on branches that are narrow relative to the size of the monkey.

Touchdown and support in digitigrade baboons (*Papio cynocephalus* and *Papio anubis)* involves only digits II-V (Fig. 4.5). Only the distal tip of the pollex intermittently contacts the substrate. Digitigrady confers an advantage by effectively lengthening the forearm. This would confer a number of advantages to a more terrestrial primate. For example, it would increase the distance the animal

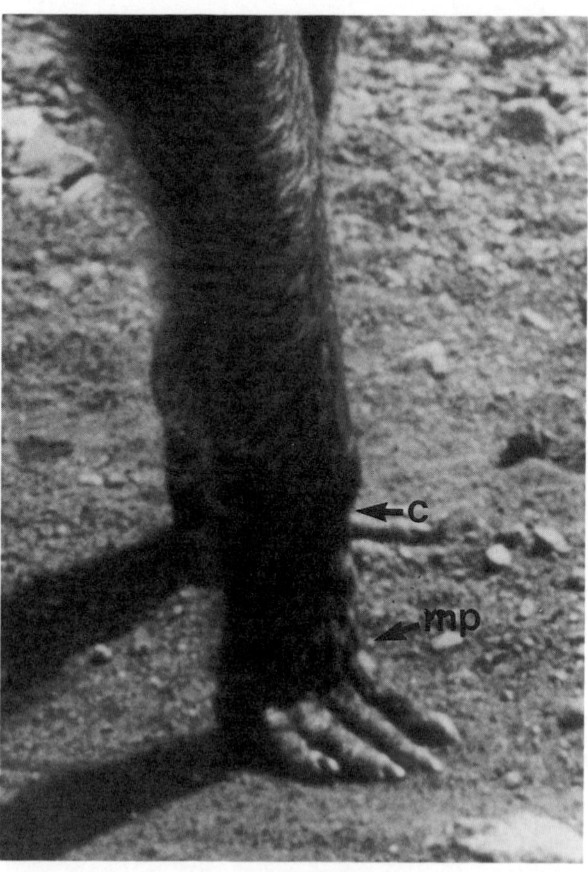

FIGURE 4.2. Left: Digitigrade hand posture in *Papio*. Thumb is not visible. Right: Weight is borne on the volar surfaces of all phalanges of digits II-V. Metacarpals are held off the substrate. Carpus = c, Metacarpophalangeal joint = *mp*.

could travel per muscle contraction, and it would increase the animal's mobility by raising the center of gravity (Lehmkuhl and Smith, 1983). The increase in distance per muscle contraction would increase the animal's potential speed (Hildebrand, 1988), and it may make each step more efficient in terms of energy utilized. This is particularly relevant to terrestrial monkeys because they often travel relatively large daily ranges (Crook and Aldrich-Blake, 1968; Kummer, 1968; Altmann and Altmann, 1970; Aldrich-Blake et al., 1971), but do not have to be concerned with balance as much when they are on the ground (Fig. 4.6).

In hominoids, knuckle-walking, a locomotor mode of African apes, has been well studied behaviorally (Tuttle, 1967, 1969), electromyographically (Tuttle, 1967, 1969; Tuttle et al., 1972; Tuttle and Basmajian, 1974; Susman and Stern, 1980), and cineradiographically (Jenkins and Fleagle, 1975). Tuttle (1967, 1969) has aptly described hand postures of African apes during knuckle-walking. The digits of fingers II through V are flexed, with the dorsal surface of the middle phalange contacting the substrate. The metacarpals are held off of the substrate and almost in line with the forearm. This position is produced by (1) hyperextension of the proximal phalanges at the metacarpophalangeal joints; (2) flexion of the proximal interphalangeal joints; and (3) flexion at the distal interphalangeal joints (Fig. 4.7). The pollex (thumb) does not touch the substrate.

Jenkins and Fleagle (1975) used cineradiography (the recording of a radiographic image with a motion picture camera [see Fodor and Malott, 1987])

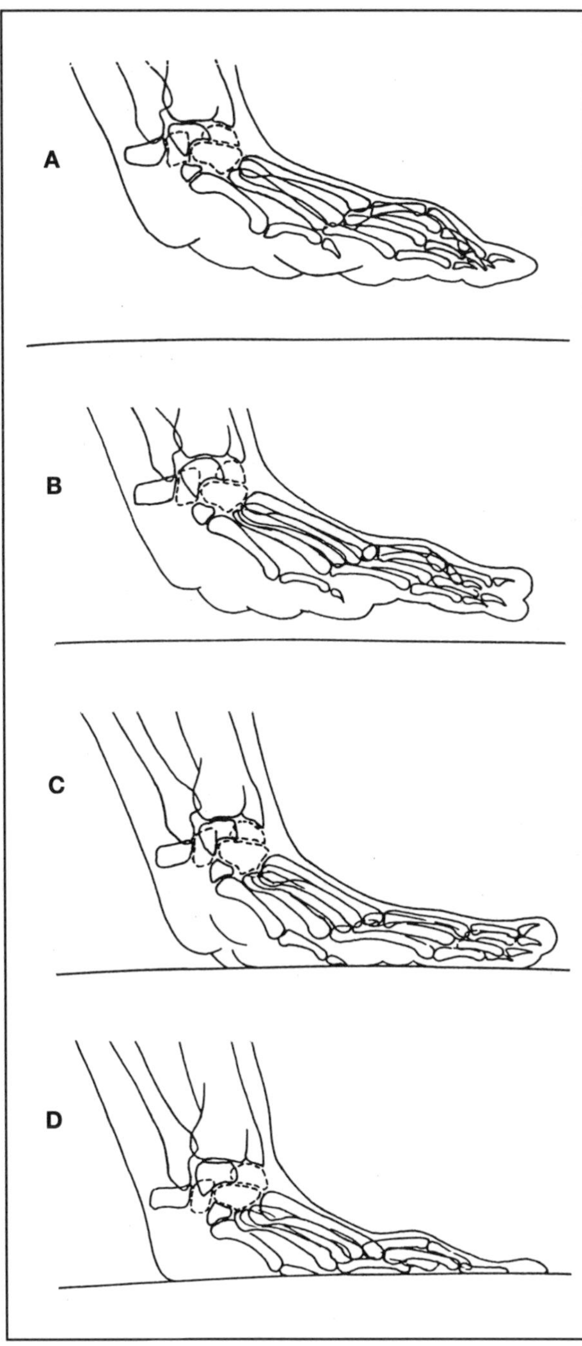

FIGURE 4.3. Touchdown in palmigrade *Cercopithecus aethiops*. The proximal manus does not contact the substrate before the distal elements. The pisiform is non-weight-bearing and maintains a horizontal orientation. Articulation of the ulna styloid with the pisiform and triquetrum is clearly visible. The articular surface of the distal radius is relatively parallel to the substrate just prior to and at touchdown.

on chimpanzees to study hand function during knuckle-walking. Jenkins and Fleagle (1975) found that the long axis of the carpals and metacarpals was between 10° and 15° to the radius. In the extended wrist, the radiocarpal joint faces dorsoproximally until the end of the support phase. The wrist begins the swing phase at 20° of flexion. The anteroposterior projection documents slight ulnar deviation, at both the proximal carpal and midcarpal joints, during weight-bearing touchdown. Jenkins and Fleagle (1975:219) stated that the joint between the antebrachium (forearm) and carpals "appears to remain static throughout a propulsive movement," and there is a "center of rotation that lies either at the level of the metacarpophalangeal joints or slightly distal to them." They did not detect flexion-extension motions at the carpometacarpal or midcarpal joints during support.

Like the African apes, various hand postures utilized by orangutans have been extensively investigated (Tuttle and Beck, 1972; Susman, 1974; Susman and Tuttle, 1976; Tuttle and Cortright, 1988). Orangutans frequently use their ulnar four digits as grasping hooks when arboreal (Rose, 1988), but when they move on the ground they often use a fist-walking hand posture (Tuttle, 1967). In fist-walking, the weight is borne on the dorsal surfaces of the proximal phalanges of digits II through V. Generally, the fist is moved rectilinear with the forearm, although ulnar deviation, supination, and volar flexion are also manifested. In modified fist-walking, the distal pollex contacts the substrate and may provide support, and the hand may be moved rectilinearly with the forearm or it may be volarflexed. In ulnar-deviated (adducted) fist-walking, the radial aspect of the hand contacts the substrate (Tuttle, 1967). In orangutan palmigrady, the wrist is dorsiflexed while the digits are extended. Tuttle (1967:179) states that the "tips of the fingers always touched the ground before the palm of the hand." This implies a similarity in hand movement to vervet palmigrady, although Tuttle (1967) ascribes it to the palmar curvature of the metacarpals and phalanges. Orangutan palmigrady is different from cercopithecoid palmigrady in that the hand is held outward, perpendicular to the forearm (Tuttle, 1970:244). In cercopithecoids, the hand is rectilinear or slightly ulnar-deviated to the forearm. In modified palmigrady, the dorsal surfaces of the distal phalanges are in contact with the substrate while

FIGURE 4.4. Vervets behaviorally lower their center of gravity when moving down an inclined substrate. Palmigrady helps to lower the center of gravity, regardless of substrate angle.

most of the individual's weight is supported by the posterior palm. Fully palmigrade hand postures are only rarely used by orangutans on the ground (Tuttle, 1970). Tuttle (1967) concludes that fist-walking is more common than palmigrady, and he proposes that orangutan hand postures are accommodations for an essentially arboreal primate.

ANATOMY OF THE WRIST AND HAND

Carpus

The anatomical region linking the hand with the forearm is the carpus, or wrist. The precise boundaries of the wrist have been a matter of some debate, because some authors include the distal radioulnar joint (Hollinshead and Jenkins, 1981) while others do not (Taleisnik, 1985). These disagreements have often been based on clinical criteria. In any case, the primate wrist includes eight or nine carpal bones (depending on the species), the distal radius, and the anatomy of the ulnocarpal space.

Primates are unusual within the Mammalia in that reductions and fusions of hand and wrist bones have not occurred as commonly as in other Orders of mammals (Bunnell, 1944; Yalden, 1970, 1971;

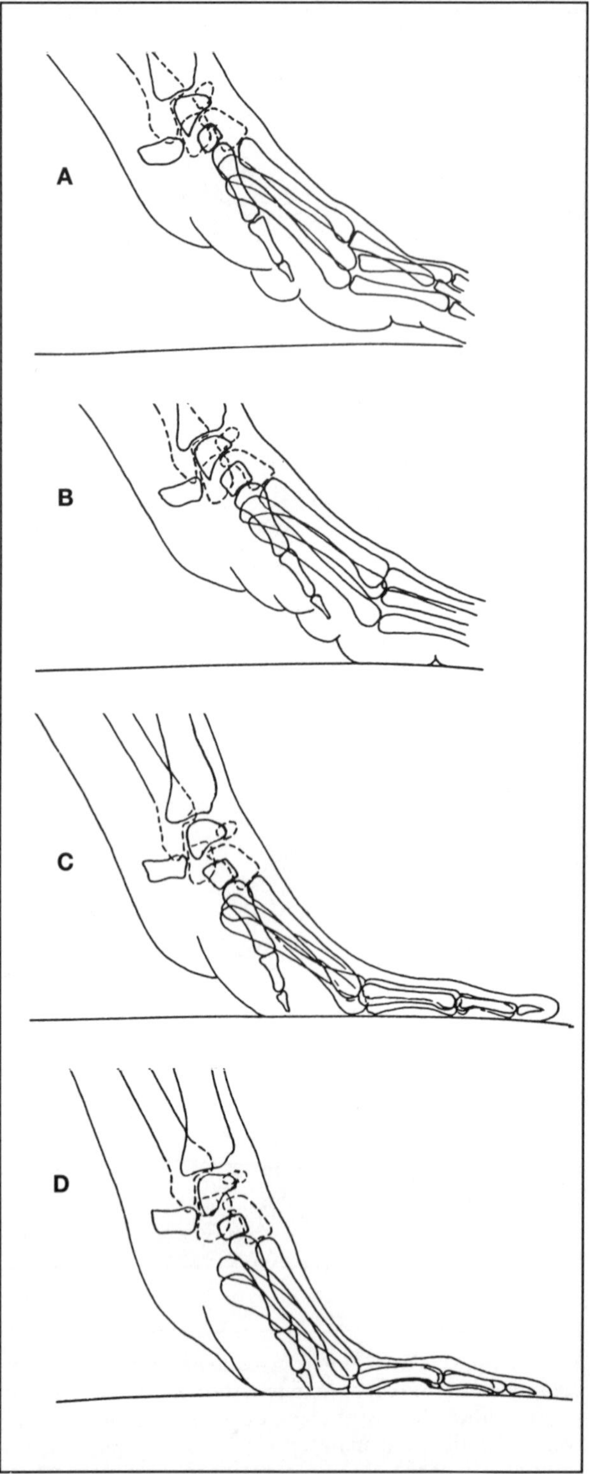

FIGURES 4.5. Digitigrade touchdown and support in a young *Papio anubis*. Touchdown includes phalanges of digits II-V. The inclination of the metacarpals is evident. The rod-like pisiform maintains a horizontal orientation.

FIGURE 4.6. Digitigrade walking in free-ranging *Papio cynocephalus*. Digitigrady lengthens the antebrachium, increasing the distance moved per muscle contraction.

Romer and Parsons, 1977). The relative lack of bone fusions is not solely due to the arboreal habitus of primate species since fusions have occurred in other arboreal mammals (e.g., marsupials; see Yalden, 1972). The maximum number of wrist bones in primates, excluding the prepollex, is nine.

The wrist or carpus is often simplistically classified as a condyloid joint (Hall, 1965; Crafts, 1985). It actually consists of several joints including the radiocarpal, intercarpal, and carpometacarpal joints (see Hollinshead and Jenkins, 1981; Lemkuhl and Smith, 1983). The standard representation of the carpus is of two transverse rows of bones: a proximal (antebrachial) row and a distal (metacarpal) row (Woodburne, 1973; Hollinshead and Jenkins, 1981; Crafts, 1985; Stern, 1988). The proximal row, from ulnar to radial sides, consists of five bones: the pisiform, the triquetrum (triangular), the lunate, the scaphoid (navicular), and the centrale. The distal row, from ulnar to radial sides, consists of the hamate, the capitate, the trapezoid, and the trapezium (Fig. 4.8). Humans and African apes lack a distinct centrale, although it is commonly present in Old and New World monkeys and occasionally in orangutans. It is also occasionally fused with the scaphoid in some individuals in the genera *Colobus* and *Cercopithecus* of the mona group (Fig. 4.9).

The traditional horizontal-row depiction of the carpus has been criticized as unrealistic in terms of wrist function. Some authors (Kapandji, 1982; Taleisnik, 1985; Weber, 1988) prefer the concept of the vertical carpus after Navarro (1935). In this view, the wrist is composed of a central column for extension and flexion flanked by two other columns. The central column is made of the lunate, capitate, and hamate. The "mobile" column is composed of the scaphoid, trapezium, and trapezoid while the triquetrum and pisiform are elements of the "rotation" column.

Hand

The Order Primates has been generally defined as possessing five rays in the hand and in the foot (Napier and Napier, 1967). Generally, this number has been considered to be primitive for tetrapods

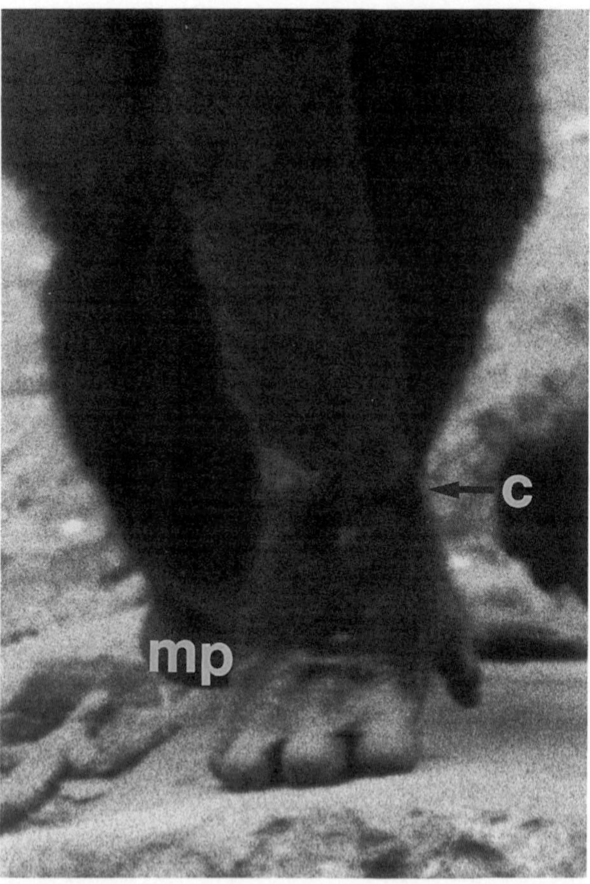

FIGURE 4.7. Left: Knuckle-walking in a gorilla. The right hand is in support. Right: The thumb is held off the substrate. The dorsal surfaces of the middle phalanges are bearing the weight, with the proximal phalanges and metacarpals held off the ground. Hyperextension of the proximal phalanges at the metacarpophalangeal (mp) joints is apparent. Carpus = c.

(Bunnell, 1944; Romer, 1966; Lewis, 1989), although recent developmental and paleontological evidence (Shubin and Alberch, 1986; Coates and Clack, 1990) appears to suggest that seven or eight may actually have been the primitive condition.

Weight-Bearing Digits

Although there have been many studies of the first carpometacarpal joint (Haines, 1944; Napier, 1955; Pieron, 1981), the joints of the principal weight-bearing digits have received much less attention. The devotion to the thumb is due largely to a preoccupation with the evolution of hominid toolmaking ability (Napier, 1962; Susman, 1988), as well as the generalization that the ulnar four carpometacarpal joints are relatively immovable mortices. The thumb contacts the substrate in palmigrade monkeys, but only the tip of the thumb intermittently contacts the substrate during digitigrade posture and locomotion (Tuttle, 1967; Whitehead, 1991). The principal line of force transmission is therefore from the phalanges of the four nonpollex digits, through the metacarpophalangeal joints and metacarpals, and finally to the carpometacarpal joints. An important difference between the hamates of digitigrade and palmigrade cercopithecoids is in the distinctness of the distal facets for the fourth and fifth metacarpals. Palmigrade species have highly distinct metacarpal facets (Fig. 4.10). These facets show a more even gradation in digitigrade species. It is, therefore, not possible to lump cercop-

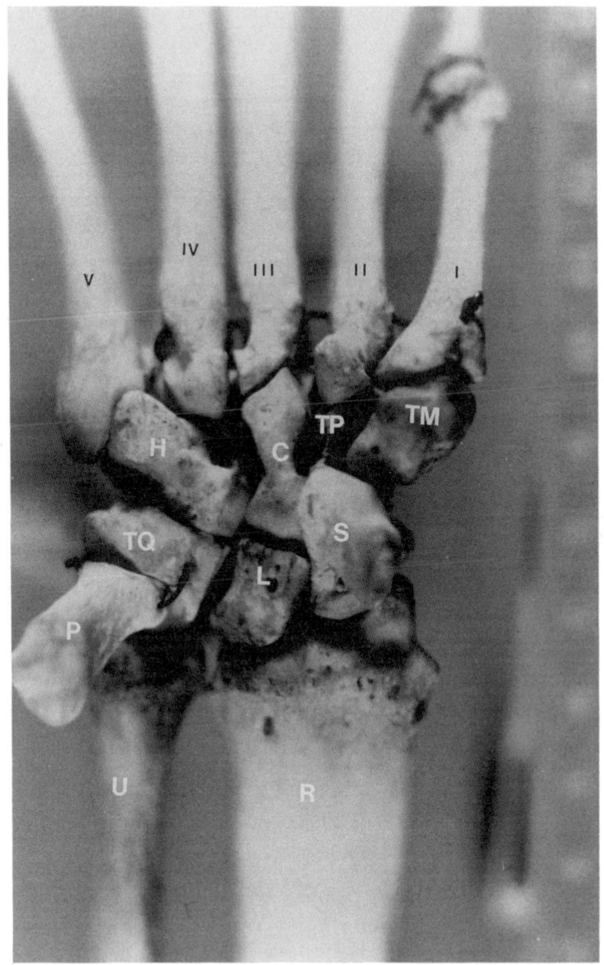

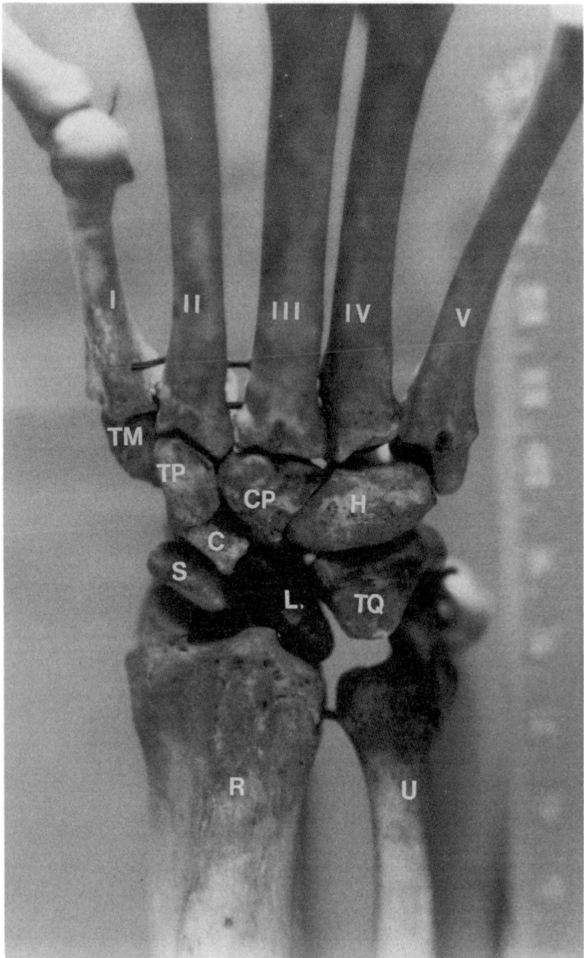

FIGURE 4.8. **A**: Palmar view of the wrist in *Papio hamadryas*. **B**: Dorsal view of the wrist in *Papio hamadryas*. C = centrate; CP = capitate; H = hamate; L = lunate; P = pisiform; R = radius; S = scaphoid; TM = trapezium; TP = trapezoid; TQ = triquitrum; and U = ulna.

ithecoids (see O'Connor, 1975) into a single "palmigrade/digitigrade" category in terms of the hametometacarpal joints.

Metacarpals and Phalanges

The five metacarpals, one for each of the five rays of the hand, articulate with the bones of the distal carpal row (Fig. 4.11). The concavoconvex base of the first metacarpal articulates with the distal saddle facet of the trapezium. The second metacarpal articulates principally with the trapezoid, with smaller areas of articulation with the trapezium radially and the capitate ulnarly. The second metacarpal is generally the longest in humans (Woodburne, 1973; Bogumill, 1988). The third metacarpal articulates with the capitate, as does the radial portion of the proximal facet of the fourth metacarpal. The radial portion of the fourth metacarpal sometimes manifests a clear, raised capitate facet in cercopithecoids but often the capitate area blends into the articular surface for the hamate. The fourth and fifth metacarpals both articulate with the hamate. Each of the fingers, with the

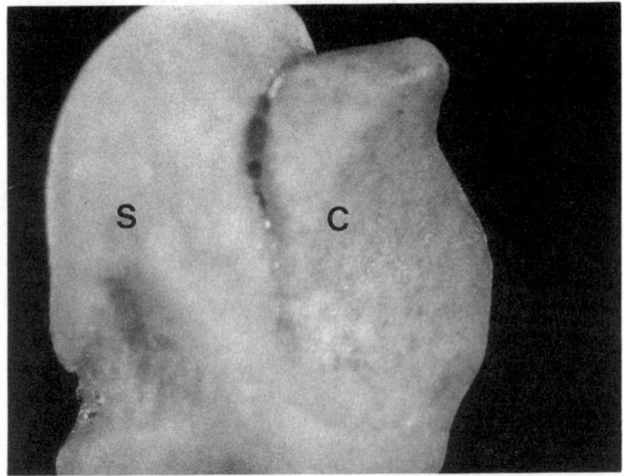

FIGURE 4.9. Fused scaphoid (S) and centrale (C) of *Colobus*. Fusion is most complete in the volar half of the contact.

FIGURE 4.10. Distal view of the hamate of *Theropithecus oswaldi*. This specimen has the distinct and separate facets for the fourth and fifth metacarpals that are found in palmigrade primates such as *Colobus guereza* and *Cercopithecus albogularis*.

exception of the pollex, has three phalanges. In the majority of primates, the pollex has two phalanges although some primates, such as *Colobus* and *Ateles*, possess only one phalanx (Tuttle, 1975; Schultz, 1986).

Capitate

The broad, concave distal facet of the capitate articulates with the entire surface of the proximal facet of the third metacarpal in anthropoids. Rose (1984) states that there is little movement and great stability of this joint in pongids. Likewise, in humans (Weber, 1988) and in cercopithecoids the capitate remains stable during both radial and ulnar deviation. The dorsoulnar corner of the distal capitate articulates with the proximal fourth metacarpal facet in anthropoids. The distoradial facets articulate with the second metacarpal. There is a radial articulation for the trapezoid. A long ulnar-oriented facet is for the hamate, and it runs proximally into the head of the capitate. The combination of proximal capitate and hamate form the ball of the midcarpal joint described by Jenkins (1981). Palmar to the hamate facet is a relatively deeply excavated area for the strong interosseous ligament. The radial aspect of the capitate head is more bulbous in humans than it is in cercopithecoids. There are three joints on the head of the capitate for articulation with the lunate, the scaphoid/centrale, and the hamate. The three facets are so different that the head of the capitate cannot be considered to be a simple sphere (Kauer and Landsmeer, 1981).

Hamate

As in modern humans, the body of the hamate is basically triangular in living apes and cercopithecoids. The anthropoid hamate articulates with the capitate, the triquetrum, the lunate, and the fourth and fifth metacarpals. The hamulus projects from the volar surface of the body. It is variable in both size and shape in different species of primates. A well-developed hamulus projects from the palmar surface of the modern human hamate and is distinctly curved radially. This surface provides sites of attachment for the flexor retinaculum, for the pisohamate ligament (which is an extension of flexor carpi ulnaris), and for the origins of opponens digiti minimi and flexor digiti minimi (Bogumill, 1988). The human hamulus projects further than does the pisiform, and is roughly equivalent in length to the body of the hamate. The curvature of the hamulus, together with the body of the hamate, forms a "concavity" that functions as a pulley for the long digital flexors (Bogumill, 1988). The distal edge of the

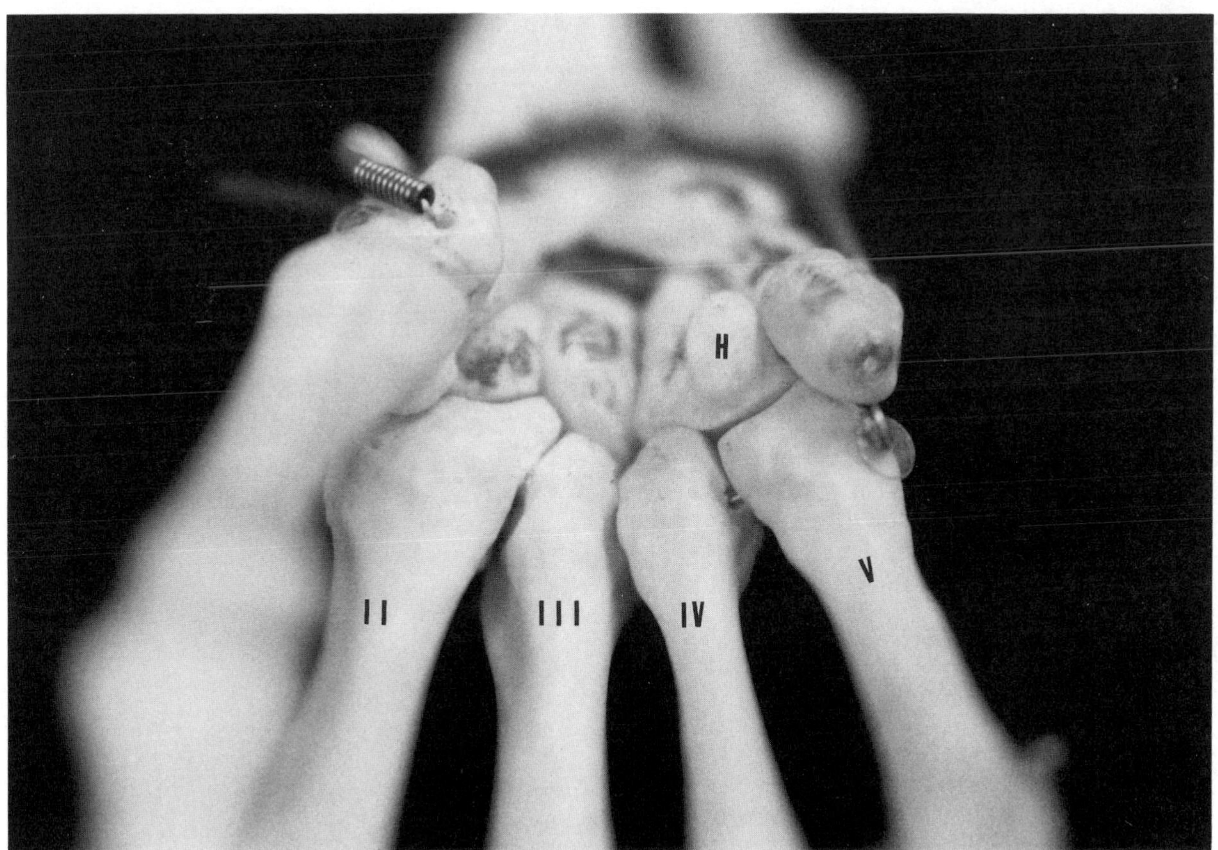

FIGURE 4.11. Articulation of the bases of the metacarpals with the distal carpal row in a modern human. The carpal tunnel is bounded by the scaphoid tubercle and trapezium on the radial side, and the pisiform and hamate hamulus on the ulnar. It is apparent that the hamulus curves principally radially in a human and does not closely girdle the volar surface of the fifth metacarpal. The styloid of the second metacarpal is prominent. Hamulus = H.

hamulus slightly crosses the volar surface of the base of the fourth and fifth metacarpals in humans. This crossing is at a distance from the volar surfaces of the metacarpals; that is, the hamulus does not contact the metacarpals' surfaces (Fig. 4.11). The volar surfaces of the proximal ends of the fourth and fifth metacarpals fit snugly against the hamulus in both *Gorilla* and *Pongo,* and the morphology of the distal hamate is more similar among gorillas and orangutans than either species is to humans. O'Connor (1975:119) illustrates a relatively long, distally projecting hamulus in gibbons and interprets it as limiting dorsiflexion.

The hamulus is variable in development in cercopithecoids. It is prominent, and hooklike in digitigrade papionines and in *Erythrocebus,* but smaller in palmigrade monkeys such as *Colobus polykomos.* Corrucini et al. (1975) did not fully appreciate this difference since they relied on an index that did not indicate the extent of overlap of the hamulus with the volar metacarpals or the direction of projection by the hamulus. The fact that cercopithecoids are not uniform (contra Schultz, 1970; Lewis, 1972, 1974; O'Connor, 1975) is underscored by the relatively short hamulus in *Theropithecus oswaldi* and the relatively longer one in the smaller *T. gelada.* Since the hamulus crosses the palmar surface of the fifth metacarpal in digitigrade monkeys, it is actually more hooklike in its longitudinal axis than in humans—contra O'Connor (1975).

Scaphoid

The scaphoid has been described as a bridge across the midcarpal joint (Taleisnik, 1985). Together with the lunate, it forms a convex surface on which articulates the radius. There is considerable variability in the shape of the human scaphoid, with a variety of bumps and notches along its contour (Poznanski, 1984). In all anthropoids, the nonarticular waist of the scaphoid divides it into dorsal and volar halves. The waist has several foramina for blood vessels, and regions for ligament attachment. The proximal surface of the dorsal half possesses the large radial facet. Tuttle (1975) believes that weight transmission is facilitated by possession of a proximal carpal row that is radioulnarly wider than it is dorsopalmarly deep in the knuckle-walking African apes.

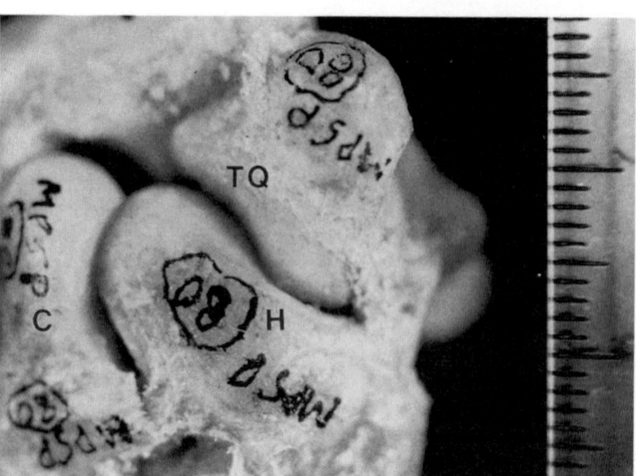

FIGURE 4.12. Helicoidal articulation between the triquetrum (TQ) and hamate (H).

Centrale

In humans, the centrale appears during the sixth week in the fetus (Taleisnik, 1985). Although it may persist as an anomality (Poznanski, 1984), it is normally fused with the scaphoid by the eighth week of development. While the centrale remains free and articulates with the trapezoid in cercopithecoids, gibbons, and orangutans, it fuses with the scaphoid by the end of infancy in knuckle-walkers (Lewis, 1989). Tuttle (1975) relates fusion of the centrale in African apes to weight transmission during knuckle-walking.

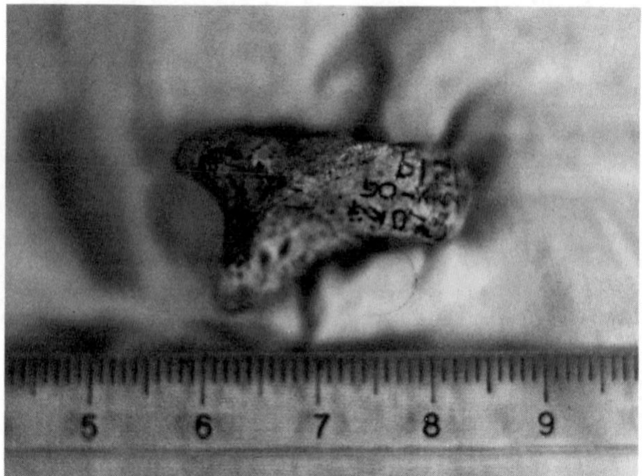

FIGURE 4.13. Hamate of *Theropithecus oswaldi* from Olorgesailie, Kenya. Specimen has a short triquetral facet. A short facet would limit ulnar deviation.

Triquetrum

The triquetrum has four facets in cercopithecoids: one for the pisiform, a second for the lunate, a third for the hamate, and a fourth for the styloid process of the ulna. Lewis (1989) writes that the ulnar styloid may be completely excluded from the wrist in orangutans, while the styloid articulates with the triquetrum in gibbons and chimpanzees. The facet for the hamate is helicoidal, and the triquetrum spirals along the articulation during radial and ulnar deviation (Fig. 4.12). This system acts as a first-class lever and is the control surface for the proximal row of carpals (Weber, 1988). The "high" position, in which the triquetrum is proximal, occurs during radial deviation. During ulnar deviation, the triquetrum spirals toward the fifth metacarpal. As a consequence, the short triquetral facet on the hamate of *Theropithecus oswaldi* would have prevented ulnar deviation to the extent found in modern baboons (Whitehead, 1990) (Fig. 4.13). The pisiform facet is often slightly concave in cercopithecoids. Taleisnik (1985) has argued that the articulated triquetrum-pisiform is functionally equivalent to the scaphoid and that the articulated lunate-triquetrum-pisiform resembles a scaphoid. Observations on the human wrist indicate that the spiral movement by the triquetrum, during ulnar deviation, causes the lunate to dorsiflex (Taleisnik, 1985).

Pisiform

The cercopithecoid pisiform is not spherical like the human pisiform. It is rod-shaped and is rarely the smallest wrist bone, as is generally true in humans. The pisiform bears two articular facets which are adjacent to each other. The first facet articulates with the triquetrum while the second articulates with the styloid process of the ulna. In humans, the pisiform bears only a single facet for articulation with the triquetrum. Lewis (1989) has demonstrated that a fibrocartilaginous intra-articular meniscus excludes the ulnar styloid from direct articulation with the pisiform in hominoids. An extra bony element, the os Daubentonii, is found in the meniscus in gibbons (Lewis, 1989). In gibbons, the pisiform articulates with the meniscus rather than with the ulnar styloid process (Lewis, 1989). The ulnar facet of the pisiform is concave in the radioulnar plane, and it

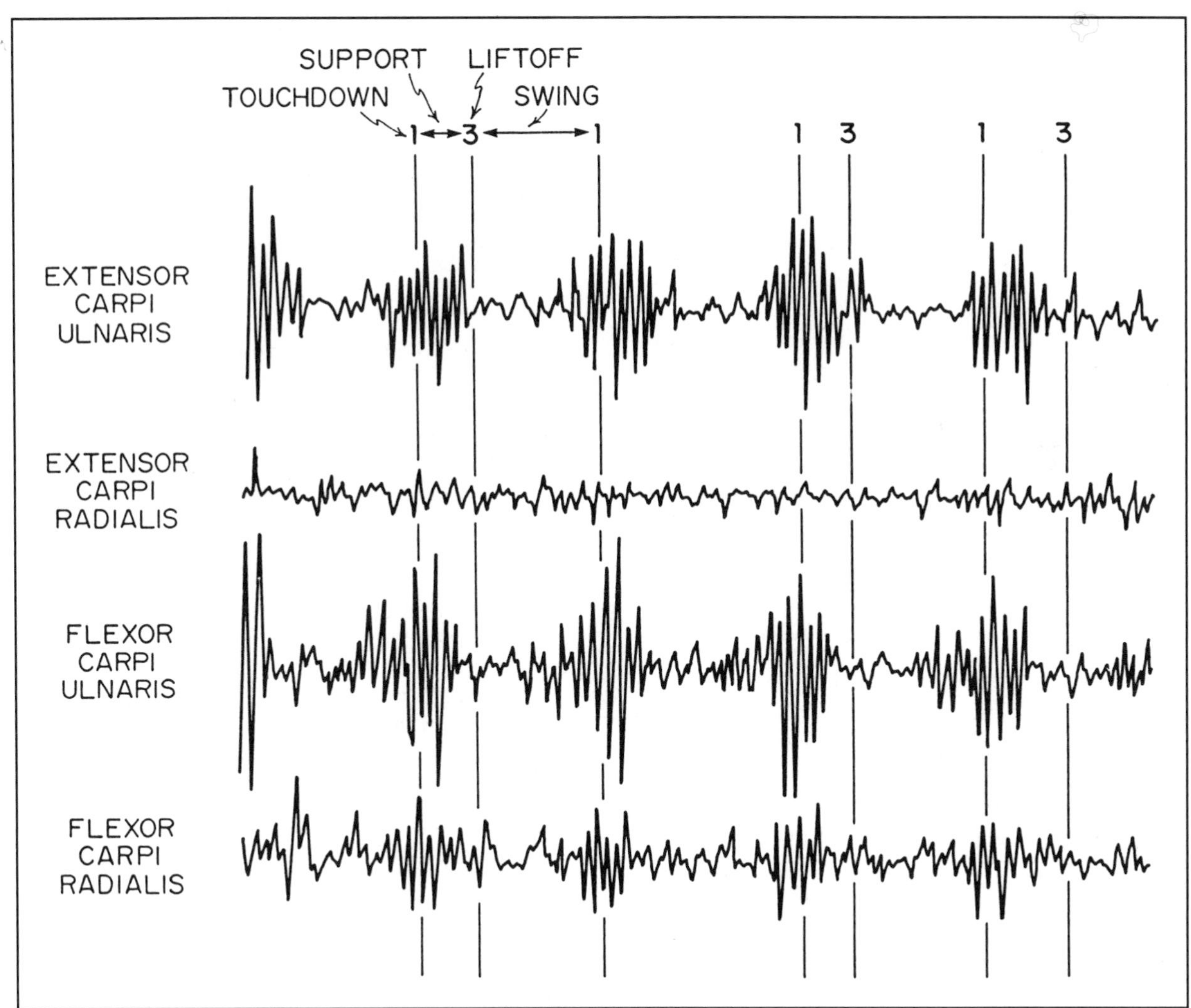

FIGURE 4.14. Electromyography of wrist extensors and flexors in an adult female *Cercopithecus aethiops*. Individual was palmigrade during the recording session and has been since her birth. Flexor carpi ulnaris begins activity prior to touchdown and continues to fire until the wrist is past the gleno-humeral joint.

faces dorsally, in a walking monkey (Figs. 4.3 and 4.5). The triquetral facet is relatively flat, and is functionally oriented distopalmarly. The base of the pisiform receives the insertion of the flexor carpi ulnaris. Although the thickness of the pisiform shaft can be variable in cercopithecoids, the shapes of the facets are relatively uniform within taxa.

Cineradiography clearly demonstrates that the pisiform is a non-weight-bearing bone in both palmigrade and digitigrade cercopithecids (Whitehead, 1991). The long axis maintains an orientation horizontal to the substrate, during the weight-bearing phase of support (Figs. 4.3 and 4.5). The palmigrade pisiform-metacarpal system is a first-class lever, analogous to the gastrocnemius-calcaneus-metatarsal system analyzed by Gowitze and Milner (1988) and Hildebrand (1988). The flexor carpi ulnaris is active during the weight-bearing phase of support (Fig. 4.14), in a manner that is analogous to the gastrocnemius muscle during human striding (Basmajian and DeLuca, 1985). Flexor carpi ulnaris provides the "in-force" for the lever. The tendon of flexor carpi ulnaris continues as the pisohamate and pisometacarpal ligaments (Weston and Kelsey, 1973; Lewis, 1989). Therefore, although the palmigrade pisiform is analogous to a calcaneus (Lewis, 1974) in the sense of muscle action, it is not analogous in terms of the weight-transmission found in humans. Rose (1988) points out that the orangutan pisiform points distally and palmarly, allowing the animal to keep its wrist flexed against extending forces while climbing.

LIGAMENTS OF THE WRIST

The movements of individual carpals, and of the wrist as a whole, are significantly influenced by two major ligament groups (Taleisnik, 1985). The intrinsic ligaments are those that run between carpal bones and control movement between carpals along a carpal row. The extrinsic ligaments are those that run between the radius and the carpus, and between the carpals and the metacarpals (Table 4.1). Weber (1988:42) points out that wrist function is dependent on a combination of joint contact surfaces and of ligaments; ligaments provide stability at the extremes of joint movement, while the joint surfaces "provide stability throughout the entire range of motion while transmitting loads."

TABLE 4.1
Summary of Wrist Ligaments[1]

EXTRINSIC
Proximal Extrinsic Ligaments
"Radial Collateral"
Volar Radiocarpal
Superficial
Deep
Radiocapitate
(Radioscaphocapitate)
Radiolunate
(Radiotriquetral)
Radioscaphoid
(Radioscapholunate)
Ulnocarpal Complex
"Triangular Fibrocartilage Complex": Ulnocarpal meniscus homologue
"Triangular Fibrocartilage Complex": Triangular fibrocartilage
Ulnolunate ligament
"Ulnar collateral ligament"
Intracapsular ulnocarpal ligament
Dorsal Radiocarpal
INTRINSIC
Short
Volar
Capitotriquetral
Dorsal
Trapeziotrapezoid
Trapeziocapitate
Capitohamate
Interosseous
Intermediate
Lunotriquetral (Radiotriquetral)
Scapholunate
Scaphotrapezium
Long
Volar intercarpal (V deltoid)
Dorsal intercarpal
RETINACULA
"Dorsal Retinaculum"
Supratendinous
Infratendinous
Flexor Retinaculum

[1] Based on Taleisnik (1988) and Mayfield (1988).

Extrinsic Ligaments

The radial collateral ligament is generally included in descriptions of human wrist anatomy (Crafts, 1985; Bogumill, 1988) as a discrete ligament but this is a matter of some controversy. Lewis et al. (1970) view the radial "collateral ligament" as a specialization of the fibrous capsule and thus do not distinguish the radial and ulnar "collateral ligaments" as separate elements. Taleisnik (1985) points out that collateral ligaments are important when motion is limited to flexion and extension. Carpal movement is not limited to the sagittal plane, but includes radial and ulnar deviation, and it has been suggested that extensor pollicis brevis and abductor pollicis longus serve as collateral support on the radial side, while extensor carpi ulnaris provides similar support on the ulnar side (Kauer, 1979).

The remaining proximal extrinsic ligaments can be divided into volar radiocarpal, ulnocarpal and dorsal radiocarpal. Mayfield (1988:55) writes that dorsiflexion, volar flexion, and radial and ulnar deviation are functionally controlled by the "specific arrangement and integrity" of the volar intracapsular ligaments. The deep volar radiocarpal ligaments are named in respect to their points of origin and insertion. They include the radiocapitate and radioscapholunate ligaments. The third ligament is called the radiotriquetral by Mayfield (1988) and this is apparently the same ligament that Taleisnik (1985) divides into two, the radiolunate and the lunotriquetral. During dorsiflexion of the scaphoid, the radioscaphoid ligament prevents excessive volar displacement of the proximal pole of the scaphoid. In volar flexion, that ligament and the radioscaphocapitate fascicle bind the scaphoid to the volar margin of the radius (Taleisnik, 1985). The superficial volar radiovolar ligaments form a complex pattern, in which most fibers form a V-shape. The apex of the V is attached to the capitate and lunate, while the ends attach proximally to the ulna and radius (Taleisnik, 1984, 1985). In humans, the dorsal radiocarpal ligament is considerably thinner than are the ligaments of the volar aspect. This dorsal ligament ties the dorsodistal radius to the lunate and triquetrum. Bogumill (1988) states that this ligament forces the wrist and hand to pronate when the forearm is pronated.

Intrinsic Ligaments

The intrinsic carpal ligaments bind the carpals. The ligaments are categorized as short, intermediate, or long on the basis of their length and on the relative degree of mobility they permit between the carpals. The short intrinsic ligaments are thick and do not permit much mobility. They also bind the distal carpals into a single functional unit (Taleisnik, 1984). The intermediate intrinsic ligaments include the lunotriquetral ligament (part of the radiotriquetral, according to Mayfield, 1988), the scapholunate ligaments, and the scaphotrapezium ligaments. The scapholunate ligaments allow motion between the scaphoid and the lunate. In humans, the scaphoid rotates about 30° and the lunate about 28° from neutral position to full dorsiflexion (Taleisnik, 1984). The scaphotrapezium ligaments allow volar-to-dorsal rocking of the distal end of the scaphoid on the biconcave surface formed by the trapezium and trapezoid (Taleisnik, 1984). Long intrinsic ligaments are found both volarly and dorsally. The volar intrinsic intercarpal ligament stabilizes the capitate. Its radial fascicle stabilizes the midcarpal joint.

Retinacula

The flexor retinaculum stretches across the wrist, on the palmar aspect, in a radioulnar direction. It connects the pisiform and hamate on the ulnar side with the scaphoid and trapezium on the radial side. Most of the extrinsic flexors pass dorsal to the flexor retinaculum, so that the ligament prevents "bowstringing" during wrist flexion (Bogumill, 1988). Stern (1988) states that the flexor retinaculum is a true transverse carpal ligament. The extensor retinaculum has a similar function relative to the flexor retinaculum, but it is actually made out of "a reinforced region of deep fascia" (Stern, 1988:566). The extensor retinaculum is composed of supratendinous and infratendinous layers in humans, and pronation-supination is facilitated by the lack of connection of the supratendinous layer on the ulna (Taleisnik, 1985). There are six septa created by the extensor retinaculum in humans that separate the extensor muscles from each other.

Extrinsic Muscles of the Wrist and Hand

The muscles of the hand and wrist can be broadly divided into those that are extrinsic and those that are intrinsic (Table 4.2). The extrinsic muscles originate at points proximal to the wrist, and insert on either wrist or hand structures. All of the muscles that move the wrist are extrinsic to it. The extrinsic muscles of the hand are pluriarticular,

TABLE 4.2
Summary of Anthropoid Forearm and Hand Muscles

LAYERS OF EXTRINSIC FLEXORS
Outer
M. flexor carpi radialis
M. palmaris longus
M. flexor carpi ulnaris
Second
M. flexor digitorum superficialis (sublimis)
Third
M. flexor digitorum profundus
M. flexor pollicis longus

LAYERS OF EXTRINSIC EXTENSORS
Superficial
M. extensor carpi radialis longus
M. extensor carpi radialis brevis
M. extensor digitorum communis
M. extensor carpi ulnaris
M. abductor pollicis longus
Deeper
M. extensor digiti proprius
M. extensor pollicis longus
M. extensor pollicis brevis (hominoids)
M. extensor digiti quarti et quinti proprius
M. extensor digiti secundi et tertii proprius
[M. extensor digiti anularis;
M. extensor digiti medius]
M. extensor digiti indicis proprius
M. extensor digiti minimi proprius

INTRINSIC FLEXORS (Palmar)
M. palmaris brevis
M. abductor pollicis brevis
M. opponens pollicis
M. flexor pollicis brevis
M. adductor pollicis
M. abductor digiti minimi (M. abductor digiti quinti manus)
M. flexor digiti minimi brevis (M. flexor digiti quinti brevis manus)
M. opponens digiti minimi (M. opponens digiti quinti manus)
Mm. lumbricales manus
Mm. contrahentes digitorum manus
Mm. interossei manus

INTRINSIC EXTENSORS (Dorsal)
Normally no intrinsic extensors

meaning that these muscles cross several joints. A pluriarticular muscle, when contracted, can produce movement in all of the joints over which it passes, although synergists usually stabilize the joint when movement is not desired (Lemkuhl and Smith, 1983). Intrinsic muscles both originate and insert within the hand.

There are three flexors and three extensors of the wrist. The flexors are flexor carpi radialis, palmaris longus, and flexor carpi ulnaris. Palmaris longus is variable in its presence in primates, being absent in most gorillas and in many humans (Swindler and Wood, 1973). All of the flexors originate from the medial epicondyle of the humerus. Stern (1988) records that there are fibrous septa that separate palmaris longus and flexor carpi radialis from the other muscles that originate from the medial epicondyle. The origin of flexor carpi radialis is largely confined to the medial epicondyle in *Papio anubis, Macaca mulatta, Alouatta seniculus,* and *Homo sapiens.* This muscle also originates from the proximal shaft of the radius in apes (Howell and Straus, 1933; Schön, 1968; Swindler and Wood, 1973) and is fused with pronator teres in rhesus monkeys (Howell and Straus, 1933). In humans, flexor carpi ulnaris possesses a humeral head, as well as an ulnar head (originating from the medial border of the olecranon, the upper part of the posterior border of the ulna, and from the area distal to the medial collateral ligament [Kaplan and Taleisnik, 1984]). With the exception of palmaris longus, the flexor tendons are deep to the flexor retinaculum. The tendon of palmaris longus is superficial to the flexor retinaculum, and becomes the palmar aponeurosis in the hand. Flexor carpi radialis inserts on the palmar aspect of the base of the second metacarpal, and sometimes on the third metacarpal. Flexor carpi ulnaris is unusual in that it inserts directly on a carpal bone, the pisiform.

The finger flexors are flexor digitorum superficialis (sublimis), flexor pollicis longus, and flexor digitorum profundus. Flexor digitorum superficialis originates from the medial epicondyle of the humerus in *Papio anubis* and *Macaca mulatta,* while it has additional origin from the coronoid process and from the radial shaft in *Homo* and *Pan* (Howell and Straus, 1933; Swindler and Wood, 1973). This muscle inserts onto the palmar surfaces of the middle phalanges of digits II through V in the rhesus monkey, olive baboons, chimpanzees, and humans. In all four of the above species, the tendons that insert on the second and fifth fingers are deep to those to the third and fourth fingers (Howell and Straus, 1933; Swindler and Wood, 1973). There are ulnar and radial heads for flexor digitorum profundus in *Papio anubis* and *Macaca mulatta,* although the tendons merge before the wrist. The radial head sends tendons to the first, second, and third fingers while the ulnar head sends tendons to the third, fourth, and fifth digits (Howell and Straus, 1933; Swindler and Wood, 1973). Schön (1968) writes that the medial head of flexor digitorum profundus inserts on digits I and II in *Alouatta seniculus,* while the ulnar head inserts on the third, fourth, and fifth digits. In humans, flexor digitorum profundus originates from the medial and anterior aspects of the ulna (Stern, 1988). It also has an ulnar origin in *Pan* (Swindler and Wood, 1973). In each of the above species, insertion is onto the palmar surface of the terminal phalange, with the deep tendons traveling through slits in the superficial tendons. In humans, flexor pollicis longus originates from the radius, travels around the trapezium, and inserts onto the distal phalange of the thumb (Valentin, 1981). In *Papio* and in *Macaca mulatta,* the long thumb flexor is part of the radial portion of flexor digitorum profundus (Howell and Straus, 1933; Swindler and Wood, 1973).

The extensor muscles of the wrist and hand consist of superficial and deep groups. Lewis (1989) states that the extensors conform to a basic therian pattern in monkeys, but exhibit a derived condition in hominoids. Although the muscles of the superficial layer are comparable among taxa, problems of homology of the deeper muscles make comparison less direct. As Lewis (1989:141) writes, the deep layer is where "the most extreme modifications of the extensor musculature have occurred, and where the most variability is found."

The three extensors of the wrist are extensor carpi radialis longus, extensor carpi radialis brevis, and extensor carpi ulnaris. The lateral epicondyle of the humerus is the source of all three muscles, with additional origin from the supracondylar ridge for extensor carpi radialis longus. The belly of extensor carpi radialis brevis is fused with that of longus in *Papio papio* (Straus, 1941), but has a distinct origin in *Macaca mulatta* (Howell and Straus, 1933). Stern (1988) reports that extensor carpi ulnaris also has an ulnar head in humans. Extensor carpi radialis longus inserts on the dorsal aspect of the base of the second metacarpal. Extensor carpi radialis brevis inserts on the dorsal surface of the base of the third metacarpal in *Macaca mulatta, Papio cynocephalus, Papio anubis, Hylobates moloch, Pan,* and *Homo sapiens,* while it inserts on both second and third metacarpals in *Cebus* (Howell and Straus, 1933; Straus, 1941; Swindler and Wood, 1973). Extensor carpi ulnaris inserts on the base of the fifth metacarpal on the dorsal surface in *Macaca,* on the ulnar side in *Papio, Hylobates moloch, Hylobates pileatus,* and *Alouatta fusca,* on the ulnopalmar surface in *Alouatta seniculus* and *Pongo,* and on the palmar surface in *Cebus* (Howell and Straus, 1933; Straus, 1941; Schön, 1968).

The origin of abductor pollicis longus is from the dorsal surfaces of the ulna and radius, and the interosseous membrane. The muscle is part of the deep layer of extensors, and manifests considerable variability in insertion among taxa. It is often stated that flexor carpi ulnaris is the only muscle that inserts directly on a carpal bone. Although this may be true in some humans, abductor pollicis longus frequently manifests an insertion on the trapezium or prepollex in a variety of primate taxa. Unfortunately, this variability among different taxa may be due to actual variability or just to differences in anatomical observation. For example, insertion on the first metacarpal with the attachment either on the radial or radiopalmar surface of the base is known to occur in *Cebus, Macaca mulatta, Papio, Hylobates, Pan, Pongo,* and *Homo sapiens.* Straus (1941) also lists an additional insertion onto the trapezium in *Cebus* and *Papio,* while Swindler and Wood (1973) describe an insertion on the prepollex for *Papio.* In *Macaca, Hylobates,* and *Pongo,* the prepollex is the second insertion with "some overspill onto the trapezium" for *Hylobates* and *Pongo* (Lewis, 1989:141). In contrast, Straus (1941) notes an insertion on the trapezium in *Pongo* rather than on the prepollex. Lewis (1989) further states that the tendon inserts onto the base of the first metacarpal and prepollex in *Pan,* with some fibers to the trapezium, while Swindler and Wood (1973) describe this insertion on the prepollex and trapezium, but not to the first metacarpal, for *Pan.* In *Alouatta seniculus* and *Alouatta caraya,* abductor pollicis longus inserts on the prepollex and trapezium, but not on the first metacarpal, while this same tendon in *Alouatta fusca* inserts only on the trapezium and the radial side of the first metacarpal (Schön, 1968). For *Gorilla,* Lewis (1989) points out that the prepollex is fused with the trapezium and that the upper tendon inserts on that composite bone. In humans, the second tendon goes to either the trapezium or to the abductor pollicis brevis (Lewis, 1989). Extensor pollicis brevis is present in hominoids but is apparently absent in monkeys.

The deep extensor stratum includes the extensor proprius (profundus). Although the proprius sends tendons to all digits in monkeys, it does not do so in hominoids (Lewis, 1989). In monkeys, the belly originating at the lateral epicondyle, which sends tendons to the fourth and fifth digits, is called extensor digiti quinti et quarti proprius (Schön, 1968; Swindler and Wood, 1973; Lewis, 1989). This muscle belly is also referred to as extensor digitorum lateralis or extensor digitorum ulnaris (Lewis, 1989). Swindler and Wood (1973) designate the muscle belly supplying the second and third digits as extensor digiti secundi et tertii proprius, although they also identify extensor digiti anularis, a muscle that inserts on the fourth digit, and extensor digiti medius, which inserts on the fifth digit. The muscle belly that sends a tendon to the thumb is separate from the muscle mass of the proprius muscle, and is the homologue of extensor pollicis longus in humans (Lewis, 1989). The tendon of extensor pollicis longus is frequently split, inserting onto the base of the first metacarpal and the trapezoid (Swindler and Wood, 1973). In great apes and humans, the proprius layer is represented by extensor pollicis longus, extensor digiti minimi, and extensor digiti indicis. The extensor digiti minimi usually sends a tendon to the fourth and fifth fingers in the orangutan, while it sends a tendon generally only to the fifth finger in African apes and humans (Lewis, 1989). Straus (1941) details the number of tendons and their distribution in a variety of genera.

Palmaris longus and extensor carpi radialis brevis are in the middle of the wrist. The other muscles are either more radial or more ulnar in their disposition, and therefore produce radial and ulnar deviation in addition to flexion and extension. Ulnar deviation of the wrist is produced by the combined synergistic action of extensor carpi ulnaris and flexor carpi ulnaris, while radial deviation is due to the actions of extensor carpi radialis longus and flexor carpi radialis (which act in concert with abductor pollicis longus and extensor pollicis brevis) (Lemkuhl and Smith, 1983).

ELECTROMYOGRAPHY OF WRIST AND HAND MUSCLES

Bäckdahl and Carlsöö (1961) found that flexion of the wrist in humans is accomplished by a combination of flexor carpi radialis, flexor carpi ulnaris, and flexor digitorum superficialis. The three muscles act synchronously in that no one muscle appears to be the prime mover. Flexor digitorum profundus does not appear to have a role in flexion of the human wrist. In extreme forced flexion of the wrist, there is contraction of extensor carpi ulnaris, but not of extensor digitorum or of extensor radialis. It has been suggested that this contraction acts to stabilize the wrist. Thus, the radial extensors of the wrist and the extensors of the fingers are passive during wrist flexion while extensor carpi ulnaris acts as an antagonist. Basmajian and DeLuca (1985) also report antagonist activity of the flexors when the wrist is extended and the metacarpophalangeal joints are hyperextended.

Tuttle and Basmajian (1974) obtained muscle activity from flexor carpi ulnaris and flexor carpi radialis during the early minutes of their experiments on gorilla knuckle-walking. However, activity decreased as the gorilla recovered from anesthesia and began to locomote normally. Neither muscle showed activity during the swing phase of knuckle-walking, or during fist-walking or "modified" palmigrady. This is particularly interesting since flexor carpi ulnaris fires prior to touchdown and during the support phase in cercopithecoids (Fig. 4.14). Tuttle et al. (1972) and Tuttle and Basmajian (1974) conclude that flexor carpi radialis aids in wrist support, despite their evidence that levels of activity decreased as the subject recovered from anesthesia.

In humans, the extensors of the wrist work synchronously with extensor digitorum to produce wrist extension (Bäckdahl and Carlsöö, 1961). During pure extension of the wrist, extensor carpi radialis brevis is more active than is extensor carpi radialis longus, but extensor carpi radialis longus is active during rapid extension (Tournay and Paillard, 1953). Both radial extensors are active during radial deviation of the wrist (Basmajian and DeLuca, 1985). Extensor carpi ulnaris is said to be relatively inactive in great apes during knuckle-walking, postural stances, fist-walking, and "modified" palmigrady (Tuttle and Basmajian, 1974), although I have found that this muscle is active in both *Cercopithecus aethiops* and in *Papio anubis* prior to and during support (Fig. 4.14).

In ulnar and radial deviation, the human flexors and extensors act reciprocally, while their antagonist muscles are relaxed. In humans, extensor digitorum contracts during radial deviation. During extremes of radial and ulnar deviation, extensor digitorum and flexor digitorum superficialis are also active.

Lemkuhl and Smith (1983) summarize information on the activity of the human muscles in grasping. The lumbricals are silent during slow, light closing of the fist. Flexor digitorum profundus is active during closure of the fist. During opening, there is synchronous activity of extensor digitorum and the lumbricals; the interossei and flexor digitorum are silent. There is considerable activity of extensor digitorum and of flexor digitorum profundus during metacarpophalangeal extension and interphalangeal flexion. There is, however, only very low or no activity of the lumbricals and interossei. In metacarpophalangeal flexion with interphalangeal extension, there is considerable activity of the lumbricals and interossei. Backhouse and Catton (1954) conclude that the lumbricals are important only in extension of the interphalangeal joints. The activity of extensor digitorum varies from high to zero, perhaps related to the amount of effort. Long (1968) suggests that the lumbricals resist the tendency of the finger to "claw," although they do not contribute to metacarpophalangeal flexion. Landsmeer and Long (1965) propose that the lumbricals contribute to metacarpophalangeal flexion by passive tension. Close and Kidd (1969) observed simultaneous lumbrical and deep flexor activity during "pinch" and "grasp." Lumbricals appear to have no effect on rotation or radial deviation of fingers during opposition with the thumb (Basmajian and DeLuca, 1985).

Long and Brown (1962) found that flexor digitorum profundus is the most consistently active finger flexor, while the superficial flexor is essentially inactive. Long (1981) reports that extensor digitorum is active during full finger flexion, and interprets this action as that of a brake.

Susman and Stern (1980) tested Preuschoft's (1973) notion that the long digital flexors maintain equilibrium at the metacarpophalangeal joints during knuckle-walking in chimpanzees. They concluded that neither flexor digitorum profundus nor flexor digitorum superficialis manifested significant activity during the stance phase. There was a burst of activity of the two muscles at the beginning of swing phase, when the hand is lifted from the substrate. Maximum activity, of the fascicules to digits III and IV, occurred during suspensory behavior. This work basically confirmed the research of Tuttle and Basmajian (1974) and Tuttle et al. (1972) on the same muscles in gorillas, although Tuttle (1975:474) later maintained that flexor digitorum superficialis helped to "safeguard the metacapophalangeal joints during overextension." Tuttle and Watts (1985:276) suggest that the more frequent muscular activity observed in gorillas may be due to their tendency to "place their knuckled hands perpendicular to the line of progression or to bear more weight on their hands."

At the beginning of interphalangeal joint extension, the human extensor digiti communis begins activity regardless of the position of the metacarpophalangeal joints. When there is extension or hyperextension of the interphalangeal joint, extensor digitorum alone is active. Long and Brown (1962, 1964) found that extensor digitorum is active during both extension and flexion of the metacarpophlangeal joint.

Hylobatids are unique among primates in possessing an accessorius interosseus muscle, a derivative of flexor brevis. Electromyography indicates that the muscle is an abductor of the "index finger" during pinch grasping, and an accessory flexor of the metacarpophlangeal joint during "whole hand grasping" (Susman et al., 1982:117).

BRACHIATION, KNUCKLE-WALKING, AND HUMAN EVOLUTION

In their classification of primate locomotion, Napier and Napier (1967) categorize gorillas, chimpanzees, and orangutans as "modified brachiators." The category is defined (Napier and Napier, 1967:388) as a "form of arboreal locomotion in which the forelimbs extended above the head play a major role in suspending the body or propelling it through space" but in which "quadrupedal walking, the weight of the forebody being taken on the knuckles or bunched fists, is commonly seen." Unfortunately, although this category was part of a behavioral classification, it was not based on actual studies of naturalistic behavior but on inferences from certain morphological similarities to gibbons. In addition, as Senut (1989) pointed out, the term *brachiation* was used differently by different authors. The important implication of the term *brachiation,* and of its subsequent inclusion of all the living apes within this category, was that human bipedality was believed to have arisen from a brachiating pre-hominid ancestor.

In 1986 McHenry (p. 177) noted: "A profound revolution occurred sometime prior to 3.5 Myr ago: the body plan of an ape-like hominoid became completely reorganized for bipedality." What did the anatomy and locomotor behavior of this prehominid look like? The morphological similarity between African apes and humans has been known for some time (Huxley, 1863). In a seminal paper in 1971, Washburn (p. 83) supported the theory that "apes and men shared an arboreal, quadrupedal way of life which evolved during the Miocene into a climbing-feeding way of life termed brachiation"; eventually, "some apes and men continued to share a ground-living, knuckle-walking adaptation, and from that point in time human ancestors first became behaviorally distinct as bipeds." This model became known as the "late divergence hypothesis" (Feldesman, 1986) and was supported by estimates of dates of divergence based on biochemistry (Sarich, 1971), although suggestions that brachiation was practiced by prehominid ancestors date from the early twentieth century (see Gregory, 1916). Washburn further contended that the trunk and arms of modern apes provided the best morphological evidence for a brachiating phase in human evolution. Although many of the anatomical characters concerned the shoulder, several features of the wrist were also used in Washburn's (1971) discussion. For example, the lack of articulation between the ulna and carpals in apes and humans was interpreted as permitting greater pronation-supination and ulnar deviation–radial deviation. The most stable position of the wrist in humans is believed to be a partially supinated one, a position "that is taken by the great apes in quadrupedal locomotion. This is clearly seen in the knuckle-walking chimpanzee" (Washburn, 1971:93). Washburn also noted that the palmar pads do not extend up the distal forearm. Likewise, the morphology of the proximal and distal phalanges, and the distal phalanx of the thumb, of the Olduvai hand bones, OH 7 (Napier, 1962) was diagnosed as being "half-way between those of the contemporary apes and modern man" (Washburn, 1971:93). Susman and Stern (1980) subsequently concluded that the OH 7 hand bones indicated climbing capabilities. Finally, Washburn (1971:97) hypothesized that, because humans often lack hair on the dorsum of the middle phalanx "it may be that the condition of mid-digital hair in man is at least a partial adaptation to knuckle-walking by our remote ancestors."

Lewis (1971, 1972), in studies of the wrist, supports the brachiationist model of Gregory (1916). Lewis (1972) argued that a suite of wrist characters resulted in the retreat of the distal ulna from direct articulation with the triquetrum and pisiform in hominoids and that the hominid intra-articular meniscus increased the range of pronation-supination. Lewis (1971:186-187) concluded that "man shares with the Pongidae a vital structural specialization which seems clearly related to brachiation; this appears to be a most forceful argument in favour of the brachiation hypothesis."

The classification of African apes as "modified brachiators" and the brachiationist hypothesis received challenges from the work of Tuttle and colleagues (1967, 1969, 1970, 1975; Tuttle and Basmajian, 1974; Tuttle and Watts, 1985; Tuttle and Cortright, 1988). Tuttle (1970) reviewed the available data on locomotor behavior and concluded that both chimpanzees and gorillas are essentially quadrupedal on both the ground and in the trees, although both retain the anatomical capability to perform suspensory postures and movements. Subsequent naturalistic fieldwork (Tuttle and Watts, 1985; Hunt, 1992) confirmed this finding, although Hunt (1992) views the selective forces that are correlated with knuckle-walking as confined to the hand. Tuttle (1967, 1969) makes a significant contribution in clearly distinguishing between fist-walking in the orangutan and knuckle-walking in African apes.

Tuttle (1975) argued that parallelism led to many of the anatomical similarities among anthropoids previously classified as brachiators. He concluded that forelimb elongation and pollical reduction evolved in parallel in *Ateles, Colobus,* and the Hominoidea. Tuttle and his colleagues (1970, 1975; Tuttle and Basmajian, 1974; Tuttle and Watts, 1985) also related a suite of hand and wrist features specifically to the knuckle-walking adaptation:

knuckle pads on the dorsal surfaces of the middle phalanges of the weight-bearing digits, transverse dorsal ridges behind the metacarpal heads, shortening and "tendonization" of the long digital flexors, well-developed ligaments supporting the palmar surfaces of digits II through V, relative length of the fifth metacarpal, morphology of the distal radius, fusion of the centrale with the scaphoid, relationship between the scaphoid and capitate in the midcarpal joint, dimensions of the radial articular surface of the proximal carpal row, and development of ligaments in the wrist. Many of these features are directly related to the limitation of dorsiflexion and radial deviation in knuckle-walking (Tuttle, 1975).

Jenkins and Fleagle (1975) discussed the presence or absence of morphological features related to knuckle-walking by comparing morphological features of the hand in the chimpanzee, gorilla, orangutan, gibbon, and rhesus macaque. They compared the relative concavity and orientation of the distal radius and concluded that the angulation of the distal radius contributes to joint stability, thus limiting shearing forces in knuckle-walking (shearing forces make one surface slide past an adjacent surface in the same plane) (Gowitze and Milner, 1988). Jenkins and Fleagle (1975) suggested that the wrists of *Pongo* and *Hylobates* have a rotary capability not found in African apes or in the rhesus monkey. This rotary capability is due to an elliptical articular surface formed by the scaphoid and lunate (in which the dorsopalmar and radioulnar axes are relatively equal in length) and a greater radioulnar curvature of those wrist bones. The possibility that there is some convergence between the knuckle-walking and cercopithecoid types of digitigrady—in which the palm is essentially part of the forearm during quadrupedal terrestrial locomotion—was not discussed by Jenkins and Fleagle (1975).

Lewis (1989) criticized Jenkins and Fleagle's interpretation of the midcarpal joint. Jenkins and Fleagle (1975) set up a trichotomy in the orientation of the articulation between the capitate and centrale-scaphoid. The centrale-articular area faces dorsally in the macaque capitate, it faces dorsally and radially in African apes, and this surface faces "almost exclusively" radially (Jenkins and Fleagle, 1975:223) in the orangutan and gibbon. Lewis (1989:87) noted that his description of the scaphoid-capitate-trapezoid mechanisms in *Pan* was misinterpreted. He (Lewis, 1989:72) stated that the "centrale portion of the scaphoid is . . . received snugly into the hollowed waist underlying the swollen part of the capitate head and the cartilage-clothed notch on the scaphoid at the rear of the centrale component locks firmly onto the articular posterior part of the capitate waist." Lewis (1989) argued that the midcarpal mechanism in *Pan* is a mirror image of the situation in prosimians and monkeys, and he interpreted the differences in apes as a mechanism for a suspensory context.

Cartmill and Milton (1977) compared galagine and lorisine wrist anatomy to that of apes and concluded that similar hominoidlike features cannot be considered indicators of an adaptation for brachiation. They examined several characters: the direction of the projection of the pisiform, the articulation of the pisiform with the ulna, the relative reduction of the ulnar styloid, the direct contact between the triquetrum and ulnar styloid, and the distal radioulnar joint. Cartmill and Milton (1977) decided that the adaptations of the lorisine wrist paralleled those of living hominoids. Yet modern lorisines are slow, deliberate climbers (Hill, 1972), so Cartmill and Milton (1977) considered the possibility of parallelism. Cartmill and Milton believed that tensile stresses related the morphological similarities in apes and lorisines. Again, Lewis (1989) challenged these views by asserting that only the demonstration of an intra-articular meniscus in lorisines would indicate parallel evolution relative to the hominoids.

Although the data on hand use, comparative anatomy, and paleontology has improved over time, arguments concerning the evolution of the hand and its role in human evolution have not diminished. It is rather clear from the debate above that the hand has played a prominent role in the widely different interpretations of the prehominid condition—from arm-swinging brachiators to ground-dwelling knuckle-walkers. There is hardly a doubt that studies on the hand will contribute greatly to many other evolutionary debates concerning primate locomotion as well.

CONCLUSION

In any discussion of the positional behavior of the hand, it is necessary to distinguish between those behaviors that are used during locomotion and those that are used during postural activities.

Prost (1965) usefully distinguishes between these positional categories: posture occurs when the ratio between positional change and time ("summary displacement") is below a threshold value, while locomotion occurs when the ratio is above the threshold value. A primate may place its digits in a variety of ways during body postures. In quadrupeds, it is crucial to characterize hand postures of species during weight-bearing, especially when the individual is passing its body weight toward and initially over the digits during actual locomotion.

Primate hands and wrists are intricate organs that serve both for manipulation and locomotion. Although it is necessary to study hands to provide a perspective on human evolution, nonhuman primate hands provide information on complex adaptations to the demands of both compression and tension during locomotion. Although morphometric studies can provide elegant and informative data on the shape of bones, it is necessary to understand the shapes of carpal bones as parts of integrated functional systems. That is, the bones act in relation to each other, and to the ligaments and muscles that control their movements. It is particularly important to apply modern, experimental, in vivo techniques to both interspecific and intraspecific comparisons, and to gather more information on the use of the hands in a naturalistic context. Too often, morphologists are constrained by anecdotal reports of hand use by workers, whose reports are focused on different types of behavioral problems. Without detailed behavioral studies, it is difficult to assign behavioral capabilities to fossil forms. A prime example of this is the early reconstruction of *Proconsul africanus*. This early Miocene hominoid was first considered to be a brachiator (Napier and Davis, 1959), because of its anatomical similarity to three anthropoid genera (*Presbytis, Ateles,* and *Pan*), but we now know that all three of these forms move with a completely different locomotor repertoire. As the data base on hand use and morphology in nonhuman primates increases, our ability to reconstruct the locomotor capabilities of prehominids and early hominids will certainly increase. Lastly, we need to understand how particular anatomical features within joint complexes permit or restrict mobility and behavior, and to investigate morphological integration between anatomical complexes and mechanical demands, rather than to simply place animals into broad locomotor categories.

Hand morphology and the neural control of hand movements during locomotion may also provide clues as to the likelihood of different models of primate evolution. For example, in cercopithecoid evolution it has been suggested that the vervet monkey, *Cercopithecus aethiops,* is derived from an ancestral stock similar in morphology to the swamp monkey, *Allenopithecus,* and that the terrestrial, patas-like characters, are secondarily acquired (Kingdon, 1988). It had previously been suggested that *Cercopithecus aethiops* had evolved in the southern African savannas (Kingdon, 1974). The extent to which the similarity in hand movement between digitigrady and *Cercopithecus aethiops* palmigrady may support either view—or may support the view that cercopithecids evolved as semiterrestrial frugivores (Leakey, 1988)—is dependent on both the relative genetic conservatism of the neurological basis of hand movement, and on the adaptive significance of palmigrady. The hand motion at touchdown may be related to a need for precise hand placement in an irregular arboreal milieu, especially one in which the animal is navigating among thorns.

Acknowlegments: I wish to thank Dr. J. Kimani and Professor G. Maloiy, of the University of Nairobi, for sponsoring my research in Kenya. Mr. R. Leakey, Dr. M. Leakey, and Dr. L. Jacobs, of the National Museums of Kenya, kindly allowed me to examine fossil and osteological material in their collections. Cineradiographic and electromyographic research was accomplished at SUNY Stony Brook, with the kind cooperation and technical assistance of Dr. J. Stern, Jr., Dr. S. Larson, Dr. W. Jungers, Mr. J. Harrison, and the staff of the Division of Laboratory Animal Resources. This research was supported by grants from the National Science Foundation (BNS-8115742), the National Academy of Sciences (Joseph Henry Fund), Boise Fund (Oxford University), Explorers Club, Sigma Xi, the Department of Biology at Yale University, and Provost's Fund for the Sciences at Yale University. Permission to conduct research in Kenya was graciously granted by the Office of the President and the National Council for Science and Technology (Res. Permit 13/001/10C 220/20). I wish to thank Dan Gebo, Mike Rose, and Charles Remington for their comments on the manuscript, and Susan Hochgraf for drawing the figures.

LITERATURE CITED

Aldrich-Blake FPG, Bunn TK, Dunbar RIM, and Headley PM (1971) Observations on baboons, *Papio anubis,* in an arid region of Ethiopia. Folia Primatol. *15:*1–35.

Altmann SA, and Altmann J (1970) Baboon Ecology: African Field Research. Chicago: University of Chicago Press.

Bäckdahl M, and Carlsöö S (1961) Distribution of the muscles acting on the wrist (an electromyographic study). Acta Morphol. Neerl. Scand. *4:*136–144.

Backhouse KM, and Catton WT (1954) An experimental study of the functions of the lumbrical muscles in the human hand. J. Anat. *88:*133–41.

Basmajian JV, and DeLuca CJ (1985) Muscles Alive: Their Functions Revealed by Electro-myography, 5th ed. Baltimore: Williams and Wilkins.

Bogumill GP (1988) Anatomy of the wrist. In DM Lichtman (ed.): The Wrist and Its Disorders. Philadelphia: W.B. Saunders, pp. 14–26.

Brooks VB (1986) The Neural Basis of Motor Control. New York: Oxford.

Bunnell S (1944) Surgery of the Hand. Philadelphia: J.B. Lippincott.

Cartmill M, and Milton K (1977) The lorisine wrist joint and the evolution of brachiating adaptations in the Hominoidea. Am. J. Phys. Anthrop. *47:*249–72.

Close JR, and Kidd C (1969) The functions of the muscles of the thumb, the index, and long fingers: Synchronous recording of motions and action potentials of muscles. J. Bone Joint Surg. (Am) *51:*1601.

Coates MI, and Clack JA (1990) Polydactyly in the earliest known tetrapod limbs. Nature *347:*66–69.

Corrucini RS, Ciochon RL, and McHenry HM (1975) Osteometric shape relationships in the wrist joint of some anthropoids. Folia Primatol. *24:*250–274.

Crafts RC (1985) A Textbook of Human Anatomy, 3d ed. New York: John Wiley and Sons.

Crook JH, and Aldrich-Blake P (1968) Ecological and behavioral contrasts between sympatric ground dwelling primates in Ethiopia. Folia Primatol. *8:*192–227.

Feldesman MR (1986) The forelimb of the newly "rediscovered" *Proconsul africanus* deom Rusinga Island, Kenya: Morphometrics and implications for catarrhine evolution. In R Singer and JK Lundy (eds.): Variation, Culture and Evolution in African Populations. Johannesburg: Witwatersrand University, pp. 179–193.

Fodor J III, and Malott JC (1987) The Art and Science of Medical Radiography, 6th ed. St Louis: Catholic Health Assoc. of the U.S.

Gowitzke BA, and Milner M (1988) Scientific Bases of Human Movement, 3d ed. Baltimore: Williams and Wilkins.

Gregory WK (1916) Studies on the Evolution of the Primates. Bull. Amer. Mus. Nat. Hist. *35:*239–355.

Grillner S (1981) Control of locomotion in bipeds, tetrapods, and fish. In VB Brooks (ed.): Handbook of Physiology. Section 1, Neurophysiology. Vol. 2, Motor Control, Part 2. Bethesda: American Physiological Society, pp. 1179–1236.

Haines RW (1944) The mechanism of rotation at the first carpo-metacarpal joint. J. Anat. *78:*44–46.

Hall MC (1965) The Locomotor System: Functional Anatomy. Springfield: Charles C. Thomas.

Hildebrand M (1988) Analysis of Vertebrate Structure, 3d ed. New York: John Wiley and Sons.

Hill WCO (1966) Primates: Comparative Anatomy and Taxonomy, Vol. 6. Edinburgh: Edinburgh University Press.

——— (1972) Evolutionary Biology of the Primates. London: Academic Press.

Hollinshead WH, and Jenkins DB (1981) Functional Anatomy of the Limbs and Back, 5th ed. Philadelphia: W.B. Saunders.

Howell AB, and Straus WL, Jr. (1933) The muscular system. In CG Hartman and WL Straus, Jr. (eds.): The Anatomy of the Rhesus Monkey *(Macaca mulatta).* Baltimore: Williams and Wilkins, pp. 89–175.

Hunt KD (1992) Positional behavior of *Pan troglodytes* in the Mahale Mountains and Gombe Stream National Parks, Tanzania. Am. J. Phys. Anthrop. *87:*83–105.

Huxley TH (1863) Evidence as to Man's Place in Nature. London: Williams and Norgate.

Jenkins FA (1981) Wrist rotation in primates: A

critical adaptation for brachiators. Symp. Zool. Soc. Lond. *48*:429–451.

Jenkins FA, and Fleagle JG (1975) Knuckle-walking and the functional anatomy of the wrists in living apes. In RH Tuttle (ed.): Primate Functional Morphology and Evolution. The Hague: Mouton, pp. 213–227.

Kapandji IA (1982) The Physiology of the Joints, 5th ed. Edinburgh: Churchill Livingstone.

Kaplan EB, and Taleisnik J (1984) The wrist. In M Spinner (ed.): Kaplan's Functional and Surgical Anatomy of the Hand, 3d ed. Philadelphia: J.B. Lippincott, pp. 153–178.

Kauer JMG (1979) The collateral ligament function in the wrist joint. Acta Morphol. Neerl. Scand. *17*:252–253.

Kauer JMG, and Landsmeer JMF (1981) Functional anatomy of the wrist. In R Tubiana (ed.): The Hand, Vol 1. Philadelphia: W.B. Saunders, pp. 142–157.

Kingdon J (1974) East African Mammals: An Atlas of Evolution in Africa, Vol. 1. Chicago: University of Chicago Press.

——— (1988) Comparative morphology of the hands and feet in the genus *Cercopithecus*. In A Gautier-Hion, F Bourliere, J-P Gautier, and J Kingdon (eds.): A Primate Radiation: Evolutionary Biology of the African Guenons. Cambridge: Cambridge University, pp. 184–193.

Kowalski K (1976) Mammals: An Outline of Theriology. Warsaw: Panstwowe Wydawnicto.

Kummer H (1968) Social Organization of Hamadryas Baboons: A Field Study. Chicago: University of Chicago Press.

Landsmeer JMF, and Long C (1965) The mechanism of finger control: Based on electromyograms and location analysis. Acta Anat. *60*:330.

Leakey M (1988) Fossil evidence for the evolution of the guenons. In A Gautier-Hion, F Bourliere, J-P Gautier, and J Kingdon (eds.): A Primate Radiation: Evolutionary Biology of the African Guenons. Cambridge: Cambridge University, pp. 7–12.

Lehmkuhl LD, and Smith LK (1983) Brunnstrom's Clinical Kinesiology, 4th ed. Philadelphia: F.A. Davis.

Lewis OJ (1971) The hominoid wrist joint. In SL Washburn and P Dolhinow (eds.): Perspectives on Human Evolution, Vol. 2. New York: Holt, Rinehart and Winston, pp. 169–191.

——— (1972) Osteological features characterizing the wrists of monkeys and apes, with a reconsideration of this region in *Dryopithecus (Proconsul) africanus*. Am. J. Phys. Anthrop. *36*:45–58.

——— (1974) The wrist articulations of the anthropoidea. In FA Jenkins, Jr. (ed.): Primate Locomotion. New York: Academic Press, pp. 143–167.

——— (1989) Functional Morphology of the Evolving Hand and Foot. Oxford: Clarendon.

Lewis OJ, Hamshere RJ, and Bucknill TM (1970) The anatomy of the wrist joint. J. Anat. *106*:532–552.

Long C II (1968) Intrinsic-extrinsic control of the fingers. J. Bone Joint Surg. *50A*:973.

——— (1981) Electromyographic studies of hand function. In R Tubiana (ed.): The Hand, Vol. 1. Philadelphia: W.B. Saunders, pp. 427–440.

Long C, and Brown ME (1962) Electromyographic kinesiology of the hand. Part 3, Lumbricalis and flexor digitorum profundus to the long finger. Arch. Phys. Med. *43*:450–460.

——— (1964) Electromyographic kinesiology of the hand: Muscles moving the long finger. J. Bone Joint Surg. *8*:1683–1706.

McHenry HM (1986) The first bipeds: A comparison of the *A. afarensis* and *A. africanus* postcranium and implications for the evolution of bipedalism. J. Hum. Evol. *15*:177–191.

Mayfield JK (1988) Pathogenesis of wrist ligament instability. In DM Lichtman (ed.): The Wrist and Its Disorders. Philadelphia: W.B. Saunders. pp. 53–73.

Napier JR (1955) The form and function of the carpo-metacarpal joint of the thumb. J. Anat. *89*:362–369.

——— (1962) Fossil hand bones from Olduvai Gorge. Nature *196*:409–411.

Napier JR, and Davis PR (1959) The forelimb skeleton and associated remains of *Proconsul africanus*. Fossil Mammals of Africa *16*:1–69.

Napier JR, and Napier PH (1967) A Handbook of Living Primates. New York: Academic Press.

Navarro A (1935) Anatomia y fisiologia del carpo. An. Inst. Clin. Quir. Cir. Exp. Montevideo.

O'Connor BL (1975) The functional morphology of the cercopithecoid wrist and inferior radioulnar joints and their bearing on some problems in the evolution of the Hominoidea. Am. J. Phys. Anthrop. *43*:113–122.

Pieron AP (1981) The first carpometacarpal joint. In R Tubiana (ed.): The Hand, Vol. 1. Philadelphia: W.B. Saunders, pp. 169–183.

Poznanski AK (1984) The Hand in Radiologic Diagnosis, with Gamuts and Pattern Profile, Vol. 1. Philadelphia: W.B. Saunders.

Preuschoft H (1973) Functional anatomy of the upper extremity. In GH Bourne (ed.): The Chimpanzee, Vol. 6. Basel: Karger, pp. 34–120.

Prost JH (1965) A definitional system for the classification of primate locomotion. Am. Anthrop. *67*:1198–1214.

Romer AS (1966) Vertebrate Paleontology, 3d ed. Chicago: University of Chicago Press.

Romer AS, and Parsons TS (1977) The Vertebrate Body, 5th ed. Philadelphia: W.B. Saunders.

Rose MD (1973) Quadrupedalism in New and Old World Monkeys. Ph.D. dissertation, University of London, London.

——— (1984) Hominoid postcranial specimens from the Middle Miocene Chinji Formation, Pakistan. J. Hum. Evol. *13*:503–516.

——— (1988) Functional anatomy of the cheiridia. In JH Schwartz (ed.): Orang-utan Biology. New York: Oxford, pp. 299–310.

Sarich V (1971) A molecular approach to the question of human origins. In P Dolhinow and V Sarich (eds.): Background for Man: Readings in Physical Anthropology. Boston: Little, Brown, pp. 60–81.

Schön MA (1968) The Muscular System of the Red Howling Monkey. U.S. National Museum Bulletin 273.

Schultz AH (1970) The comparative uniformity of the Cercopithecoidea. In JH Napier and PH Napier (eds.): Old World Monkeys: Evolution, Systematics, and Behavior. New York: Academic Press, pp. 39–51.

Schultz M (1986) The forelimb of the Colobinae. In DR Swindler and J Erwin (eds.): Comparative Primate Biology. Vol. 1, Systematics, Evolution and Anatomy. New York: Alan R. Liss, pp. 559–669.

Senut B (1989) La locomotion des pré-hominides. In G Giacobini (ed): Hominidae. Proc. 2nd Int. Congress of Human Paleontology. Milan: Jaca Book, pp. 53–60.

Shubin NH, and Alberch P (1986) A morphogenetic approach to the origin and basic organization of the tetrapod limb. In MK Hecht, B Wallace, and GT Prance (eds): Evolutionary Biology, Vol. 20. New York: Plenum Press, pp. 319–387.

Stern JT, Jr. (1988) Essentials of Human Anatomy. Philadelphia: F.A. Davis.

Straus WL, Jr. (1941) The phylogeny of the human forearm extensors. Human Biology *13*:23–50, 203–238.

Susman RL (1974) Facultative terrestrial hand postures in an orangutan and pongid evolution. Am. J. Phys. Anthrop. *40*:27–38.

——— (1988) New postcranial remains from Swartkrans and their bearing on the functional morphology and behavior of *Paranthropus robustus*. In FE Grine (ed.): Evolutionary History of the "Robust" Australopithecines. New York: Aldine de Gruyter, pp. 149–172.

Susman RL, Jungers WL, and Stern JT, Jr. (1982) The functional morphology of the accessory interosseous muscle in the gibbon hand: Determination of locomotor and manipulatory compromises. J. Anat. *134*:111–120.

Susman RL, and Stern JT, Jr. (1980) EMG of the interosseous and lumbrical muscles in the chimpanzee (*Pan troglodytes*) hand during locomotion. Am. J. Anat. *157*:389–397.

Susman RL, and Tuttle RH (1976) Knuckling behavior in captive orangutans and a wounded baboon. Am. J. Phys. Anthrop. *45*:123–124.

Swindler DR, and Wood CD (1973) An Atlas of Primate Gross Anatomy: Baboon, Chimpanzee and Man. Seattle: University of Washington.

Taleisnik J (1984) Current concepts of the anatomy of the wrist. In MS Spinner (ed.): Kaplan's Functional and Surgical Anatomy of the Hand, 3d ed. Philadelphia: J.B. Lippincott, pp. 179–201.

——— (1985) The Wrist. New York: Churchill Livingstone.

Tournay A, and Paillard J (1953) Electromyographie des muscles radiaux à l'état normal. Rev. Neurol. *89*:277–279.

Tuttle RH (1967) Knuckle-walking and the evolution of hominoid hands. Am. J. Phys. Anthrop. *26*:171–206.

——— (1969) Knuckle-walking and the problem of

human origins. Science *166:*953–961.

——— (1970) Postural, propulsive, and prehensile capabilities in the cheiridia of chimpanzees and other great apes. In GH Bourne (ed.): The Chimpanzee, Vol 2. Basel: Karger, pp. 167–253.

——— (1975) Parallelism, brachiation, and hominoid phylogeny. In WP Luckett and FS Szalay (eds.): Phylogeny of the Primates: A Multidisciplinary Approach. New York: Plenum Press, pp. 447–480.

Tuttle RH, and Basmajian JV (1974) Electromyography of forearm musculature in *Gorilla* and problems related to knuckle-walking. In FA Jenkins, Jr. (ed.): Primate Locomotion. New York: Academic Press, pp. 293–347.

Tuttle RH, Basmajian JV, Regenos E, and Shine G (1972) Electromyography of knuckle-walking: Results of four experiments on the forearm of *Pan gorilla.* Am J. Phys. Anthrop. *37:*255–266.

Tuttle RH, and Beck BB (1972) Knuckle-walking hand postures in an orangutan (*Pongo pygmaeus).* Nature *236:*33–34.

Tuttle RH, and Cortright GW (1988) Positional behavior, adaptive complexes, and evolution. In JH Schwartz (ed.): Orang-utan Biology. New York: Oxford, pp. 311–330.

Tuttle RH, and Watts DP (1985) The positional behavior and adaptive complexes of *Pan gorilla.* In S Kondo (ed.): Primate Morphophysiology, Locomotor Analyses and Human Bipedalism. Tokyo: University of Tokyo, pp. 261–288.

Valentin P (1981) Extrinsic muscles of the hand and wrist: An introduction. In R Tubiana (ed.): The Hand, Vol. 1. Philadelphia: W.B. Saunders, pp. 237–243.

Washburn SL (1971) The study of human evolution. In P Dolhinow and V Sarich (eds.): Background for Man: Readings in Physical Anthropology. Boston: Little, Brown, pp. 82–117.

Weber ER (1988) Wrist mechanics and its association with ligamentous instability. In DM Lichtman (ed.): The Wrist and Its Disorders. Philadelphia: W.B. Saunders, pp. 41–52.

Weston WJ, and Kelsey CK (1973) Functional anatomy of the pisi-cuneiform joint. British J. Radiology *46:*692–694.

Whitehead PF (1990) Anatomy of the wrist in *Theropithecus oswaldi* from Olorgesailie, Kenya. J. Vertebrate Paleon. *10*(Suppl.): 48A.

——— (1991) Hand postures during quadrupedal walking in *Cercopithecus aethiops* and *Papio anubis:* Cineradiographic and cinematographic analysis. Am. J. Phys. Anthrop. Suppl. *12:*183.

Woodburne RT (1973) Essentials of Human Anatomy, 5th ed. New York: Oxford.

Yalden DW (1970) The functional morphology of the carpal bones in carnivores. Acta Anat. *77:*481–500.

——— (1971) The functional morphology of the carpus in ungulate mammals. Acta Anat. *78:*461–487.

——— (1972) The form and function of the carpal bones in some arboreally adapted mammals. Acta Anat. *82:*383–406.

CHAPTER FIVE

Functional Morphology of the Vertebral Column in Primates

Liza Shapiro

Over the past two decades, functional studies of primate postcranial morphology incorporating comparative anatomy, naturalistic observation, and various experimental approaches have become increasingly abundant (e.g., Jenkins, 1974a; Morbeck et al., 1979; Tuttle et al., 1979; Kondo, 1985; Jouffroy et al., 1990). These analyses have been successful in clarifying the complex relationship between postcranial form and function within and among diverse groups of extant primates. Such information also has been crucial in attempts to reconstruct the evolutionary development of primate postcranial adaptation (e.g., Strasser and Dagosto, 1988).

The spine is the central element of the locomotor skeleton, acting as a partially flexible, partially rigid link between the head and limbs. The spine's important role in the locomotor system is attested by its morphological variation among primates that rely on different positional behaviors (e.g., Keith, 1902, 1923, 1940; Vallois, 1927, 1928; Reynolds, 1931; Winckler, 1936; Schultz, 1938, 1961; Schultz and Straus, 1945; Slijper, 1946; Washburn and Buettner-Janusch, 1952; Erikson, 1960, 1963; Washburn, 1963; Benton, 1967, 1974; Ankel 1972; Donisch, 1973; Clauser, 1975, 1980; Rose, 1975; Yang, 1981; Jungers, 1984; Filler, 1986; Kelley, 1986; Kelley and Pilbeam, 1986; Hurov, 1987; Sanders, 1990, 1991; Shapiro, 1990, 1991a; Ward, 1990, 1991; Ward and Latimer, 1991).

Traditionally, considerations of primate axial anatomy have emphasized human morphology in an effort to understand the evolutionary development of bipedal posture and locomotion. For example, a consideration of axial anatomy was a key point in Sir Arthur Keith's (1923) contention that a common ancestry involving "brachiation" and "orthograde" posture could explain the morphological similarities characterizing humans and other hominoids. Keith's work on vertebral morphology was a major contribution to research on evolutionary adaptation, but he blurred possible functional and morphological distinctions between orthogrady, brachiation, and bipedalism. In addition, Keith did not address the variable extent to which each of the nonhuman hominoids engages in brachiation, nor did he acknowledge aspects of trunk morphology that are shared by hominoids and nonhominoid primates that engage in a variety of positional behaviors (Erikson, 1963; Benton, 1967, 1974; Cartmill and Milton, 1977; see also Fleagle et al., 1981). Of course, Keith's perspective was hindered in the sense that he lacked data on the behavior of primates in their natural habitats.

Subsequent researchers have identified many important and functionally relevant distinctions in vertebral anatomy among various groups of primates (see references above) but less attention has been given to vertebral diversity *within* these groups. With the increase in field studies addressing positional

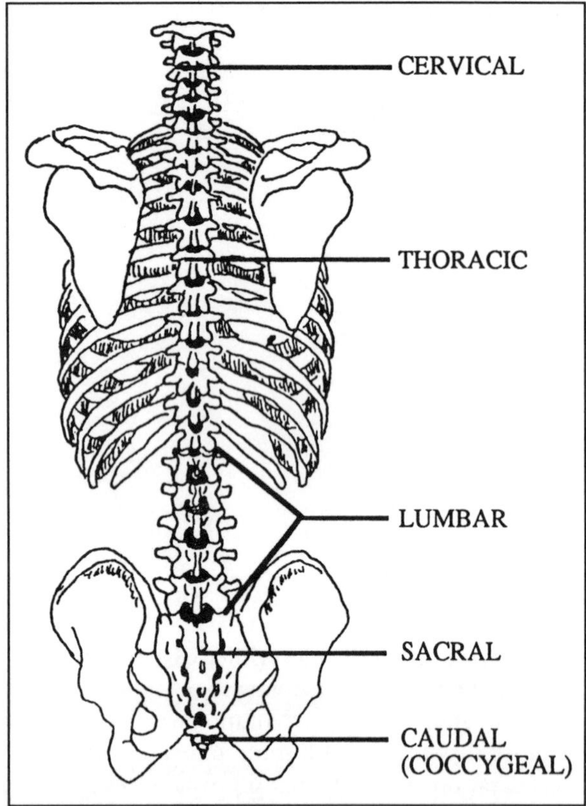

FIGURE 5.1. Dorsal view of the human vertebral column illustrating the five types of vertebrae. After Kapandji (1974). By permission of Churchill Livingstone.

behavior and the ever-expanding application of quantitative and highly technical methodologies to the study of primate morphology and locomotion (Fleagle, 1979; Tuttle et al., 1979; Martin, 1989; Jouffroy et al., 1990), we are in a better position than ever before for elucidating the functional morphology and evolutionary development of the primate back. Indeed, new interest in primate vertebral anatomy is being generated among anthropologists (e.g., Shapiro and Jungers, 1988, 1989; Sanders, 1990, 1991; Shapiro, 1990, 1991a, 1991b; Ward, 1990, 1991; Ward and Latimer, 1991), but many questions remained unanswered. This discussion is a review of prior and current vertebral research, but it also serves to identify gaps in our knowledge and to suggest new directions in which we need to explore.

Each of the five regions (cervical, thoracic, lumbar, sacral, caudal) (Fig. 5.1) of the vertebral column has been shown to vary morphologically among primates (e.g., Schultz, 1961; Ankel 1972). Morphological specializations of each type of vertebra are for the most part functionally associated with posture and locomotion, but they may also be related to respiration (thoracic; Kapandji, 1974) or obstetrics (sacrum; e.g., Tague and Lovejoy, 1986; but see Abitbol, 1987a, 1987b).

Although all regions of the vertebral musculoskeletal system are functionally important, variation in the primate spine is most prominently displayed by the lumbar region. Accordingly, more research has been carried out on lumbar comparative anatomy and function in primates than on any other type of vertebra. Therefore, in this chapter, a brief discussion of the nonlumbar portions of the column will be followed by a detailed treatment of the comparative functional morphology of the lumbar vertebrae in primates. The emphasis on lumbar vertebrae here should not imply that there is little to discover about the other regions of the column. Rather, this discussion should serve to exemplify the complexity of vertebral function when considering only one of its many components, and is meant to stimulate further research on all aspects of the vertebral column.

The danger inherent in isolating segments from a functional system is an overemphasis on a particular part of the body, or the tendency to forget that anatomical segments function together rather than independently. If this tendency can be overcome, however, the benefit of this approach is a more detailed view of an anatomical region, leading ultimately to a better understanding of its relationship to the functional system as a whole.

CERVICAL VERTEBRAE

Cervical vertebrae (Figs. 5.1 and 5.2A) function primarily in support and movement of the head and neck. Variation among primates in the morphology of the cervical region has been documented with respect to several parameters. One of the most prominent differences in cervical structure among primates relates to the relative elongation of the spinous processes. For example, the cervical spinous processes of the great apes (especially those of male gorillas and orangutans) are exceptionally elongated. Elongation of the spinous processes improves the leverage of the neck musculature necessary to

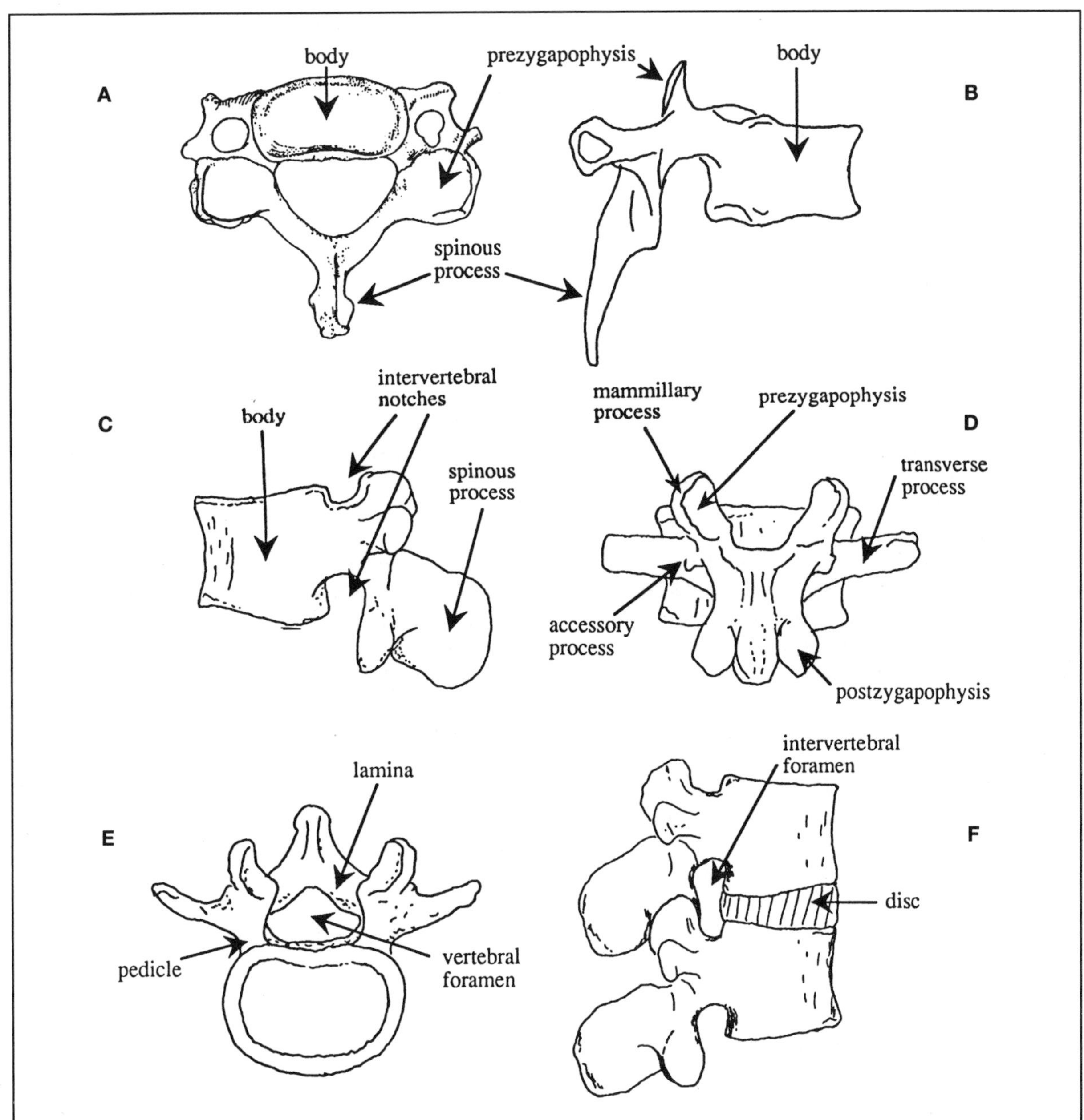

FIGURE 5.2. Human vertebrae. A: cranial view of cervical; B: lateral view of thoracic; C: lateral view of lumbar; D: dorsal view of lumbar; E: cranial view of lumbar; and F: lateral view of two articulated lumbar vertebrae. After Kapandji (1974), Pegington (1985), and Bogduk and Twomey (1987). By permission of Churchill Livingstone.

support the large, heavy heads of these primates (Slijper, 1946; Schultz, 1961; Ankel, 1972). In addition, Slijper (1946) found that *Loris tardigradus* shares with great apes relatively long cervical spinous processes, despite the fact that *Loris* does not have a relatively large head. Slijper (1946) suggested that in both great apes and *Loris*, the elongation of the cervical spines compensates for head-support leverage lost by the reduction of the postcondylar portion of the occiput. On the other hand, the lengthened spinous processes (last two cervicals and first three thoracics) of *Perodicticus* have been associated with the unique hairless, cornified epithelium that covers their tips and are said to function as a "sensory zone" in "peaceful social interactions" (Walker, 1970). Other functionally important variations in cervical anatomy among primates relate to the shape of the vertebral bodies, the orientation of the articular processes (zygapophyses), and the orientation of the odontoid process of the second cervical vertebra (axis) (Schultz, 1961; Ankel,1972).

THORACIC VERTEBRAE

Most research on the thoracic region (Figs. 5.1 and 5.2B) in primates has focused on variation in number of vertebrae, the proportional length of the region, and the shape of the rib cage.

In contrast to the constant number of cervical vertebrae (seven) in primates and all mammals except manatees and sloths (Flower, 1876; Todd, 1922; Gadow, 1933; Schultz, 1961) there is considerable variability in the number of thoracic (rib-bearing) vertebrae between groups of primates and even within a genus or species (Schultz and Straus, 1945; Schultz, 1961; Erikson, 1963; Abitbol, 1987b). Variation in the number of thoracic vertebrae among primate taxa is closely associated with variation in the number of lumbar vertebrae. Primates also vary with respect to the proportional length of each of these regions (Schultz, 1961; Jungers, 1984). These relationships are discussed in detail in the lumbar section, as is the fact that the lorisines have the most numerous thoracic vertebrae of all primates (Schultz, 1961).

Hominoids have a mediolaterally broad, dorsoventrally flattened rib cage with relatively ventrally placed thoracic vertebrae and curved ribs. This pattern contrasts sharply with the narrow, deep rib cage of almost all nonhominoids (Keith, 1923; Schultz, 1961; Ankel, 1972). Exceptions include *Perodicticus, Tarsius, Ateles, Brachyteles,* and *Colobus,* which tend toward the hominoid condition (Ankel, 1972). The differences in thoracic shape are functionally related to the placement of the scapula on the rib cage, excursion of the shoulder, and/or control of the trunk during movements such as arboreal "bridging" (Keith, 1923; Schultz, 1961; Cartmill and Milton, 1977; Jenkins et al., 1978). The distinctive anteroposteriorly expanded ribs of *Arctocebus* may provide the thoracolumbar stability required during slow arboreal climbing (Jenkins, 1970).

In comparison to other types of vertebrae, very little is known about functional variation among primates in details of thoracic vertebral structure. Further research in this area is clearly needed.

SACRAL AND CAUDAL VERTEBRAE

The sacrum (Fig. 5.1) is a single bone formed by the fusion of several sacral vertebrae, and forms the dorsal wall of the pelvic girdle. Hominoids have more sacral vertebrae than do most nonhominoids. An increase in the number of sacral vertebrae is generally accompanied by a reduction in the number of caudal (tail) vertebrae (e.g., most lorisines, *Cacajao)* (Schultz, 1961; Ankel, 1972). In hominoids, an increased number of sacral vertebrae is accompanied by a reduction in the number of lumbar vertebrae (Keith, 1902; Schultz, 1961; Ankel, 1972; Abitbol, 1987b). The unique aspects of human sacral structure are well linked to bipedalism (see lumbar section and Abitbol, 1987a, 1987b).

Ankel's (1962, 1965, 1972) detailed work on the comparative anatomy of the caudal vertebrae and the sacral canal (particularly in the prehensile-tailed species) has been unequalled. Prehensile tails are distinguished from nonprehensile tails by a number of morphological features. For example, prehensile tails are characterized by relatively high neural arches, and a greater number of caudal vertebrae with neural and ventral arches when compared to nonprehensile tails. These features are associated with the well-developed musculature and neurovascular system of prehensile tails. The mobility of prehensile tails is enhanced by the more distal placement of the longest caudal vertebra. In addition, in prehensile-tailed primates, the distal sacral aperture of the neural canal is larger relative to the proximal sacral aperture than in non-prehensile-tailed primates. The relatively large distal sacral aperture provides space for the well-developed nervous system required to innervate the complex musculature of the prehensile tail (Ankel, 1962, 1965, 1972). Future research on these well-documented structural attributes of caudal and sacral vertebrae would benefit from an in vivo analysis of tail use in various primates.

LUMBAR VERTEBRAE: BASIC ANATOMY

The lumbar vertebrae are those that lie between the thorax and the pelvis (Fig. 5.1). There are several basic components to each lumbar vertebra (Fig. 5.2C–5.2F).

Vertebral Body

The vertebral body (Fig. 5.2C) is the main weight-bearing unit of the vertebra. Its weight-bearing function is facilitated by the body's flat articular surface and by the crosshatched design of the trabeculae that make up the body's internal structure (Bogduk and Twomey, 1987). Interposed between each two bodies is a deformable disc (Fig. 5.2F); these three structures together form an "interbody joint." The interbody joint is named as such to avoid confusion with the other two intervertebral joints, the zygapopophyseal joints (Bogduk and Twomey, 1987). The presence of the disc permits flexibility between vertebral bodies without compromising their weight-bearing function or stability (Bogduk and Twomey, 1987).

Neural Arch (Pedicles and Laminae)

Projecting dorsally from the vertebral body is the neural arch, formed by the pedicles and laminae (Fig. 5.2E). The two pedicles project from the dorsal aspect of the vertebral body on each side, and are continuous with the two laminae. Each lamina curves dorsomedially to fuse at the midline, completing the neural arch. The fused bone is usually referred to in the singular form, "lamina."

Vertebral and Intervertebral Foramina

The space surrounded by the dorsal aspect of the vertebral body, the pedicles, and the lamina is the vertebral foramen (Fig. 5.2E). The vertebral foramina of successive vertebrae form the vertebral canal, through which the spinal cord passes.

In lateral view, there is a space or notch caudal to each pedicle, and a shallower notch cranial to each pedicle. These notches are referred to as intervertebral notches (Fig. 5.2C). The intervertebral notches of two successive vertebrae form an intervertebral foramen (Fig. 5.2F), through which a spinal nerve passes.

Zygapophyses

The zygapophyses are the synovial joints of the vertebral column. On each side of the superior aspect of the neural arch (near the junction between the pedicle and lamina) lies a prezygapophysis (Fig. 5.2D), also referred to as the superior articular process. The postzygapophyses (Fig. 5.2D), or inferior articular processes, are continuous with the inferior edge of the lamina. The postzygapophyses of the first lumbar vertebra articulate with the prezygapophyses of the second lumbar vertebra, and so on.

Spinous and Transverse Processes

The spinous and transverse processes (Figs. 5.2C, 5.2D) are the primary bony levers for the back and abdominal muscles that attach to lumbar vertebrae. On each vertebra, a single spinous process projects dorsally from the midline of the lamina. There are two transverse processes, one on each lateral aspect of the vertebra. The particular placement of the transverse processes on the vertebra varies markedly among primates. This point will be discussed in further detail later in the chapter.

Mammillary and Accessory Processes

The mammillary processes (Figs. 5.2D, 5.3) are small tubercles located on the dorsal edge of each prezygapophysis. They serve as attachment sites for multifidus, an intrinsic back muscle.

The accessory processes (Figs. 5.2D, 5.3) serve as attachment sites for the lumbar portion of extensor caudae lateralis (a tail muscle), and/or longissimus and the intertransversarii muscles (back muscles). In most primates, the accessory processes are pointed, quite prominent, and are located on the dorsal aspect of the lamina just lateral to the postzygapophysis (Fig. 5.3). They are usually quite prominent on the first few lumbar vertebrae, decreasing in size or disappearing toward the caudal end of the lumbar region. In hominoids, the accessory processes are located on the dorsomedial aspect of the lumbar transverse processes (Fig. 5.2D). Although variable in size, they usually appear as a small bump rather than as a pointed projection. Like the accessory processes of other primates, they diminish or disappear toward the last lumbar vertebra.

COMPARATIVE ANATOMY AND FUNCTIONAL MORPHOLOGY: LUMBAR VERTEBRAE

Counting Lumbar Vertebrae

A fairly obvious distinction among primates is the number of vertebrae that comprise the lumbar region, and its length proportional to the rest of the column (e.g., Schultz, 1938, 1961; Schultz and Straus, 1945; Hill, 1953-1974; Clauser, 1975, 1980). There are two widely accepted methods by which one can define and count lumbar vertebrae: the "rib" definition and the "zygapophyseal" definition.

Rib Definition

This method of counting lumbar vertebrae is predominant in the literature. Using this definition, lumbar vertebrae are defined as those vertebrae that lie between the thorax and the sacrum/pelvis and do not bear ribs. The range of variation among primates in number of lumbar vertebrae using this definition is summarized in Table 5.1 (Schultz, 1961).

Zygapophyseal Definition

Lumbar vertebrae can be easily distinguished if one chooses to define them by their lack of ribs. It is more functionally relevant to define thoracic versus lumbar vertebrae with respect to the zygapophyses, however, because the types of movements permitted between vertebrae are dependent on the shape and orientation of these structures (Washburn and Buettner-Janusch, 1952; Erikson, 1960, 1963; Washburn, 1963; Clauser, 1975, 1980). According to the zygapophyseal definition, a thoracic vertebra is one in which the articular surfaces of the prezygapophyses are flat and face dorsally or dorsolaterally, while those of the postzygapophyses face ventrally or ventromedially (i.e., the articular surfaces lie in or close to a coronal plane). In contrast, the articular surfaces of lumbar prezygapophyses are concave and mostly medially directed; those of the lumbar postzygapophyses are convex and face mostly laterally (i.e., the articular surfaces lie closer to a sagittal plane than do those of thoracic vertebrae) (Fig. 5.4).

TABLE 5.1
Number of Lumbar Vertebrae Defined by Lack of Ribs[1]

TAXON	NUMBER OF LUMBAR VERTEBRAE (MODE) RIB DEFINITION
Prosimians	
Daubentonia, Galago, Tarsius, Varecia	6
Arctocebus	6,7
Lemur, Hapalemur, Cheirogaleus, Microcebus, Nycticebus, Perodicticus	7
Avahi, Propithecus, Loris	8
Indri	8, 9
Lepilemur	9
Anthropoids	
Platyrrhines	
Ateles, Brachyteles, Lagothrix	4
Alouatta	5
Cebus, Pithecia, Callithrix	6
Cacajao	6,7
Leontopithecus, Callimico, Aotus, Callicebus, Saimiri	7
Catarrhines	
Papio, Theropithecus	6
Macaca, Cercocebus, Cercopithecus, Erythrocebus, Presbytis, Rhinopithecus, Nasalis, Colobus	7
Hylobates, Homo	5
Pongo, Pan, Gorilla	4

[1] Numbers represent mode; two numbers listed indicate equal frequencies of each in sample. All data are from Schultz (1961); not all primate genera available.

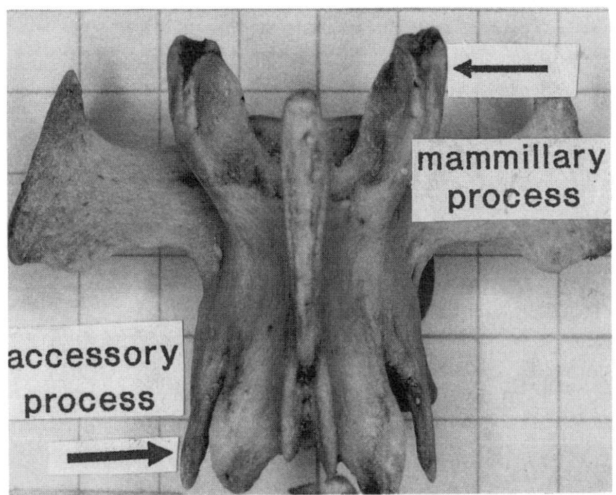

FIGURE 5.3. Lumbar vertebra of *Alouatta* in dorsal view. Note the prominent accessory processes. Background scale is four squares per inch.

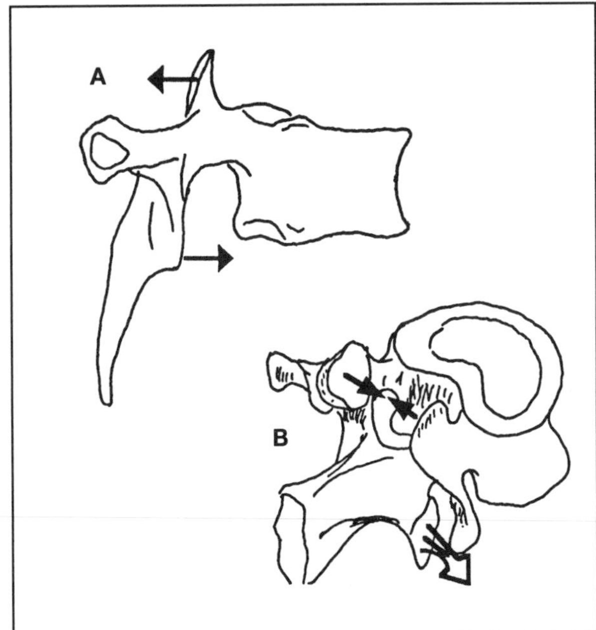

FIGURE 5.4. Thoracic (A) and lumbar vertebrae (B) (zygapophyseal definition), indicating the differences in shape and orientation of their zygapophyses. After Kapandji (1974). By permission of Churchill Livingstone.

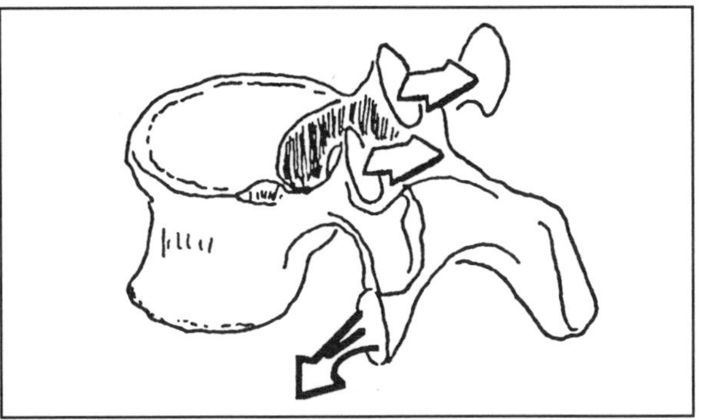

FIGURE 5.5. Diaphragmatic vertebra in which the prezygapophyses are thoracic-like and the postzygapophyses are lumbar-like. After Kapandji (1974). By permission of Churchill Livingstone.

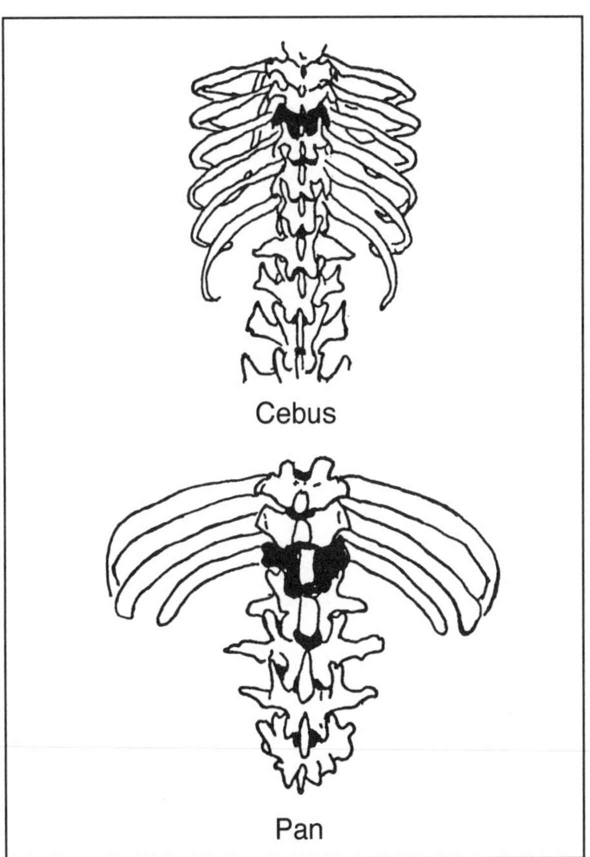

FIGURE 5.6. Dorsal views of the thoracolumbar transition in *Cebus* versus *Pan*. The diaphragmatic vertebra is shaded. In *Cebus*, there are three rib-bearing vertebrae with lumbar-like zygapophyses caudal to the diaphragmatic vertebra. In *Pan*, the diaphragmatic vertebra is the last rib-bearing vertebra. After Erikson (1963). By permission of Academic Press Inc. (London) Ltd.

Diaphragmatic Vertebra

The abrupt change in zygapophyseal shape between thoracic and lumbar regions necessitates the existence of a vertebra that is transitional with respect to its zygapophyses. This vertebra is referred to as the diaphragmatic vertebra. It is usually a rib-bearing vertebra, but its prezygapophyses are thoracic-like (for articulation with the thoracic vertebra that lies immediately cranial to it), and its postzygapophyses are lumbar-like (for articulation with the lumbar vertebra that lies immediately caudal to it) (Fig. 5.5).

In great apes and humans, the diaphragmatic is almost always the last rib-bearing vertebra. In all other primates there are usually one to three additional rib-bearing vertebrae caudal to the diaphragmatic vertebrae (Fig. 5.6). Maximum mobility in the sagittal plane is said to occur at the diaphragmatic vertebra (thoracolumbar junction) in humans (Kashimoto et al., 1982), but this may or may not be true for other primates or nonprimate mammals (Jenkins, 1974b).

With respect to the zygapophyseal definition, in any primate, the vertebra subjacent to the diaphragmatic vertebra will always be the first lumbar vertebra whether or not it bears a rib.

Lumbar Vertebral Bodies, Length of the Lumbar Region

If lumbar vertebrae are defined by their zygapophyseal orientation, rather than by their lack of ribs, their number increases (and the number of thoracic vertebrae decreases by the same amount) (Table 5.2). In other words, some primates may have from one to three rib-bearing vertebrae that also have the lumbar type of zygapophyses. Accordingly, the joints of these vertebrae are capable of moving like those of lumbar vertebrae, but they extend into the rib-bearing thorax, essentially extending the functional length of the lumbar column.

Tables 5.1 and 5.2 reveal that the primates with the most numerous lumbar vertebrae (by either definition) are the leaping prosimians such as *Lepilemur, Indri, Propithecus,* and *Avahi,* but also *Loris,* a slow-moving quadrupedal prosimian (see discussion below). Although the cercopithecoids are fairly uniform in their number of lumbar vertebrae, those that leap frequently (e.g., *Presbytis* and *Colobus;* Fleagle, 1977a, 1977b; Rose, 1978) have relatively elongated lumbar regions for their size compared to their more exclusively quadrupedal counterparts. This elongation is not due to an increase in number of vertebrae, however, but to an increase in the craniocaudal length of each lumbar vertebral body (Ward, 1991).

Rapid flexion and extension of a long lumbar region could be expected to provide an extra propulsive force to a leap (Erikson, 1963; Fleagle, 1988) especially in conjunction with the increase in back muscle mass that characterizes leaping primates (Fleagle, 1977a; Emerson, 1985). Emerson (1985) has noted, however, that most mammalian leapers (i.e., her "jumpers") shorten the presacral vertebral column thereby shifting the center of mass caudally. Presumably, this minimizes clockwise (viewed from the right) torques placed on the body by the propulsive thrust of the hind limbs. Emerson's observations would seem to contradict the long lumbar region/leaping association put forth for primates. It is unclear, however, whether the "presacral" reduction of leapers is restricted to a particular vertebral region. The example offered by Emerson concerns the shortening of the neck in jumping rodents (Hatt, 1932), and she does not indicate whether the same is true of all other mammalian leapers. Further research on the role of the lumbar region in leaping and in other locomotor behaviors is needed to clarify this issue.

The hominoid lumbar region is distinctive for two reasons. First, because the diaphragmatic vertebra is also the last rib-bearing vertebra in great apes and humans, these hominoids have the same number of lumbar vertebrae by either definition (i.e., if the diaphragmatic is counted as a thoracic vertebra; see Table 5.2). Second, the hominoids are notable for their reduction of the proportional length of the lumbar region (relative to trunk length and length of the presacral spine; see Schultz, 1961). This reduced length is a result of a decrease in the number of vertebrae (compared to other primates) as well as a relative reduction in the craniocaudal length of each vertebral body. In other words, the bodies of hominoid lumbar vertebrae are shorter relative to their width than are those of nonhominoids (Schultz, 1961; Benton, 1967, 1974; Ankel, 1972; Rose, 1975; Clauser, 1980; Ward, 1991). Although lumbar vertebral body length scales negatively

TABLE 5.2
Comparison of the Two Methods for Counting Thoracic and Lumbar Vertebrae in Selected Primates[1]

	RIB DEFINITION				ZYGAPOPHYSEAL DEFINITION			
TAXON	R	N	THOR	LUMB	R	N	THOR	LUMB
Indri	P	5	12 (12–14)	9 (8–9)	P	10	11 (11–13)	10 (9–10)
Propithecus	S	20	12 (12–13)	8 (7–8)	P	9	11 (10–11)	9 (9–10)
Varecia	P	7	13 (12–13)	6 (6–7)	P	8	10 (10–11)	9 (8–9)
Ateles	E	45	14 (13–15)	4 (4–5)	E	27	12 (12–13)	6 (5–6)
Brachyteles	E	5	14 (13–14)	4 (4–5)	E	4	12 (12–13)	6 (5–7)
Lagothrix	E	57	14 (13–15)	4 (4–5)	E	48	11 (10–12)	7 (6–8)
Alouatta	E	29	14 (13–16)	5 (4–6)	E	21	11 (10–13)	8 (7–9)
Cebus	S	32	14 (13–15)	6 (4–7)	P	24	11 (10–12)	8 (7–10)
Cercopithecus aethiops	W	60	12 (12–13)	7 (6–7)	W	60	10 (9–10)	9 (8–10)
Papio	W	11	13 (12–13)	6	W	11	10 (9–11)	9
Hylobates lar	W	159	13 (12–14)	5 (4–6)	W	159	13 (12–14)	6 (5–7)
Hylobates syndactylus	S	29	13 (11–14)	5 (3–5)	P	2	12.5 (12–13)	5
Pongo	S	127	12 (11–13)	4 (3–5)	P	27	12 (12–14)	4 (3–4)
Pan troglodytes	S	162	13 (12–14)	4 (3–4)	P	65	13 (12–13)	4 (3–5)
Pan paniscus	P	9	13 (12–14)	4	P	9	13 (12–14)	4
Gorilla	S	81	13 (12–14)	4 (3–4)	P	72	13 (12–14)	4 (3–4)
Homo	S	125	12 (11–13)	5 (4–6)	S P	34	12 (11–13)	5 (4–6)

[1] Individual numbers represent mode; numbers in parentheses represent range.

N = sample size; R = reference; THOR = thoracic; LUMB = lumbar; P = personal observations; S = Schultz (1961); E = Erikson (1963); and W = Washburn (1963). For the zygapophyseal definition, diaphragmatic vertebra (both types of zygapophyses) is included in the thoracic count.

allometrically within both hominoids and monkeys, lumbar vertebral bodies are also shorter in hominoids relative to overall body weight than are those of nonhominoids (Ward, 1991). The most pronounced expression of reduction in number of lumbar vertebrae among hominoids is found in *Pan, Gorilla,* and *Pongo* (Schultz, 1938, 1961; Schultz and Straus, 1945; Ward, 1991).

Among nonhominoids, the atelines (*Ateles, Brachyteles, Lagothrix, Alouatta;* Rosenberger and Strier, 1989) have the fewest number of lumbar vertebrae, but their reduction in number is less extreme than that of the great apes (especially if one compares the number of zygapophyseal lumbar vertebrae) (Table 5.2). Relative to other atelines, *Ateles* and *Brachyteles* have relatively short vertebral bodies for their overall body size (personal observations; see also Ward, 1991). This contradicts Benton's (1967, 1974) contention that the atelines' reduction in the proportional length of the lumbar region relative to other monkeys is not a result of a shortening of the craniocaudal length of vertebrae, but is due solely to a decrease in number of lumbar vertebrae. Benton did not consider each genus separately, but based his conclusion on a mean value for the four genera.

Keith (1923) suggested that the reduction of the lumbar region in hominoids is related to their upright or orthograde posture and brachiating ancestry. According to Keith, this lumbar reduction is part of a wider complex of traits also attributable to orthogrady that includes dorsal extension and widening of the iliac crest, loss of the tail producing increased sacral area for back extensor muscle attachment, lateral lengthening of the ribs, and ventral displacement of thoracic vertebrae relative to the rib cage.

At least some of the orthograde/brachiating features Keith (1923) ascribed to hominoids are exhibited by other nonhominoid primates that engage in a variety of positional behaviors (Erikson, 1963; Benton, 1967, 1974; Cartmill and Milton, 1977). For example, Erikson (1963) described in detail the various similarities between the atelines and the hominoids in the morphology of the thorax, vertebral column, and forelimb. He attributed these similarities to brachiation, while admitting that *Alouatta* and *Lagothrix* are at best, "incipient" brachiators.

An alternative explanation for the morphological features shared by hominoids and atelines has been presented by Cartmill and Milton (1977). These authors propose that lumbar reduction enables these primates to control movements between the thorax and pelvis during bridging behaviors employed to cross gaps in trees. Specifically, with a short lower back and a relatively long, stiff thoracic region, the trunks of these primates are long enough to facilitate reaching across gaps, but rigid enough for the maintenance of controlled movements.

Cartmill and Milton (1977) reinforce their interpretation with a consideration of the anatomy and behavior of the lorisines (*Nycticebus, Perodicticus, Arctocebus, Loris).* The posture of these primates is not orthograde, nor do they brachiate (Walker, 1969). They do, however, rely on bridging in their locomotor repertoire and, like hominoids, the lorisines are characterized by a widened rib cage, ventral displacement of the thoracic vertebrae, and by an elongated sacrum (except *Loris).* Moreover, although the lumbar region of lorisines is not reduced, their thorax is quite elongated and, like that of the hominoids and atelines, "contributes disproportionately to total trunk length" (Cartmill and Milton, 1977:268; see also Straus and Wislocki, 1932). Presumably, such vertebral proportions enable primates to bridge effectively, and may have nothing specifically to do with brachiation or with orthogrady.

Jungers (1984) notes that the bridging explanation is quite consistent with the behavior and trunk proportions of lesser apes (and perhaps *Pongo).* That is, bridging is a significant component of their positional behavior, the length of their thoracic column is positively allometric, and their lumbar length is negatively allometric. With respect to the larger bodied hominoids who bridge infrequently, however, Jungers (1984) proposes that lumbar reduction may act instead to prevent buckling of the trunk in response to the large forces transmitted to the vertebral column during vertical climbing.

The dramatic reduction of the lumbar region characterizing the great apes is responsible for the extreme negative allometry of its length among catarrhines as a group (Jungers, 1984). Accordingly, compared to the great apes, the cercopithecoids have a relatively long lumbar region that most likely facilitates leaping or running by increasing vertebral mobility in the sagittal plane (Jungers, 1984; Ward, 1991; see also Hildebrand, 1959). Nevertheless, within cercopithecoids, the lumbar region exhibits negative allometry in length; that is, larger cercopithecoids have relatively shorter lumbar regions (Jungers, 1984). Jungers' (1984) suggestion that relative shortening of the lumbar region in large cercopithecoids may be associated with the rigidity of their spine during quadrupedal locomotion is consistent with data reported by Hurov (1987), who demonstrated that vervets use spinal flexibility to increase stride, while larger-bodied patas monkeys do not. Further research is needed to supplement what little comparative data is available concerning primate spinal flexibility.

Although humans are not unique with respect to *number* of lumbar vertebrae, human lumbar vertebral bodies do reflect several unique adaptations to bipedalism. For example, among hominoids, consecutive increases in transverse diameters of lumbar vertebral bodies (moving caudally) are highest in humans (Rose, 1975). In addition, although lumbar intervertebral surface area scales isometrically in nonhuman hominoids (Shapiro, 1991a; Ward, 1991), the area of this weight-bearing surface is larger in humans than one would expect based on body weight alone, especially in the lower lumbar region (Shapiro, 1991a). These two aspects of human lumbar vertebral body anatomy are most likely a response to the relatively large, cumulative loads transmitted through the lumbar column during bipedal stance and locomotion (Schultz, 1961).

The most obvious uniquely human adaptation is our lumbar lordosis, or dorsal concavity of the lumbar region (Fig. 5.7). Lordosis is primarily a result of the wedging of the discs. That is, the ventral edge of each disc is longer craniocaudally than is its dorsal edge. In conjunction with the sharp angle between the last lumbar vertebra and the dorsally tilted sacrum (*lumbosacral angle),* wedging is especially pronounced at the lumbosacral junction. At this level, both the lumbosacral disc and the vertebral body of L5 are wedged (Kapandji, 1974; Lindh, 1980; Nissan and Gilad, 1986; Bogduk and Twomey, 1987; cf. Ward and Latimer, 1991). Various other primates, especially hominoids, have been said to have a lumbar lordosis and lumbosacral angle, but one that is much less pronounced than that characterizing humans (Cunningham, 1886; Schultz, 1961; Abitbol, 1987).

The ontogenetic development of lordosis and the lumbosacral angle coincides with the adoption of bipedalism (Abitbol, 1987). Lumbar lordosis promotes stability during bipedal stance and locomotion by placing the body's center of gravity directly over the hip joints. It is interesting that lordosis can develop if an otherwise pronograde primate is trained to walk bipedally early in ontogeny (Preuschoft et al., 1988).

Zygapophyseal Joints

The zygapophyseal joints work in conjunction with the interbody joints to control the overall range of movement between vertebrae. Vertebral movements that occur at the zygapophyses are guided by the shape and orientation of these joints. In all primates (and in mammals in general), the lumbar zygapophyses are distinctive for their relatively sagittal orientation, and for the concave/convex shape of their articular surfaces (Rockwell et al., 1938).

Zygapophyseal Orientation

Zygapophyseal orientation can be defined as the angle formed between a sagittal plane and the chord of the arc formed by the curved articular surface (Fig. 5.8). This angle would measure 0° in a completely sagittally oriented joint, and 90° in a horizontally/coronally oriented joint.

The more sagittally oriented the zygapophysis (i.e., the extent to which the articular surfaces of the prezygapophyses face medially, or the postzygapophyses face laterally), the better the joint is designed to resist rotation (Rockwell et al., 1938). On the other hand, the more the joint surfaces tend to incline toward a horizontal plane (i.e., face dorsally or ventrally), the more easily ventral displacement of one vertebra on another is resisted. An intermediate position is resistant to both rotation and ventral displacement (Fig. 5.9; Bogduk and Twomey, 1987).

In humans, the association between the structure and mobility of the zygapophyseal joints has been tested in vitro (e.g., Lewin et al., 1962; Markolf, 1972; Kashimoto et al., 1982; Haher et al., 1989; Singer, 1989), in vivo (e.g., Gregersen and Lucas, 1967; Kenesi and Lesur, 1985; Van Schaik et al., 1985; Singer, Breidahl, and Day, 1989; Singer, Day, and Breidahl, 1989), and by computer modeling (Scholten and Veldhuizen, 1985). The majority of these studies have demonstrated that very little rotation occurs in

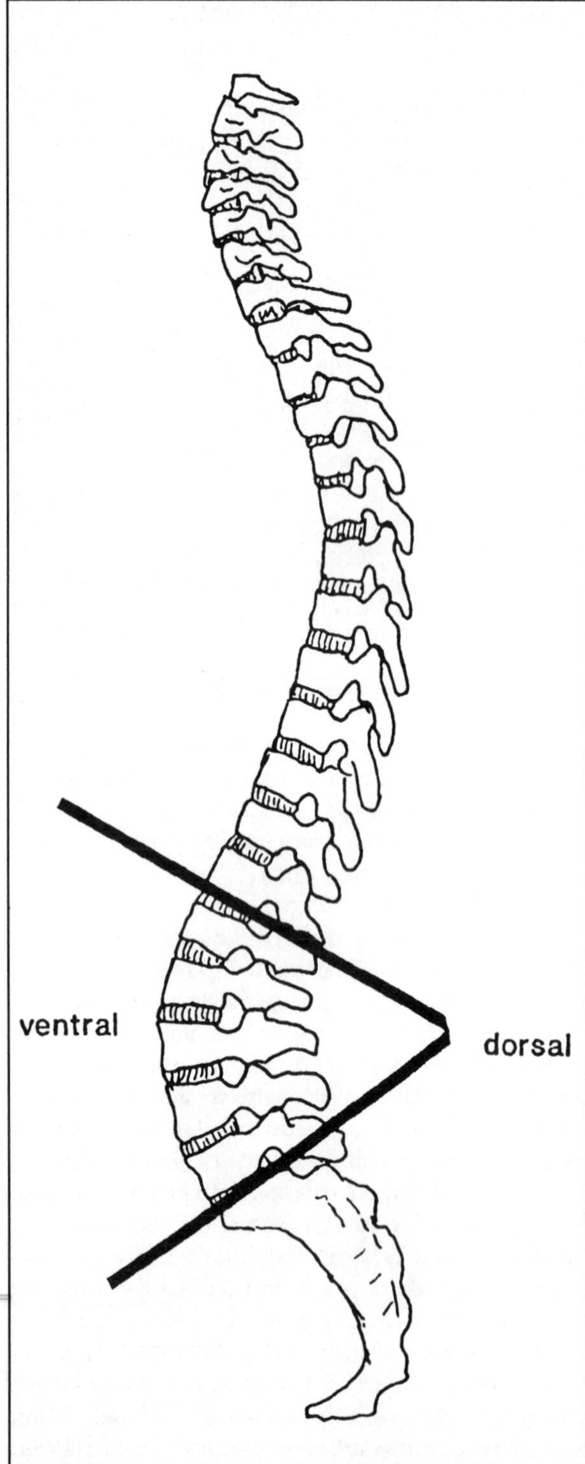

FIGURE 5.7. Lateral view of the human vertebral column indicating the lordosis (dorsal concavity) of the lumbar region. After Kapandji (1974). By permission of Churchill Livingstone.

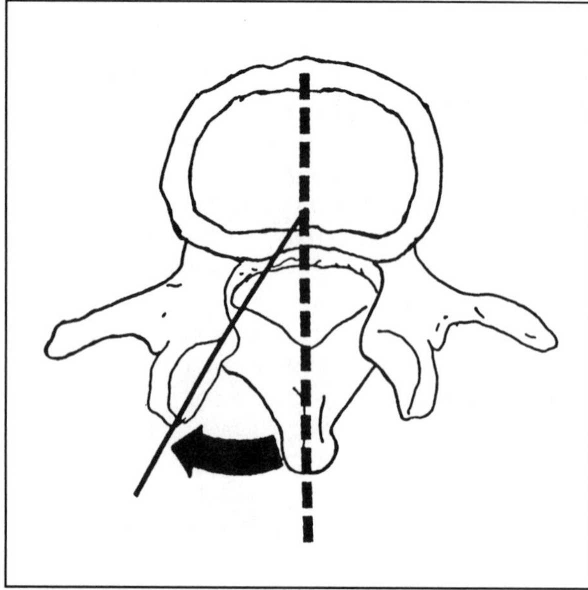

FIGURE 5.8. Cranial view of a human lumbar vertebra indicating orientation of the left prezygapophysis. Orientation refers to the angle formed between a sagittal plane and the chord of the arc formed by the concave articular surface of the prezygapophysis.

the lumbar region, and that the upper and midlumbar prezygapophyses (and accordingly, the articulating postzygapophyses) are designed to resist rotation as a result of their "sagittal" (actually, oblique or dorsomedial) orientation of less than 45°. Due to this orientation, impaction of the zygapophyses prevents excessive rotation at the joints. Without this orientation, the zygapophyseal joints would be susceptible to a significant amount of rotation, because the axis of rotation lies at the posterior aspect of the intervertebral disc (Cossette et al., 1971; Kenesi and Lesur, 1985; Bogduk and Twomey, 1987).[1]

When lumbar prezygapophyseal orientation is compared (at corresponding vertebral levels) among closely related nonhuman primates, little variation is found despite prominent differences in positional behavior (Clauser, 1980; Shapiro, 1991a). If one considers the change in prezygapophyseal orientation *along* the lumbar column (i.e., between the upper lumbar vertebrae and the first sacral vertebra), however, the human pattern is quite distinctive. Humans are the only hominoids in which the prezygapophyses of the last lumbar and first sacral vertebrae are oriented more obliquely than are the

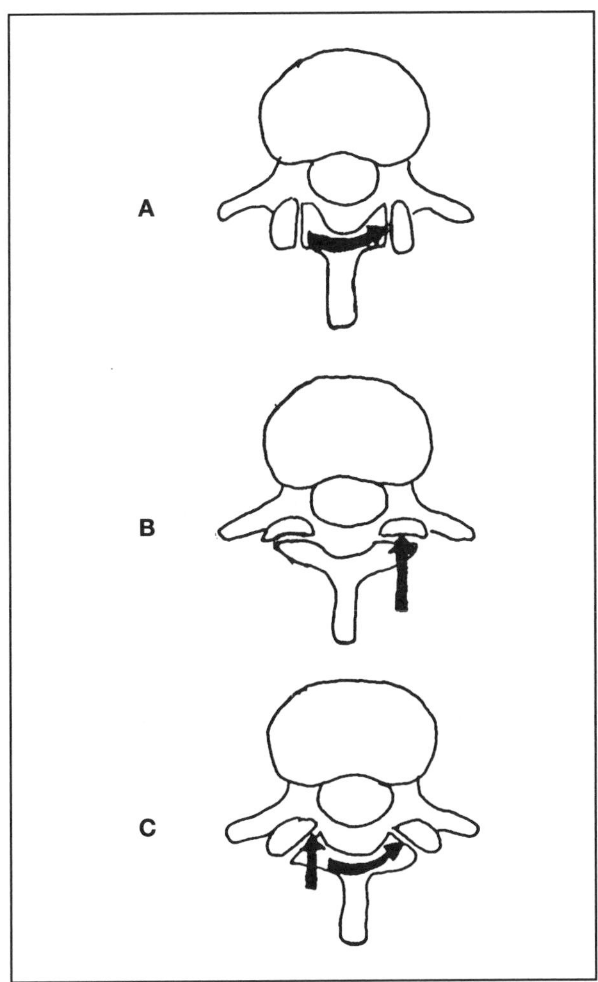

FIGURE 5.9. Lumbar vertebrae in cranial view demonstrating three hypothetical articulations of flat zygapophyseal joints. The postzygapophyses of one vertebra are shown articulating with the prezygapophyses of the subjacent vertebra. A: Joints oriented in a sagittal plane (0°) resist rotation but not ventral displacement; B: joints oriented in a horizontal plane (90°) resist ventral displacement but not rotation; C: obliquely oriented joints (45°) resist both rotation and ventral displacement. After Bogduk and Twomey (1987). By permission of Churchill Livingstone.

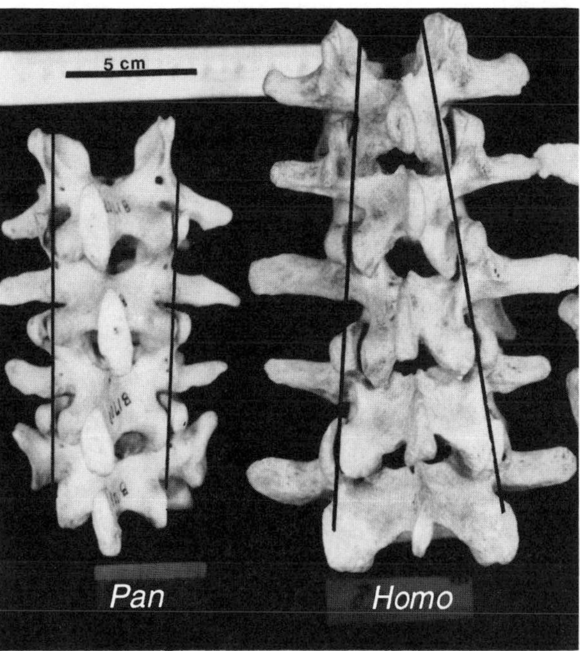

FIGURE 5.10. Dorsal views comparing the lumbar vertebrae of *Pan* and *Homo* illustrate the increasing obliquity of the zygapophyses and the widening of the lamina found only in humans.

upper and midlumbar prezygapophyses. Generally, the opposite pattern characterizes nonhuman hominoids (Fig. 5.10). This basic distinction between humans and great apes was recognized qualitatively as early as 1892 by Struthers and was later quantified by Odgers (1933). Subsequent research has demonstrated that the human pattern is unique with respect to nonhominoids as well (Clauser, 1980; Shapiro, 1991a). It is likely that the distinctive widening of the lamina, the increasing obliquity of the zygapophyses, and the widening of the vertebral bodies are functionally related (Shapiro, 1991a; Ward and Latimer, 1991; see also Cihak, 1970; Van Schaik et al., 1985).

The fact that the largest angular values in humans (i.e., the most oblique angles) are located at the last lumbar and first sacral vertebra is most likely attributable to bipedalism and to our unique lumbar lordosis. The sharp lumbosacral angle associated with lordosis produces a tendency for the last lumbar vertebra (L5) to slide forward on the sacrum, as well as a tendency for L4 to slide forward (ventrally) on L5 (Davis, 1961; Bogduk and Twomey, 1987) (Fig. 5.11). As mentioned above, zygapophyses oriented at or near 45° are in the best position to resist such forward displacement while simultaneously providing resistance to axial rotation (Bogduk and Twomey, 1987). Accordingly, the values at L5 and S1 (for articulation with L4 and

L5, respectively) measure at or near 45° in humans (Horwitz and Smith, 1940; Shapiro, 1991a).

Zygapophyseal Shape

The curved shape of lumbar zygapophyseal joints provide resistance to both ventral displacement and axial rotation. The ventromedial end of the prezygapophysis faces dorsally, resisting ventral displacement. At the same time, the placement of the dorsolateral end assures full impact of the opposite zygapophysis as it swings into rotation (Bogduk and Twomey, 1987) (Fig. 5.12).

To date, there has been no comprehensive analysis of the shape (e.g., curvature) of prezygapophyseal articular surfaces in nonhuman primates. Given the important contribution curvature has been shown to make toward resisting torsion and ventral displacement in humans (Lewin et al., 1962; Bogduk and Twomey, 1987; Yunliang et al., 1990), comparative research on this aspect of vertebral morphology should prove enlightening.

Any comparative investigation of zygapophyseal joint shape should include allometric considerations. For example, Halpert et al. (1987) demonstrated that among African bovids, larger animals have more curved zygapophyses in response to increasing demands for spinal rigidity. Gorillas appear to have much flatter articular surfaces than do smaller hominoids (personal observations). Evaluation of this apparent dissimilarity between primates and bovids awaits an in-depth analysis of primate zygapophyseal joint shape.

Other Functions of the Zygapophyses

In addition to their role in movement, there is evidence (at least for humans) that the lower lumbar zygapophyses act as load-bearing structures during extension, sustained compression, or prolonged standing with a lordotic spine. In these postures, weight is transmitted as the postzygapophysis contacts the subjacent prezygapophysis or lamina (Markolf, 1972; Hutton and Cyron, 1978; Lin et al., 1978; Shah et al., 1978; Adams and Hutton, 1980, 1983; Jayson, 1983; Lorenz et al., 1983; Dunlop et al., 1984; Yang and King, 1984; Pal and Routal, 1987; El-Bohy et al., 1989). These data exemplify a large body of clinical and biomechanical research addressing the role of the posterior elements of the spine (i.e., the neural arch and its associated ligaments and bony processes) in load bearing (e.g., see references already cited and also White and Hirsch, 1971; Denis, 1983; Louis, 1985; Pal and Routal, 1986; Pal et al., 1988; Sanders, 1990; Shapiro, 1990, 1991a), and stand in contrast to the traditional view of the spine, in which the anterior elements (the bodies and discs) are solely responsible for bearing the load passing through the column.

Spinous Processes

The lumbar spinous processes act as bony levers for the muscles capable of extending (straightening) the lower back. The most prominent muscle that acts on the spinous processes in the lumbar region is multifidus. Multifidus (i.e., its largest portion) is composed of groups of fascicles that arise from the spinous process of each lumbar vertebra. The fascicles extend caudally across two or three vertebrae, and insert on lumbar mammillary processes. The more caudally situated fascicles reach to portions of the iliac crest and the sacrum (Fig. 5.13). In primates with tails, multifidus continues caudally as the tail muscle extensor caudae medialis (George, 1967; Bogduk, 1973; Macintosh et al., 1986). The principal action of multifidus is extension of the lower back (Macintosh and Bogduk, 1986). Multifidus is an important stabilizer of the trunk in human bipedalism (e.g., Waters and Morris, 1972; Thorstensson et al., 1982; Thorstensson et al., 1984; Dofferhof and Vink, 1985; Vink and Karssemeijer, 1988) and also plays a prominent role in nonhuman primate locomotor behaviors, especially quadrupedalism and bipedalism (Shapiro and Jungers, 1988; Shapiro, 1991a, 1991b). The lever arm for multifidus, as expressed by the dorsal projection of the spinous processes, is longer (relative to body size) in indrids than it is in prosimians who leap less frequently, or when compared to hominoids or atelines (Shapiro, 1991a). This indicates more effective back muscle leverage in indrids, and is most likely attributable to the powerful extension of the back accompanying leaping (Fleagle, 1977a; Shapiro, 1991a). Generally, however, relative dorsal projection of the spinous processes exhibits little difference in value among closely related primates with diverse positional behaviors (Shapiro, 1991a; cf. Ward, 1991).

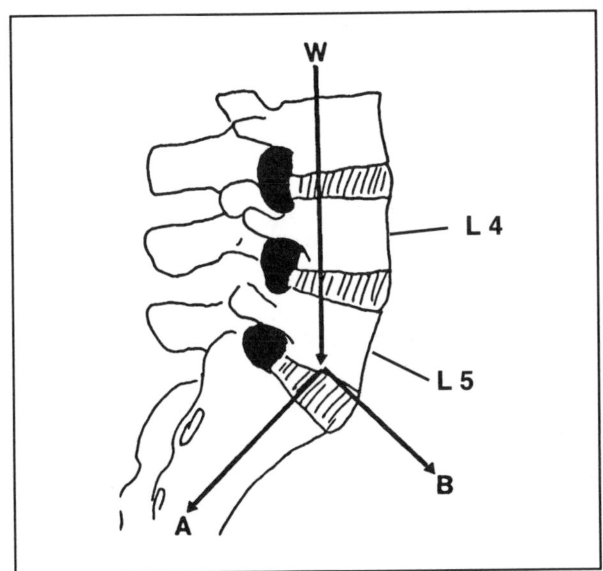

FIGURE 5.11. Lateral view of the lower lumbar vertebrae and sacrum of a human indicating the lines of force passing through the lumbar region as weight (W) is transmitted from the upper trunk to the legs (A). Component B, brought about by lordosis and the sharp lumbosacral angle, produces a tendency for the last lumbar vertebra to slide forward on the sacrum (and to a lesser extent, for L4 to slide forward on L5). These movements are resisted by the oblique orientation of the zygapophyses. After Davis (1961). By permission of Cambridge University Press (New York).

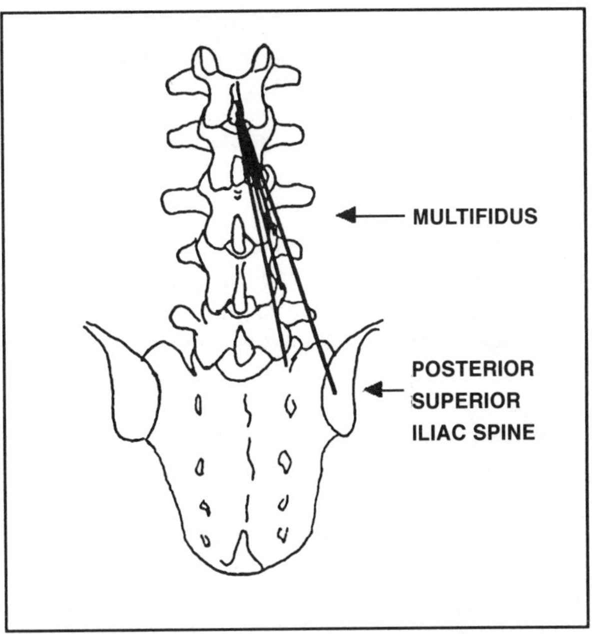

FIGURE 5.13. Dorsal view of the human lumbar region, with lumbar attachments of multifidus on the spinous and mammillary processes. After Macintosh et al. (1986). By permission of the publishers Butterworth-Heinemann Ltd.©

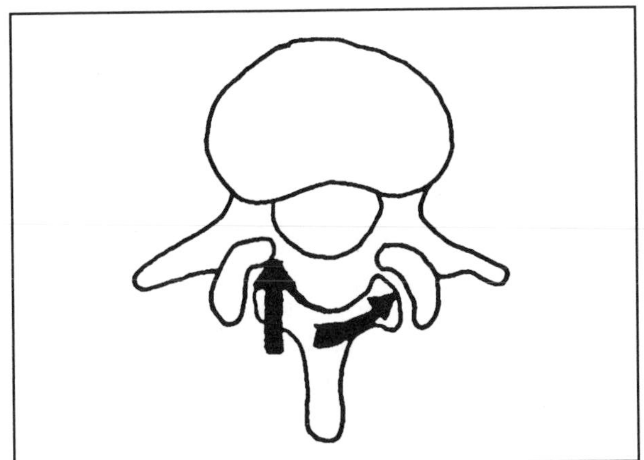

FIGURE 5.12. Cranial view of curved lumbar zygapophyseal joints. Curved joints resist both rotation and ventral displacement. After Bogduk and Twomey (1987). By permission of Churchill Livingstone.

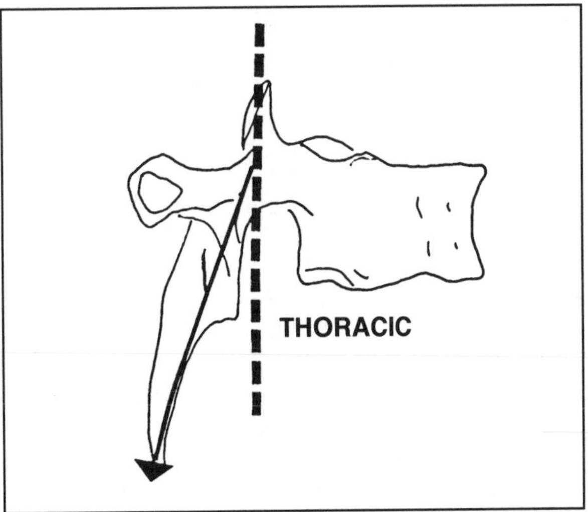

FIGURE 5.14. Lateral view of a human thoracic vertebra indicating the orientation of the spinous process measured from a coronal plane. A spine at an angle less than 90° is caudally oriented (depicted here), a spine at an angle of 90° is upright or perpendicular, and a spine at an angle greater than 90° is cranially oriented.

The craniocaudal orientation of the spinous processes changes from one vertebral region to the next. Craniocaudal orientation of spines refers to the angle formed between a coronal plane and a line passing from the base to the tip of the spinous process (Fig. 5.14). Transitions in orientation are most apparent between the thoracic and lumbar regions. The spines of thoracic vertebrae (rib-definition) generally point caudally in primates (Mivart, 1865; Slijper, 1946; Hill, 1953-1974; Ankel, 1972), and it has been noted that this caudal inclination is the most pronounced in the hominoids (Slijper, 1946).

In the lumbar region, spinous process inclination varies among primates. For example, the lumbar spinous processes of most primates are usually referred to as cranially oriented (Mivart, 1865; Slijper, 1946; Hill, 1953-1974; Ankel, 1972; Clauser, 1980). This categorization can be based on the overall shape of the spines; that is, they often have a small tip on the craniodorsal edge that points cranially (e.g., *Alouatta* in Fig. 5.15). Otherwise, it refers to the cranial orientation of the cranial and/or caudal edges of the spinous processes (e.g., *Cebus* and *Varecia* in Fig. 5.15). The cranially inclined lumbar spines that characterize many primates (like most mammals) meet the caudally inclined thoracic spines at the anticlinal vertebra (Fig. 5.16). The anticlinal vertebra has a spinous process that is oriented perpendicular to the vertebral body (Slijper, 1946). The anticlinal vertebra may or may not coincide with the diaphragmatic vertebra (Slijper, 1946; Clauser, 1975).

The spinous processes of hominoids are qualitatively different from those of most monkeys and prosimians—their dorsal edges are squared off, with no craniodorsal point, and they appear to be oriented perpendicularly or caudally relative to the vertebral body. Interestingly, the hominoid-type of squared-off spines also characterize *Ateles, Brachyteles, Indri,* and *Propithecus* (Fig. 5.15).

Although it is not difficult to see a basic qualitative difference in lumbar spinous process orientation between hominoids and most other primates, it is more difficult to quantify the inclination of lumbar spines than that of thoracic spines. This is due to the fact that (1) lumbar spinous processes are longer craniocaudally and, accordingly, the dorsal tip or edge of the spine is longer, and (2) the inclination of the cranial and caudal edges of the spinous process may differ. Consequently, there are many lines that could be drawn between the base and the tip of the spinous process, making the true orientation of the spine elusive (Clauser, 1980) (Fig. 5.17). Not surprisingly, the method of quantification of this aspect of vertebral morphology has been inconsistent among researchers (e.g. Lanier, 1939; Slijper, 1946; Clauser, 1980; Ward, 1991).

Although several workers have noted the variation in spinous process orientation among primates (e.g., Mivart, 1865; Slijper, 1946; Hill, 1953-1974; Ankel, 1972), functional explanations for this variation are rarely offered. One exception can be found in Slijper's (1946) extensive analysis of the mammalian vertebral column. Slijper noted that each spinous process is oriented perpendicular to the line of force of the most "important" muscle acting on it, providing maximum leverage with the least amount of material (Fig. 5.18). If more than one muscle of equal "importance" is attached to a spinous process, it is oriented in a position intermediate to the perpendiculars of each muscle.[2] For example, Slijper (1946) proposed that cranially oriented "postdiaphragmatic" spinous processes (which characterize most primates) reflect the importance of a muscle such as the thoracic portion of longissimus. He attributed more "upright" or caudally inclined spines to the importance of multifidus, and also observed that upright or caudal spines are associated with craniocaudal shortening of the postdiaphragmatic vertebral bodies. This association appears to be true for hominoids.

Another functionally important aspect of the lumbar spinous process is the relative expansion of its dorsal edge ("tip") in the craniocaudal plane. For example, Erikson (1963) attributed the squared-off, "broad interlocking" spines of *Ateles* as a mechanism for limiting mobility of the spine (Fig. 5.19). Presumably, the wider the expansion, the less interspinal distance there is available for the supraspinous or interspinous ligaments or muscles, and the more rigid is the lumbar region (as seen in ungulates; Gambaryan, 1974). In contrast, narrowed spinous process tips are associated with an increase in sagittal flexibility of the lumbar region due to the expansion of the ligaments or muscles (as seen in carnivores; Gambaryan, 1974) (Fig. 5.20).

Other than Erikson's (1963) brief consideration, little is known about variation among primates with respect to interspinal distances and the expansion of

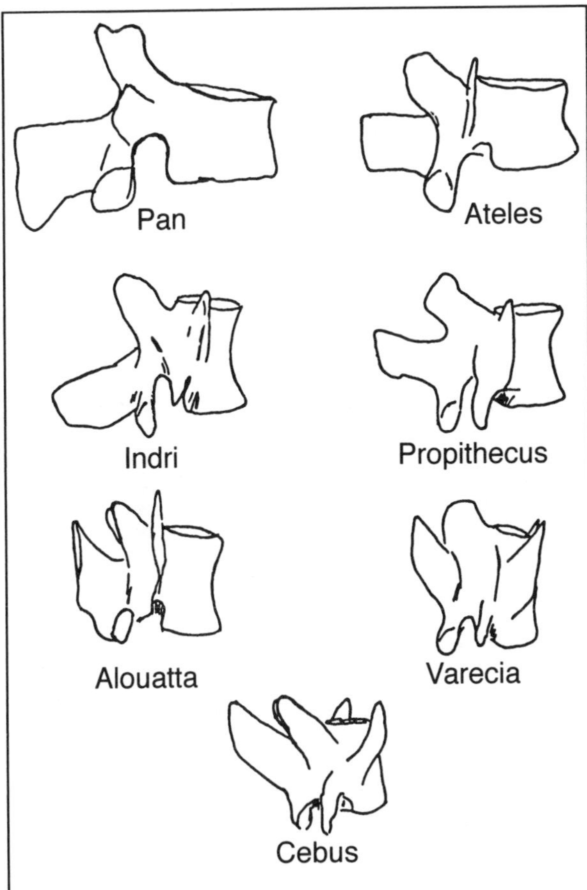

FIGURE 5.15. Lateral views of the lumbar vertebrae of various primates, exemplifying differences in spinous process shape and orientation. Note the squared-off, perpendicularly/caudally oriented spines of *Pan, Ateles, Indri,* and *Propithecus.* In contrast, the spine of *Alouatta* has a craniodorsal point, while those of *Cebus* and *Varecia* are angled cranially as a whole (not to scale).

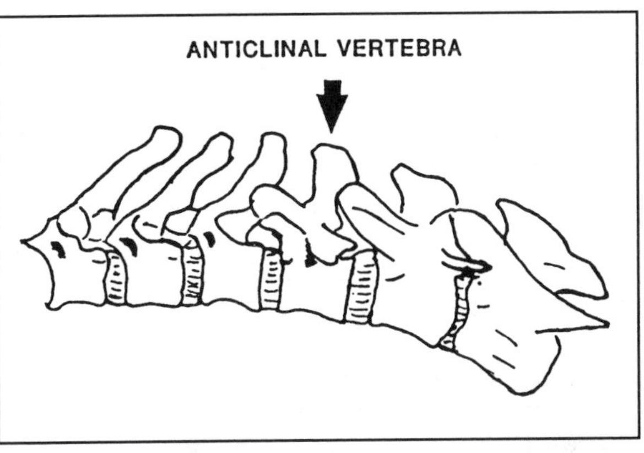

FIGURE 5.16. The anticlinal vertebra is the transitional vertebra between thoracic and lumbar regions with respect to the orientation of its spinous process. In the drawing, cranial is to the left. After Kimura et al. (1987).

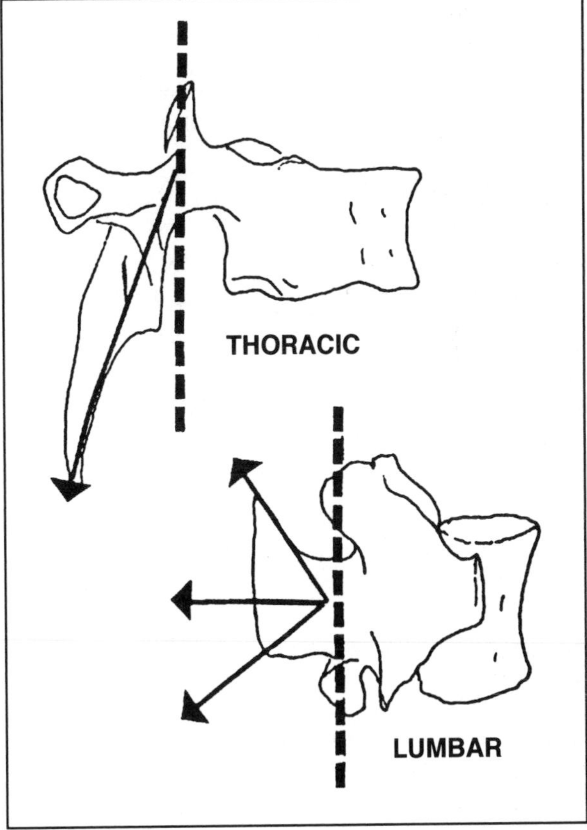

FIGURE 5.17. Orientation of the spinous process in thoracic (human) and lumbar (nonhuman) vertebrae (lateral view). The multiple lines drawn through the lumbar spinous process demonstrate the difficulty in quantifying this aspect of lumbar vertebral morphology (not to scale).

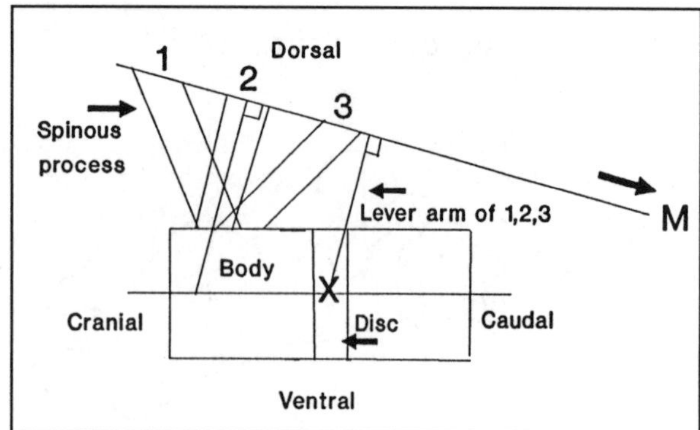

FIGURE 5.18. Schematic lateral view of two articulating lumbar vertebrae illustrating three possible orientations of the spinous process relative to the line of muscle force (M). The muscle would have the same lever arm (perpendicular distance between the line of muscle force and X, the axis of rotation) by attaching to each spine, but spine no. 2 provides the same leverage with the least amount of material, that is, the shortest spine. After Slijper (1946).

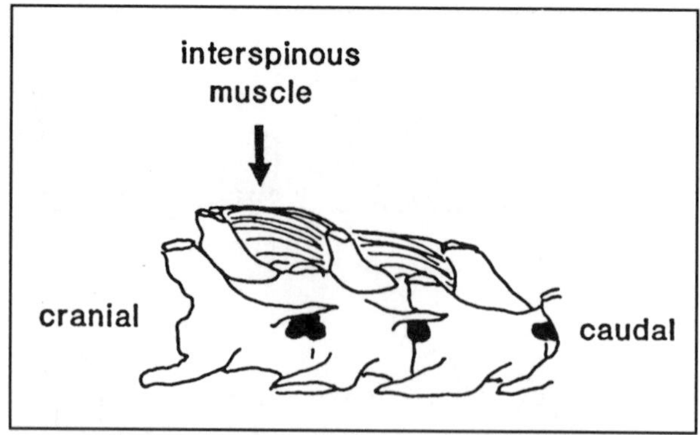

FIGURE 5.20. Lateral view of three articulating lumbar vertebrae of a nonprimate mammal (mustelid). The narrow tips of the spinous processes, the expansion of the interspinal space, and the well-developed interspinous muscles presumably account for the sagittal flexibility of the lumbar spine of this mammal. After Gambaryan (1974). By permission of Keter Publishing House Ltd.

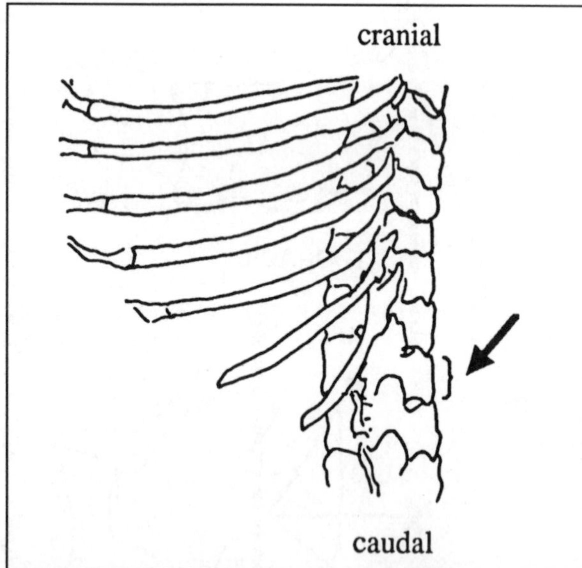

FIGURE 5.19. Lateral view of the (partial) thoracolumbar region in *Ateles*. Note the craniocaudal expansion of the spinous processes. Illustration as depicted by Erikson (1963) by permission of Academic Press Inc. (London) Ltd.

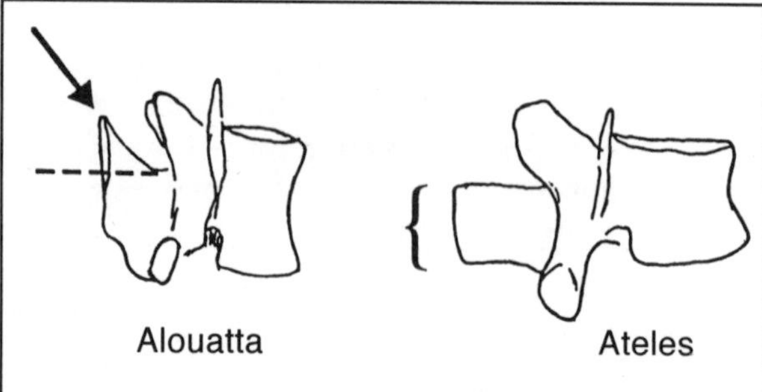

FIGURE 5.21. Lateral view of two lumbar vertebrae demonstrating two ways by which the dorsal edge of the spinous process may expand in the craniocaudal plane: expansion by a craniodorsal "point" (*Alouatta*) or expansion along the entire "squared-off" dorsal edge (*Ateles*).

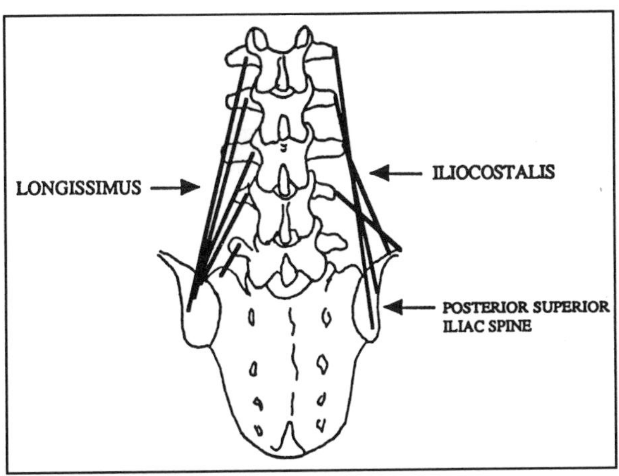

FIGURE 5.22. Dorsal view of the human lumbar region illustrating the attachments of longissimus on the dorsomedial aspects of the transverse processes and iliocostalis on the lateral tips of the transverse processes. After Macintosh and Bogduk (1987). By permission of J. B. Lippincott Co.

the spinous process tip. A thorough comparative and quantitative analysis of this feature would further our understanding of the role of spinal flexibility/rigidity in primate locomotion. However, it is probably important to consider not only expansion of the tip, but its shape as well. For example, the difference between craniocaudal expansion of a squared-off edge and extension of a craniodorsal point might be functionally relevant (Fig. 5.21).

Transverse Processes

Like the spinous processes, the lumbar transverse processes act as bony levers for the muscles that move the lumbar column. There is a considerable amount of ambiguity in the human anatomical literature as to the specific attachments of the intrinsic back muscles in the lumbar region (e.g., longissimus and iliocostalis). Most authors treat the lumbar portion of these muscles as a single muscle mass called the erector spinae, and most descriptions emphasize the individual divisions of these muscles in the thoracic and cervical regions. Several workers have documented separate lumbar attachments for longissimus and iliocostalis in humans as well as in nonhuman primates. According to these studies, longissimus and iliocostalis attach to the dorsomedial and dorsolateral aspects of lumbar transverse processes, respectively (Fig. 5.22) (George, 1967; Bogduk, 1973, 1980; Sato and Nakazawa, 1982; Kimura et al., 1987; Macintosh and Bogduk, 1987; but see Bustami, 1986).

Due to their location, longissimus and iliocostalis can effect extension or lateral flexion of the lumbar region. They stabilize the trunk during quadrupedalism (Shapiro, 1991a, 1991b) and bipedalism (e.g., Waters and Morris, 1972; Thorstensson et al., 1982; Thorstensson et al., 1984; Dofferhof and Vink, 1985; Shapiro and Jungers, 1988; Vink and Karssemeijer, 1988), and are also active during armswinging and climbing (Shapiro and Jungers, 1988, 1989; Shapiro, 1991a, 1991b).

Portions of psoas major and quadratus lumborum insert on the ventromedial and ventrolateral aspects of the transverse processes, respectively (Fig. 5.23). In conjunction with multifidus, longissimus, and iliocostalis, quadratus lumborum functions to stabilize the trunk during bipedalism (Waters and Morris, 1972). It also attaches to and stabilizes the last rib, thereby serving a respiratory function by creating a fixed point from which the diaphragm can act (Boyd et al., 1965). Psoas major extends beyond the lumbar region and combines with iliacus to form iliopsoas, which attaches to the lesser trochanter of the femur. It is thus principally a flexor of the hip. During bipedal walking in humans and in nonhuman primates, psoas major flexes the hip as the leg begins to swing forward (Basmajian and DeLuca, 1985; Stern, 1988; S. Larson, personal communication). In human bipedalism, it also regulates extension of the hip at the end of support phase (Basmajian and DeLuca, 1985; Stern, 1988). If the hip is stabilized, the vertebral portion of psoas major can flex the lumbar spine forward, and it also contributes to the maintenance of upright postures (Basmajian and DeLuca, 1985; Bogduk and Twomey, 1987).

The leverages of iliocostalis, longissimus, psoas major, and quadratus lumborum are dependent on the shape, placement, and orientation of the lumbar transverse processes. The farther dorsal the transverse process lies relative to the axis of extension, the more efficient are the leverages of longissimus and iliocostalis when they contract bilaterally. The farther lateral the tip of the transverse process lies relative to the axis for lateral flexion, the better are the leverages of iliocostalis and quadratus lumborum for lateral flexion when they contract unilaterally.

The lumbar transverse processes vary among primates in three respects (Fig. 5.24): (1) the location at which they originate on the vertebra, (2) their

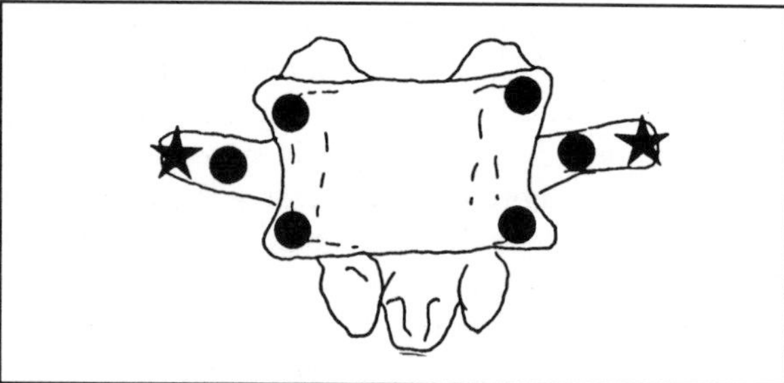

FIGURE 5.23. Ventral view of lumbar vertebra illustrating the attachment sites of psoas major (circles) on the vertebral body and ventromedial aspect of the transverse processes, and quadratus lumborum (stars) on the ventrolateral aspects of the transverse processes.

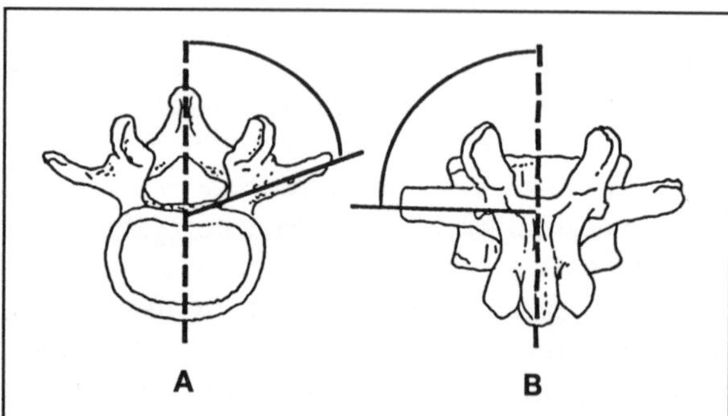

FIGURE 5.24. A: cranial view of a human lumbar vertebra illustrating ventrodorsal orientation of the transverse process measured from a sagittal plane (orientation depicted here is dorsal); B: dorsal view of a human lumber vertebra illustrating craniocaudal orientation of the transverse process measured from a sagittal plane (orientation depicted here is lateral, that is, neither cranial nor caudal).

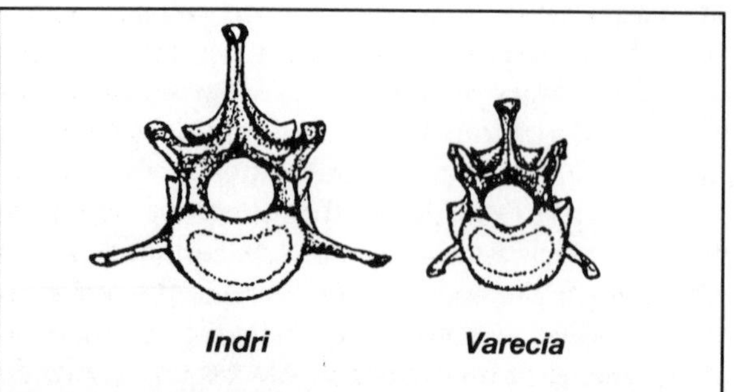

FIGURE 5.26. Cranial views of midlumbar vertebrae in *Indri* and *Varecia*. Note that in both species the transverse processes are oriented ventrally, but the angle of orientation is less acute and the tips of the transverse processes lie further dorsally relative to the ventral aspect of the vertebral body in *Indri*.

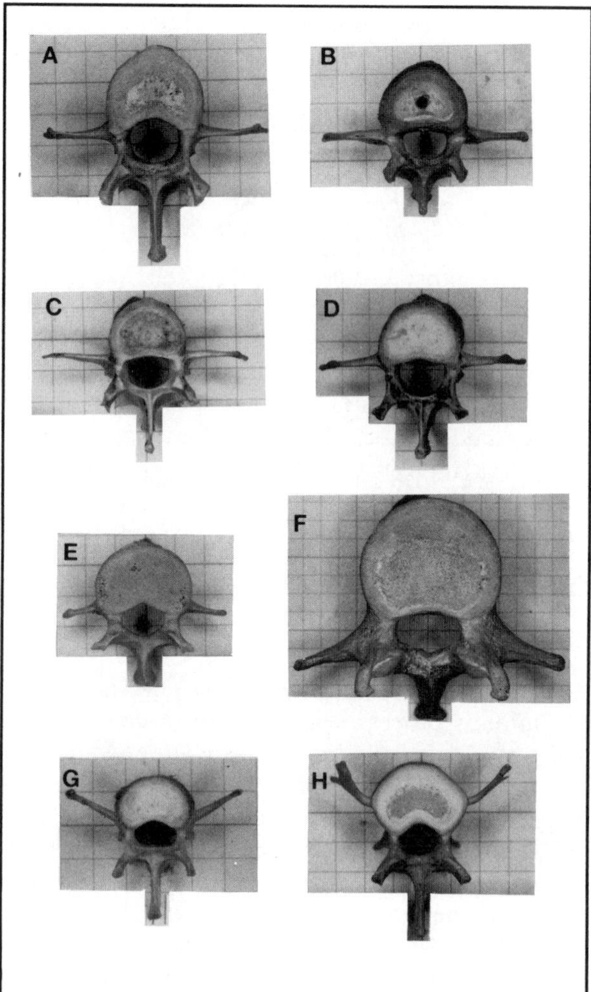

FIGURE 5.25. Cranial views of lumbar vertebrae of various primates illustrating the ventrodorsal placement and orientation of the transverse processes. Transverse processes are situated at the junction of the vertebral body and pedicles and project laterally in *Ateles* (A), *Brachyteles* (B), *Lagothrix* (C), *Alouatta* (D), and *Hylobates* (E). In great apes and humans, they arise from the pedicle-lamina junction and project dorsally, as in *Gorilla* (F). In most other primates, the transverse processes arise from the vertebral body and point ventrally, as in *Cebus* (G) and *Cercopithecus* (H). Background scale is four squares per inch.

ventrodorsal orientation—angular orientation relative to a coronal plane (measured from a sagittal plane), and (3) their craniocaudal orientation—angular orientation relative to a transverse plane (measured from a sagittal plane).

Location on the Vertebra

Among most primates, the lumbar transverse processes arise from the broadest part of the vertebral body. In *Alouatta, Lagothrix, Brachyteles, Ateles,* and *Hylobates,* they arise more dorsally from the roots of the neural arch, where the pedicles join the bodies. In the great apes and humans, they arise even further dorsally, at the pedicle-lamina junction (Benton, 1967, 1974; Ankel, 1972; Filler, 1986; Kelley, 1986) (Figs. 5.24, 5.25). The implications of these differences are discussed below in conjunction with ventrodorsal orientation.

Ventrodorsal Orientation

The transverse processes of *Hylobates* and the atelines project laterally; that is, the tips and bases of the transverse processes lie in line with each other and with the dorsal aspect of the vertebral body. In the great apes and humans, the transverse processes are oriented dorsally (Figs. 5.24, 5.25) (Mivart, 1865; Benton, 1967; Filler, 1986; Kelley, 1986; Shapiro, 1991a).

The dorsal placement and orientation of the transverse processes in great apes and humans are related to the wide flare of the iliac blades and the dorsal expansion of the iliac crest and ribs that characterize these primates. These anatomical relationships help to maintain a consistent line of action for muscles such as longissimus and iliocostalis (Keith, 1923; Waterman, 1929; Reynolds, 1931; Ward, 1991). In addition, the structural arrangement of the transverse processes of great apes and humans provides efficient leverage for extension. That is, their attachments are relatively far from the axis of rotation located in the intervertebral disc.

The lumbar transverse processes of most primates are oriented ventrally, (i.e., the distal tips lie ventral relative to the base of the process), which is also true of most mammals (Mivart, 1865; Benton, 1967; Gambaryan, 1974; Filler, 1986; Kelley, 1986; Shapiro, 1991a). Generally, the ventral position and orientation of the transverse processes characterizing most primates creates a relatively deep compartment for the erector spinae muscles (i.e., the compartment that is formed by the spinous process, lamina, and transverse process). Therefore, the erector spinae muscles are generally more massive in cross section (relative to body size) in these primates than in the hominoids or the atelines (Keith, 1940; Benton, 1967, 1974).

"Ventral" orientation is not necessarily equivalent in extent among these various primates. For example, although both the indriids and *Varecia* have ventrally oriented transverse processes, the angle of orientation is less acute, and the tips of the processes lie relatively further dorsally in the former than in the latter (Fig. 5.26) (Shapiro, 1991a). This would place the back muscles relatively farther dorsally, improving leverage for extension, and is most likely attributable to the indriids' emphasis on leaping and orthograde posture in comparison to the more quadrupedal and pronograde *Varecia* (Petter et al., 1977; Walker, 1979; Gebo, 1987; Dagosto, 1989). Although the ventrodorsal orientation of the transverse process is correlated with body size in African bovids (i.e., as species get larger, their transverse processes become more perpendicularly oriented relative to a sagittal plane), this does not appear to be true within related groups of primates (Shapiro, 1991a; Ward, 1991).

Craniocaudal Orientation

In addition to differences in ventrodorsal placement and orientation, the primate lumbar transverse process varies with respect to its craniocaudal orientation as well. The evaluation of the craniocaudal orientation of the transverse process is difficult for reasons similar to those mentioned concerning the spinous process. That is, one must consider the shape and orientation of the lateral tip of the transverse process, as well as the orientation of the cranial and caudal edges (Fig. 5.27).

Qualitative descriptions of this orientation are inconsistent in the literature. Some authors refer to the lateral tip specifically (e.g., "their lateral ends turn cranially"; Benton, 1967:202), while others indicate that the process as a whole points "forward" (e.g., Mivart, 1865:559). Nevertheless, these observations have identified a basic difference between the transverse processes of hominoids and those of most other primates. Specifically, the latter tend to have "cranially" oriented processes (which resemble those of many other mammals), in contrast to the

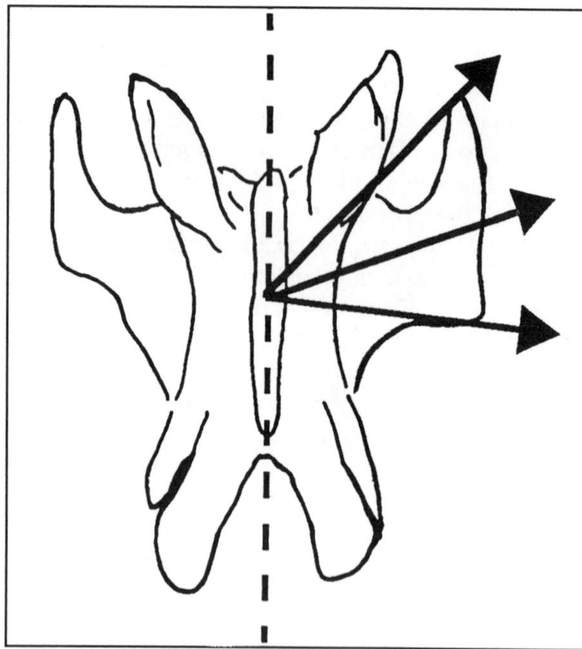

FIGURE 5.27. Dorsal view of a lumbar vertebra of *Cercopithecus*. The multiple lines drawn through the transverse process demonstrate the difficulty in quantifying its craniocaudal orientation.

more "lateral" orientation of those of hominoids (Fig. 5.28).

The transverse processes of nonprimate mammals exhibit some of the same types of morphologies characterizing those of primates. Therefore, despite the lack of functional analyses offered in the literature for primate transverse process orientation (especially craniocaudal), one can gain some functional insight by considering transverse process morphology in nonprimate mammals (albeit qualitatively). For example, Gambaryan (1974) associated the lumbar rigidity characterizing ungulates with the fact that their transverse processes are situated horizontally and widened at the lateral ends, where the intertransversal ligament is attached (Fig. 5.29). In contrast, the ventrocranially (Gambaryan's "forward and downward") oriented processes of carnivores with their narrow tips and strongly developed intertransversal muscles lends great lateral flexibility to the lumbar spine. In lagomorphs, the craniocaudal expansion at the lateral ends of the transverse processes presumably functions to restrict lateral movements, while also promoting sagittal plane flexion and extension by increasing attachment area for the flexor and extensor muscles (Gambaryan, 1974; see also Hatt, 1932).

A thorough understanding of the functional morphology of the transverse processes in primates as well as in nonprimates is dependent upon an adequate quantification of transverse process orientation, which has rarely been attempted. Otherwise, it remains unclear in what respect the transverse process truly varies, and interspecific variation becomes difficult to verify, let alone interpret functionally.

Lateral Projection

The extent to which the transverse process projects from the median sagittal plane is an indication of the leverage for lateral flexion by muscles such as iliocostalis, quadratus lumborum, and to a lesser extent, longissimus (Fig. 5.30). If one compares closely related primates that differ in positional behavior, however, there is little difference in the relative lateral projection of their lumbar transverse processes and, therefore, the muscular leverage for lateral flexion (Shapiro, 1991a). This is not to say, however, that leverages for lateral flexion do not differ among primates at all. Iliocostalis and quadratus lumborum for example, attach to the ribs and iliac crest as well as to the lumbar transverse processes. A more lateral placement of these muscles on the rib cage or the iliac crest most likely has a more pronounced effect on lateral flexion than do their transverse process attachments.

CONCLUSION

It is clear that we cannot fully understand the functional morphology of primate positional behavior without a consideration of the vital role played by the vertebral column and its musculature. The contribution of the spine to postural and locomotor function is attested to by its morphological variability among primates that move in different ways. Future research should concentrate on quantifying the morphological variation that as of yet has been described only qualitatively. Even more importantly, reliable links between vertebral form and function depend upon in vivo investigations of the movements of the spine during various positional behaviors; this includes naturalistic observations as well as more physiological approaches (Fleagle, 1979).

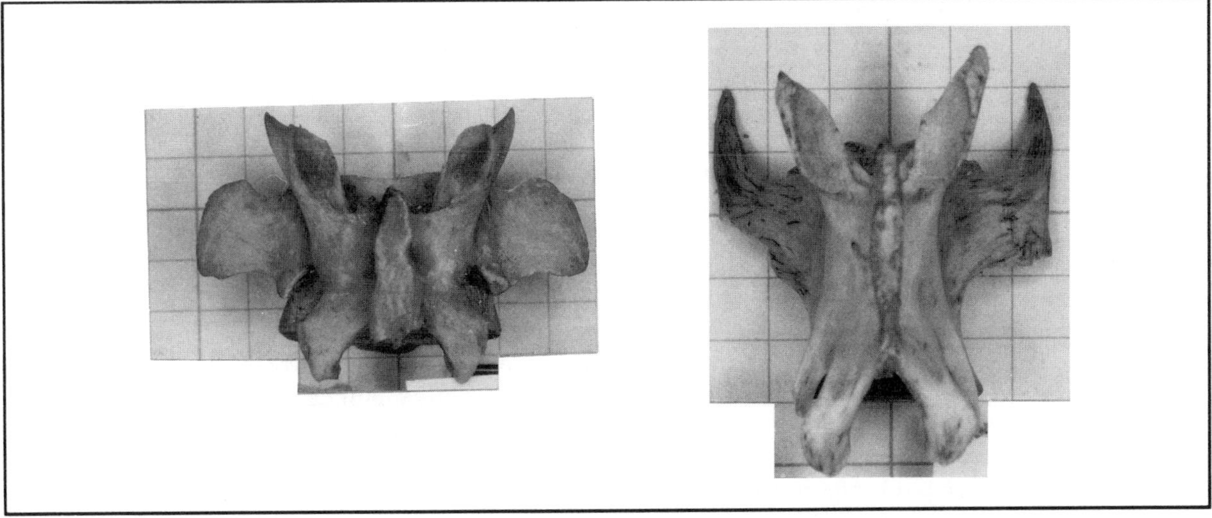

FIGURE 5.28. Dorsal views of lumbar vertebrae of *Hylobates syndactylus* (left) and *Cercopithecus aethiops* (right). Compare the lateral orientation of the transverse processes of the hominoid to their cranial orientation in the cercopithecoid (and most other primates). Background scale is four squares per inch.

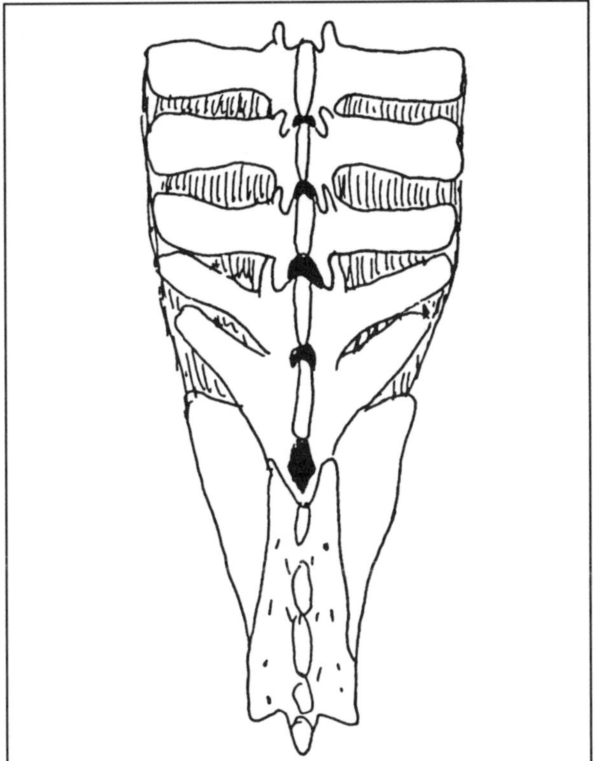

FIGURE 5.29. Dorsal view of the lower lumbar region in an ungulate (*Equus*). Lumbar rigidity is achieved by transverse processes that are oriented laterally and widened at their lateral ends. The intertransversal ligament is depicted with vertical hatching. After Gambaryan (1974). By permission of Keter Publishing House Ltd.

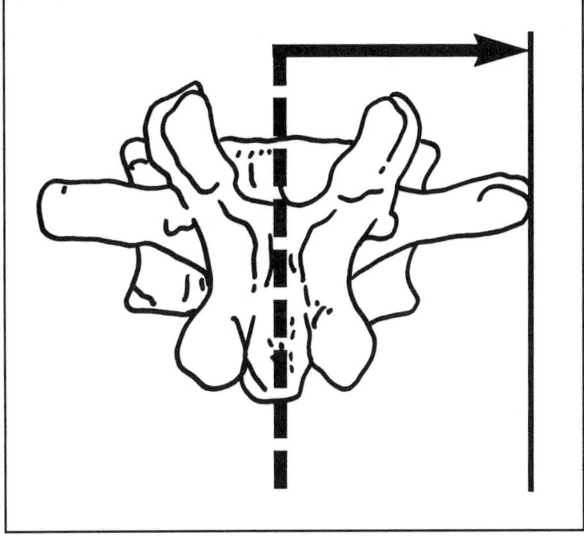

FIGURE 5.30. Lateral projection of the transverse process from a median sagittal plane (human).

Previous research has emphasized vertebral distinctions across broad taxonomic groups of primates. These distinctions are functionally relevant, but they should also be treated as a baseline from which to inspect axial function in more detail. For example, functional explanations for variation in the relative length of the lumbar region usually refer to the long spine's "flexibility" in leaping or running, or the short spine's "rigidity" in bridging or climbing. Biomechanically speaking, it is a reasonable assumption that a longer spine would be more "flexible" and vice versa, but such assumptions need to be verified objectively and quantitatively with kinematic analyses of the spine. Only with such methods can we move beyond broad taxonomic categories and address the functional variability that might characterize members within these groups. The same holds true for other aspects of vertebral anatomy that differ between "long-backed" and "short-backed" primates (Benton, 1967, 1974), such as the shape and orientation of transverse and spinous processes. These parameters have been noted by several workers, but their influence on axial function has yet to be fully understood. The morphology of these bony levers varies both among and within diverse groups of primates, which suggests functional differences above and beyond a simple "long-backed"/"short-backed" dichotomy, and warrants further investigation (e.g., using three-dimensional morphometrics, electromyography, etc.). In addition, although zygapophyseal orientation appears to vary little among nonhuman primates, a detailed inspection of other aspects of zygapophyseal morphology (e.g., shape, size, curvature) might provide us with additional functional insight.

A thorough understanding of the primate spine is dependent upon the investigation of vertebral form and function in a wide variety of primates that engage in a diversity of positional behaviors. It is time to give primate axial morphology attention comparable to that which has been paid to other aspects of postcranial anatomy. The most reliable approach is a multifaceted one, combining comparative anatomy, naturalistic observation, and in vivo studies in a controlled setting (Fleagle, 1979). With such an approach, form and function associations become testable, and ultimately can be applied to the fossil record in order to reconstruct the evolution of primate locomotor adaptations. The lumbar region, as well as the rest of the vertebral musculoskeletal system, holds much promise toward this purpose.

NOTES

Acknowledgments: I would like to thank Dan Gebo for the opportunity to contribute to this book. This chapter has benefited greatly from discussions with William Jungers, Jack Stern, Jr., John Fleagle, Farish Jenkins, Jr., and Susan Larson. Thanks also go to Kay Clark and especially Alex Duncan for their work on the figures, and to John Kappelman, Claud Bramblett, Alex Duncan, Carol Ward, and two anonymous reviewers for their helpful suggestions. I am grateful to the curators and staff of the following museums for access to collections in their care: the American Museum of Natural History, the Cleveland Museum of Natural History, the Field Museum of Natural History, the Museum of Comparative Zoology (Harvard University), the United States National Museum of Natural History, the Musée Nationale de l'Histoire Naturelle, the Rijksmuseum van Natuurlijke Historie, the Powell-Cotton Museum, the Musée Royal de l'Afrique Centrale, the Institut Royal des Sciences Naturelles de Belgique, and the British Museum of Natural History. Financial support for my research summarized here was provided by NSF grants BNS 8519747 and 8819621 to Jack Stern, Jr., an NSF Dissertation Improvement Grant BNS 8823083 (supervised by W. Jungers), the Doctoral Program in Anthropological Sciences, S.U.N.Y. at Stony Brook, and a Herbert and Mildred Weisinger Fellowship.

1. A common assumption is that the axis of rotation lies at or near the zygapophyses (Gregersen and Lucas, 1967; Ankel, 1972; Haher et al., 1989). This is not likely to be true, because such an axis would subject the intervertebral disc to intolerable shearing moments. Rather, the zygapophyses and the annulus fibrosus of the disc act together to resist torsion around an axis located in the disc (Lewin et al., 1962; Singer, Day, and Briedahl, 1989). An axis of rotation within the zygapophyseal joint occurs only after rotation occurs around the axis located in the disc, and after a postzygapophysis has impacted its corresponding prezygapophysis. This secondary rotation surpasses the critical range beyond which the risk of torsional injury increases (Bogduk and Twomey, 1987).

2. Slijper (1946) defines the most "important" muscle as the one with the highest point of intersection with a line perpendicular to the longitudinal axis of the vertebra. This makes his argument potentially circular because the method by which he determines "importance" is not independent of spinous process orientation.

LITERATURE CITED

Abitbol M (1987a) Evolution of the lumbosacral angle. Am. J. Phys. Anthrop. *72*:361–372.

——— (1987b) Evolution of the sacrum in hominoids. Am. J. Phys. Anthrop. *74*:65–81.

Adams MA, and Hutton WC (1980) The effect of posture on the role of the apophysial joints in resisting intervertebral compressive forces. J. Bone Joint Surg. *62B*(3):358–362.

——— (1983) The mechanical function of the lumbar apophysial joints. Spine *8*(3):327–330.

Ankel F (1962) Vergleichende untersuchungen über die skelettmorphologie des greifschwanzes südamerikanischer affen (Platyrrhina). Z. Morph. Ökil. Tiere *52*:131–170.

——— (1965) Der canalis sacralis als indikator für die länge der caudalregion der primaten. Folia Primatol. *3*:263–276.

——— (1972) Vertebral morphology of fossil and extant primates. In R Tuttle (ed.): The Functional and Evolutionary Biology of Primates. Chicago: Aldine, pp. 223–240.

Basmajian J, and DeLuca C (1985) Muscles Alive: Their Functions Revealed by Electromyography, 5th ed. Baltimore: Williams and Wilkins.

Benton RS (1967) Morphological evidence for adaptations within the epaxial region of the primates. In H Vagtborg (ed.): The Baboon in Medical Research, Vol. 2. Austin: University of Texas Press, pp. 201–216.

——— (1974) Structural patterns in the Pongidae and Cercopithecidae. Yrbk. Phys. Anthrop. *18*:65–88.

Bogduk N (1973) Aspects of Comparative Anatomy of the Dorsal Lumbosacral Region. Thesis, submitted for admission to the degree of Bachelor of Medical Science, University of Sydney, Australia.

——— (1980) A reappraisal of the anatomy of the human lumbar erector spinae. J. Anat. *131*:525–540.

Bogduk N, and Twomey L (1987) Clinical Anatomy of the Lumbar Spine. New York: Churchill Livingstone.

Boyd W, Blincoe H, and Hayner JC (1965) Sequence of action of the diaphragm and quadratus lumborum during quiet breathing. Anat. Rec. *151*:579–582.

Bustami F (1986) A new description of the lumbar erector spinae muscle in man. J. Anat. *144*:81–91.

Cartmill M, and Milton K (1977) The lorisiform wrist and the evolution of "brachiating" adaptations in the Hominoidea. Am. J. Phys. Anthrop. *47*:249–272.

Cihak R (1970) Variations of lumbosacral joints and their morphogenesis. Acta Univer. Carolin. Med. *16*:145–165.

Clauser DA (1975) The numbers of vertebrae in three African cercopithecine species. Folia Primatol. *23*:308–319.

——— (1980) Functional and Comparative Anatomy of the Primate Spinal Column: Some Postural and Locomotor Adaptations. Ph.D. dissertation, University of Wisconsin, Milwaukee.

Cossette JW, Farfan HF, Robertson GH, and Wells RV (1971) The instantaneous center of rotation of the third lumbar intervertebral joint. J. Biomech. *4*:149–153.

Cunningham DJ (1886) The lumbar curve in man and apes, with an account of the topographical anatomy of the chimpanzee, orangutan and gibbon. Cunningham Memorial, Royal Irish Acad. No. 2.

Dagosto M (1989) Locomotion of *Propithecus diadema* and *Varecia variegata* at Ranomafana National Park, Madagascar. Am. J. Phys. Anthrop. *78*:209.

Davis PR (1961) Human lower lumbar vertebrae: Some mechanical and osteological considerations. J. Anat. *95*:337–344.

Denis F (1983) The three column spine and its significance in the classification of acute thoracolumbar spinal injuries. Spine *8*(8):817–831.

Dofferhof ASM, and Vink P (1985) The stabilizing function of the mm. iliocostalis and the mm. multifidi during walking. J. Anat. *140*(2):329–336.

Donisch E (1973) A comparative study of the back muscles of gibbon and man. In DM Rumbaugh (ed.): Gibbon and Siamang, Vol. 2. Basel: Karger, pp. 96–120.

Dunlop RB, Adams MA, and Hutton WC (1984) Disc space narrowing and the lumbar facet joints. J. Bone Joint Surg. *66B*(5):706–710.

El-Bohy AA, Yang KH, and King AI (1989) Experimental verification of facet load transmission by direct measurement of facet lamina contact pressure. J. Biomech. *22*:931–941.

Emerson S (1985) Jumping and leaping. In M Hildebrand, D Bramble, K Liem, and D Wake (eds.): Functional Vertebrate Morphology. Cambridge: Harvard University Press, pp. 58–72.

Erikson GE (1960) The vertebral column of New World primates. Anat. Rec. *138*:346.

——— (1963) Brachiation in New World monkeys and in anthropoid apes. Symp. Zool. Soc. Lond. *10*:135–164.

Filler AG (1986) Axial Character Seriation in Mammals: An Historical and Morphological Exploration of the Origin, Development, Use, and Current Collapse of the Homology Paradigm. Ph.D. dissertation, Harvard University.

Fleagle JG (1977a) Locomotor behavior and muscular anatomy of sympatric Malaysian leaf monkeys (*Presbytis obscura* and *Presbytis melalophos*). Am. J. Phys. Anthrop. *46*(2):297–308.

——— (1977b) Locomotor behavior and skeletal anatomy of sympatric Malaysian leaf-monkeys. Yrbk. Phys. Anthrop. *20*:440–453.

——— (1979) Primate positional behavior and anatomy: Naturalistic and experimental approaches. In ME Morbeck, H Preuschoft, and N Gomberg (eds.): Environment, Behavior and Morphology: Dynamic Interactions in Primates. New York: Gustav Fischer, pp. 313–325.

——— (1988) Primate Adaptation and Evolution. New York: Academic Press.

Fleagle JG, Stern JT, Jr., Jungers WL, Susman RL, Vangor AK, and Wells JP (1981) Climbing: A biomechanical link with brachiation and with bipedalism. In MH Day (ed.): Vertebrate Locomotion. Sympos. Zool. Soc. Lond. No. 48. London: Academic Press, pp. 359–375.

Flower W (1876) Introduction to the Osteology of the Mammalia. London: Macmillan and Company.

Gadow H (1933) Evolution of the Vertebral Column. London: Cambridge University Press.

Gambaryan P (1974) How Mammals Run. New York: John Wiley and Sons.

Gebo DL (1987) Locomotor diversity in prosimian primates. Am. J. Primatol. *13*:271–281.

George RM (1967) The intrinsic back musculature of *Macaca mulatta*. Masters thesis, Medical College of Virginia, Richmond.

Gregersen G, and Lucas D (1967) An in vivo study of the axial rotation of the human thoracolumbar spine. J. Bone Joint Surg. *49A*:247–262.

Haher T, Felmy W, Baruch H, Devlin V, Welin D, O'Brien M, Ahmad J, Valenza J, and Parish S (1989) The contribution of the three columns of the spine to rotational stability: A biomechanical model. Spine *14*:663–669.

Halpert AP, Jenkins FA, Jr., and Franks H (1987) Structure and scaling of the lumbar vertebrae in African bovids (Mammalia: Artiodactyla). J. Zool. Lond. *211*:239–258.

Hatt R (1932) The vertebral columns of ricochetal rodents. Bull. Amer. Mus. Nat. Hist. *63*:599–738.

Hildebrand M (1959) Motions of the running cheetah and horse. J. Mammal. *40*:481–495.

Hill WCO (1953–1974) Primates: Comparative Anatomy and Taxonomy. Vols. 1–8. Edinburgh: Edinburgh University Press.

Horwitz T, and Smith RM (1940) An anatomical, pathological and roentgenological study of the intervertebral joints of the lumbar spine and of the sacroiliac joints. Am. J. Roentgenol. *43*:173–186.

Hurov J (1987) Terrestrial locomotion and back anatomy in vervets (*Cercopithecus aethiops*) and patas monkeys (*Erythrocebus patas*). Am. J. Primatol. *13*:297–311.

Hutton WC, and Cyron BM (1978) Spondylolysis: The role of the posterior elements in resisting the intervertebral compressive force. Acta Orthop. Scand. *49*:604–609.

Jayson M (1983) Compression stresses in the posterior elements and pathologic consequences. Spine *8*(3):338–339.

Jenkins FA, Jr. (1970) Anatomy and function of expanded ribs in certain edentates and primates. J. Mamm. *51*:288–301.

——— (1974a) Primate Locomotion. New York:

Academic Press.

——— (1974b) Tree shrew locomotion and the origins of primate arborealism. In F Jenkins, Jr. (ed.): Primate Locomotion. New York: Academic Press, pp. 85–116.

Jenkins FA, Jr., Dombrowski P, and Gordon E (1978) Analysis of the shoulder in brachiating spider monkeys. Am. J. Phys. Anthrop. *48:*65–76.

Jouffroy F, Stack M, and Niemitz C (1990) Gravity, Posture and Locomotion in Primates. Il Sedicesimo: Firenze.

Jungers WL (1984) Scaling of the hominoid locomotor skeleton with special reference to lesser apes. In H Preuschoft, D Chivers, W Brockelman, and N Creel (eds.): The Lesser Apes. Edinburgh: Edinburgh University Press, pp. 146–169.

Kapandji IA (1974) The Physiology of the Joints. Vol. 3, The Trunk and Vertebral Column. London: Churchill Livingstone.

Kashimoto T, Yamamuro T, and Hatakeyama K (1982) Anatomical and biomechanical factors in the curve pattern formation of idiopathic scoliosis. Acta Orthop. Scand. *53:*361–368.

Keith A (1902) The extent to which the posterior segments of the body have been transmuted and suppressed in the evolution of man and allied primates. J. Anat. Lond. *37:*18–40.

——— (1923) Man's posture: Its evolution and disorders. 2. The evolution of the orthograde spine. British Medical Journal *1:*499–502.

——— (1940) Fifty years ago. Am. J. Phys. Anthrop. *26:*251–267.

Kelley JJ (1986) Paleobiology of Miocene Hominoids. Ph.D. dissertation, Yale University, New Haven.

Kelley JJ, and Pilbeam D (1986) The dryopithecines: Taxonomy, anatomy and phylogeny of Miocene large hominoids. In DR Swindler and J Erwin (eds.): Comparative Primate Biology, Vol. 1. New York: Alan R. Liss, pp. 316–411.

Kenesi C, and Lesur C (1985) Orientation of the articular processes at L4, L5, S1: Possible role in pathology of the intervertebral disc. Anat. Clin. *7:*43–47.

Kimura K, Konishi M, Takahashi Y, and Iwamoto S (1987) The skeletal system of *Macaca fascicularis*—Description and measurement. Part 4, Vertebral column. J. Natl. Def. Med. Coll. *12:*183–194.

Kondo S (1985) Primate Morphophysiology, Locomotor Analyses and Human Bipedalism. Tokyo: University of Tokyo Press.

Lanier R (1939) The presacral vertebrae of American white and negro males. Am. J. Phys. Anthrop. *25:*341–420.

Lewin T, Moffett B, and Viidik A (1962) The morphology of the lumbar synovial intervertebral joints. Acta. Morph. Neer. Scand. *4:*299–319.

Lin HS, Liu YK, Adams KH (1978) Mechanical response of the lumbar intervertebral joint under physiological (complex) loading. J. Bone Joint Surg. *60A*(1):41–55.

Lindh M (1980) Biomechanics of the lumbar spine. In V Frankel and M Nordin (eds.): Basic Biomechanics of the Skeletal System. Philadelphia: Lea & Febiger, pp. 255–289.

Lorenz M, Patwardhan A, and Vanderby R (1983) Load-bearing characteristics of lumbar facets in normal and surgically altered spinal segments. Spine *8*(2):122–130.

Louis R (1985) Spinal stability as defined by the three-column spine concept. Anat. Clin. *7:*33–42.

Macintosh J, and Bogduk N (1986) The biomechanics of the lumbar multifidus. Clin. Biomech. *1:*205–213.

——— (1987) The morphology of the lumbar erector spinae. Spine *12:*658–668.

Macintosh J, Valencia F, Bogduk N, and Munro R (1986) The morphology of the human lumbar multifidus. Clin. Biomech. *1:*196–204.

Markolf KL (1972) Deformation of the thoracolumbar intervertebral joints in response to external loads. J. Bone Joint Surg. *54A*(3):511–533.

Martin RD (1989) (ed): New Quantitative Developments in Primatology and Anthropology. 1. Schultz-Biegert Symposium in Kartause Ittingen. Folia Primatol. Vol. 53. Basel: Karger.

Mivart St. G (1865) Contributions towards a more complete knowledge of the axial skeleton in the primates. Proc. Zool. Soc. Lond., pp. 545–592.

Morbeck ME, Preuschoft H, and Gomberg N (1979) Environment, Behavior, and Morphology: Dynamic Interactions in Primates. New York: Gustav Fischer.

Nissan M, and Gilad I (1986) Dimensions of human lumbar vertebrae in the sagittal plane. J. Biomech. *19:*753–758.

Odgers PNB (1933) The lumbar and lumbosacral diarthrodial joints. J. Anat. *67:*301–317.

Oxnard C, and Yang H (1981) Beyond biometrics: Studies of complex biological patterns. Symp. Zool. Soc. Lond. *46:*127–167.

Pal G, Cosio L, and Routal R (1988) Trajectory architecture of the trabecular bone between the body and the neural arch in human vertebrae. Anat. Rec. *222:*418–425.

Pal G, and Routal R (1986) A study of weight transmission through the cervical and upper thoracic regions of the vertebral column in man. J. Anat. *148:*245–261.

——— (1987) Transmission of weight through the lower thoracic and lumbar regions of the vertebral column in man. J. Anat. *152:*93–105.

Pegington J (1985) Clinical Anatomy in Action: The Vertebral Column and Limbs. Edinburgh: Churchill Livingstone.

Petter J, Albignac R, and Rumpler Y (1977) Faune de Madagascar. Vol. 44, Mammifères Lemuriens (Primates, Prosimiens). Paris: Orstom, CNRS.

Preuschoft H, Hayama S, and Gunther MM (1988) Curvature of the lumbar spine as a consequence of mechanical necessities in Japanese macaques trained for bipedalism. Folia Primatol. *50:*42–58.

Reynolds E (1931) The evolution of the human pelvis in relation to the mechanics of the erect posture. Papers of the Peabody Museum of Am. Archaeology and Ethnology *21*(*5*):255–334.

Rockwell H, Gaynor Evans F, and Pheasant H (1938) The comparative morphology of the vertebrate spinal column. Its form as related to function. J. Morph. *63:*87–117.

Rose M (1975) Functional proportions of primate lumbar vertebral bodies. J. Hum. Evol. *4:*21–38.

——— (1978) Feeding and associated positional behavior of black and white colobus monkeys (*Colobus guereza).* In G Montgomery (ed.): The Ecology of Arboreal Folivores. Washington, D.C.: Smithsonian Institution Press.

Rosenberger AL, and Strier KB (1989) Adaptive radiation of the ateline primates. J. Hum. Evol. *18:*717–750.

Sanders W (1990) Weight transmission through the lumbar vertebrae and sacrum in australopithecines. Am. J. Phys. Anthrop. *81:*289.

——— (1991) Comparative study of hominoid lumbar neural canal dimensions. Am. J. Phys. Anthrop. Suppl. *12:*157.

Sato T, and Nakazawa S (1982) Morphological classification of the muscular tubercles of the vertebrae. Okaj. Folia Anat. Jpn. *58:*1167–1186.

Scholten P, and Veldhuizen A (1985) The influence of spine geometry on the coupling between lateral bending and axial rotation. Engin. Med. *14:*167–171.

Schultz AH (1938) The relative length of the regions of the spinal column in Old World primates. Am. J. Phys. Anthrop. *24:*1–22.

——— (1961) Vertebral column and thorax. Primatologia, Vol. 4. Basel: Karger, pp. 1–66.

Schultz AH, and Straus WL, Jr. (1945) The numbers of vertebrae in primates. Proc. Am. Phil. Soc. *89*(4):601–626.

Shah JS, Hampson WGJ, and Jayson MIV (1978) The distribution of surface strain in the cadaveric lumbar spine. J. Bone Joint Surg. *60B*(2):246–251.

Shapiro LJ (1990) Vertebral morphology in the Hominoidea. Am. J. Phys. Anthrop. *81:*294.

——— (1991a) Functional Morphology of the Primate Spine with Special Reference to Orthograde Posture and Bipedal Locomotion. Ph.D. dissertation, State University of New York, Stony Brook.

——— (1991b) Quadrupedalism and back muscle function in hominoids. Am. J. Phys. Anthrop. Suppl. *12:*160.

Shapiro LJ, and Jungers WL (1988) Back muscle function during bipedal walking in chimpanzee and gibbon: Implications for the evolution of human locomotion. Am. J. Phys. Anthrop. *77:*201–212.

——— (1989) Back muscle function during suspensory behaviors in chimpanzee, orangutan and gibbon. Am. J. Phys. Anthrop. *78:*300.

Singer KP (1989) Thoracolumbar mortice joint: Radiological and histological observations. Clin. Biomech. *4:*137–143.

Singer KP, Breidahl PD, and Day RE (1989) Posterior element variation at the thoracolumbar transition: A morphometric study using computed tomography. Clin. Biomech. *4:*80–86.

Singer KP, Day RE, and Breidahl PD (1989) In vivo

axial rotation at the thoracolumbar junction: An investigation using low dose CT in healthy male volunteers. Clin. Biomech. *4:*145–150.

Slijper E (1946) Comparative biologic-anatomical investigations on the vertebral column and spinal musculature of mammals. Verh. K. Ned. Akad. Wet. *42:*1–128.

Stern JT, Jr. (1988) Essentials of Gross Anatomy. Philadelphia: F.A. Davis Company.

Strasser B, and Dagosto M (1988) The Primate Postcranial Skeleton: Studies in Adaptation and Evolution. London: Academic Press.

Straus WL, Jr., and Wislocki GB (1932) On certain similarities between sloths and slow lemurs. Bull. Mus. Comp. Zool. Harv. *74:*45–56.

Struthers J (1892) On the articular processes of the vertebrae in the gorilla compared with those in man and on costo-vertebral variation in the gorilla. J. Anat. Lond. *27:*131–138.

Tague RG, and Lovejoy CO (1986) The obstetric pelvis of A.L. 288–1 (Lucy). J. Hum. Evol. *15:*237–255.

Thorstensson A, Carlson H, Zomlefer M, and Nilsson J (1982) Lumbar back muscle activity in relation to trunk movements during locomotion in man. Acta Physiol. Scand. *116:*13–20.

Thorstensson A, Nilsson J, Carlson H, and Zomlefer M (1984) Trunk movements in human locomotion. Acta Physiol. Scand. *121:*9–22.

Todd TW (1922) Numerical significance in the thoracicolumbar vertebrae of the Mammalia. Anat. Rec. *24:*261–286.

Tuttle RH, Buxhoeveden DP, and Cortwright GW (1979) Anthropology on the move: Progress in experimental studies of nonhuman primate positional behavior. Yrbk. Phys. Anthrop. *22:*187–214.

Vallois HV (1927) Les variations des muscles spinaux chez les primates supérieurs. Comp. Rend. Acad. Sci. *184:*232–234.

——— (1928) Les muscles spinaux chez l'homme et les anthropoides. Ann. Sci. Nat. Zool. *10:*1–65.

Van Schaik JPJ, Verbiest H, and Van Schaik FDJ (1985) The orientation of laminae and facet joints in the lower lumbar spine. Spine *10*(1):59–63.

Vink P, and Karssemeijer N (1988) Low back muscle activity and pelvic rotation during walking. Anat. Embryol. *178:*455–460.

Walker A (1969) The locomotion of the lorises, with special reference to the potto. E. Afr. Wildl. J. *7:*1–5.

——— (1970) Nuchal adaptations in *Perodicticus potto.* Primates *11:*135–144.

——— (1979) Prosimian locomotor behavior. In GA Doyle and RD Martin (eds.): The Study of Prosimian Behavior. New York: Academic Press, pp. 543–565.

Ward CV (1990) The lumbar region of the Miocene hominoid *Proconsul nyanzae.* Am. J. Phys. Anthrop. *81*(2):314.

——— (1991) Functional anatomy of the lower back and pelvis of the Miocene hominoid *Proconsul nyanzae* from Mfangano Island, Kenya. Ph.D. dissertation, Johns Hopkins University, Baltimore.

Ward CV, and Latimer B (1991) The vertebral column of *Australopithecus.* Am. J. Phys. Anthrop. Suppl. *12:*180.

Washburn SL (1963) Behavior and human evolution. In SL Washburn (ed.): Classification and Human Evolution. Chicago: Aldine, pp. 190–203.

Washburn SL, and Buettner-Janusch J (1952) The definition of thoracic and lumbar vertebrae. Am. J. Phys. Anthrop. *10:*251.

Waterman H (1929) Studies on the evolution of the pelves of man and other primates. Bull. Am. Mus. Nat. Hist. *58:*585–642.

Waters RL, and Morris JW (1972) Electrical activity of muscles of the trunk during walking. J. Anat. *111*(2):191–199.

White A, and Hirsch C (1971) The significance of the vertebral posterior elements in the mechanics of the thoracic spine. Clin. Orth. Rel. Res. *81:*2–14.

Winckler, G (1936) Le muscle ilio-costal. Arch. Anat. Hist. Embr. *21:*143–251.

Yang KH, and King AI (1984) Mechanism of facet load transmission as a hypothesis for low-back pain. Spine 9(6):557–565.

Yunliang G, Biaoming H, Yunxi T, and Guangting L (1990) The shape and curvature of the lumbar zygapophyseal joints and their mechanical analysis. Acta Anthrop. Sinica 9:254–259.

C H A P T E R   S I X

The Functional Anatomy of the Hip and Thigh in Primates

Robert L. Anemone

Due to its important role in the positional behavior[1] of primates, the hindlimb has been the subject of much attention in recent years (see Sigmon and Farslow, 1986, for a recent review). My own interests have mainly concerned questions of form and function of the hindlimb among living prosimian primates (Anemone, 1988, 1990), in particular among those forms that practice a specialized form of saltatory locomotion known as vertical clinging and leaping (Napier, 1967; Napier and Walker, 1967a, 1967b; Walker, 1967, 1974, 1979). Detailed comparative studies of the functional anatomy and positional behavior of living taxa are essential to the successful reconstruction of the behavior of extinct forms, which are often represented solely by fragmentary fossil remains (Bock and Von Wahlert, 1965; Fleagle and Simons, 1983; Kay and Covert, 1984). In this chapter I discuss the hip and thigh among living primates and tree shrews in an attempt to elucidate form-function relationships involving musculoskeletal anatomy and positional behavior. After reviewing the basic comparative anatomy of the hip and thigh, discussing some principles of biomechanical analysis, and classifying the living prosimian primates on the basis of locomotor behavior, I present some results of my work on the functional anatomy of the hindlimb among extant prosimians and tupaiids. In this way, I hope to demonstrate one possible approach to the general problem of understanding the relationship of primate form and function, and to lay further groundwork for the reconstruction of the positional behavior of extinct primates on the basis of their fossilized bony remains. The tree shrews (Family Tupaiidae) are included in this review in spite of their nonprimate status (see Luckett, 1980), because they may closely resemble the ancestral primate morphotype with respect to postcranial morphology and positional behavior (e.g., Jenkins, 1974).

OSTEOLOGY

The primate bony pelvis is formed by the sacrum and the left and right os coxae or innominate bones, and plays several important roles in reproduction (e.g., forming the birth canal) and locomotion (e.g., providing origin to many important hindlimb muscles). Each os coxae is composed of three separate bones, the ilium, the ischium, and the pubis, which fuse in the acetabulum in the adult (Figs. 6.1 and 6.2). There are three important joints in the pelvis. The sacro-iliac joint is a slightly movable synovial joint between left and right ilia and a variable number of sacral vertebrae. The left and right pubes are joined in the midline at the pubic symphysis, a fibrocartilaginous joint which may fuse in older nonhuman primates. The acetabulum and the femoral head form the hip joint proper, a deep ball-and-socket joint that allows a considerable degree of mobility in a very stable synovial joint.

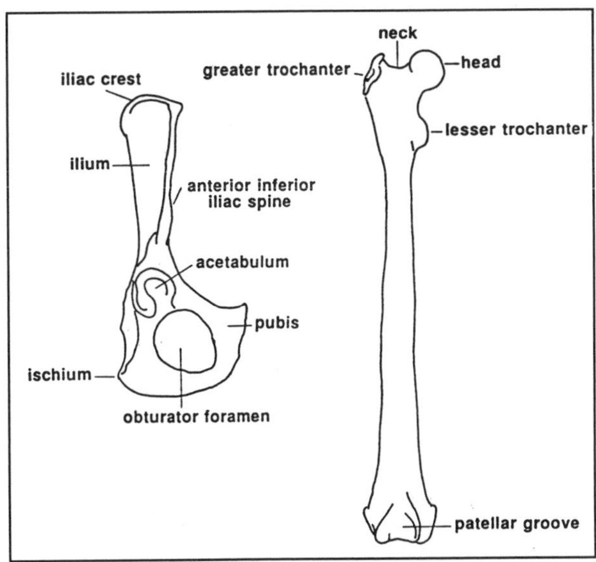

FIGURE 6.1. Osteological points mentioned in the text labeled on the right femur and innominate of *Saguinus labiatus*.

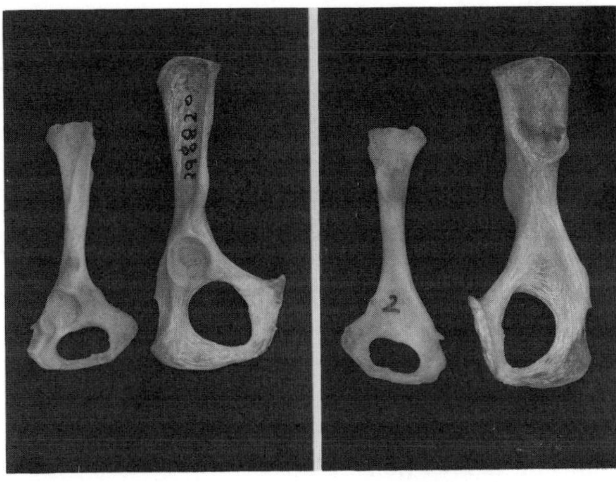

FIGURE 6.2. Ventral and dorsal views of right innominate of *Galago crassicaudatus* (left) and *Saguinus labiatus* (right).

The ilium assumes a dorsal position in the primate os coxae and has three distinct surfaces: The gluteal surface faces laterally and dorsally and provides origin to the gluteal musculature; the iliac surface faces ventromedially and provides origin to the iliacus muscle; and the medially facing sacral surface forms the articulation for the sacro-iliac joint. Most nonhuman primates possess long and narrow iliac blades with some degree of expansion of the cranial edge, known as the iliac crest. A raised part of the ventral edge of the ilium cranial to the acetabulum is the anterior inferior iliac spine, and provides origin to the rectus femoris muscle, part of the important knee extensor group known as the quadriceps femoris. The ischium is positioned posteriorly, and is composed of a relatively short and stout, caudally directed bar of bone ending in the ischial tuberosity, as well as a variable contribution to the thin, ventrodorsally running ischio-pubic ramus. The ischial tuberosity provides origin to the hamstring muscles, powerful extensors of the hindlimb. The pubis is a thin, ventrally positioned bone composed of a superior pubic ramus, a pubic symphysis, and a contribution to the ischio-pubic ramus. The adductor muscles arise from much of the ischio-pubic ramus, as well as from the symphyseal part of the pubis. The pubis and the ischium enclose the obturator foramen which, during life, is covered by the obturator membrane. The acetabulum has a horseshoe-shaped or lunate articular surface for the femoral head. The nonarticular, ventral opening of the lunate surface is known as the acetabular notch, and the (also nonarticular) center of the lunate surface is called the acetabular fossa. The ligamentum teres or round ligament of the head of the femur runs from the acetabular fossa to the fovea capitis femoris of the femoral head.

Among primates, the femur is the largest of the long bones (Figs. 6.1 and 6.3). Important landmarks on its proximal end include the femoral head, neck, and trochanters. The head is spherical to cylindrical in shape and, forming part of the hip joint, bears the articular surface for the acetabulum. The head faces medially and is connected to the femoral shaft by the neck. On its medial end is a small pit, the fovea capitis femoris, for attachment of the ligamentum teres. The trochanters are bony ridges and rugosities resulting from the pull of certain muscles attaching to the proximal femur. The greater trochanter bears the insertion of the gluteus medius and gluteus minimus muscles and the origin of the vastus lateralis. On the distal aspect of the greater trochanter lies the trochanteric fossa, a small recess for attachment of the lateral rotators of the femur (the gemelli and obturators). Also on the lateral aspect of the femur, but distal to the greater trochanter, lies the third trochanter. Both the gluteus

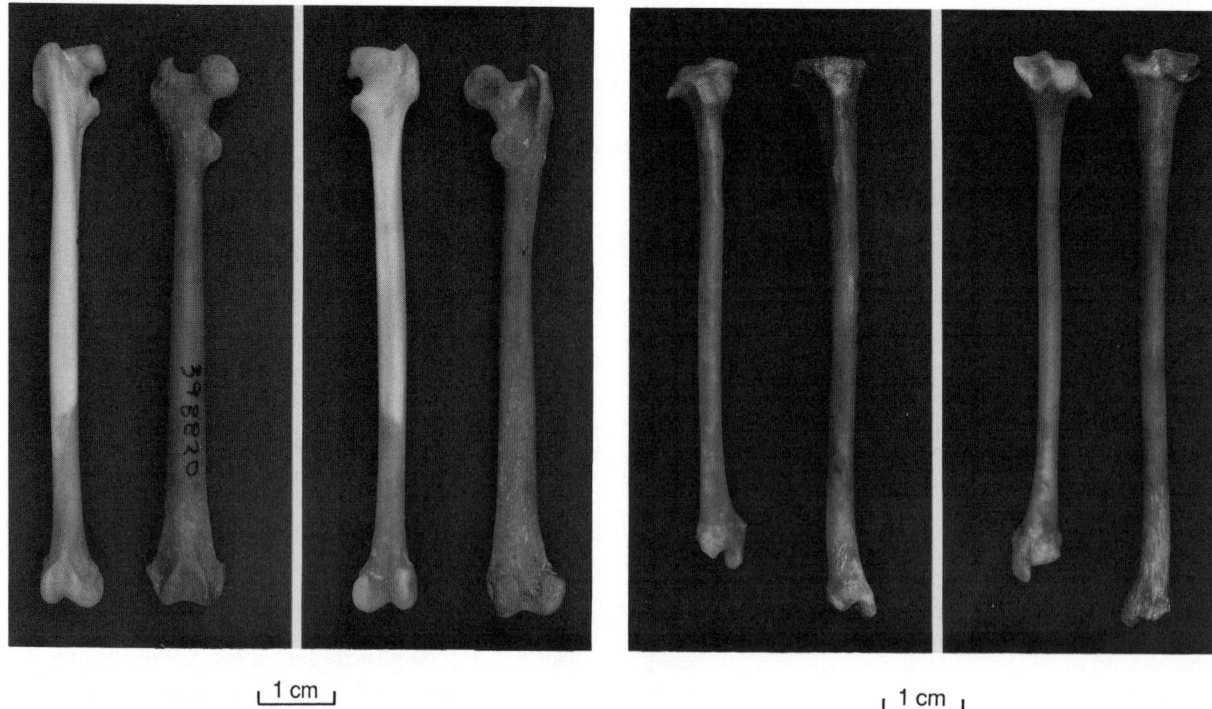

FIGURE 6.3. Ventral and dorsal views of right femur of *Galago crassicaudatus* (left) and *Saguinus labiatus* (right).

FIGURE 6.4. Ventral and dorsal views of right tibia of *Galago crassicaudatus* (left) and *Saguinus labiatus* (right).

superficialis (pars anterior) and tensor fasciae femoris insert onto the third trochanter. The lesser trochanter is on the medial aspect of the proximal femur and bears the insertion of the iliopsoas. Although the greater and lesser trochanters are found among all primates, the third trochanter is absent from some lorisids and all anthropoid primates.[2] Although it provides surface area for muscular attachments along much of its length, the femoral shaft has little in the way of distinctive morphology. The distal end of the femur, however, contains the articular areas for both the tibia and the patella. The patellar surface is a midline groove on the ventral aspect of the distal femur on which the patella slides during flexion and extension at the knee. The width, length, and depth of this groove vary, but it is always bounded by raised lateral and medial ridges. Distally and posteriorly are the femoral condyles, which articulate with the tibial condyles in the knee joint. On the distal aspect, the femoral condyles are separated by the intercondylar fossa, where the cruciate ligaments of the knee attach. The collateral ligaments of the knee attach on the lateral and medial epicondyles, immediately adjacent to the condyles.

The medial and lateral bones of the leg are, respectively, the tibia and the fibula (Fig. 6.4). The distal part of the knee joint is formed by the tibial condyles, the flat superior-facing articular surfaces, which articulate with the femoral condyles. The medial and lateral tibial condyles are separated by a raised, nonarticular area called the intercondylar eminence. Articular fibrocartilages separate the femoral and tibial condyles, and the bones are connected by two sets of internal (cruciate) and external (collateral) ligaments. The anterior aspect of the tibial shaft is medio-laterally flattened, and its roughened proximal surface bears the marks of the insertion of the patellar ligament of the quadriceps femoris muscle. The lateral surface of the tibia articulates with the fibula superiorly and inferiorly among all primates except for the tarsiers, which have a distally fused tibia-fibula.

MYOLOGY

Published descriptions of prosimian hip and thigh musculature used in this review include Woolard (1925), Jouffroy (1962, 1975), Grand and Lorenz (1968), George (1977), McArdle (1981), and Stevens et al. (1981). In addition, my own dissection notes on Galagidae, Tupaiidae, and Tarsiidae were consulted. The major published works on higher primate hip and thigh musculature include Hartman and Straus (1933), Gregory (1950), Schön (1968), Uhlmann (1968), Stern (1971), and Swindler and Wood (1973). The major muscles of the primate hip and thigh are depicted in lateral and medial views in Figures 6.5 and 6.6, respectively.

Hip Flexors

As indicated by its name, the iliopsoas is a compound muscle composed of the iliacus and psoas major muscles. These important hip flexors vary little among primates. The former arises on the medial surface of the blade of the ilium, while the latter takes its origin from the intervertebral discs, vertebral bodies, and transverse processes of the lumbar vertebrae. The two muscles unite and insert by a common tendon into the lesser trochanter of the femur. The psoas minor is a rather insignificant muscle that runs on the ventral surface of psoas major. It arises from intervertebral discs and vertebral bodies in the lumbar region and inserts by a small tendon into the ilio-pectineal eminence on the pubis.

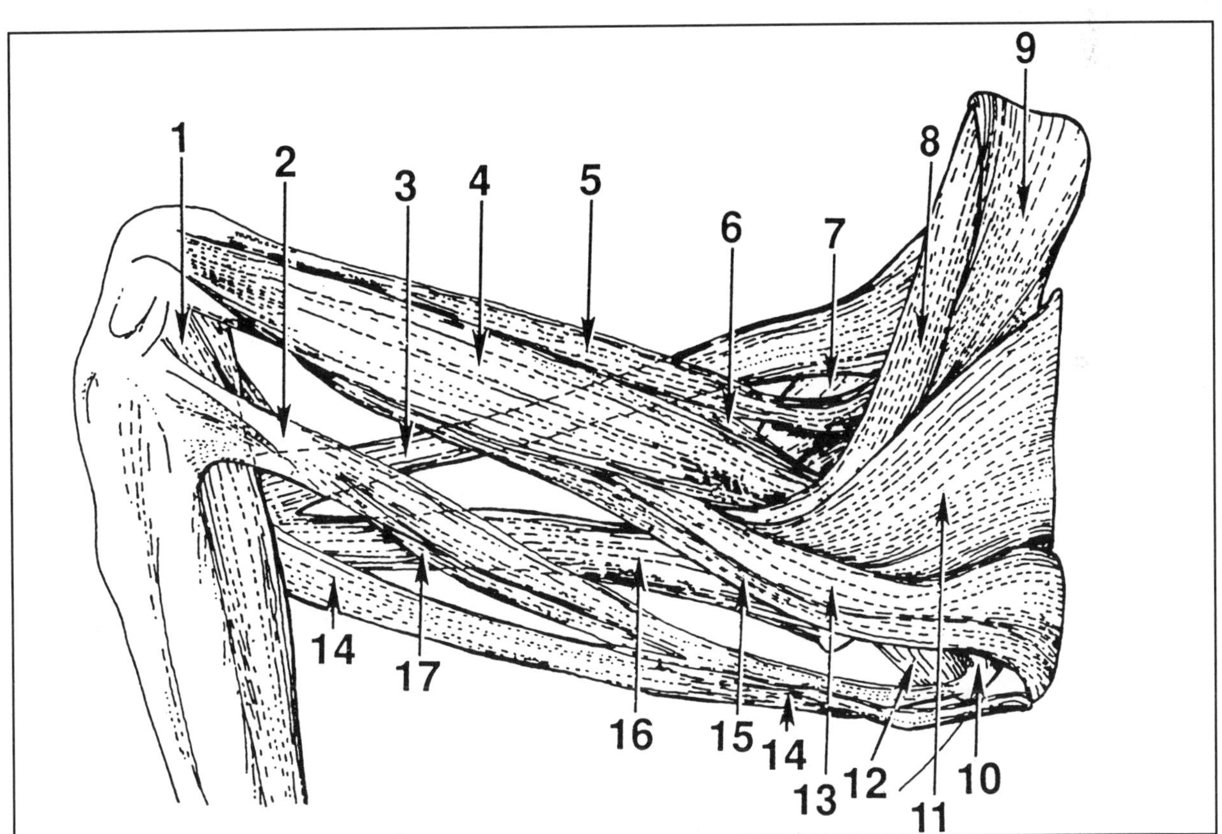

FIGURE 6.5. Musculature of the left hip and thigh of *Varecia variegata,* lateral aspect: 1 = gastrocnemius; 2 = flexor cruris lateralis (biceps femoris); 3 = sartorius; 4 = vastus lateralis; 5 = rectus femoris; 6 = vastus medialis; 7 = iliopsoas; 8 = tensor fasciae femoris; 9 = gluteus medius; 10 = gemellus inferior; 11 = gluteus superficialis pars anterior; 12 = quadratus femoris; 13 = gluteus superficialis pars posterior (femorococcygeus); 14 = semitendinosus; 15 = caudofemoralis; 16 = gracilis; and 17 = semimembranosus. Modified from Jouffroy (1975). By permission of Plenum Press.

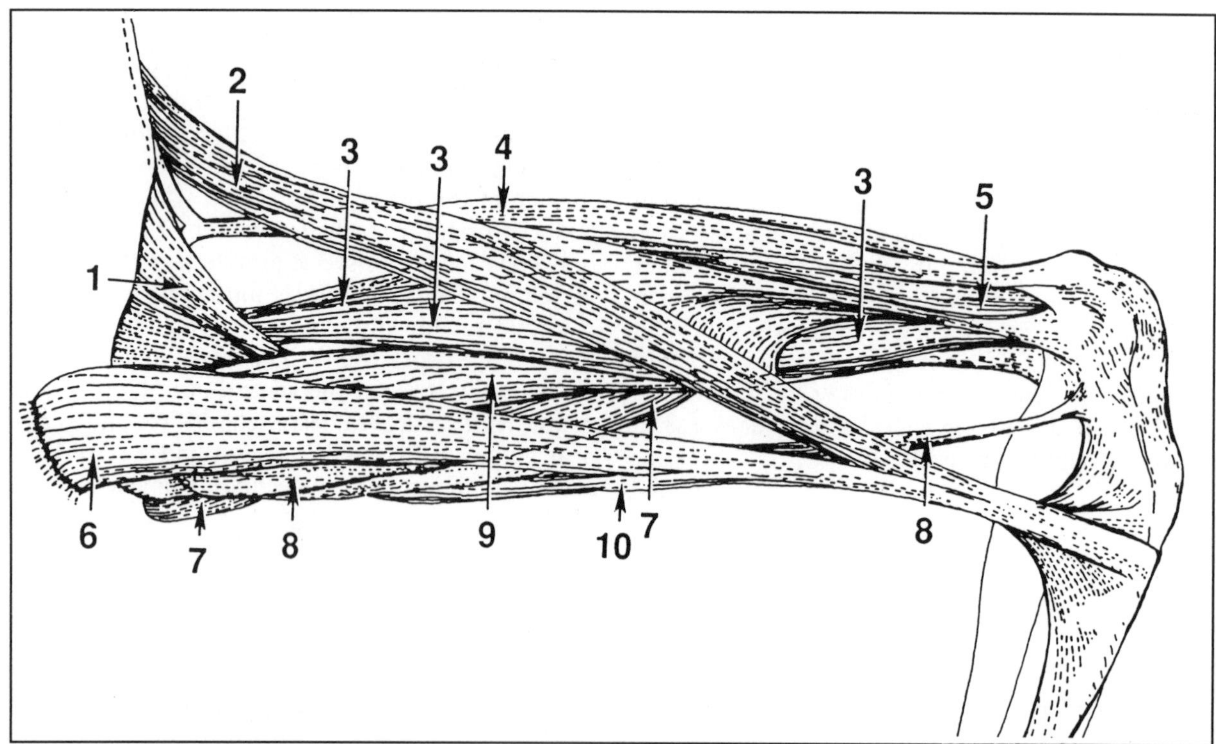

FIGURE 6.6. Musculature of the left hip and thigh of *Propithecus verreauxi,* medial aspect: 1 = iliopsoas; 2 = sartorius; 3 = vastus medialis; 4 = rectus femoris; 5 = vastus lateralis; 6 = gracilis; 7 = femorococcygeus; 8 = semimembranosus; 9 = adductor magnus; and 10 = semitendinosus. Modified from Jouffroy (1975). By permission of Plenum Press.

Gluteal Muscles

Gluteus superficialis—known as gluteus maximus in human anatomy—has two variably separated parts among nonhuman primates, the pars anterior and pars posterior. The pars posterior lies dorsal to the sciatic nerve, and is the femorococcygeus of some authors. The pars anterior arises from the gluteal fascia overlying gluteus medius and/or the caudal fascia overlying abductor caudae medialis. The pars anterior inserts via its "ascending tendon" (Stern, 1971) into the third trochanter. This insertion is in common with that of the tensor fasciae femoris. The pars posterior of the gluteus superficialis arises from the transverse processes of the first two or three caudal vertebrae and inserts for a variable distance on a line distal to the third trochanter. The gluteus superficialis serves both nonhuman and human primates as an extensor of the thigh. Stern (1972) cogently argues that the anatomical specializations that make the gluteus superficialis "maximus" in humans are confined to the cranial portion of the muscle, and serve to increase lateral stability during bipedalism (i.e., abduction). But see Robinson et al. (1972) and Sigmon (1974) for an opposing viewpoint.

Situated along the anterior border of the gluteus superficialis (from which it is often difficult to separate), the tensor fasciae femoris arises by muscular fibers from part or all of the iliac crest cranial to the origin of gluteus medius, and also from the gluteal fascia overlying gluteus medius. It inserts with the ascending tendon from gluteus superficialis pars anterior into the third trochanter. This muscle is called tensor fasciae latae by human anatomists because it inserts into the fascia lata in humans. Its functions are flexion and medial rotation of the leg.

There is little variation in the gluteus medius among the prosimians. It arises from much of the gluteal aspect of the ilium (cranial to the gluteal line if present) and inserts into the greater trochanter of the femur. It is the largest and most powerful of the

gluteal muscles among nonhuman primates. Its functions are medial rotation and especially extension of the thigh. The gluteus minimus is part of the deepest layer of gluteal musculature and, like gluteus superficialis, is often divided into two separate parts. The pars anterior arises from the inferior ventral edge of the ilium, slightly cranial to the acetabulum. This is the "scansorius" muscle of some authors (see Sigmon, 1969, for a review). The pars posterior arises from the dorsal edge of the ilium, both dorsal and cranial to the acetabulum. In all primates there is a single insertion into the greater trochanter. Its function is mainly as a medial rotator of the thigh. Stern (1971) states that gluteus medius and minimus act jointly as a "regulator femoris," assuring the integrity of the hip joint during a wide variety of movements and postures.

Femoral Rotators

The obturator externus and internus muscles are lateral rotators of the femur that arise from the external and internal surfaces, respectively, of the ischial and pubic rami around the obturator foramen and from the obturator membrane. They insert with the gemelli into the trochanteric fossa. Gemellus superior and inferior are also lateral rotators of the hip joint. They both arise from the ischium, the former from the ischial spine and the latter from the ischial tuberosity. They join obturator internus as it crosses the lesser sciatic notch and insert with this muscle into the trochanteric fossa. The piriformis is a small muscle that runs from sacral vertebrae and the dorsal edge of the ilium to the greater trochanter.

Extensors of the Leg

The sartorius arises from the ventral border of the blade of the ilium, usually in the vicinity of the anterior superior iliac spine. The insertion is into the anterior tibial crest, often in common with the tendons of gracilis and semitendinosus. This common insertion is the pes anserinus of human anatomy. The major functions of the sartorius include flexion of the thigh and extension of the leg.

The quadriceps femoris muscle group is composed of the rectus femoris and the three vasti muscles, vastus lateralis, vastus medialis, and vastus intermedius. Rectus femoris is the only part of the quadriceps femoris to cross both the hip and the knee joints. Its double origin is similar in all primates. The straight or iliac head arises from the anterior inferior iliac spine, and the reflected or acetabular head arises from the ilium just cranial to the acetabulum. All four parts of the quadriceps fuse above the knee to form the quadriceps tendon, which inserts into the patella. Distally, the patellar tendon runs from the patella to insert into the tibial tuberosity. This four-headed muscle is the major extensor of the leg at the knee joint in primates. Rectus femoris has the additional function of flexing the femur at the hip. The deepest and usually the smallest member of the quadriceps group is the vastus intermedius. This muscle arises by short muscle fibers that insert into a tendon running longitudinally along its superficial aspect. It arises for a variable distance from the ventral aspect of the femoral shaft. Both the vastus medialis and the vastus lateralis are large muscles arising, respectively, from the ventral aspect of the neck of the femur and the anterior crest of the greater trochanter.

Adductors

A relatively uniform muscle among primates, gracilis arises mainly from the symphyseal ramus of the pubis, as well as from a variable part of the adjacent horizontal and vertical pubic rami. It inserts into the anterior tibial crest, often in common with sartorius and semitendinosus. It is mainly an adductor of the thigh. Other functions include flexion and medial rotation of the leg. Another very uniform muscle among primates, pectineus typically arises from the superior pubic ramus and inserts to a variable extent on the medial and dorsal aspect of the femur, distal to the lesser trochanter. The pectineus also adducts and flexes the thigh.

The adductor group proper includes adductor magnus, adductor longus, and adductor brevis. They each arise from the pubis and part of the ischium and insert along the dorsal surface of the femur to a variable extent. The entire group is relatively uniform among primates, serving mainly in adduction but also in flexion and sometimes in extension of the thigh.

Hip Extensors

The quadratus femoris arises from the body of the ischium and from the ischio-pubic ramus. Its insertion extends a variable distance distally on the posterior aspect of the femur from the level of the third trochanter. It is an extensor of the thigh.

The group of muscles known as the hamstrings is composed of semitendinosus, semimembranosus, and flexor cruris lateralis (the biceps femoris of human anatomy). These muscles arise together from the ischial tuberosity and insert on the posteromedial aspect of the proximal tibia (semitendinosus and semimembranosus) and the lateral tibial condyle and/or head of the fibula (flexor cruris lateralis). There is a good deal of intertaxonal variation among primates with respect to details of origins, insertions, and internal divisions of these muscles. Among many prosimians, for example, semitendinosus has two heads of origin, a caudal head arising from the transverse processes of the first few caudal vertebrae, and an ischial head arising from the ischial tuberosity. Similarly, flexor cruris lateralis has two heads of origin (long head arising from the ischial tuberosity and short head arising from the lateral lip of the linea aspera) among all hominoids and some ceboids. Similar variation exists in the distal extent and degree of fusion or separation of the tendinous insertion of the hamstrings. The functions of the hamstrings include flexion of the leg and extension of the thigh.

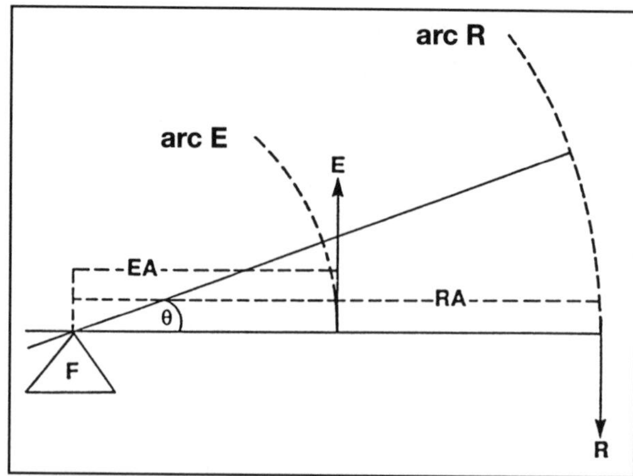

FIGURE 6.7. Biomechanics of a simple lever of the third class. *F* = fulcrum; *EA* = effort arm; *E* = effort; *RA* = resistance arm; *R* = resistance; θ = angle through which the lever moves; arc *E* = the path along which the point of application of the effort force moves; arc *R* = the path along which the point of application of the resistance force moves. Arrows denote vector quantities. Modified from Gowitzke and Milner (1980). © 1980 by Williams & Wilksons Co., Baltimore.

BIOMECHANICS

A brief consideration of the basics of biomechanics—the application of Newtonian physics to problems involving animal and human movement—is a necessary prerequisite to understanding the functional anatomy of the primate hip and thigh. Much of the following discussion comes from Gowitzke and Milner (1980). A lever is a device, usually a rigid bar, that can transmit force and efficiently do work when work is done upon it (Fig. 6.7). A lever rotates around a fixed point (fulcrum) in response to an applied force. Forces that act upon a lever to overcome a resistance can be termed effort forces *(E)*, and the moment arm of the effort force, or more simply, the effort arm *(EA)*, is the perpendicular distance between the line of action of the effort force and the axis of rotation or fulcrum. The load or the resistance *(R)* is a force that acts to rotate the lever in the opposite direction of the effort force. Similarly, the perpendicular distance from the point of application of the load to the fulcrum is known as the moment arm of the resistance, or, the resistance arm *(RA)*. The moment of force *(M)* resulting from a force acting at some distance from a fulcrum is the product of the force and the perpendicular distance between the force and the fulcrum. Thus, the moment of an effort force *(Me)* is the product of that force *(E)* and the effort arm *(EA)*. Similarly, the moment of a resistance force *(Mr)* is the product of that force *(R)* and the resistance arm *(RA)*.

$$Me = E \times EA \tag{1}$$

$$Mr = R \times RA \tag{2}$$

A joint system is at equilibrium when the effort moment equals the resistance moment, or $Me = Mr$, and therefore $E \times EA = R \times RA$. This can be simplified to

$$R/E = EA/RA \tag{3}$$

The ratio of the effort arm to the resistance arm is defined as the mechanical advantage *(MA)* of the lever. When the mechanical advantage of a lever system is large, little effort is required to balance a large resistance, and conversely, when mechanical advantage is small, a large effort is required to attain equilibrium. Furthermore, the angular displacement *(S)* of a lever can be calculated as the product of the angle (ϕ) through which the lever moves, and the perpendicular distance to the fulcrum. Thus,

$$Sr = RA\phi \tag{4}$$

and

$$Se = EA\phi \tag{5}$$

The ratio of the distance traveled by the two ends of a bony lever is then $Sr/Se = RA\phi/EA\phi$, which simplifies to

$$Sr/Se = RA/EA = 1/MA \tag{6}$$

The significance of this result is that it explains the inverse relationship between mechanical advantage and velocity of movement in anatomical systems. High mechanical advantages are advantageous because they allow relatively large resistance forces to be overcome with relatively small muscular forces. The angular displacement or distance through which the distal end of the lever moves is, however, correspondingly small, and the velocity low. Lever systems with low mechanical advantage have the disadvantage of requiring large muscular forces to overcome resistance, but they have the advantage of allowing a great distal displacement and a resulting great velocity of movement, for a given muscular contraction.

The speed-power dichotomy can be traced back at least to Gregory (1912), and has been accepted by many other authors, including Elftman (1929), Waterman (1929), Howell (1932, 1944), Smith and Savage (1956), Walker (1967), Stern (1974), and McArdle (1981). The general theory accepted by all of these authors is that short muscular moment arms are optimally designed to provide rapid movements, while long moment arms sacrifice speed for powerful contractions. Using the illustrative example of the teres major muscle, an important retractor (extensor) of the mammalian forelimb, Smith and Savage (1956) nicely illustrated the osteological and myological differences between the cursorial horse *(Equus)* and the fossorial nine-banded armadillo *(Dasypus)*. In the armadillo, both the inferior angle of the scapula, which provides attachment to teres major, and the attachment on the humerus for teres major, are distally prolonged, resulting in a very large effort arm for this muscle (Fig. 6.8). In the horse, the scapula is narrow and bladelike and the humeral attachment for teres major is proximally placed, resulting in a short effort arm for teres major. The relatively longer and more vertically oriented forelimb of the horse increases the distance of the glenoid fossa (fulcrum) from the point of application of force (resistance) at the ground. The overall result is that the cursorial horse has a much smaller mechanical advantage in humeral retraction than the fossorial armadillo, resulting in a speed-adapted limb in the former and a power-adapted limb in the latter mammal. In a series of papers, Hall-Craggs (1964, 1965a, 1965b, 1966a, 1966b, 1974) argued that tarsal elongation among the bushbabies (Galagidae) is an adaptation for the rapid plantarflexion needed to attain the acceleration required by the rapid leaping locomotion of these animals. Tarsal elongation among these animals is restricted to the calcaneus and navicular bones; occurs entirely distal to the fulcrum of plantarflexion, the tibio-talar joint, and as a

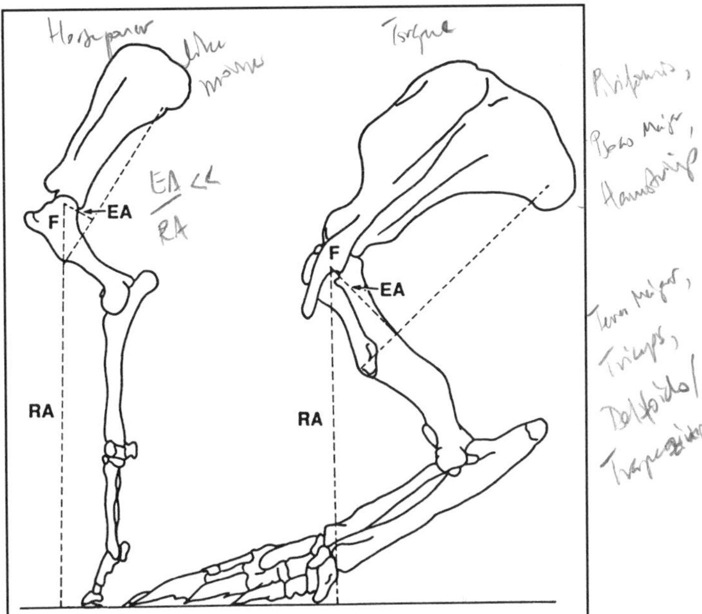

FIGURE 6.8. Musculoskeletal adaptations for speed and power in the mammalian forelimb. **Left:** forelimb of *Equus*, the modern horse; **Right:** forelimb of *Dasypus*, the nine-banded armadillo. Abbreviations defined in Figure 6.7. Modified from Smith and Savage (1956). By courtesy of the Linnean Society of London.

result greatly increases the length of the resistance arm. Hall-Craggs realized that the resultant low mechanical advantage in plantarflexion among bushbabies necessitates enormous muscular forces, but that these forces were generated by the multipinnate and hypertrophied triceps surae muscles (soleus and especially gastrocnemius), and allow the rapid acceleration required in bipedal saltation.

Stern (1974) demonstrated that the speed-power dichotomy is more complex than recognized by most previous authors. Incorporating insights from research in muscle physiology, Stern derived a mathematical model to simulate the effects of changed sites of muscular attachment on the dynamic characteristics of bone-muscle systems. Comparing physiologically identical muscles (same length of contractile tissues, cross-sectional areas, rate of contraction, muscle fiber angle), Stern found that muscles that attach close to joints (and hence have short effort arms) allow greater velocity and less power than muscles that attach at a greater distance from the joint. However, these results were qualified in several important ways. First, Stern suggested that there is an optimum moment arm length for physiologically identical muscles such that either increasing or decreasing it will result in lessened dynamic performance. Therefore, although it is true that shorter moment arms in general yield faster movements, it is not true that reduction of muscular moment arms will always produce greater velocity. Second, Stern (1974:422) found that "a muscle ideally suited to produce high velocity of movement at a certain position of the limb is not best designed to bring the limb to this position most quickly." These results indicate that larger moment arms will bring the limb to a given position in the shortest time, due to the greater initial acceleration attained, while smaller moment arms produce greater velocities at a given point in the limb movement. Third, Stern concluded "that these muscles which are ideally suited for producing high velocity at some determined position retain their ability to supply further torque at this point whereas those muscles designed to bring the limb to the same position most quickly are forced to contract so rapidly when the position is reached that they can no longer exert force" (Stern, 1974:422).

Another relevant biomechanical concept for the analysis of the functional anatomy of the hindlimb is moment of inertia *(I)*. The moment of inertia is, for rotational motion, the analog of mass in linear motion. As such, it is a measure of the resistance of a body at rest to change in angular motion, and is proportional both to the mass of the body and to the distribution of mass about the axis of angular movement (Gowitzke and Milner, 1980). Using the concept of the radius of gyration *(K)*, which is defined as "the distance from the axis of rotation of a point at which the total mass *(M)* of a body might be concentrated without changing the moment of inertia of the body" (Gowitzke and Milner, 1980:78), moment of inertia can be calculated as

$$I = MK\,2 \qquad (7)$$

Thus, the moment of inertia of a limb increases with increasing distribution of its mass at a distance from the center of rotation. With the muscular mass proximally restricted and the moment of inertia minimized, the limb can be rotated at a greater rate by a given muscular force. This is the same principle that allows an ice skater to rotate rapidly or slowly depending upon the distance the arms are held from the body—closely for speed (decreasing *K* and *I;* increasing angular velocity) or further from the body to slow down (increasing *K* and *I:* decreasing angular velocity)—and it is clearly of value to cursors (Howell, 1944) as well as to saltatorial animals whose hindlimbs need to attain relatively high velocities during takeoff (Hall-Craggs, 1965a, 1965b). In his analysis of the functional anatomy of *Megaladapis,* an extinct Malagasy subfossil lemur, Jungers (1976:513) states: "The greater Io (moment of inertia) becomes, the greater the force needed to produce a given angular velocity. Conversely, the rate at which a limb may be rotated, or the cadence, is dependent not only upon the total mass of the limb, but upon the distribution of this mass as well."

LEAPING AMONG PROSIMIAN PRIMATES: A CASE STUDY

It is a well-known fact that some of the most proficient leapers among the primates are those prosimians labeled by Napier and Walker (1967a) as "vertical clingers and leapers" (VCLs). This group includes several of the smaller African bushbabies (Galagidae), the Indriidae and Lepilemuridae of Madagascar, and the Asian tarsiers (Tarsiidae). VCLs typically engage in a mode of arboreal loco-

TABLE 6.1
Classification of Prosimian Positional Behavior

ACTIVE QUADRUPEDALISM

These animals are primarily rapid, arboreal quadrupedal runners and climbers, although some are fairly proficient leapers. They are all capable of some vertical climbing.

Lemuridae	*Lemur, Varecia*
Cheirogaleidae	*Cheirogaleus, Microcebus, Phaner*
Daubentoniidae	*Daubentonia*
Galagidae	*Galago crassicaudatus*

SLOW-CLIMBING QUADRUPEDALISM

These animals are all slow and cautious arboreal climbers that rely on bridging behaviors rather than leaping in order to cross gaps in the forest canopy. They use suspensory movements more than any other prosimians.

Lorisidae	*Loris, Nycticebus, Arctocebus, Perodicticus*

VERTICAL CLINGING AND LEAPING

These rapid, saltatorial animals usually rely on leaping to cross gaps in the forest, may preferentially use vertical supports during postural and locomotor behavior, and are often bipedal hoppers on the ground. Extensive variation exists within this group in the amount of leaping and quadrupedalism utilized, and in the size and orientation of preferred supports.

Galagidae	*Galago* (except *G. crassicaudatus*)
Tarsiidae	*Tarsius*
Indriidae	*Indri, Avahi, Propithecus*
Lepilemuridae	*Lepilemur, Hapalemur*

motion in which the hindlimbs act together to provide the substantial propulsive force necessary for the exceptional leaping abilities of these animals. Using high-speed cinematography, Hall-Craggs (1964, 1965a) demonstrated that *Galago senegalensis* is easily able to leap from the ground to a vertical height of greater than 7 feet, a distance equal to 14 times its body length (excluding its tail). The magnitude of this jump is better appreciated when one realizes that at the start of the jump the animal's center of gravity is approximately one and a half inches off the ground (Hall-Craggs, 1965a). Although the Madagascan and Asian VCL taxa have not yet been subjected to this kind of detailed locomotor analysis, there is little doubt that they are equally proficient saltators in the field (Jolly, 1966; Bearder and Doyle, 1974; Pollock, 1975, 1977; Charles-Dominique, 1977; Crompton, 1980, 1984; Niemitz, 1984). Other prosimians can be classified as either active quadrupeds or slow-climbing quadrupeds. See Table 6.1 for one classification of prosimian positional behaviors.

Although Napier and Walker (1967a) stressed the primary importance of the vertical substrate, both as takeoff and landing platforms during leaps and as preferred postural supports, different VCL taxa have diverse preferences for vertical, horizontal, and oblique supports (Oxnard et al., 1990). These preferences are often heavily influenced by the relative abundance of supports of differing orientation and diameter in particular habitats (Jolly, 1966; Sussman, 1974; Tattersall, 1977, 1982; Richard, 1978). Rather than being a unique postural adaptation of VCL prosimians, Gebo (1987:279) states that "all prosimians use the posture of vertical clinging, which in reality is nothing more than a stopping position in a vertical climbing sequence." Interspecific differences in details of leaping behavior are also evident among VCL taxa (Stern and Oxnard, 1973), with some taxa (e.g., tarsiers and bushbabies) using "curled-up positions," while others (e.g., indriids) use "stretched-out positions" during the flight phase of the leap (Oxnard et al., 1990:19). In their discussion of the pedal grasping mechanism among primates, Szalay and Dagosto (1988:27) suggest that "rapid, successive, leaping and landing with a habitual grasp (i.e., grasp leaping)" may be a critical euprimate adaptation, which is still of great importance to many living saltatory prosimians.

In spite of the many problems with the "VCL hypothesis" (recently reviewed in Anemone, 1990), an analysis of the hip and thigh among living prosimian primates that focuses upon leaping adaptations can provide striking insights into the nature of functional adaptation, as well as yield important baseline data for reconstructing the positional

TABLE 6.2
Relative Limb Proportions among Prosimian Primates and Tupaiids[1]

Family	IM Index[2]		Crural Index		Hindlimb L.		Femoral L.		Tibial L.	
Lorisidae[3]	88.1	(2.2)[4]	95.0	(2.5)	83.1	(9.1)	42.6	(4.5)	40.5	(4.6)
Galagidae	60.6	(7.7)	95.6	(6.6)	109	(9.7)	56.2	(5.7)	53.3	(4.7)
Tarsiidae	57.7	(1.5)	101	(1.8)	148	(5.9)	73.8	(2.7)	74.4	(3.3)
Indriidae	59.2	(3.3)	86.3	(1.7)	132	(6.4)	71.1	(3.4)	61.3	(3.1)
Lepilemuridae	60.7	(2.0)	95.4	(2.7)	101	(3.8)	51.9	(1.8)	49.5	(2.3)
Lemuridae	70.2	(1.0)	94.2	(1.4)	93.0	(4.0)	47.8	(1.8)	45.1	(1.8)
Cheirogaleidae	70.6	(1.4)	108	(6.1)	78.5	(6.1)	37.7	(2.3)	40.8	(4.0)
Tupaiidae	72.5	(1.3)	108	(2.8)	72.8	(2.9)	34.9	(1.3)	37.8	(1.8)
Species	**IM Index**		**Crural Index**		**Hindlimb L.**		**Femoral L.**		**Tibial L.**	
Galago alleni	63.4	(0.9)	94.8	(5.0)	116	(2.5)	59.6	(1.4)	56.4	(2.0)
G. senegalensis	52.5	(1.9)	92.6	(4.5)	119	(4.0)	61.8	(1.5)	57.1	(3.1)
G. crassicaudatus	68.3	(1.4)	94.0	(1.8)	100	(3.8)	51.7	(1.9)	48.6	(2.0)
G. demidovii	67.0	(1.5)	108	(4.2)	103	(5.9)	49.2	(2.5)	53.4	(3.7)

[1] All indices are defined in the text.
[2] Intermembral index.
[3] Total sample size = 296. See Anemone (1988) for sample sizes for individual taxa.
[4] Data include mean and standard deviation.

behavior of extinct primates.

One outstanding characteristic of the morphology of predominantly leaping prosimian primates is their "hindlimb dominance" (Martin, 1972). This refers to the fact that the hindlimbs of leaping prosimians are absolutely longer than their own forelimbs, and are relatively longer than the hindlimbs of quadrupedal and slow-climbing prosimians. Osteometric data in support of this assertion have been collected on hundreds of prosimian skeletons by Walker (1967), Jouffroy (1975; Jouffroy and Lessertisseur, 1979), Jungers (1979), McArdle (1981), and Anemone (1988, 1990). The presence of elongated limbs is a common trait found in many cursorial mammals, serving to increase the stride length and, other things being equal, the velocity of quadrupedal locomotion (Howell, 1944; Gambaryan, 1974). The functional significance of long hindlimbs for saltatory animals is readily explainable because the hindlimb provides the lever with which muscular forces are generated to enable the individual to reach takeoff velocity. "Since the force required for a particular leap is inversely proportional to the distance (and time) over which the force is applied, longer hindlimbs allow animals of similar size either to decrease the force required to leap a specific distance or to increase the distance leaped with a particular set of musculature" (McArdle,1981:114).

Table 6.2 includes the results of five different measures of relative hindlimb length among prosimian families and tupaiids. The intermembral index is a standard and widely used measure of forelimb length (humerus + radius) relative to hindlimb length (femur + tibia). Values greater than 100 indicate that an animal's forelimbs are longer than its hindlimbs, while hindlimb-dominant animals, with hindlimbs absolutely longer than forelimbs, have intermembral indices lower than 100. The data clearly show that specialized leapers have the lowest intermembral indices among the prosimians, ranging from 57.7 among Tarsiidae to 60.7 among Lepilemuridae, while all other prosimian families and tupaiids have intermembral indices at or above 70. It is interesting to note that, strictly speaking, there are no "forelimb dominant" prosimians: All prosimian taxa have intermembral indices substantially lower than 100. The suborder of higher primates or Anthropoidea, on the other hand, includes several genera with forelimbs and hindlimbs of approximately equal length (e.g., *Alouatta, Lagothrix,* and *Papio* all have intermembral index between 95 and 98), as well as many other genuinely forelimb dominant taxa (e.g., *Hylobates* and *Pongo* with intermembral indices of 129 and 144, respectively) (Napier and Napier, 1967).

Because by itself intermembral index cannot distinguish between taxa with long hindlimbs and those with short forelimbs, several other osteometric measures are necessary to distinguish between these two possibilities. Relative hindlimb length (Table 6.2) relates the length of the hindlimb (femur + tibia) to that of the skeletal trunk length (STL), a body size variable (Biegert and Maurer, 1972). These data make it clear that the low intermembral index among leapers is explainable on the basis of lengthening of the hindlimb, since leapers have the relatively longest hindlimbs among prosimians. Note, however, the large amount of interfamilial variation in relative hindlimb length among leapers, ranging from 148.2 in Tarsiidae to 101.3 in Lepilemuridae. We will return to this question of metric variation among leapers later. Schultz (1930, 1933) presents data on the range in relative hindlimb length among Anthropoidea that indicates that cercopithecoid monkeys as well as great and especially lesser apes all have relatively long hindlimbs.[3] The cercopithecoid sample (N = 10) averages 106.5, while the anthropoid apes include *Pongo* at 119.2 (N = 5), *Pan* at 128.0 (N = 7), *Gorilla* at 123.8 (N = 5), and *Hylobates* (including *Symphalangus)* at 138.4 (N = 15) (Schultz, 1933).

The next two ratios in Table 6.2, relative femoral length and relative tibial length, allow us to determine in which bone(s) of the hindlimb of prosimian leapers the observed elongation occurs. Examination of these ratios indicates that leapers have longer femora and tibiae than other prosimians, and that (with two exceptions) leapers have relatively slightly longer femora than tibiae. The exceptions are Tarsiidae, which have slightly longer tibiae (74.4) than femora (73.8), and Indriidae, which have substantially longer femora (71.1) than tibiae (61.3). Few comparable data on higher primates have been published. Schultz (1933) provides raw data for *Pan, Pongo,* and *Gorilla,* from which relative femoral and tibial lengths can be calculated, remembering that Schultz uses a different skeletal measure of trunk length than the one employed here. Relative femoral length is shortest in *Pongo* (58.7, N = 5), longest in *Pan* (64.3, N = 7), and intermediate in *Gorilla* (61.4, N = 4). Relative tibial length averages 48.8 in *Gorilla,* 50.4 in *Pongo,* and 53.0 in *Pan.*

Examination of the data in Table 6.2 yields some interesting conclusions concerning patterns of variation in these measures. Perhaps the most important point to note is the large degree of variation among predominantly leaping taxa in most of these measures of the hindlimb. For example, relative femoral length varies among leapers between 73.8 (Tarsiidae) and 51.9 (Lepilemuridae). Although there can be little doubt that these data clearly establish the hindlimb dominance of leaping prosimians, it is also clear that all leapers do not share hindlimb dominance in equal proportion. Consideration of these data for the four families of Madagascan primates (Fig. 6.9) indicates a spectrum of values running from the most (Indriidae and Lepilemuridae) to the least hindlimb dominant (Lemuridae and Cheirogaleidae). Although the endpoints of this spectrum (i.e., Indriidae and Cheirogaleidae) are always extremely divergent, adjacent taxa are quite similar, and sometimes show some overlap. A similar spectrum can be observed among the species in the genus *Galago* (Fig. 6.10), in which *G. alleni* and *G. senegalensis* are in all measures more hindlimb dominant than *G. demidovii* and *G. crassicaudatus.* Not surprisingly, this anatomical spectrum mirrors a behavioral spectrum within this genus in that the former two species are much more frequent and proficient leapers than the latter two (Charles-Dominique, 1977; McArdle, 1981; Oxnard et al., 1990).

Another useful approach to the relationship between morphology and positional behavior relies on the osteological evidence for muscular mechanics. Because bone is a living tissue and reacts to the stresses created by the pull of attached muscles, the

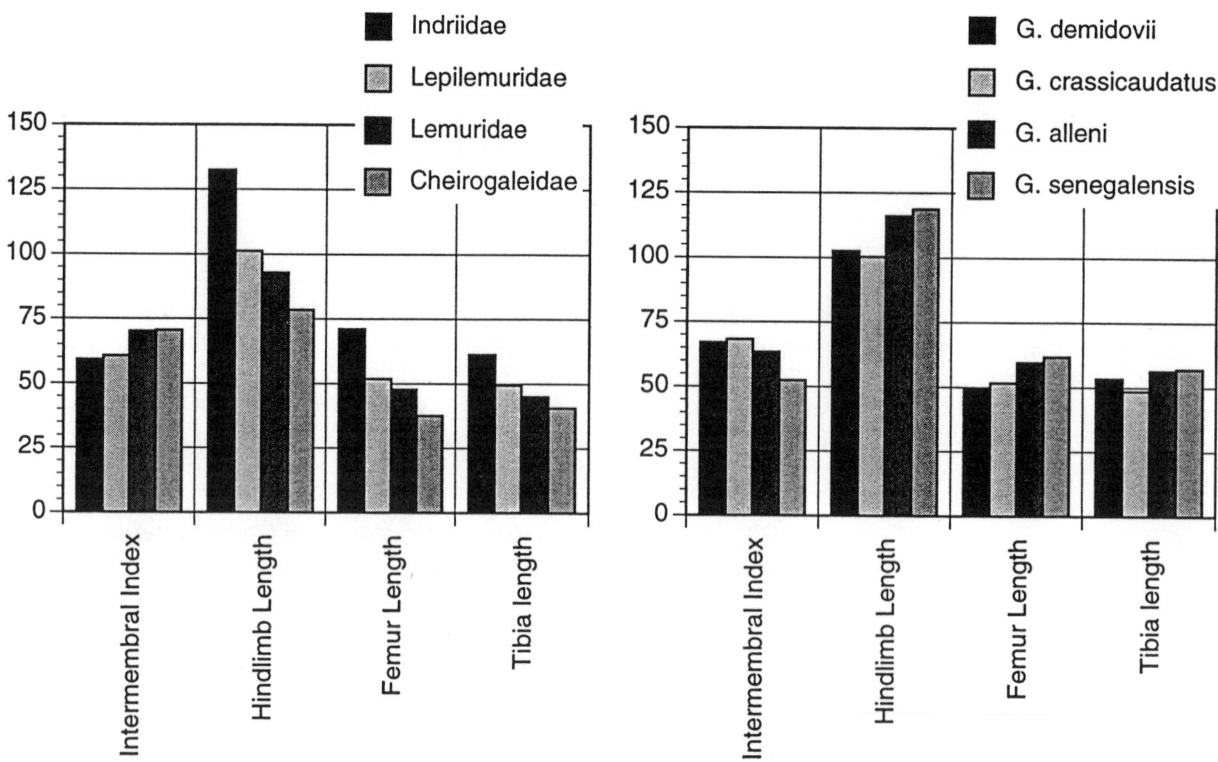

FIGURE 6.9. Limb proportions among Malagasy Lemuriformes. Intermembral Index = forelimb length/hindlimb length × 100; Hindlimb Length = hindlimb length/skeletal trunk length × 100; Femur Length = femur length/skeletal trunk length × 100; Tibia Length = tibia length/skeletal trunk length × 100.

FIGURE 6.10. Limb proportions among species of the genus *Galago*. Intermembral = forelimb length/hindlimb length × 100; Hindlimb Length = hindlimb length/skeletal trunk length × 100; Femur Length = femur length/skeletal trunk length × 100; Tibia Length = tibia length/skeletal trunk length × 100.

sites of muscle origins and insertions are often marked on the bone by raised surfaces, rugosities, or tuberosities. These osteological markers allow the functional morphologist to investigate muscular mechanics on the basis of preserved bony morphology. The proximal femur is a particularly interesting anatomical region for this kind of analysis because it bears the easily visible attachments of several very important muscles of the locomotor system. As discussed previously, the femoral trochanters mark the insertions of the iliopsoas (lesser trochanter), the gluteus medius and minimus (greater trochanter), and the gluteus superficialis and tensor fasciae femoris (third trochanter). Measuring the distance from the hip joint to the distal point of muscle insertion at each of the femoral trochanters is easily accomplished, and can be considered an approximation of the muscular moment arm or effort arm, although this is probably only strictly true when the limb is flexed 90° (Stern, 1974; Jungers, 1976). Similarly, the length of the hindlimb or of the relevant hindlimb segment can be considered an approximation of the resistance arm (Gregory, 1912; Elftman, 1929; Smith and Savage, 1956; Jungers, 1976; McArdle, 1981). In this way we can compare the mechanical advantage of bone-muscle systems among animals of differing behavioral repertoires, and in so doing, test hypotheses of functional adaptation. The pelvis also lends itself to an analysis of this kind since its three constituent bony parts are attachment sites for important hindlimb muscle groups. The length of the ilium, the ischium, and the pubis are good approximations of the effort arms of the tensor fasciae femoris, the hamstring muscles, and of the adductors respectively.

Table 6.3 presents a series of ratios that, among

TABLE 6.3
Femoral Ratios among Prosimian Primates[1]

Family	Hip Flexor		Gluteal Index		Gr. Trochanter		Condylar	
Lorisidae[2]	18.1	(3.3)	15.0	(2.1)	13.2	(2.0)	84.0	(5.1)
Galagidae	16.1	(1.8)	12.5	(2.0)	11.5	(1.8)	108	(5.7)
Tarsiidae	13.6	(0.6)	10.9	(0.6)	12.4	(0.6)	117	(6.0)
Indriidae	12.9	(1.3)	12.9	(1.7)	11.2	(0.6)	112	(10.4)
Lepilemuridae	14.8	(0.9)	14.0	(1.2)	12.7	(0.5)	110	(3.0)
Lemuridae	16.7	(1.2)	16.6	(1.2)	13.9	(0.7)	101	(3.5)
Cheirogaleidae	18.0	(1.2)	13.6	(0.8)	13.6	(0.8)	103	(3.6)
Tupaiidae	18.8	(1.4)	34.2	(2.8)	16.6	(1.1)	91.4	(4.0)
Species	**Hip Flexor**		**Gluteal Index**		**Gr. Trochanter**		**Condylar**	
Galago alleni	16.5	(0.9)	13.1	(1.3)	11.3	(0.3)	111	(1.1)
G. senegalensis	14.5	(1.3)	10.8	(0.8)	11.2	(2.1)	113	(3.2)
G. crassicaudatus	18.2	(0.9)	14.5	(1.3)	11.6	(1.9)	102	(2.0)
G. demidovii	16.1	(1.0)	12.1	(1.0)	12.0	(0.9)	104	(3.4)

[1] All ratios are defined in the text.
[2] Toatal sample size = 296. See Anemone (1990) for sample sizes for individual taxa.

other things, approximate the mechanical advantage (*EA/RA*) of some hindlimb muscle groups. The Hip Flexor Index is a measure of the distal extent of the lesser trochanter relative to overall femoral length, and thus of the mechanical advantage of hip flexion by the iliopsoas muscle. A proximally positioned lesser trochanter would shorten the moment arm of hip flexion, reduce the mechanical advantage of the muscle, and maximize speed. An additional adaptive rationale for proximal placement of this (and other) muscle insertions on the femur is a reduction in the moment of inertia of the limb. In both its effect on the length of the flexor moment arm and on the moment of inertia of the hindlimb as a whole, a more proximal position of the lesser trochanter can be seen as an adaptation likely to be found in leaping prosimians. The prediction is confirmed in broad outline. Indriidae and Tarsiidae show the most proximally placed lesser trochanters, followed by Lepilemuridae and Galagidae. Interspecific variation within the Galagidae is in the predicted direction in light of the behavioral propensities of the various species. *G. senegalensis,* the most specialized leaper in the genus, has the lowest hip flexor index, while the mostly quadrupedal *G. crassicaudatus* has the highest. The most distally positioned lesser trochanters in the sample are to be found in Tupaiidae, Cheirogaleidae, and Lorisidae.

The Gluteal Index measures the distal extent of the third trochanter relative to total femoral length. This bony landmark is the attachment site for the tendons of pars anterior of the gluteus superficialis, whose much-debated functions (see Stern, 1971, 1972 for excellent reviews) must include some components of both abduction and extension of the hip, and of the tensor fasciae femoris, an important hip flexor. Distal prolongation of the third trochanter has the effects of increasing the moment arms of these two muscles, and of increasing the mass moment of inertia of the hindlimb as a whole. As for the lesser trochanter, the three families with the most proximally positioned third trochanter are Tarsiidae, Galagidae, and Indriidae. Among the Malagasy genera, *Lepilemur* and *Microcebus* are most similar to the three leaping families, as might be predicted on the basis of the large leaping component of locomotor behavior within these two genera. Tupaiidae is most clearly distinguished from all prosimians on the basis of an extremely distally positioned third trochanter. Among the galagos, *G. senegalensis* diverges from *G. crassicaudatus* in the direction of Tarsiidae and Indriidae, while both *G. alleni* and *G. demidovii* are intermediate.

The Greater Trochanter Index measures the distance from the hip joint to the lateral edge of the greater trochanter relative to femoral length. Because the gluteus medius and minimus insert into this region of the greater trochanter, it represents an approximation of the moment arm of these muscles in extension of the thigh. Due to the importance of rapid hip extension in leaping, one might predict that specialized leapers would have lower values for these ratios than would quadrupeds. The taxa with the lowest mechanical advantage in gluteal hip extension are Tarsiidae, Galagidae, and Indriidae. Conversely, those taxa with relatively large values, and hence longer moment arms, are the quadrupedal Tupaiidae, Cheirogaleidae, Lemuridae, and Lorisidae. Among the Malagasy taxa, there is a clear spectrum from the leaping Indriidae with the lowest values, through the progressively less saltatorial Lepilemuridae, Lemuridae, and Cheirogaleidae. This ratio also indicates minor differences between the leaping and quadrupedal species of *Galago,* but always in the predicted direction (i.e., smaller values for *G. senegalensis* than for *G. crassicaudatus).*

In his study of the hip and thigh of Lorisiformes, McArdle (1981) calculated similar ratios to some of those used here. Although my results are essentially similar to his, he offered the caveat that the use of femoral length as the denominator in these ratios may have skewed the results, because taxa with apparently more proximally situated trochanters usually have the longest femora. There is, however, every reason to believe that a lineage could increase the length of the femur and maintain high mechanical advantage in the muscles crossing the hip joint. Therefore, the fact that some prosimian taxa (i.e., most leapers) have reduced their muscular moment arms in conjunction with an increased functional length of the hindlimb is significant. The fact that femoral elongation appears to have taken place distal to the femoral trochanters is of functional importance, and should not be ignored.

Some distinctive qualitative traits of the proximal femur of Galagidae and Tarsiidae include a cylindrical head with large posterior expansion of its articular surface onto a thick and short neck that is nearly perpendicular to the shaft. These traits appear to be related and to allow easy separation of Galagidae and Tarsiidae (and to a lesser extent, Lorisidae) from all Lemuriformes and Tupaiidae. The crucial trait in this suite of characteristics is the posterior expansion of the articular surface onto the neck; it is this trait that gives the head a cylindrical shape and makes the neck appear short, thick, and perpendicular. The proximal femoral anatomy among all Lemuriformes includes a spherical femoral head with little or no posterior expansion of articular surface onto the neck. The articular expansion seen in galagos and tarsiers is a postural adaptation of importance during vertical clinging behavior in small-bodied prosimians. While vertically clinging to an arboreal support, the femora are strongly flexed, abducted, and laterally rotated. In this position, the posterior expansion of articular surface comes into contact with the distal articular region of the acetabulum, allowing the necessary posture preparatory to leaping. Jenkins and Camazine (1977) similarly related the degree of expansion of the articular surface of the femoral head onto the neck among carnivores to the relative importance of the use of the femur in abducted postures. The absence of this trait among Indriidae may be the result of subtle differences in clinging behavior of these larger animals, and requires further study. All leaping prosimians are joined by the hindlimb-dominant Malagasy quadrupeds in possessing greater trochanters that overhang the ventral aspect of the shaft. The significance of this trait appears to be related to the direction of pull or leverage of the vastus lateralis, which arises from the ventral aspect of the greater trochanter.

The Condylar Index (Table 6.3) is a ratio of the anterior-posterior depth of the femoral condyles to their medio-lateral width. One of the original morphological traits of vertical clingers and leapers noted by Walker (1967, 1974; Napier and Walker, 1967a, 1967b) was the deep antero-posterior dimension of the femoral condyles. Femoral condyles with a large antero-posterior diameter were thought to act as a pulley to increase the leverage (i.e., mechanical advantage) of knee extension by the quadriceps femoris. By increasing the distance from the quadriceps tendon to the center of rotation of the knee joint, the moment arm of this important extensor is increased, allowing a more powerful action. The Condylar Index was calculated to directly test the claim that vertical clingers and leapers can be distinguished from other prosimians on the basis of this trait. The data indicate no clear difference between leapers and active quadrupedal prosimians. The only separation effected by the Condylar Index is between the Lorisidae and Tupaiidae, both of whom have femoral condyles that are wider (mediolaterally) than deep (dorso-ventrally), and all other prosimians, where these relations are reversed. Both the quadrupeds and leapers of the Malagasy Republic, and the African and Asian leapers have femoral condyles that are slightly deeper than wide. While Walker's (1967, 1974; Napier and Walker, 1967a, 1967b) functional analysis of the femoral condyles acting as a "pulley" for the action of the quadriceps femoris in knee extension is certainly correct, the presence of this trait does not distinguish between active quadrupedal prosimians and specialized leapers.

Visual examination of the distal femur among prosimians indicates that a ventrally raised patellar articular surface is found only among the most specialized leapers (Galagidae, Tarsiidae, Indriidae, and Lepilemuridae). This feature serves to further increase the antero-posterior dimension of the femoral condyles, and the mechanical advantage of femoral extension (Walker, 1967; Napier and Walker, 1967a; Anemone, 1988, 1990). Several other traits of distal femoral anatomy are found among all hindlimb-dominant taxa (i.e., all groups except Lorisidae and Tupaiidae) and appear to be related to extension of the leg by the quadriceps femoris. These animals all share an especially prominent lateral patellar ridge that counteracts the pull of the hypertrophied vastus lateralis and thus guards against patellar dislocation during contraction of that muscle. A narrow and deep patellar groove with a large arc of curvature is a further indication of the emphasis on flexion-extension at the knee (Tardieu, 1981) among hindlimb-dominant prosimians. Although these traits—as well as posteriorly facing tibial articular surfaces—are in sharp contrast to the condition seen in the specialized slow-climbing Lorisidae, they do not allow active quadrupedal and leaping prosimians to be successfully distinguished.

Dissection of the quadriceps femoris muscle group reveals two specializations that have been interpreted by some authors as adaptations to leaping. These are the presence of a large and powerful vastus lateralis (Murie and Mivart, 1872; Alezais, 1900; Jouffroy, 1962, 1975; Stern, 1971) and the presence of a fibrocartilaginous superior patella in the tendon of vastus intermedius (Retterer and Vallois, 1912; Jouffroy, 1962, 1975). The vastus lateralis is hypertrophied among all Malagasy prosimians, and Jouffroy (1962, 1975) suggests that this is a critical adaptation that allows all Lemuriformes some degree of proficiency in arboreal leaping. My own dissections of the quadriceps femoris among Galagidae *(G. senegalensis, G. demidovii,* and *G. crassicaudatus)* and Tarsiidae reveal that vastus lateralis routinely comprises more than 60 percent of the total quadriceps mass, with vastus medialis accounting for an additional 20-30 percent (Anemone, 1988). Hypertrophy of vastus lateralis among saltatorial mammals has been confirmed by McArdle (1981) for Galagidae, Woolard (1925) for Tarsiidae, Stern (1971) and Plaghki et al. (1981) for Cebidae, Alezais (1900) for Rodentia, and Hopwood and Butterfield (1976) for Macropodidae. In addition to its large size, the origin of vastus lateralis is restricted to the proximal femur among both nonprimate (Alezais, 1900; Hopwood and Butterfield, 1976) and primate leapers (Woolard, 1925; Jouffroy, 1962, 1975; Stern, 1971; McArdle, 1981; Plaghki et al., 1981). This proximally restricted origin is seen also in vastus medialis, and appears to be related to the possession of long and parallel muscle fibers among Lemuridae (Jungers et al., 1980), Galagidae, and Tarsiidae (Anemone, 1988). In fact, prosimian leapers appear to be maximizing both the physiological cross section of vastus lateralis (and vastus medialis) and the length of its constituent muscle fibers. Stern (1971) has shown that the combination of a large cross-sectional area and long fiber length allows a muscle to develop maximal force generation at high velocities of contraction, and would appear to be well adapted to the needs of leaping animals. Jungers et al. (1980) have confirmed the importance of vastus lateralis in the leap of *Lemur fulvus* with telemetered electromyography. Their results indicate that vastus lateralis consistently reaches maximal burst levels as it initiates the leap in this taxon. They further suggest that the amount of force generated by this muscle is enhanced by active stretching (i.e., eccentric contraction) when the femur is hyperflexed in the crouch

TABLE 6.4
Pelvic Ratios among Prosimian Primates[1]

Family	Ischium 1	Ischium 2	Ilium L.	Pubis 1	Pubis 2	Iliac Crest
Lorisidae[2]	25.7 (2.0)	34.9 (3.4)	73.7 (2.6)	44.3 (3.0)	60.2 (4.3)	23.7 (2.8)
Galagidae	27.6 (1.5)	37.3 (2.6)	74.0 (1.9)	33.2 (2.4)	44.9 (3.7)	28.6 (3.5)
Tarsiidae	25.6 (1.2)	33.0 (1.8)	77.5 (1.3)	38.0 (3.0)	49.0 (3.9)	23.6 (2.1)
Indriidae	28.1 (1.3)	38.4 (2.4)	73.3 (2.1)	29.2 (2.0)	39.9 (2.7)	43.6 (5.2)
Lepilemuridae	30.0 (1.3)	41.4 (2.1)	72.4 (1.7)	29.4 (3.3)	39.0 (4.5)	31.4 (6.2)
Lemuridae	29.4 (1.3)	41.5 (1.9)	71.0 (1.0)	28.3 (1.3)	41.4 (2.0)	31.3 (3.0)
Cheirogaleidae	32.8 (1.7)	46.9 (3.3)	70.0 (2.1)	35.1 (2.9)	50.2 (3.6)	20.8 (1.9)
Tupaiidae	40.2 (1.1)	64.8 (2.3)	62.1 (1.0)	30.4 (2.1)	48.9 (3.1)	37.7 (4.0)
Species	**Ischium 1**	**Ischium 2**	**Ilium L.**	**Pubis 1**	**Pubis 2**	**Iliac Crest**
Galago alleni	27.2 (0.1)	36.6 (0.9)	74.4 (1.7)	33.0 (2.9)	56.4 (2.0)	25.0 (2.9)
G. senegalensis	26.5 (1.3)	35.2 (1.7)	75.2 (1.2)	32.0 (2.1)	57.1 (3.1)	28.3 (2.7)
G. crassicaudatus	28.7 (0.9)	39.7 (1.7)	72.3 (1.3)	34.3 (2.3)	48.6 (2.0)	29.8 (3.9)
G. demidovii	27.8 (1.0)	36.9 (1.4)	75.3 (0.9)	34.1 (1.6)	53.4 (3.7)	27.0 (3.4)

[1] All ratios are defined in the text.
[2] Total sample size = 296. See Anemone (1988) for sample sizes for individual taxa.

that typically precedes the leap. Stern's (1971) theoretical work and the experimental results of Jungers et al. (1980) strongly support the importance of vastus lateralis in the leaping of prosimian primates. Rather than characterizing only VCL prosimians, however, the described morphology of vastus lateralis appears to link both quadrupedal and saltatorial Lemuriformes with Tarsiidae and Galagidae.

A functional relationship between the morphology of vastus intermedius, the presence of a superior patella within its tendon, and leaping behavior (Retterer and Vallois, 1912; Jouffroy, 1962, 1975) among primates has recently been questioned (Stern, 1971; Jungers et al., 1980). The internal architecture of vastus intermedius appears to be relatively uniform among prosimians. It is mainly composed of short, fleshy fibers arising from much of the ventral femoral shaft to insert into a tendon running superficially along the length of the muscle. Stern (1971:244) suggests that "the only advantage associated with short-fibered vasti is that they probably manage to maintain some superiority in the number of fibers and, therefore, are able to produce more powerful extension at slow velocities of contraction." This morphology appears better suited to slow postural activities and maintenance of the integrity of the knee joint than for the propulsive necessities of leaping. Jungers et al. (1980) have confirmed electromyographically the primary postural activity of vastus intermedius in *Lemur fulvus*. Their results indicate that vastus intermedius undergoes no force enhancement through active stretching prior to the leap, fails to show increased electrical activity during the leap, and is the only part of the quadriceps that is electrically active during resting postures with flexion at the knee. Stern (1971) found a superior patella in the tendon of vastus intermedius in a number of cebid genera, including *Cebus* and *Alouatta*. Because these two genera leap, respectively, the most and the least frequently among the Cebidae, Stern (1971) denied any special relationship between the superior patella and leaping (contra Retterer and Vallois, 1912; Jouffroy, 1962, 1975). Among prosimian primates, a superior patella is found in Lemuriformes (Jouffroy, 1962, 1975) and Galagidae and Tarsiidae (Anemone,

1988), while it is absent in Tupaiidae. It is unclear whether the Lorisidae possess a superior patella: Jouffroy (1962) states that it is lacking, while McArdle (1981) claims that it is present in all four genera. Because it only articulates with the patellar surface of the femur during extreme flexion at the knee, Jungers et al. (1980:287) suggest that the presence of a superior patella is most likely "related to the complex tensile and compressive stresses generated in the tendon during the completely hyperflexed phase of leaping." In any event, the presence of the superior patella does not link the specialized leaping prosimians to the exclusion of the hindlimb-dominant quadrupeds of Madagascar.

Table 6.4 presents ratios concerning the relative lengths of the ilium and iliac crest, ischium, and pubis. The denominators in these ratios were total pelvic length (Ischium 1, Ilium L., and Pubis 1) and ilium length (Ischium 2, Pubis 2, and Iliac Crest). Both measures of relative ischial length indicate that leapers and slow climbers possess shorter ischia than do active quadrupedal prosimians, and that Tupaiidae have relatively longer ischia than any prosimian. Interspecific variation in *Galago* indicates a longer ischium among the more quadrupedal taxa and a shorter ischium among the more specialized leapers.

As a result of the origin of the hip extensor muscles (flexor cruris lateralis, semitendinosus, semimembranosus, adductor magnus, quadratus femoris) from its ramus and tuberosity, the length of the ischium is widely accepted as a good osteological approximation of the moment arm of these muscles (Howell, 1932, 1944; Smith and Savage, 1956; Walker, 1967, 1974; McArdle, 1981). Among some nonprimates, the simple speed-power dichotomy appears to accurately explain relative ischial length. Smith and Savage (1956) note the long ischium in powerful aquatic mammals like *Phoca,* and the short ischium found among cursorial Equiidae. Since bipedal saltators, like cursors, require great accelerations in hip extension, we can expect to find a short ischium among these taxa. Indeed, Walker (1967) first noted the unusually short ischium among prosimian saltators. Zuckerman et al. (1973) further noted that a short ischium distinguished Lorisiformes (both Galagidae and Lorisidae) from all other primate taxa in their study. These results have since been confirmed by Godfrey (1977), McArdle (1981), and Anemone (1988, 1989). Interestingly, nonprimate terrestrial saltators, both marsupial (Elftman, 1929) and placental (Howell, 1932), have long ischia compared to their quadrupedal relatives. The difference in ischial length between arboreal primate and terrestrial nonprimate saltators may have something to do with the presence of a large "counterbalancing" tail among marsupial (Elftman, 1929) and rodent (Howell, 1932; Bartholomew and Caswell, 1951) saltators. Howell (1944:47) has gone so far as to state that "saltation in mammals requires a heavy tail as a counterbalance." Yet a tail does not appear to be an important element in prosimian leaping and is, indeed, almost totally lacking in *Indri* (Tattersall, 1982). In addition, in light of Stern's (1974) findings, we can hypothesize that the two groups are maximizing different kinds of velocity, resulting in different optimal muscular effort arm lengths. Although acknowledging the significant behavioral differences between primate and nonprimate leapers, Godfrey (1977) suggests that the shortened ischium among primates may be associated with habitual femoral extension at the hip, and not specifically with leaping. In this way, she explains the similarly short ischium of both bipedal *Homo* and of the specialized hanger *Paleopropithecus.* McArdle (1981) notes that the prosimian leapers are also characterized by proximal insertions of the hamstring muscles into the tibia. This would seem to be another adaptation for rapid extension at the hip (Gregory, 1912; Stern, 1974). It seems safest to state that the short ischium among prosimian leapers is an adaptation that allows a very rapid extension of the hip with little loss of power. It is unclear why the Lorisidae are also characterized by a very short ischium.

In a recent paper, Fleagle and Anapol (1992) suggest that the short ischium among nonhuman primate leapers is best explained by analogy with the human ischium. Both human bipedalism and vertical leaping require extreme ranges of femoral excursion, and both humans and primate leapers have short and dorsally projecting ischia. This dorsal reorientation of the origin of the hamstring muscles allows the length of the effort arm of hip extension to be maintained when the hip is fully extended, and thus affords effective leverage to the hip extensor musculature during the typically extreme femoral excursions that are an integral part of bipedal walking and vertical leaping.

The length of the ilium (Table 6.4) approximates the moment arm of the tensor fasciae femoris, an important flexor of the femur at the hip joint during the recovery stroke of locomotion. In addition, it is a good measure of the length of attachment of the gluteus medius, which arises along much of its lateral surface. The data clearly indicate that leaping taxa have significantly longer ilia than do quadrupedal taxa, specifically that tarsiids have the longest ilia among prosimians, followed closely by galagids and indriids. The shortest ilia are found among Cheirogaleidae and Tupaiidae. Among the Malagasy prosimians there is a clear spectrum of increasing relative iliac length from Cheirogaleidae to Lemuridae, Lepilemuridae, and finally Indriidae. The Lorisidae are intermediate between leapers and unspecialized quadrupeds in relative iliac length. There is considerable interspecific variation in iliac length among the species of the genus *Galago*. *Galago senegalensis* and *G. alleni* have significantly longer ilia than do their more quadrupedal relatives *G. demidovii* and especially *G. crassicaudatus*.

As noted previously, the length of the ilium is greatest among the leaping prosimians and smallest among unspecialized quadrupeds (e.g., Tupaiidae and Cheirogaleidae). Several important muscles of the hip and thigh arise along the ilium, both the gluteal group of extensors and a number of important hip flexors (e.g., tensor fasciae femoris, sartorius, rectus femoris, and iliacus). The length of the ilium affects the dynamic characteristics of these muscles in a number of ways. Because the gluteus medius, a powerful femoral extensor, arises along much of the length of the lateral surface of the iliac blade, lengthening of the ilium increases the area of origin and the average length of the fibers of this muscle. Because all skeletal muscle fibers contract approximately one-third of their length, longer fibers contract an absolutely greater distance, and thus bring the distal end of a bone through a greater distance. Because similar muscle fibers, short or long, contract at a uniform rate, longer fibers confer greater velocity than do physiologically similar short muscle fibers (Gowitzke and Milner, 1980). With respect to the femoral flexors, it is clear that an increase in length of the iliac blade increases the length of the moment arm of the tensor fasciae femoris, which arises from the iliac crest and inserts into the femoral fascia. The moment arms of the other femoral flexors would probably also increase with a longer ilium, but not to the same degree as that of tensor fasciae femoris. An increase in iliac length, then, would appear to have two different effects, namely an increase in the velocity of gluteal extension of the femur, and increased power of femoral flexion with the added benefit (according to Stern, 1974) of flexing the extended femur during the recovery stroke of locomotion in a shorter time. A survey of the relative length of the ilium among nonprimates yields some interesting results. Gregory (1912) and Smith and Savage (1956) argue that long and narrow ilia are characteristic of cursorial quadrupedal mammals, while graviportal mammals tend to have short and broad iliac blades. Smith and Savage (1956) go on to hypothesize that two different muscle groups can be functionally differentiated in femoral extension at the hip among mammals. The gluteal group, with short moment arms but extensive origins along the iliac blade are the "fast" extensors that provide a late burst of acceleration, while the ischial group, arising from a longer ischial moment arm, provide the power needed in early stages of leaping or running to overcome inertia. Howell (1932) found that the saltatorial rodents he investigated were characterized by short ilia, especially in the distal part of the ilium between the sacral articulation and the acetabulum. Again, we can only note the apparent contradiction between terrestrial, nonprimate saltators and arboreal, prosimian leapers in this trait and suggest that the significant biomechanical differences in leaping must explain this seeming paradox. Howell later disagreed with Gregory (1912) and stated that the "length of the ilium . . . is not a cursorial character, although a long ilium is frequently . . . present in cursorial species" (Howell, 1944:167). Rather, Howell (1944) argued that a long ilium is a cushioning mechanism that acts by means of the attachment of ligaments and back muscles along its length to transmit stresses from the femur incurred during locomotion. Although cursorial and saltatorial animals certainly incur large stresses during locomotion, other, mainly large-bodied quadrupeds, would also be expected to have long ilia if Howell (1944) were correct. Stern (1974) and Waterman (1929) both suggest that the functional significance of the long ilium in cursorial mammals—and, by extension, in saltatorial prosimians—lies in its effect on the moment arm for flexion of the femur on the hip. Femoral flexion mainly comes into play during the recovery stroke of mammalian locomotion, after the extensors have provided the main propulsion. Both cursorial quadrupeds and bipedal saltators need to flex the extended hindlimb in the shortest possible time: cursors in preparation for the next propulsive stage, and saltators in preparation for landing. As a result, we would expect a greater moment arm for femoral flexion than for extension. Indeed, Stern (1974) states that the moment arm of the tensor fasciae femoris, an important femoral flexor, is the longest in the hindlimb. The long ilial moment arm (flexion) in combination with the short ischial moment arm (extension) appear to be equally well adapted for rapid running on all fours (Stern, 1974) and for rapid bipedal leaping in the trees.

In his original diagnosis of the morphology of vertical clingers and leapers, Walker stated that a narrow, bladelike ilium was characteristic of indriids, tarsiids, and galagids (Walker, 1967; Napier and Walker, 1967a). The results presented here (Table 6.4) agree with respect to Tarsiidae and Galagidae. These latter families are joined by Lorisidae and Cheirogaleidae in having very narrow iliac blades. Contrary to Walker, however, indriids are distinguished from all other prosimians by virtue of their very broad ilia (Godfrey, 1977; Tattersall, 1982). Lemuridae, Lepilemuridae, and Tupaiidae also have broad ilia, although consistently less wide than those of Indriidae. Both Tattersall (1982) and Godfrey (1977) state that the function of the broadened indriid ilium is to provide a broad area of attachment for the large gluteus medius and iliacus, which arise from its lateral and medial surfaces, respectively. Howell (1944:166) adds the effects of abdominal muscles and "ligamentous factors, which are hard to evaluate" in determining the breadth of the ilium. The powerful erector spinae muscles in saltatorial macropodids (Elftman, 1929) and in indriids (Tattersall, 1982) may also affect the width of the ilium. These intrinsic muscles of the back are important in hyperextension of the vertebral column during leaping, and arise above the proximal ilium, along the lumbar and sacral vertebrae.

The Lorisidae have by far the longest relative pubic length among prosimian primates (Table 6.4). They are most closely approximated in this trait by the Cheirogaleidae, Tupaiidae, and Tarsiidae. Galagidae are intermediate in relative pubic length, while the shortest pubes are found in the Lemuridae, Lepilemuridae, and Indriidae. Among the Malagasy taxa there is a clear spectrum of decreasing pubic length from the generalized quadrupedal Cheirogaleidae, through the active quadrupedal Lemuridae and Lepilemuridae, to the leaping Indriidae. This same trend is seen among the species of *Galago*. At the familial level, relative pubic length does not successfully distinguish between active quadrupeds and leapers, nor does it link all three families of leapers.

Few authors have discussed or presented data on the relative length of the pubis among mammals. The pubis provides attachment for the adductor musculature of the hindlimb, and pubic length can be considered an approximation of the length of the adductor moment arm. Howell (1944) states that pubic length is shortest among taxa, which confine their hindlimb movements to the parasagittal plane of flexion and extension and thus deemphasize the importance of adduction. Quadrupedal cursorial animals would be expected to have short pubes, as would terrestrial bipedal saltators. Prosimian saltators—although they strongly emphasize flexion-extension in locomotion—also rely on femoral adduction while clinging in vertical postures to trunks or branches. There must, therefore, be a compromise in the length of their adductor moment arms. Among prosimian primates, the longest pubis is found among the slow-climbing Lorisidae, followed closely by the quadrupedal Cheirogaleidae and Tupaiidae. The small-bodied leapers (Galagidae and Tarsiidae) have intermediate pubic lengths, while the indriids, lepilemurids, and lemurids have the shortest pubes in the suborder. These results make at least partial sense in that the taxa with few or no cursorial habits retain the longest pubes. Yet rather than link all leapers, pubic length clearly distinguishes galagos and tarsiers from indriids. Further research might indicate that the significant differences in pubic length among prosimian leapers may be related to differences in body size and associated differences in clinging behavior suggested by Cartmill's (1974) vertical support model (Jungers, 1976).

SUMMARY

In their original statement of the VCL hypothesis, Napier and Walker (1967a; Walker, 1967) suggested a list of morphological traits that could be found among all extant vertical clingers and leapers.

Many, although by no means all of these traits are osteological characteristics of the hindlimb: long ilia and short ischia; a long and straight femur with a cylindrical head placed upon a short neck; posterior expansion of the femoral head's articular surface; narrow and deep patellar groove with a prominent lateral patellar ridge; and posteriorly facing femoral and tibial condyles. They further suggested that this same set of traits can be found among most Eocene fossil primates, and that vertical clinging and leaping might "possibly be regarded as the earliest locomotor specialization of primates and therefore preadaptive to some or all of the later patterns of primate locomotion" (Napier and Walker, 1967a:204). Although these ideas were quickly subjected to intense critical scrutiny by Cartmill (1972), Stern and Oxnard (1973), and Martin (1974), the VCL hypothesis retains heuristic value as an eminently testable hypothesis dealing with important functional and evolutionary questions (Anemone, 1990; Fleagle and Anapol, 1992).

My work clearly indicates a more complex morphological situation in the hindlimb skeleton among extant prosimians and, by extension, argues against an ancestral and preadaptive role for vertical clinging and leaping among euprimates. On the basis of the functional anatomy of the hip and thigh, and fully supported by further comparisons from other anatomical regions (Anemone, 1988), at least two—and more likely three—different morphological solutions to the demands of an arboreal, saltatory way of life among prosimian primates seemingly exist (Oxnard, German, and McArdle, 1981; Oxnard et al., 1990). Indriids are clearly distinct from both tarsiids and galagids in many aspects of hip and thigh morphology: the similarities between these groups are most likely convergences and hence provide evidence of independent acquisition of saltatory habits and adaptations among Madagascan and African-Asian prosimians. In spite of the greater similarities between galagids and tarsiids, I am convinced that they are also convergent, and again indicate separate evolutionary histories of leaping among African and Asian prosimians. This viewpoint has also been suggested by Oxnard (1984) and coworkers (Oxnard et al., 1990) and by the detailed comparative anatomical analysis of *Tarsius* and *Galago* by Niemitz and Hollihn (1985).

The comparative anatomical investigations reported here have revealed a number of adaptive resemblances in the hip and thigh among vertical clinging and leaping prosimians. All have long hindlimbs, proximally positioned femoral trochanters, low mechanical advantage in many muscle groups crossing the hip, long ilia, and short ischia, and raised patellar articular surfaces with prominent lateral patellar ridges. The functional significance of these traits is for the most part well understood, and can easily be related to the physiological demands of a rapid saltatory mode of locomotion. Because rapid cursorial locomotion requires many of the same biomechanical and physiological adaptations, most of these traits fail to consistently distinguish between saltatorial and rapid quadrupedal prosimians. These difficulties are compounded by a realization of the variability of most prosimian locomotor behavior, and in particular the observation that many of the taxa usually labeled as rapid or active quadrupeds engage in some degree of leaping. There is a clear need on the part of functional morphologists for the collection of more and better quantitative field data on variability in positional behaviors among living primates. Only in the light of improved data on behavioral variation will further comparative analysis of the functional morphology of extant primates, among whom both form (i.e., morphology) and function (i.e., behavior) may be studied, improve our ability to reconstruct the behavior of fossil primates.

NOTES

1. I use "positional behavior" in the sense of Prost (1965) to refer to both locomotor and postural behaviors.

2. The third trochanter is usually found in all prosimians although it may be very weakly developed or absent among lorisids. It is usually absent among anthropoid primates, although Aiello and Dean (1990) state that it occurs infrequently in *Homo sapiens* and *Pan*. I find little support for Jouffroy's (1975) suggestion that, among primates, the third trochanter is present only in *Perodicticus, Galago,* and callitrichids.

3. Schultz (1930, 1933) used the anterior trunk height as the denominator of his relative limb measurements. This measurement is taken between suprasternale and symphysion on living or preserved specimens or on articulated skeletons.

LITERATURE CITED

Aiello L, and Dean C (1990) Human Evolutionary Anatomy. London: Academic Press.

Alezais H (1900) Le quadriceps fémoral des sauteurs. C. R. Soc. Biol. *52:*510–511.

Anemone RL (1988) The Functional Morphology of the Prosimian Hindlimb: Some Correlates between Anatomy and Positional Behavior. Ph.D. dissertation, University of Washington, Seattle.

——— (1990) The VCL hypothesis revisited: Patterns of femoral morphology among quadrupedal and saltatorial prosimian primates. Am. J. Phys. Anthrop. *83:*373–393.

Bartholomew GA, and Caswell HH (1951) Locomotion in kangaroo rats and its adaptive significance. J. Mammal. *32:*155–169.

Bearder SK, and Doyle GA (1974) Ecology of bushbabies, *Galago senegalensis* and *Galago crassicaudatus,* with some notes on their behavior in the field. In RD Martin, GA Doyle, and AC Walker (eds.): Prosimian Biology. New York: Academic Press, pp. 109–130.

Biegert J, and Maurer R (1972) Rumpfskelettlange, Allometrien und Korperproportionen bei catarrhinen Primaten. Folia Primatol. *17:*142–156.

Bock WJ, and von Wahlert G (1965) Adaptation and the form-function complex. Evolution *19:*269–299.

——— (1972) Arboreal adaptations and the origin of the Order Primates. In R Tuttle (ed.): The Functional and Evolutionary Biology of Primates. Chicago: Aldine-Atherton, pp. 97–122.

Cartmill M (1974) Pads and claws in arboreal locomotion. In FA Jenkins (ed.): Primate Locomotion. New York: Academic Press, pp. 45–83.

Charles-Dominique P (1977) Ecology and Behavior of Nocturnal Primates. Prosimians of Equatorial West Africa. New York: Columbia University Press.

Crompton RH (1980) A Leap in the Dark: Locomotor Behavior and Ecology in *Galago senegalensis* and *Galago crassicaudatus.* Ph.D. dissertation, Harvard University, Cambridge.

——— (1984) Foraging, habitat structure, and locomotion in two species of *Galago.* In P Rodman, and JGH Cant (eds.): Adaptations for Foraging in Nonhuman Primates. New York: Plenum Press, pp. 73–111.

Elftman HO (1929) Functional adaptations of the pelvis in marsupials. Bull. Amer. Mus. Nat. Hist. *58:*189–232.

Fleagle JG, and Anapol FC (1992) The indriid ischium and the hominid hip. J. Hum. Evol. *22:*285–305.

Fleagle JG, and Simons EL (1983) The tibio-fibular articulation in *Apidium phiomense,* an Oligocene anthropoid. Nature *301:*238–239.

Gambaryan PP (1974) How Mammals Run. New York: John Wiley and Sons.

Gebo DL (1987) Locomotor diversity in prosimian primates. Am. J. Primatol. *13:*271–281.

George RM (1977) The limb musculature of the Tupaiidae. Primates *18:*1–34.

Godfrey L (1977) Structure and Function in *Archaeolemur* and *Hadropithecus* (Subfossil Malagasy Lemurs): The Postcranial Evidence. Ph.D. dissertation, Harvard University.

Gowitzke BA, and Milner M (1980) Understanding the Scientific Bases of Human Movement. Baltimore: Williams and Wilkins.

Grand TL, and Lorenz R (1968) Functional analysis of the hip joint in *Tarsius bancanus* (Horsfield, 1821) and *Tarsius syrichta* (Linnaeus, 1758). Folia Primatol. *9:*161–181.

Gregory W (ed.) (1950) The Anatomy of the Gorilla. New York: Columbia University Press.

Gregory WK (1912) Notes on the principles of quadrupedal locomotion and on the mechanism of the limbs in hoofed animals. Ann. N.Y. Acad. Sci. *22:*267–294.

Hall-Craggs ECB (1964) The jump of the Bush Baby. Med. Biol. Illus. *14:*170–174.

——— (1965a) An analysis of the jump of the lesser galago *(Galago senegalensis).* J. Zool. *147:*20–29.

——— (1965b) An osteometric study of the hindlimb of the Galagidae. J. Anat. *99:*119–126.

——— (1966a) Muscle tension relationships in *Galago senegalensis.* J. Anat. *100:*699–700.

——— (1966b) Rotational movements in the foot of *Galago senegalensis.* Anat. Rec. *154:*287–294.

——— (1974) Physiological and histochemical

parameters in comparative locomotor studies. In RD Martin, GA Doyle, and AC Walker (eds.): Prosimian Biology. London: Duckworth, pp. 829–845.

Hartman C, and Straus W (eds.) (1933) The Anatomy of the Rhesus Monkey *(Macaca mulatta)*. Baltimore: Williams and Wilkins.

Hopwood PR, and Butterfield RM (1976) The musculature of the proximal pelvic limb of the eastern grey kangaroo *Macropus major* (Shaw) and *Macropus giganteus* (Zimm). J. Anat. *121:*259–272.

Howell AB (1932) The saltatorial rodent *Dipodomys:* Functional and comparative anatomy of its muscular and osseous systems. Proc. Am. Acad. Arts and Sci. *67:*377–536.

——— (1944) Speed in Animals. Chicago: University of Chicago Press.

Jenkins, FA (1974) Tree shrew locomotion and the origins of primate arborealism. In FA Jenkins (ed.): Primate Locomotion. New York: Academic Press, pp. 85–115.

Jenkins FA, and Camazine SM (1977) Hip structure and locomotion in ambulatory and cursorial carnivores. J. Zool. *181:*351–370.

Jolly A (1966) Lemur Behavior: A Madagascar Field Study. Chicago: University of Chicago Press.

Jouffroy FK (1962) La musculature des membres chez les lemuriens de Madagascar. Etude descriptive et comparative. Mammalia *26:*1–326.

——— (1975) Osteology and myology of the Lemuriform postcranial skeleton. In I Tattersall, and RW Sussman (ed.): Lemur Biology. New York: Plenum Press, pp. 149–192.

Jouffroy FK, and Lessertisseur J (1979) Relationships between limb morphology and locomotor adaptations among prosimians: An osteometric study. In ME Morbeck, H Preuschoft, and N Gomberg (ed.): Environment, Behavior, and Morphology: Dynamic Interactions in Primates. New York: Gustav Fischer, pp. 143–181.

Jungers WL (1976) Hindlimb and pelvic adaptations to vertical climbing and clinging in *Megaladapis,* a giant subfossil prosimian from Madagascar. Yrbk. Phys. Anthrop. *20:*508–524.

——— (1979) Locomotion, limb proportions, and skeletal allometry in lemurs and lorises. Folia Primatol. *32:*8–28.

Jungers WL, Jouffroy FK, and Stern JT (1980) Gross structure and function of the Quadriceps femoris in *Lemur fulvus:* An analysis based on telemetered electromyography. J. Morph. *164:*287–299.

Kay RF, and Covert HH (1984) Anatomy and behavior of extinct primates. In DJ Chivers, BA Wood, and A Bilsborough (eds.): Food Acquisition and Processing in Primates. New York: Plenum Press, pp. 467–508 .

Luckett, WP (ed.) 1980 Comparative Biology and Evolutionary Relationships of Tree Shrews. New York, Plenum Press.

McArdle JE (1981) Functional morphology of the hip and thigh of the Lorisiformes. Contrib. Primatol. *17:*1–132.

Martin RD (1972) Adaptive radiation and behaviour of the Malagasy lemurs. Phil. Trans. Roy. Soc. Lond. Soc. *185:*295–352.

Murie J, and Mivart SG (1872) On the anatomy of the Lemuroidea. Trans. Zool. Soc. Lond. *7:*1–113.

Napier JR (1967) Evolutionary aspects of primate locomotion. Am. J. Phys. Anthrop. *27:*333–342.

Napier JR, and Napier PH (1967) A Handbook of Living Primates. London: Academic Press.

Napier JR, and Walker AC (1967a) Vertical clinging and leaping—a newly recognized category of primate locomotion. Folia Primatol. *6:*204–219.

——— (1967b) Vertical clinging and leaping in living and fossil primates. In D Starck, R Schneider, and HJ Kuhn (eds.): Neue Ergebnisse der Primatologie. Stuttgart: Fischer Verlag, pp. 66–69.

Niemitz, C (1984) Locomotion and posture of *Tarsius bancanus.* In C Niemitz (ed.): Biology of Tarsiers. Stuttgart: Fischer Verlag, pp. 191–225.

Niemitz C, and Hollihn KU (1985) Semiquantitative Untersuchung zur Druckbelastung des Knochengewebes bei springender Fortbewegung. Z. Morph. Anthrop. *75:*273–285.

Oxnard CE (1984) The place of *Tarsius* as revealed by multivariate statistical morphometrics. In C Niemitz (ed.): Biology of Tarsiers. Stuttgart: Fischer Verlag, pp. 17–32.

Oxnard CE, Crompton RH, and Lieberman SS

(1990) Animal Lifestyles and Anatomies: The Case of the Prosimian Primates. Seattle: University of Washington Press.

Oxnard CE, German R, Jouffroy FK, and Lessertisseur J (1981) A morphometric study of limb proportions in leaping prosimians. Am. J. Phys. Anthrop. *54:*421–430.

Oxnard CE, German R, and McArdle J (1981) The functional morphometrics of the hip and thigh in leaping prosimians. Am. J. Phys. Anthrop. *54:*481–498.

Plaghki L, Goffart M, Beckers-Bleukx G, and Moureau-Lebbe A (1981) Some characteristics of the hind limb muscles in the leaping night monkey *Aotus trivirgatus* (Primates, Anthropoidea, Cebidae). Comp. Biochem. Physiol. *70:*341–349.

Pollock J (1975) Field observations on *Indri indri:* A preliminary report. In I Tattersall, and R Sussman (eds.): Lemur Biology. New York: Plenum Press, pp. 287–311.

——— (1977) The ecology and socioecology of feeding in *Indri indri:* In TH Clutton-Brock (ed.): Primate Ecology: Studies of Feeding and Ranging Behavior in Lemurs, Monkeys and Apes. London: Academic Press, pp. 37–69.

Prost, JH (1965) A definitional system for the classification of primate locomotion. Amer. Anthrop. *67:*1198–1214.

Retterer E, and Vallois H (1912) De la double rotule de quelques primates. C.R. Soc. Biol. (Paris) *73:*379–382.

Richard A (1978) Behavioral Variation: Case Study of a Madagascar Lemur. Lewisburg, PA: Bucknell University Press.

Robinson JT, Freedman L, and Sigmon BA (1972) Some aspects of pongid and hominid bipedality. J. Hum. Evol. *1:*361–369.

Schön M (1968) The muscular system of the red howling monkey. U.S. Nat. Mus. Bull. *273:*1–185.

Schultz AH (1930) The skeleton of the trunk and limbs of higher primates. Hum. Biol. *2:*303–438.

——— (1933) Die Korperproportionen der erwachsenen catarrhinen Primaten, mit spezieller Berucksichtigung der Menschenaffen. Anthropol. Anz. *10:*154–185.

Sigmon BA (1969) The scansorius muscle in primates. Primates *10:*247–261.

——— (1974) A functional analysis of pongid hip and thigh musculature. J. Hum. Evol. *3:*161–185.

Sigmon BA, and Farslow DL (1986) The Primate Hindlimb. In DR Swindler, and J Erwin (eds.): Systematics, Evolution and Anatomy. New York: Alan R. Liss, pp. 671–718.

Smith JM, and Savage RJG (1956) Some locomotory adaptations in mammals. J. Linn. Soc. Lond. *42:*603–622.

Stern JT (1971) Functional myology of the hip and thigh of Cebid monkeys and its implications for the evolution of erect posture. Biblio. Primatol. *14:*1–319.

——— (1972) Anatomical and functional specializations of the human gluteus maximus. Am. J. Phys. Anthrop. *36:*315–340.

——— (1974) Computer modeling of gross muscle dynamics. J. Biomech. *7:*411–428.

Stern JT, and Oxnard CE (1973) Primate locomotion: Some links with evolution and morphology. Primatologia *4:*1–93.

Stevens JL, Edgerton VR, Haines DE, and Meyer DM (1981) An Atlas and Source Book of the Lesser Bushbaby, *Galago senegalensis.* Boca Raton: CRC Press.

Sussman RW (1974) Ecological distinctions in sympatric species of *Lemur.* In RD Martin, GA Doyle, and AC Walker (eds.): Prosimian Biology. New York: Academic Press, pp. 75–108.

Swindler D, and Wood C (1973) An Atlas of Primate Comparative Anatomy. Baboon, Chimpanzee, and Man. Seattle: University of Washington Press.

Szalay FS, and Dagosto M (1988) Evolution of hallucial grasping in the primates. J. Hum. Evol. *17:*1–33.

Tabachnick BG, and Fidell LS (1983) Using Multivariate Statistics. New York: Harper and Row.

Tardieu C (1981) Morpho-functional analysis of the articular surfaces of the knee-joint in Primates. In AB Chiarelli, and RS Corruccini (eds.): Primate Evolutionary Biology. Berlin: Springer Verlag, pp. 68–80.

Tattersall I (1977) Ecology and behavior of *Lemur fulvus mayottensis* (Primates, Lemuriformes). Anthro. Pap. Amer. Mus. Nat. Hist. *54:*421–482.

——— (1982) The Primates of Madagascar. New York: Columbia University Press.

Uhlmann K (1968) Huft- und Oberschenkelmuskulatur: Systematische und vergleichende Anatomie. Primatologia *4*:1–442.

Walker AC (1967) Locomotor Adaptations in Recent and Fossil Madagascan Lemurs. Ph.D. dissertation, University of London.

——— (1974) Locomotor adaptations in past and present prosimian primates. In FA Jenkins (ed.): Primate Locomotion. New York: Academic Press, pp. 349–381.

——— (1979) Prosimian locomotor behavior. In GA Doyle, and RD Martin (eds.): The Study of Prosimian Behavior. New York: Academic Press, pp. 543–565.

Waterman HC (1929) Studies on the evolution of the pelvis of man and other primates. Bull. Amer. Mus. Nat. Hist. *58*:585–642.

White JF, and Gould SJ (1965) Interpretation of the coefficient in the allometric equation. Amer. Natur. *99*:5–18.

Woollard HH (1925) The anatomy of *Tarsius spectrum*. Proc. Zool. Soc. Lond. *70*:1071–1184.

Zuckerman S, Ashton EH, Flinn RM, Oxnard CE, and Spence TF (1973) Some locomotor features of the pelvic girdle in primates. Symp. Zool. Soc. Lond. *33*:71–165.

C H A P T E R S E V E N

Functional Morphology of the Foot in Primates

Daniel L. Gebo

The grasping foot has long been recognized as one of the most fundamental of all adaptations for the entire Order Primates (e.g., Mivart, 1873; Wood Jones, 1916; Gregory, 1920; Morton, 1924b; Clark, 1959; Napier, 1967; Martin, 1968, 1986; Cartmill, 1972, 1974; Szalay and Decker, 1974; Szalay and Dagosto, 1988; Dagosto, 1988). This novel adaptation allowed primates to successfully compete within the small branch milieu of the arboreal environment. The basic form of the grasping foot has been further modified from its original design into a variety of different types, and in this chapter I briefly describe and explain the functional significance of several of these novel foot adaptations in living primates. I begin by assessing the different types of foot positions and grasps that primates utilize, and go on with a description of the external characteristics of primate feet, the function of the grasping big toe, bone and joint rotations within the foot—a few of the functional varieties within the Order—and finally, a brief comment concerning the adaptive changes of prosimian and anthropoid foot morphology.

FOOT POSITIONS AND GRASPING FEET

All primates, with the exception of the great apes and humans, utilize a heel-elevated or semiplantigrade foot position when they move (e.g., climb, leap, or move quadrupedally) (Figs. 7.1 and 7.2; Weidenreich, 1923; Morton, 1924a, 1924b; Keith, 1929; Tuttle, 1970; Jouffroy, 1975; Susman, 1983; Gebo, 1986a, 1987, 1992; Meldrum, 1991).[1] On a horizontal support, the foot is inverted and slightly abducted; the big toe is curled around the support and opposes the lateral digits while the proximal end of the calcaneus is elevated above the support. Thus, only bones distal to the transverse tarsal joint contact the support (Fig. 7.1). Weight is borne by the plantar surfaces of the cuboid, navicular, entocuneiform, and the digits. When a primate foot grasps a vertical support, the foot assumes a more abducted position but is essentially positioned the same as described above. Similarly, when a primate foot is placed on the ground, the foot is in an everted and adducted position with the heel elevated above the support (Fig. 7.1). All primates have the ability to place their heels upon a support or the ground and do so often while resting or in certain postures, but with the exception of the great apes and humans, heel contact during movement sequences is rare, if it ever occurs.

In the African apes, the foot is fully plantigrade during terrestrial locomotion in that the heel actually contacts the surface of the ground at the end of swing phase (Fig. 7.1). When African apes knuckle-walk on a terrestrial substrate, the lateral side of the heel contacts the ground first, then the lateral side of the foot makes contact, and finally there is a medial shift in weight transference such that the entire

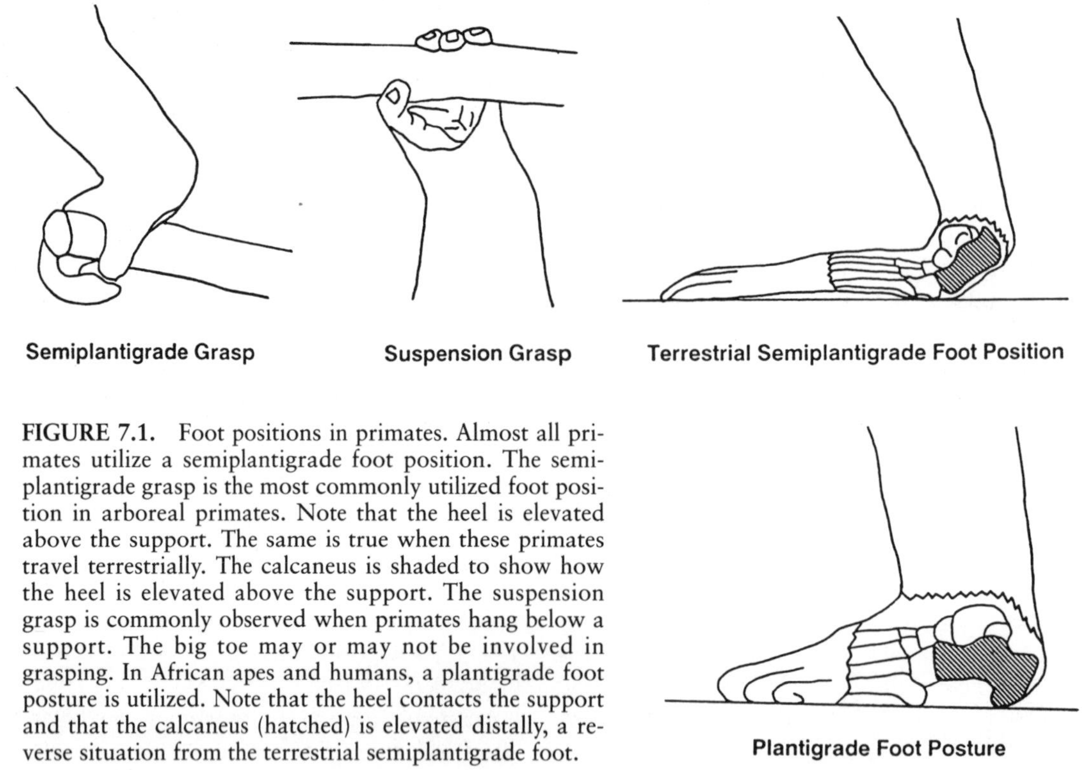

Semiplantigrade Grasp

Suspension Grasp

Terrestrial Semiplantigrade Foot Position

Plantigrade Foot Posture

FIGURE 7.1. Foot positions in primates. Almost all primates utilize a semiplantigrade foot position. The semiplantigrade grasp is the most commonly utilized foot position in arboreal primates. Note that the heel is elevated above the support. The same is true when these primates travel terrestrially. The calcaneus is shaded to show how the heel is elevated above the support. The suspension grasp is commonly observed when primates hang below a support. The big toe may or may not be involved in grasping. In African apes and humans, a plantigrade foot posture is utilized. Note that the heel contacts the support and that the calcaneus (hatched) is elevated distally, a reverse situation from the terrestrial semiplantigrade foot.

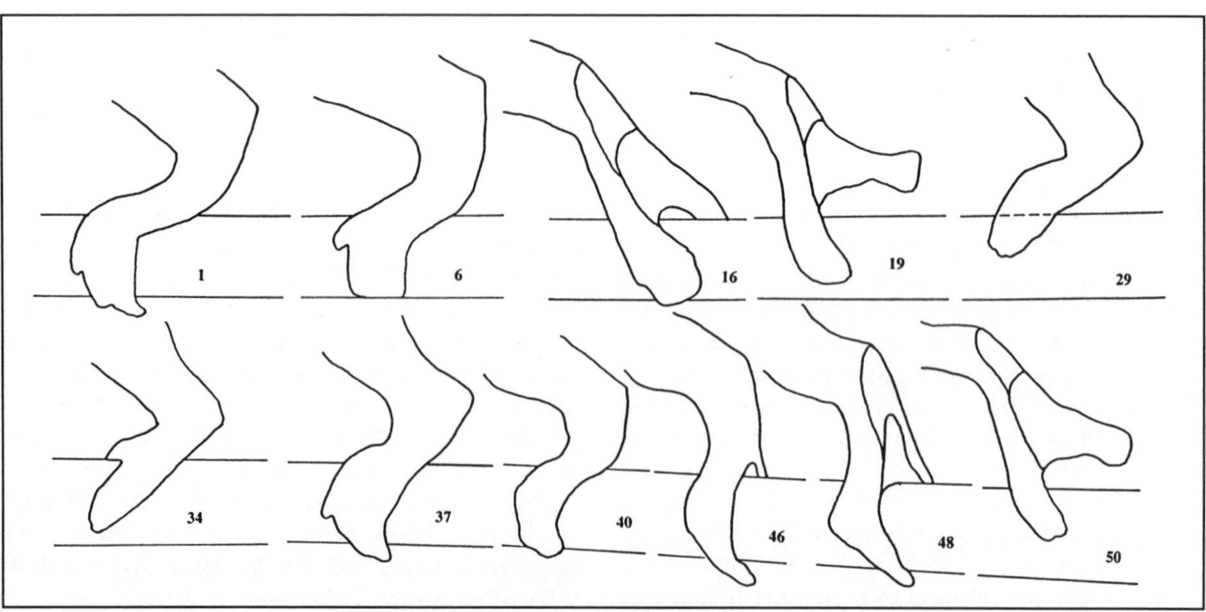

FIGURE 7.2. Arboreal foot sequence. This is a drawing of a frame-by-frame slow-motion film sequence showing the foot positions of *Lemur fulvus* traveling quadrupedally across a horizontal support. Note that the heel never touches the support during the swing or stance phases of arboreal quadrupedalism.

foot contacts the ground surface (Elftman and Manter, 1935; Tuttle, 1969; Susman, 1983).

When a primate hangs below a support using only its feet, it uses a different type of grasp, the suspension grasp, where the phalanges of digits two through five are hooked above the horizontal support (Morton, 1924a; Sarmiento, 1983). The big toe is flexed and it may or may not participate in the grasp (Fig. 7.1).

Orangutans often use their feet to suspend their bodies from a horizontal or oblique support or they place their feet in a grasping position along the side of a vertical support (Morton, 1924a; Tuttle, 1970; Cant, 1987; Rose, 1988). These foot positions are similar to those of the heel-elevated primates in that the heel does not touch the surface of an arboreal support during movement sequences. Because orangutans have greatly reduced the relative length of their big toe by greatly lengthening their lateral digits, they frequently use the suspension grasp. When orangutans move terrestrially, they place the entire lateral side of their foot against the surface of the ground (Tuttle, 1970, 1972) giving them a highly inverted foot set. This terrestrial foot position is an unusual one for the foot of an orangutan and should not be viewed as equivalent to the plantigrade foot position of African apes.

VARIETIES OF PRIMATE FEET

The presence or absence of nails, relative calcaneal length, the position of the big toe, and the degree to which hair covers the bottom surface of primate feet are four features that help to sort out the varieties of primate feet. (Figure 7.3 and Table 7.1 document these varieties.)

Most primates have rounded and blunt nails instead of the more primitive mammalian claw. This novel adaptation facilitates grasping in primates, but many primates have redeveloped pointed or keeled nails, which look clawlike in appearance (Clark, 1936, 1959; Hill, 1953; Cartmill, 1974; Garber, 1980, 1984) (Table 7.1). The clawlike nails of primates probably give the same type of adaptive advantage as the claws of other arboreal mammals. That is, claws and clawlike nails allow small mammals to cling to large-diameter supports that cannot be accommodated with the grasping span of a primate hand or foot. In fact, to a small-sized primate, a very large support appears as a flat wall, making it difficult to grasp any part of this surface (Cartmill, 1974). Therefore, nailed primates are mechanically not equipped to utilize the varied food resources located on large vertical supports like the trunks of trees. Primates that have redeveloped clawlike nails (e.g., *Daubentonia* and the tamarins and marmosets) can and do exploit this surface (see Garber, 1980, 1992); the pointed or clawlike nails act like the primate equivalent of crampons.

The relative length of the foot and its constituent parts has received a lot of attention in the literature, with particular emphasis being paid to the extremely long feet of galagos and tarsiers (e.g., Schultz, 1963; Jouffroy and Lessertisseur, 1979). A long foot increases the length of the leg and thus the lever arm for stride distance or for leaping (Hall-Craggs, 1965a, 1965b; McArdle, 1981; Anemone, 1990). This is not unusual among other leaping or cursorial mammals. What is unusual about the long feet of primates is which bones have been elongated. Morton (1924a) argued that the reason the tarsals—especially the calcaneus and the navicular—rather than the metatarsals are elongated in leaping primates such as galagos and tarsiers is a morphological compromise between the mechanical demands of leaping and grasping. In addition to increasing lever arms for a better force or gear ratio to achieve greater leaping distances, the feet of arboreal primates must also be able to grasp and function to help primates climb. By elongating the distal segments of the calcaneus and the navicular, primates have been able to maintain joint mobility in the tarsus and the grasping digits distally while still being able to increase foot length.

Although all primates, except hominids, have a grasping big toe, its relative length and position are highly variable (Fig. 7.3). For example, lorises have modified the position of the big toe by greatly abducting this digit relative to the lateral digits. Indriids and gibbons have very deep clefts between the big toe and the second digit. All three obviously expand the grasping span of the foot and one might infer from this that these primates are frequent climbers and use relatively large supports. Baboons, on the other hand, possess a very short big toe and, although they are capable climbers, they are most often observed on the ground.

The plantar surface of primate feet comes in two varieties (Fig. 7.3). The first type is found in all tooth-combed prosimians and tarsiers where the

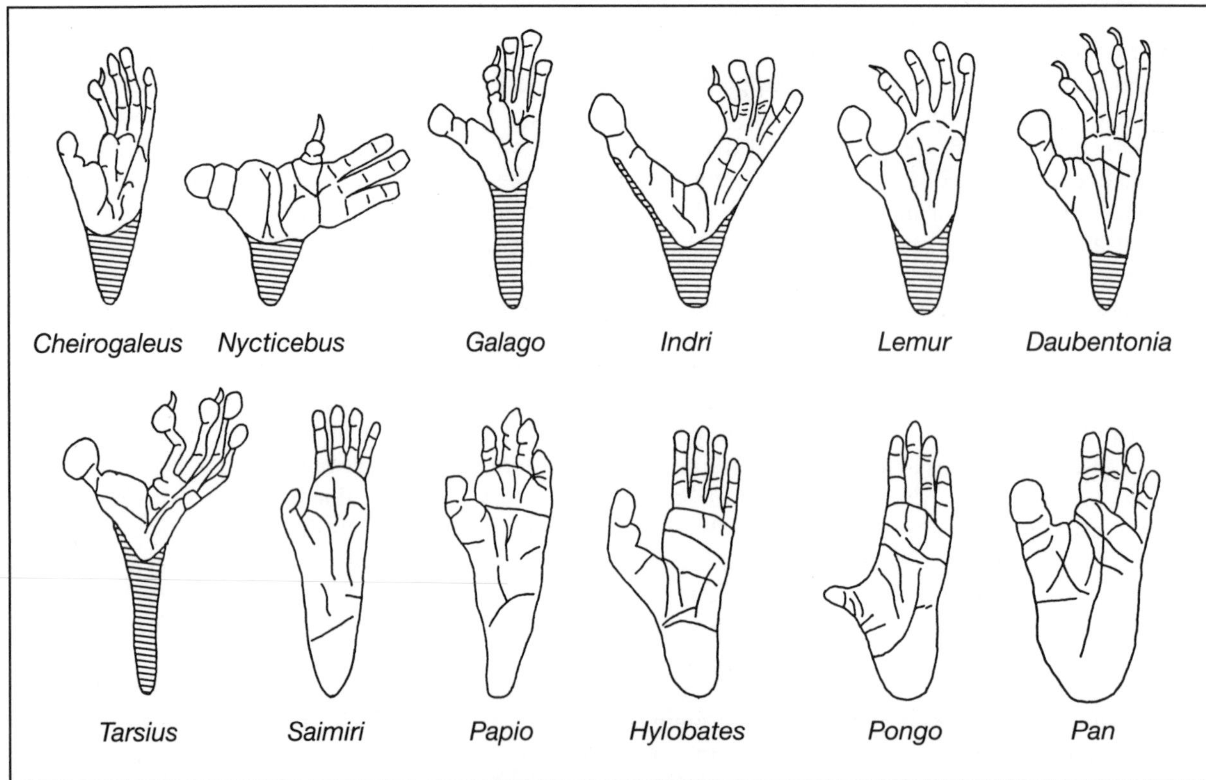

FIGURE 7.3. Varieties of primate feet. Note that all the prosimians have hairy heels (horizontal hatching) while all anthropoids possess a frictional surface on the heel. Note the grooming claws on digit two for *Cheirogaleus, Nycticebus, Galago, Indri, Lemur, Daubentonia,* and *Tarsius;* the extra grooming claw on digit three for *Tarsius*; and the clawlike nails on digits three through five for *Daubentonia.* Redrawn from Hill (1960) and Biegert (1963).

proximal end is hairy and a frictional or palmlike surface extends distally (distal to the transverse tarsal joint).[2] This means that the tarsal segment (the heel region) of the foot is covered with hair (Hill, 1953), is not a frictional surface at all, and in fact, does not participate in touching a support during movements and most postures (see discussion that follows). The second type of primate foot, found in all anthropoids, is entirely covered with a frictional surface from heel to toe. Thus, anthropoids have increased the frictional surface of the plantar side of the foot relative to prosimians. The reason for the expansion of the frictional surface in anthropoids is still unclear but it is most likely due to a postural rather than a locomotor explanation and may in fact, be coupled to the loss in anthropoids of the more forceful grasping big toe of the prosimians.

FOOT ANATOMY AND GRASPING

Primate feet possess four large tarsal bones (the talus, calcaneus, navicular, and cuboid) and three smaller tarsal bones (the three cuneiforms—entocuneiform, mesocuneiform, and ectocuneiform), five metatarsals, three phalanges on each lateral digit, and two on the big toe (Fig. 7.4). The digits and the entocuneiform are involved in grasping while the metatarsals (except for the first) are firmly attached to their respective tarsal elements. The navicular, cuboid, talus, and calcaneus are especially important for the complicated bone rotations that occur during foot inversion and eversion.

Grasping in the primate foot is accomplished by simple flexion of the lateral digits at the metatarsal-phalangeal and interphalangeal joints with metatarsal torsion facilitating flexion around curved

TABLE 7.1
Varieties of Primate Feet

	Calcaneal Length	Nail Type[1]	Big Toe Span	Sole
Cheirogaleids	long	pointed	average divergence	proximal half is hairy
Lorisids	short	blunt	widely abducted	proximal half is hairy
Galagids	very long	blunt[2]	average divergence	proximal half is hairy
Lemurids	moderate	pointed[2]	average divergence	proximal half is hairy
Indriids	short	pointed	deep cleft	proximal half is hairy
Daubentoniids	moderate	clawlike	average divergence	proximal half is hairy
Tarsiids	very long	blunt[3]	average divergence	proximal half is hairy
Callitrichids	moderate	clawlike	average divergence	proximal half is frictional
Cebids	moderate	blunt	average divergence	proximal half is frictional
Atelids	moderate	blunt	average divergence	proximal half is frictional
Cercopithecids	moderate	blunt	average divergence	proximal half is frictional
Hylobatids	moderate	blunt	deep cleft	proximal half is frictional
Pongids	short	blunt	average divergence	proximal half is frictional

[1] All tooth-combed prosimians possess a narrow grooming "claw" on their second digit while tarsiers possess two, one on their second and one on their third digits.

[2] Clawlike for L. rubriventer, especially pointed for Hapalemur, Lepilemur, Euoticus elegantulus.

[3] Pointed for Tarsius spectrum and T. pumilis.

surfaces (Morton, 1924a). The movements of the big toe are more complex due to its abducted position within the foot and its position of opposition to the lateral digits during grasping. Like the other digits, one component of grasping is flexion of the metatarsal-phalangeal and interphalangeal joints. But an additional component of motion, the swing of the big toe, which allows digital opposition, is accomplished at the entocuneiform first metatarsal joint (Fig. 7.5). The shape of this joint varies greatly in primates (Szalay and Dagosto, 1988) but all have a saddle-shaped joint surface, which allows for mostly medio-lateral (i.e., the swing) movements of the big toe. It is the position of the entocuneiform within the foot as well as the proximodistal torsion of the first metatarsal, which allows the big toe to be set away (abducted) from the other digits. The orientation of the distal facet of the entocuneiform has been modified in lorises leading to a far more abducted position of the big toe compared to other primates (Fig. 7.3).

The swing movements of the first metatarsal (i.e., big toe) are accomplished by contraction of the peroneus longus muscle and adductor hallucis, for adduction, while flexion for grasping is due chiefly to flexor hallucis longus and brevis contractions. For the lateral digits, flexor digitorum longus and brevis are, for the most part, the digital flexors to the digits. In most primates, both of the long flexor muscles to the digits actually send a large or a small tendon to each of the five digits and typically fuse tendons together after sending their respective tendon to the first digit. Thus, contraction of either long flexor can actually produce flexion in all of the five digits. The extensor musculature straightens the digits and repositions the big toe (abducts it) when a grasp is released.

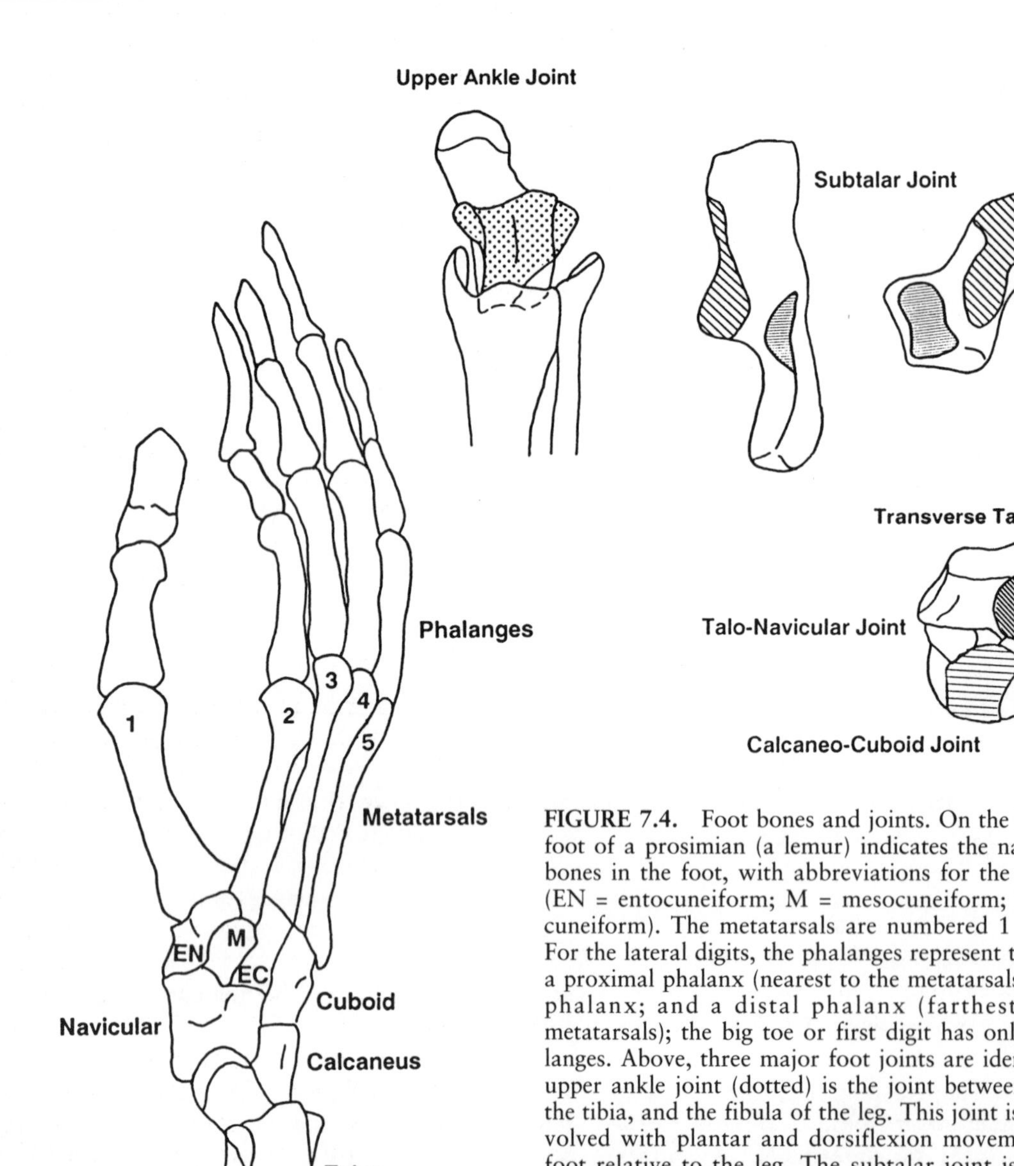

FIGURE 7.4. Foot bones and joints. On the left, a right foot of a prosimian (a lemur) indicates the names of the bones in the foot, with abbreviations for the cuneiforms (EN = entocuneiform; M = mesocuneiform; EC = ectocuneiform). The metatarsals are numbered 1 through 5. For the lateral digits, the phalanges represent three bones: a proximal phalanx (nearest to the metatarsals); a middle phalanx; and a distal phalanx (farthest from the metatarsals); the big toe or first digit has only two phalanges. Above, three major foot joints are identified. The upper ankle joint (dotted) is the joint between the talus, the tibia, and the fibula of the leg. This joint is mainly involved with plantar and dorsiflexion movements of the foot relative to the leg. The subtalar joint is where the talus and the calcaneus meet. Note that the posterior calcaneal facet and the posterior talar facet (horizontal hatching) articulate together, while the anterior calcaneal facet and the anterior talar facet (oblique hatching) articulate together. The transverse tarsal joint is composed of the talo-navicular joint (oblique hatching)—where the talus and the navicular articulate—and the calcaneo-cuboid joint (horizontal hatching)—where the calcaneus and the cuboid articulate. The subtalar and transverse tarsal joints are mainly involved with foot rotations during inversion and eversion movements.

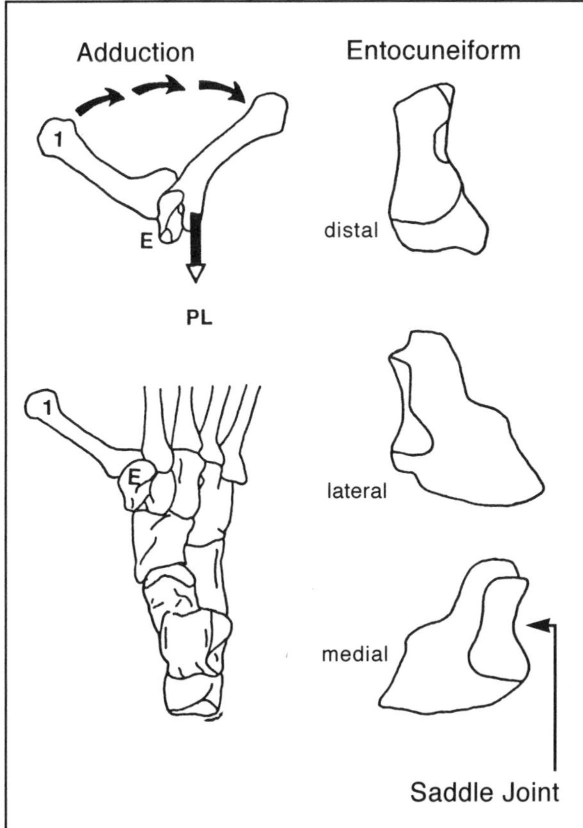

FIGURE 7.5. Adduction of the first digit. On the left, the entocuneiform (E) and the first metatarsal (1) are identified and show the adductive swing of the first digit via the pull from the peroneus longus muscle (PL) on the peroneal tubercle. On the right, three different views of the entocuneiform are shown. The lateral and medial views show the saddle-shaped joint surface for the entocuneiform.

FOOT MUSCULATURE, EXTRINSIC AND INTRINSIC

Extrinsic muscles originate outside the foot (i.e., from the femur, the tibia, and the fibula) but send their tendons to insert on the foot. Five extrinsic muscles (gastrocnemius, flexor hallucis longus, flexor digitorum longus, tibialis anterior, and peroneus longus) are especially important for foot function (Fig. 7.6). The gastrocnemius is important for plantarflexion of the foot, an important movement for propulsion, and is especially important and therefore larger in leaping primates. The two long flexors flex the big toe and the lateral digits and are important for grasping. Likewise, the peroneus longus muscle adducts the big toe and is considered an important muscle for grasping. The tibialis anterior muscle is important for foot inversion because it pulls on and rotates the navicular laterally. Thus, only the gastrocnemius is especially important for rapid forward propulsion while the other extrinsic muscles are mostly involved in grasping or climbing activities.

Intrinsic muscles originate within the foot and in general these muscles are all very small relative to the large extrinsic muscles. Only adductor hallucis is especially prominent, and this muscle helps to adduct the big toe during grasping. Each of the small intrinsic muscles are involved with subtle movements of the digits.

JOINT MOVEMENTS

Foot Inversion

Joint movements within the many and closely situated bones of the foot are both subtle and complicated (see Decker and Szalay, 1974; Szalay and Decker, 1974; Lewis, 1980a, 1980b, 1980c, 1981, 1989), but certain key features can be addressed here (Fig. 7.7). First, the invertor musculature (tibialis anterior, flexor hallucis longus, and flexor digitorum longus) contracts and this exerts a pull on the medial part of the plantar surface of the foot. The pull by tibialis anterior rotates the navicular in a lateral direction, so that the navicular comes to lie on top of (dorsal to) the cuboid rather than on its medial side (Fig. 7.8). This is the beginning of foot inversion. Meanwhile, the long flexor tendons are pulling the distal parts of the foot toward the talus and the calcaneus at the transverse tarsal joint. This movement forces the cuboid (i.e., the cuboid process) to screw into the calcaneus (i.e., at the calcaneal pit) at the calcaneal-cuboid joint (Fig. 7.9). As the cuboid screws into the calcaneus, the cuboid is also moving proximally and slides closer to the talo-navicular joint, placing the calcaneal-cuboid joint more in line (transverse) with the talo-navicular joint for greater ease in rotation. The proximal displacement of the cuboid allows the cuboid to also slide along the cuboid-navicular and the cuboid-ectocuneiform facets. Thus, the ectocuneiform protrudes distally to the cuboid at the end of these bone movements (Fig. 7.9).

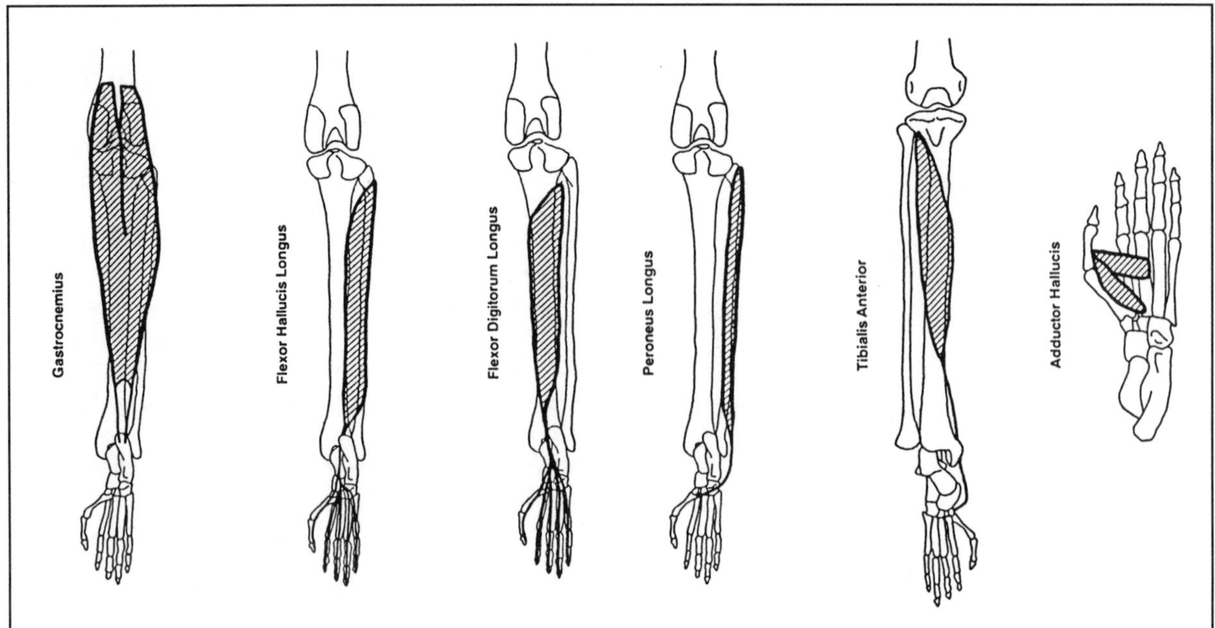

FIGURE 7.6. Extrinsic and intrinsic muscles. Extrinsic muscles for the foot on a generalized primate leg include the gastrocnemius, the flexor hallucis longus, flexor digitorum longus, peroneus longus, and tibialis anterior. Adductor hallucis is the only intrinsic foot muscle on this figure.

The long flexor tendons also rotate the calcaneus laterally because they pull up from below the sustentaculum tali while the short calcaneal-cuboid ligament keeps the calcaneus and the cuboid coupled together in close association. Thus, lateral rotation of the cuboid necessitates rotation of the calcaneus as well (since no muscles attach to the calcaneus to bring about this rotation).

As the calcaneus rotates, the talus is shifted backward (proximal) along the curvature of the posterior calcaneal facet in a helical path of motion (Lewis, 1980b). The talar neck is lifted above and medially away from the sustentaculum tali and the spring ligament during foot inversion (Fig. 7.9). The talar head shifts medially and dorsally to correspond with the lateral direction of movement by the navicular. The talar head and neck point in a proximal-distal line of direction during foot inversion with the navicular above the rim of the talar head. Thus, the helical motion of the talus helps to contribute to the rotational movements occurring at the transverse tarsal joint in allowing the primate foot to become fully aligned in an inverted foot posture.

The cuneiforms are firmly held in place by ligaments and, although some sliding movements can occur at the intercuneiform joints, these are essentially unimportant for foot inversion. All three of the cuneiforms hold the second metatarsal firmly in a notch, however, and together they form a stable arch within the foot in which all other tarsal bones rotate around.

Foot Eversion

In eversion (Figs. 7.7 and 7.9), the peroneal musculature pulls on the first metatarsal tubercle (via peroneus longus) on the medial side and the fifth metatarsal tubercle (via peroneus brevis) on the lateral side of the foot. The navicular and cuboid rotate medially as does the calcaneus. The dorso-lateral ligament prevents extensive medial rotation between the calcaneus and the cuboid. At the subtalar joint, the talus slides forward and locks with the bony ridges along the dorsal surface of the distal calcaneus. The talar head and neck are positioned downward (plantad) and medially with the talar neck contacting the sustentaculum tali and the talar head lying above the rim of the navicular.

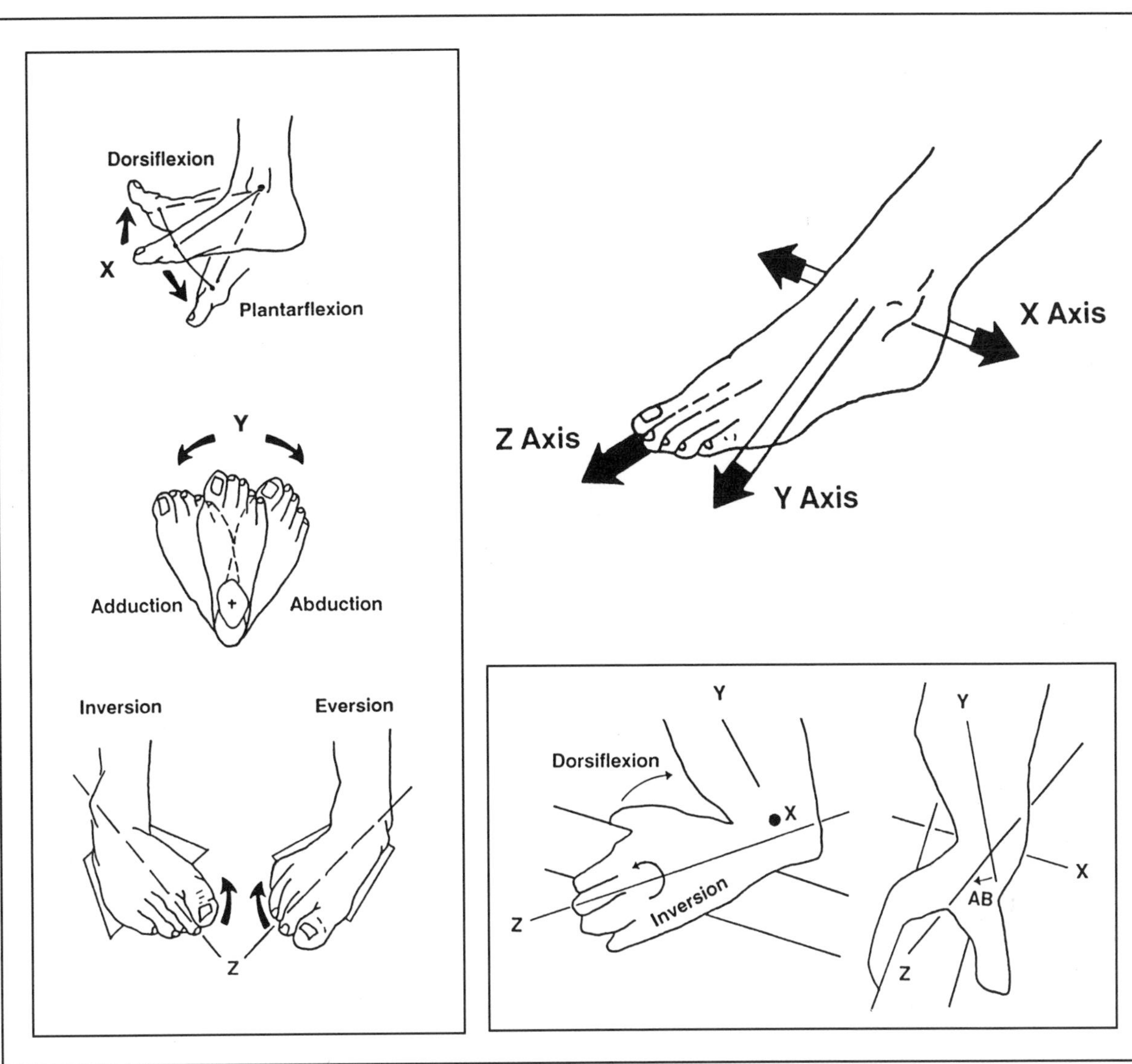

FIGURE 7.7. Foot movements. The illustration of the human foot (top upper right) identifies the three main axes of the foot. The X, Y, and Z axes (dark arrows) represent the transverse axis of the ankle, the long axis of the leg, and the long axis of the foot, respectively. The left column shows the major movements of the human foot broken down along these three axes of the foot and leg. The X axis (top left), the transverse axis of the ankle that runs through the upper ankle joint, allows dorsiflexion and plantarflexion movements. The Y axis (middle left) is the long axis of the leg and shows movement away from the midline—abduction and adduction. The Z axis (bottom left) is the long axis of the foot and shows the rotation movements involved in inversion and eversion. The two illustrations at the lower right show a primate foot grasping a horizontal branch in conjunction with the three axes. Note that the foot is dorsiflexed and inverted while grasping with the heel above the support and that the foot is abducted (AB) relative to the Y axis. Modified from Kapanji (1970).

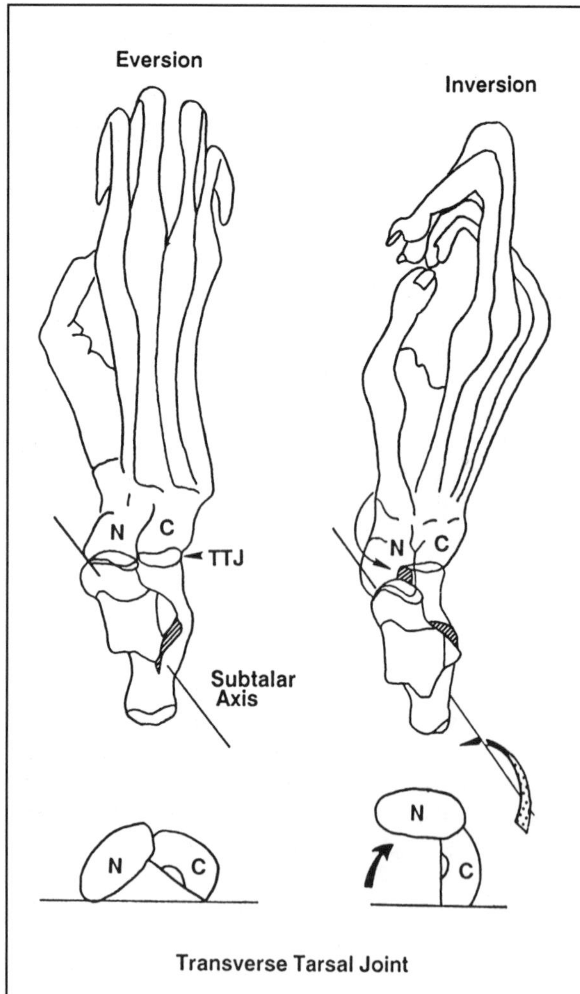

FIGURE 7.8. Foot inversion and eversion. The foot of a primate is positioned in everted and inverted postures. Note the positions of the big toe and forefoot relative to the tarsals in each foot position; the position of the navicular as it rotates upward (dorsal and lateral) from an everted position (see lower illustrations); and the position of the talus and calcaneus in both eversion and inversion relative to the subtalar joint (hatched). The dotted arrow (inversion) shows the rotation or twisting of the calcaneus as it rotates laterally. The smaller dark arrows show the dorsolateral rotation of the navicular. The dark line represents the subtalar axis. N = navicular; C = cuboid; TTJ = transverse tarsal joint. Modified from Lewis (1980b).

Plantarflexion and Dorsiflexion of the Foot

In dorsiflexion (Fig. 7.7), the extensor musculature contracts to pull the dorsal surface of the foot toward the tibia and fibula. The trochlea of the talus slides proximally toward the tibia until it locks, due to the distally wider joint surface or by contact with the talar neck. The tibial malleolus contacts the talo-tibial cup in full dorsiflexion. While the foot is being dorsiflexed, it is also being abducted and everted due to the curvature of the medial trochlear rim. During dorsiflexion, the talar neck shifts medially as does the talar head, which bulges beyond the navicular, and there is a large gap between the navicular and cuboid during dorsiflexion since the foot is abducted.

In plantarflexion (Fig. 7.7), the talus slides distally to the tibia and fibula and stops when contact with the posterior tubercles of the talus is made. The foot is also adducted and inverted in this position. Plantarflexion is particularly important in propulsion and especially so for leaping primates. This is evident by the sizes of the superficial flexors of the foot (i.e., gastrocnemius, soleus, and plantaris) relative to the deep flexors (i.e., flexor hallucis longus, flexor digitorum longus, and posterior tibialis). For example, two frequent leapers like *Tarsius syrichta* and *Galago senegalensis* possess superficial to deep flexor ratios of 1.96 and 1.43, respectively, while *Nycticebus coucang*, a frequent climber, has a ratio of 0.22.

FUNCTIONAL VARIETIES OF PRIMATE FEET

Lorises

Lorises have modified their foot relative to other prosimians in several interesting ways. First, lorises have shifted the position of the big toe medially and greatly shortened the second digit (Fig. 7.3; Hill, 1953). This has increased the span of the grasp relative to foot size and is an adaptation for grasping larger diameter supports (Gebo, 1989b). This novel foot adaptation is accomplished by reorienting the distal navicular facets, which realigns the entocuneiform, and thus widely abducts the big toe relative to the other digits (Dagosto, 1986; Gebo, 1989b).

Lorises have also modified their foot joints in such a way that they possess the most mobile feet of all primates. Their mean ranges of motion far ex-

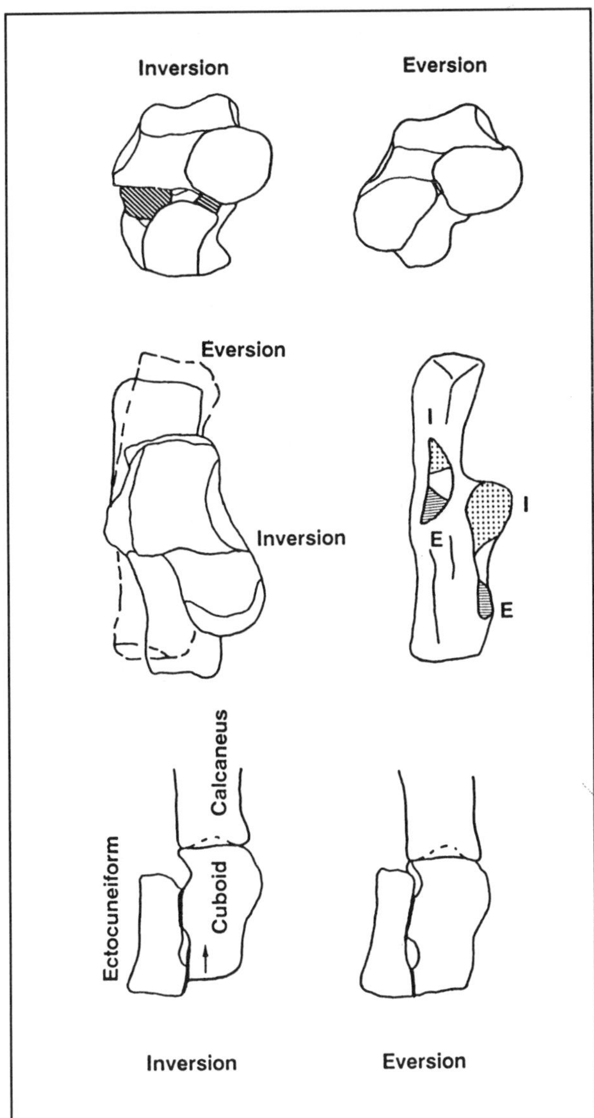

FIGURE 7.9. Bone movements in eversion and inversion. Top: In inversion, the talus has slid back along the posterior calcaneal facet exposing the front half (oblique hatching), and the talar head has lifted above the anterior calcaneal facet (oblique hatching). In eversion, the talus has slid forward covering the surface of the posterior calcaneal facet, and the talar head is pointing downward. Middle: On the left, a standardized talar position shows the two different positions of the calcaneus during inversion and eversion. On the right, the anterior and posterior facets of the calcaneus are shaded to indicate which part is utilized by the talus during inversion (I, dotted surface) and eversion (E, horizontal hatching). Bottom: The position of the cuboid and ectocuneiform are shown in both inversion and eversion. Note that in inversion, the cuboid has moved into the calcaneal pivot region and slid (arrow) along the edge of the ectocuneiform relative to eversion.

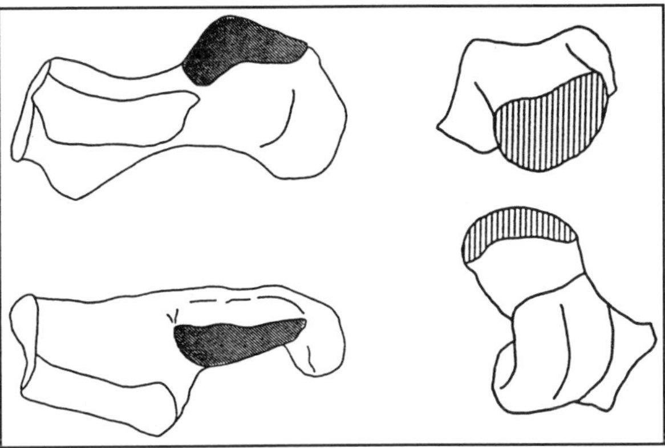

FIGURE 7.10. Calcaneus and talus from a loris foot. Note the unusual shape of the calcaneus (left) and talus (right), the long posterior calcaneal facet (oblique hatching), and the wide talar head (vertical hatching).

ceed that of other prosimians (Table 7.2). This mobility is evident in many bony features. For example, lorises possess a very deep pivot at the calcaneo-cuboid joint, a long posterior calcaneal facet for the talus to slide across (Fig. 7.10), a broad talar head for the navicular to rotate across (Fig. 7.10), and a short distal calcaneal length, which brings the transverse tarsal joint into a more transverse alignment facilitating rotation.

Lorises have increased the sizes of certain muscles responsible for grasping and climbing activities. The deep flexors (i.e., flexor hallucis longus and flexor digitorum longus) are huge muscles in lorises. Flexor digitorum longus is especially large and even originates from the femur (Murie and Mivart, 1872; Jouffroy, 1962; Grand, 1967; Jolly and Gorton, 1974; Gebo, 1989b). Similarly, soleus is well developed and is even larger than the gastrocnemius in lorises.

Lemurs and Indriids

In lemurids and indriids, the foot looks more similar to other primates, but in fact, it is also highly modified (Gebo, 1985; Gebo and Dagosto, 1988). The foot of lemurids and indriids is folded (not flat-lying) with the third, fourth, and fifth digits being reoriented (i.e., arched and lying on top of themselves) (Fig. 7.11). This is reflected in or caused by several modifications of the tarso-metatarsal joints and the joint between the second and third

TABLE 7.2
Foot Mobility in Prosimians (degrees)

	n	mean	range
Plantar and Dorsiflexion at the Upper Ankle Joint			
Lorisids	3	160	155–168
Tarsiids	1	149	—
Cheirogaleids	5	148	128–167
Galagids	7	135	127–142
Indriids	1	132	—
Lemurids	10	129	116–151
Abduction and Adduction at the Upper Ankle Joint			
Lorisids	3	75	63–95
Cheirogaleids	5	65	62–70
Galagids	7	63	60–64
Tarsiids	1	63	—
Lemurids	10	55	31–75
Indriids	1	50	—
Inversion and Eversion at the Transverse Tarsal Joint			
Lorisids	3	180	175–185
Indriids	1	170	—
Cheirogaleids	5	162	140–170
Galagids	7	158	147–180
Lemurids	10	131	105–170
Tarsiids	1	120	—
Plantar and Dorsiflexion at the Transverse Tarsal Joint			
Tarsiids	1	76	—
Galagids	7	67	40–95
Lemurids	10	48	30–67
Lorisids	3	45	35–57
Cheirogaleids	5	41	35–50
Indriids	1	39	—

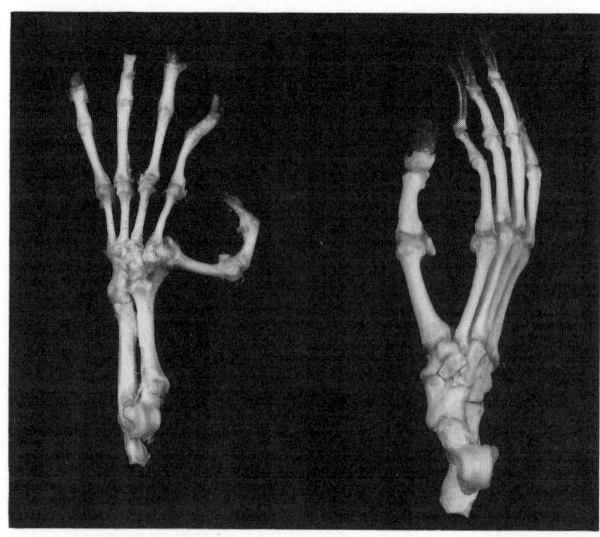

FIGURE 7.11. Folding of lemurid and indriid feet. Left = *Galago;* right = *Lemur.* Note that the galago foot lays flat along the surface (key in on the metatarsals) while the digits of the lemur are stacked upon one another. Lemurid and indriid feet are always folded.

metatarsal (see Gebo, 1985). Lemurids and indriids have also greatly increased the size of their adductor hallucis muscle relative to other intrinsic foot muscles and have also fused both heads of this muscle together (Fig. 7.12). In fact, in lemurs and indriids, this muscle is twice its size relative to other prosimians. Thus, lemurs and indriids possess a specialized adductor grasping foot with greater force production between the first and second digits relative to other primates. The significance of these modifications appears to be related to their use of vertical supports and their relatively larger body sizes (Gebo, 1985).

Indriids have gone even further than lemurs in adapting their feet for more powerful grasping and frequent climbing by increasing the cleft of the big toe relative to the lateral digits and by greatly increasing the size of their deep flexor musculature, especially flexor hallucis longus. Although indriids are known as spectacular leapers, their propulsive musculature (the superficial flexors of the foot) is less developed relative to their deep flexors, and thus, indriids are best thought of as thigh-powered rather than foot-powered leapers (see Gebo and Dagosto, 1988).

Galagos

Next to tarsiers, galagos have the longest distal calcaneal length of any primate (see Figs. 7.11 and 7.13). This great length has led to modifications along the calcaneus as well as the transverse tarsal and navicular-cuneiform joints (see Schultz, 1963;

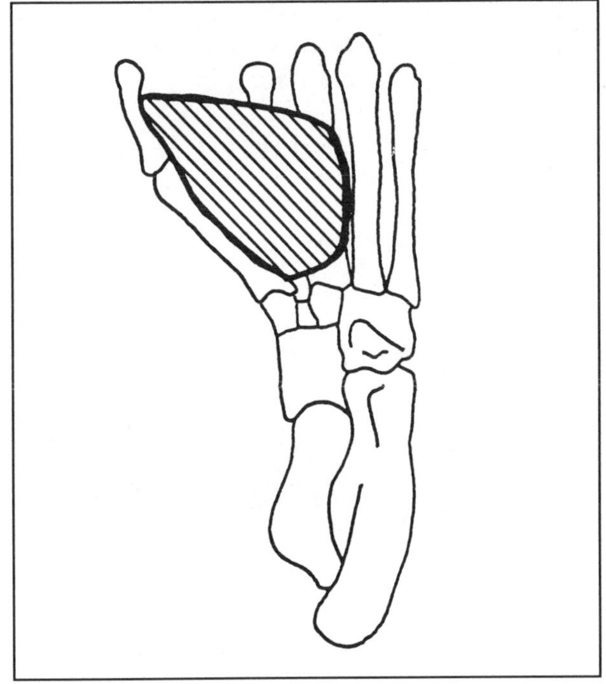

FIGURE 7.12. Adductor hallucis (hatched area) is especially prominent and both heads of this muscle are fused in lemurids and indriids.

Hall-Craggs, 1965b, 1966; Jouffroy and Lessertisseur, 1979; McArdle, 1981; Dagosto, 1986; Gebo, 1987, 1989a). First, the transverse tarsal joint is offset, with the calcaneo-cuboid joint being displaced well distally in the foot relative to the talo-navicular joint. The distal calcaneal joint surface has a very deep pivot region and the cuboid correspondingly possesses a prominent process. Both features help to maximize and stabilize the rotations at the calcaneo-cuboid joint. Second, a synovial joint between the calcaneus and the very elongated navicular has been developed (Fig. 7.13) and Hall-Craggs (1966) has explained how this synovial joint increases the strength of the distal tarsus, but does not reduce mobility. He noted that galagos are able to rotate the navicular around the calcaneus, which acts like a stable axis. Third, the navicular-cuneiform joints prohibit rotational mobility (Gebo, 1987). In galagos, the joint surface for the distal navicular is composed of three prongs, one for each of the cuneiforms, and these prongs form an irregular joint contour. This irregular contour severely prohibits rotational movements at this joint. In most other ways, like muscle development or basic joint function, galago feet are very similar to those of lemurids and cheirogaleids. In all, the great elongation of tarsal elements in the galago foot has led to several modifications in the bones and joints, which are responsible for foot rotation during inversion and grasping activities.

Tarsiers

Tarsiers possess the most specialized foot within the entire Order of Primates. Not only is the tarsier foot elongated and specialized for leaping (i.e., tarsiers possess a fused fibula, very large superficial flexor musculature for propulsion, and long tarsal elements, especially the calcaneus and navicular; see

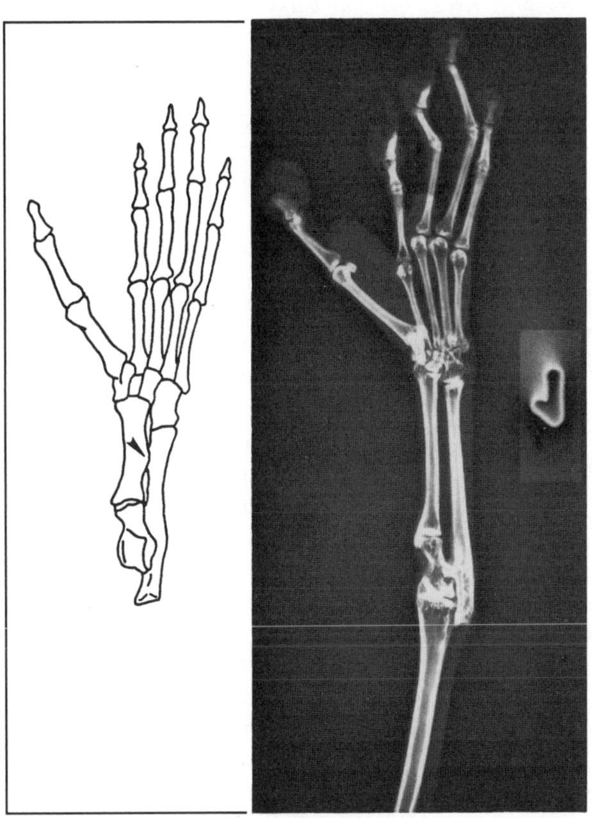

FIGURE 7.13. Galago and tarsier feet. On the left is a line drawing (dorsal view) of an everted galago foot while on the right is an x-ray of an everted tarsier foot. Note the long navicular and calcaneus in both and the unusual calcaneo-navicular synovial joint (arrow) in the galago. Compare with Figure 7.11.

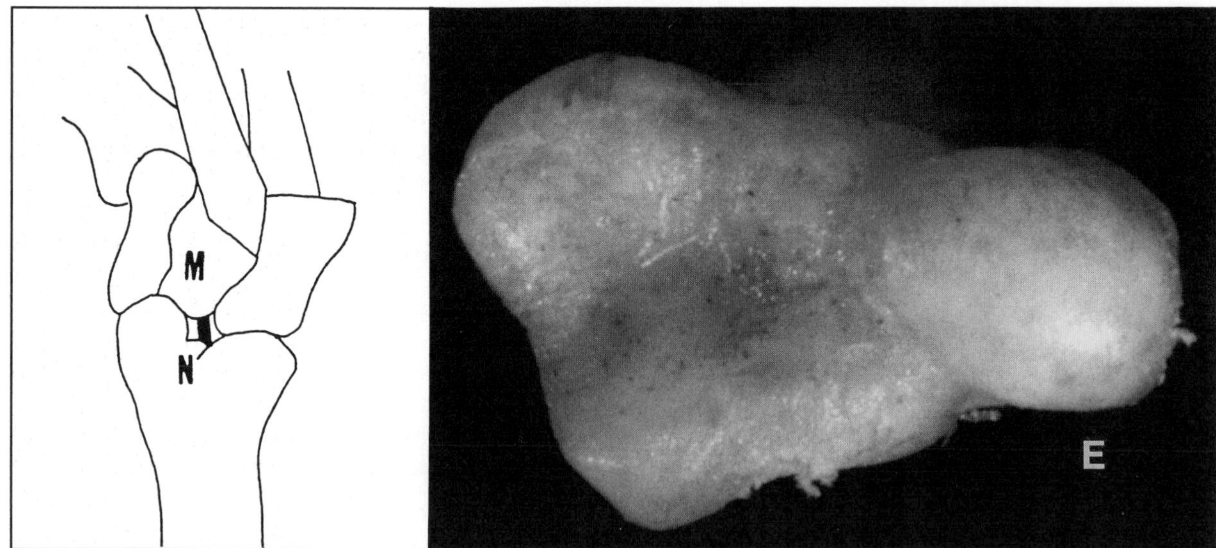

FIGURE 7.14. Tarsier navicular. The photograph on the right shows the distal surface of a tarsier navicular. Note the large round ball-like surface on the distal navicular (E). This surface articulates with the entocuneiform and because of its ball-like shape allows tarsiers to rotate the navicular at this distal joint surface. The line drawing on the left (a dorsal view of the joints between the distal navicular surface and the three cuneiforms) shows that the distal navicular joint surface does not contact the mesocuneiform as in other primates but has added a ligament (black bar) to connect these two bones, which contributes to the freedom of movement at this joint. N = navicular; M = mesocuneiform.

Fig. 7.13), it has actually undergone several more significant modifications in its musculoskeletal anatomy (see Gebo, 1987). First, tarsiers have greatly expanded toe pads and have the ability to arch their digits at the proximal and middle interphalangeal joints. Both of these adaptations, along with a specialized contrahentes musculature, allow tarsiers to put more pressure over a greater area at the tips of their toes, a helpful adaptation for slick vertical surfaces like bamboo. Second, tarsiers have very well-developed intrinsic foot muscles. Third, tarsiers have highly modified their foot joints for inversion. For example, the calcaneo-cuboid joint lacks a pivot in tarsiers (Hafferl, 1932; Szalay, 1976), the talar head is triangular in shape rather than round, and the distal navicular joint surface is highly modified (see discussion that follows). The first two features reflect a more limited range of mobility at the transverse tarsal joint, while the distal navicular joint surface is constructed for great mobility (Fig. 7.14). For example, the mesocuneiform facet of the navicular does not contact the mesocuneiform, and a ligament holds these two bones in close association (Fig. 7.14). Similarly, the entocuneiform facet is very round and ball-like in tarsiers (Hafferl, 1932) whereas the ectocuneiform joint surface is large and dished like an open C. Both of these distal navicular facets are unusual in their shape relative to other primates (Fig. 7.14). Thus, tarsiers have the ability (during foot inversion) to not only rotate the navicular at its proximal end (talo-navicular joint) but to rotate the distal end of the navicular (navicular-cuneiform joints) at the same time (Gebo, 1987). This proximal and distal rotation causes the entire navicular to slightly spin at both ends. This type of navicular rotation is highly unusual relative to other primates since the distal navicular joint surface normally locks up so that rotations at the talo-navicular joint may proceed. Thus, tarsiers possess a unique method of bone rotation within the foot and one that has evolved only once within the Order.

Old World Monkeys

In Old World monkeys, the toes have become shorter (including the big toe), and the key joints involved with bone rotations for foot inversion have

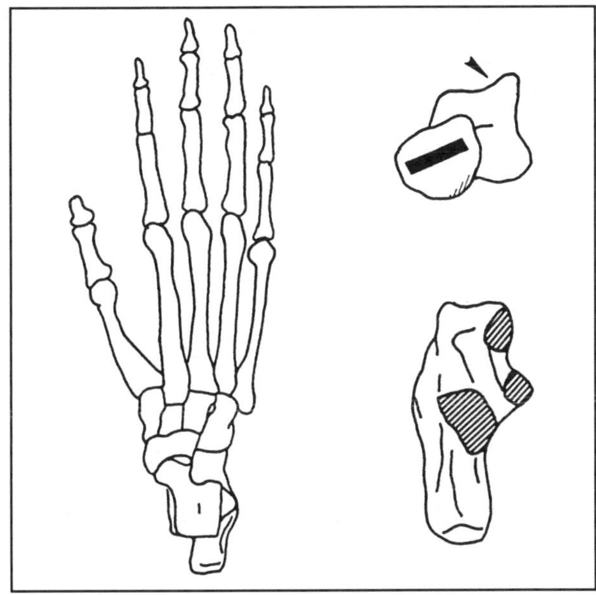

FIGURE 7.15. Old World monkey foot bones. Note the short digits in the foot (left); the asymmetrical upper ankle joint (arrow) with the high lateral edge (talus, upper right); the twisted talar head (black bar) and the plantar facet (hatching on the talar head), which is visible due to the twisting of the talar head; and the separated anterior calcaneal facets and the short and downward-angled posterior calcaneal facet (oblique hatching) on the calcaneus (lower right).

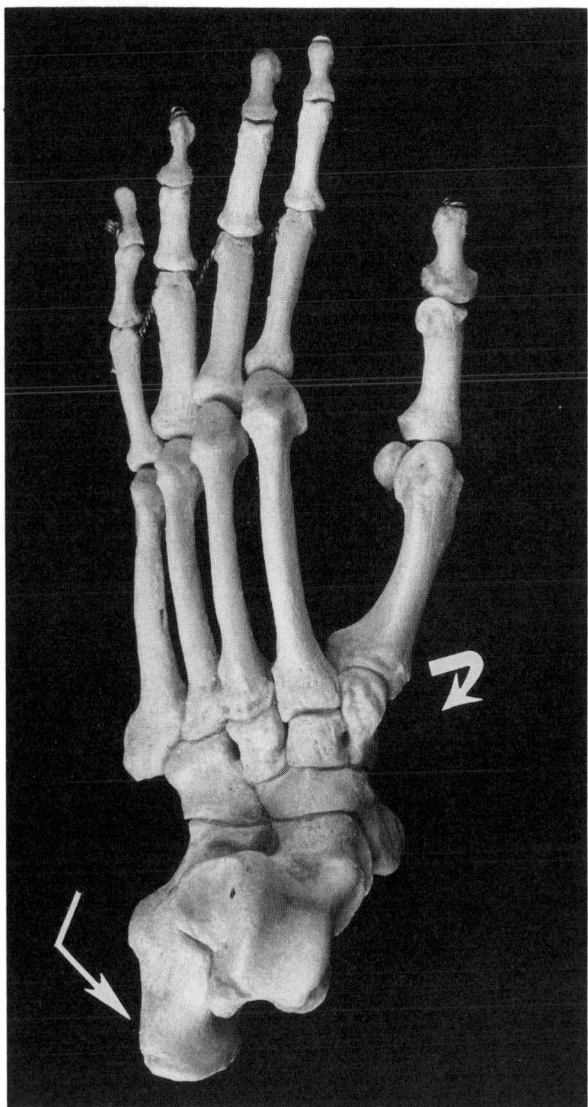

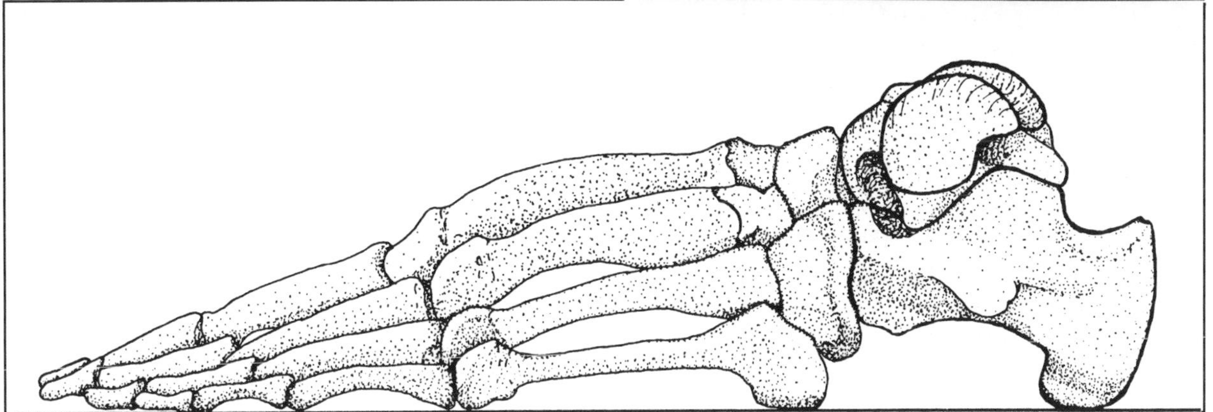

FIGURE 7.16. Top right: Dorsal view of an articulated chimpanzee foot. The white arrows illustrate the counter-rotation of the fore and aft parts of the foot during the stance phase of terrestrial knuckle-walking. Bottom: Lateral view of a chimpanzee foot illustrating heel contact with the substrate and elevation of the distal calcaneal end.

been modified to limit mobility (Fig. 7.15). Short toes, especially the big toe, imply a lessened emphasis on the ability to grasp branches. Reduced joint mobility similarly implies less capability to cope within an arboreal environment. For example, the upper ankle joint is asymmetical (Fig. 7.15), and this limits abduction of the foot, an important movement during inversion of the foot and when grasping vertical supports. Similarly, at the subtalar joint, the posterior calcaneal facet is short and bent downward (plantad) while the anterior (distal) calcaneal facet is separated into two facets with the most distal facet facing more medially than dorsally. These features reduce the distance for bones to slide across and thus limit inversion at the subtalar joint. At the transverse tarsal joint, the talar head is twisted (laterally) (Fig. 7.15) and this brings part of the plantar surface of the talar neck into view (anteriorly) while the calcaneo-cuboid joint surface is flat with a shallow pivot on the calcaneus and a correspondingly small cuboid process. Again, these features point to lessened abilities to rotate bones at the transverse tarsal joint, especially during foot inversion.

Although the feet of all Old World monkeys show reduced joint motion capabilities and toe lengths in comparison to other arboreal primates, many species of Old World monkeys are still highly arboreal and frequent climbers (e.g., Napier and Napier, 1967; Fleagle, 1977; Morbeck, 1979; Rose, 1979; Morbeck, 1979; Cant, 1988; Strasser, 1988; Gebo, 1989a). The only reasonable interpretation of this situation is to go beyond a basically mechanical "current use" view and to explain this phenomenon as a historical event. It seems clear that cercopithecoids went through a terrestrial phase in their evolution (e.g., Delson, 1975; Andrews and Aiello, 1984; Strasser, 1988; Gebo, 1989a), in which they lost their long toes and reduced mobility in their foot joints. Thus, all of the highly arboreal recent Old World monkeys are secondarily arboreal. This reinvasion of the arboreal niche must have been recent enough in evolutionary time that very few modifications of the foot reflecting arboreality can be detected. Relative to cercopithecines, however, the arboreal colobines have lengthened their fourth digit and have increased the size of the flexor digitorum longus muscle, as well as elongating the posterior calcaneal facet to facilitate their secondarily adapted life in the trees (see Strasser, 1988).

African Apes

The African apes use an unusual foot posture, plantigrady (Fig. 7.16), rather than utilizing the more common heel-elevated foot posture of other primates (Gebo, 1992). This novel method of placing the foot onto the ground with a lateral heel strike is only found to occur in the African apes and humans. This plantigrade foot position necessitates a complete reorientation of the angle of the calcaneus relative to the forefoot and thus reorients the upper ankle, and the subtalar and transverse tarsal joints as well (Fig. 7.16). Because the heel (i.e., the proximal end of the calcaneus) contacts the ground at the end of swing phase while the foot is inverted, the calcaneus is pinned under the weight of the body as the animal moves forward over its foot. Thus, the calcaneus is laterally rotated (inverted and pinned to the ground) while the forefoot is rotating medially (being everted and moving in the opposite direction of the calcaneus). Several bony modifications have occurred in the talus and the calcaneus of African apes to allow this unusual sequence of foot positions (see Gebo, 1992). For example, the talar head is angled downward (plantad) relative to the talar body, the sustentaculum tali has shifted proximally, the subtalar joint has lost some mobility, and the lateral side of the posterior end of the calcaneus (heel breadth) has expanded. This unusual foot adaptation in chimpanzees and gorillas appears to be related to terrestriality and is an adaptation that African apes share with humans (Gebo, 1992).

FOOT FUNCTION: PROSIMIANS VERSUS ANTHROPOIDS

In tooth-combed prosimians, the upper ankle joint is modified relative to tarsiers and anthropoids. The talar facet for the tibia is long and extends to the edge of the plantar talar body. These primates have also rotated the medial malleolus of the tibia (Dagosto, 1985). The talar facet is long in tarsiers, and both tarsiers and anthropoids possess a "straight" medial malleolus. Anthropoids, on the other hand, possess a dorsoplantarly shortened talar facet for the tibia (Gebo, 1986b). On the other side of the talus, tooth-combed prosimians possess a talar facet for the fibula that extends outward in an oblique direction away from the talar body, and this surface lacks a crease to separate the lateral process from the dorsal surface (Dagosto, 1986, 1988; Gebo, 1986a, 1986b, 1988; Beard et al., 1988). In tarsiers and anthropoids, this joint surface is steep-sided with a crease and a separate lateral process. On the posterior side of the talus, tooth-combed prosimians possess a trochlear joint surface that is offset from the groove for the flexor hallucis longus tendon whereas tarsiers and anthropoids possess a midline position for this groove relative to the trochlea (Fig. 7.17) (Gebo, 1986b).

In anthropoids, the reduced talar facet for the tibia and the steep lateral facet for the fibula indicate shorter bony prongs to hold the talus secure and lessened capabilities for foot abduction relative to tooth-combed prosimians. The shift in the digital axis in anthropoids from the fourth digit (in living prosimians) to the third indicates less peroneal development of the foot and thus lessened usage of vertical supports (Walker, 1974). A foot with a longer third digit is symmetrical and better for use on horizontal supports. Likewise, the decrease in the length of the peroneus longus tubercle on the first metatarsal along with a decrease in length and robusticity in anthropoids indicates a decrease in forceful adduction of the big toe during grasping. Less powerful grasping is adaptive for frequent use of horizontal supports rather than the more "gravity burdened" vertical support holds. Similarly, the midtrochlear position of the groove for flexor hallucis longus suggests a foot posture that is less vertically oriented. Thus, anthropoids possess a foot adapted for horizontal supports and above-branch quadrupedalism rather than for the vertical climbing of tooth-combed prosimians (Szalay and Dagosto, 1980; Rollinson and Martin, 1981; Gebo, 1986b).

Tarsiers are intermediate in morphology. They, like anthropoids, lack talar features to facilitate foot abduction but have other prosimian features like a long fourth digit and a large peroneal tubercle on the first metatarsal, plus several specialized features such as large toe pads, well-developed intrinsic foot muscles (Gebo, 1986b), and a transverse set of the talus upon the calcaneus (Jouffroy et al., 1984) that allows tarsiers to utilize vertical supports. Thus, although tarsier feet are highly specialized, they are adaptively most similar to the feet of tooth-combed prosimians.

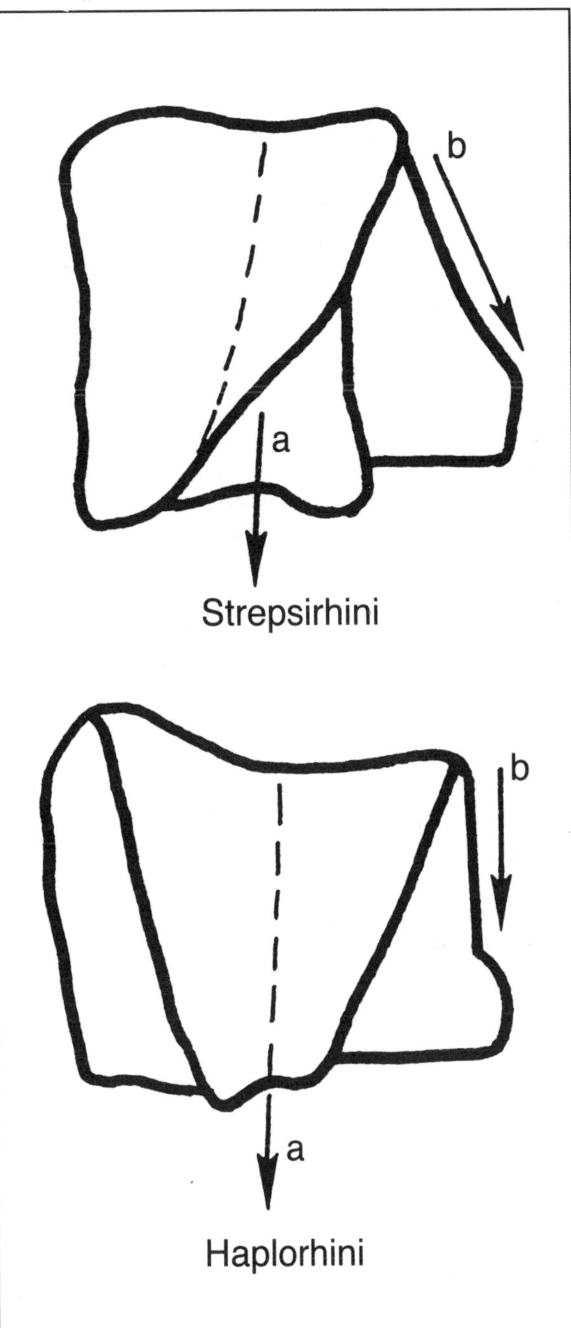

FIGURE 7.17. Posterior view of a primate talus. Note the position of the groove for flexor hallucis longus (a) and the slope of the talofibular facet (b). In tooth-combed prosimians or strepsirhini, the flexor groove is offset relative to the posterior surface of the trochlea, and the talofibular facet slopes outward. In tarsiers and anthropoids, or haplorhines, the groove is in a midline position relative to the trochlea, and the talofibular facet is steep-sided.

CONCLUSION

In the early 1900s, Gregory (1920), Morton (1922, 1924a, 1924b, 1927), Keith (1923, 1929), and Weidenreich (1923) began seriously to consider foot morphology as a tool to decipher primate and, in particular, human evolution. Since this early work, researchers such as Straus (1930), Hafferl (1932), Jouffroy (1962, 1975), Schultz (1963), Hall-Craggs (1965a, 1965b, 1966), Grand (1967), Cartmill (1972, 1974, 1979), Jenkins (1974), Jouffroy and Lessertiseur (1979), Jenkins and McClearn (1984), Szalay and his colleagues (Decker and Szalay, 1974; Szalay and Decker, 1974; Szalay, 1976; Szalay and Drawhorn, 1980, Szalay and Dagosto, 1988), Lewis (1980a, 1980b, 1980c, 1981, 1989), and others have continued this tradition until now we have quite a number of workers whose research focus is to understand foot anatomy, foot function, and primate and human evolution. In fact, studies on foot morphology are intricately involved in most, if not all, major debates concerning primate and human evolution today, which is a far cry from past decades.

In this chapter, I have attempted to briefly review the functional morphology of the primate foot. Almost all primates utilize a semiplantigrade grasping foot, with modifications from this pattern in orangutans, African apes, and humans. All prosimians, save one, have haired heels, and all anthropoids have extended the frictional surface of the forefoot to the heel. The importance of the primate grasping big toe has been widely acclaimed in the literature and is virtually without exception in the Order (i.e., out of over two hundred species of primates, only humans lack it). This novel adaptation can be traced back to the earliest part of the early Eocene and has been documented in a variety of adapid and omomyid taxa (Gregory, 1920; Simpson, 1940; Decker and Szalay, 1974; Dagosto, 1983, 1988; Rose and Walker, 1985; Covert, 1988; Gebo, 1988; Szalay and Dagosto, 1988; Gebo et al., 1991). Wood Jones (1916), Gregory (1920), Clark (1959), Martin (1968, 1986), Cartmill (1972, 1974), Szalay (1977), Szalay and Dagosto (1988), Rasmussen (1990), and Sussman (1991) have fully discussed the evolutionary importance of this novel adaptation and its importance to the successful radiation and survival of primate species. Similarly, primates have evolved foot musculature and joint anatomies to complement the grasping big toe, and these concordant adaptations allow the primate foot to grip and to hold onto a variety of curved surfaces within the arboreal environment. This innovative grasping foot has further been modified in a number of living lineages, and it is our knowledge of these patterns that has richly contributed to our understanding of primate locomotor adaptation and evolution. The next logical step is to apply this information to the known morphological record of extinct species. Fortunately, foot bones—and in particular tarsals—are moderately common in the primate fossil record. In fact, foot bones of fossil primates are known from the Paleocene to the Pliocene and are documented in a variety of taxa in each of these epochs (e.g., Gregory, 1920; Simpson, 1940; Walker, 1970; Oxnard, 1972; Day, 1973; Decker and Szalay, 1974; Wood, 1974; Conroy, 1976a, 1976b; Szalay, 1976; Lewis, 1981; Harrison, 1982; Latimer et al., 1982; Susman and Stern, 1982; Conroy and Rose, 1983; Dagosto, 1983, 1988; Fleagle, 1983; Langdon, 1984; Ford, 1986, 1988; Rose, 1986; Szalay and Langdon, 1986; Gebo, 1986c, 1988; Gebo and Simons, 1987; Latimer et al., 1987; Covert, 1988; Latimer and Lovejoy, 1989; Gebo et al., 1990, 1991; Meldrum, 1990). Thus, one of the major goals of functional morphology is better realized in this anatomical system than in other areas of the postcranial skeleton. Although considerable research has been conducted on this fossil material and major questions still remain, it is quite obvious from this body of work that our understanding of foot morphology has grown over the decades and in the process has made significant contributions to phylogenetic studies. This trend will continue, and I look forward to the new adaptive and phylogenetic hypotheses generated from study of this region of the body.

NOTES

1. Patas monkeys (*Erythrocebus patas*) possess digitigrade feet.

2. The only exception to this is *Lemur catta*, which has extended the frictional surface to its heel region.

LITERATURE CITED

Andrews P, and Aiello L (1984) An evolutionary model for feeding and positional behaviour. In DJ Chivers, BA Wood, and A Bilsborough (eds.): In Food Acquisition and Processing in Primates. New York: Plenum Press, pp. 429–466.

Anemone RL (1990) The VCL hypothesis revisited: Patterns of femoral morphology among quadrupedal and saltatorial prosimian primates. Am. J. Phys. Anthrop. *83*:373–393.

Beard KC, Dagosto M, Gebo DL, and Godinot M (1988) Interrelationships among primate higher taxa. Nature *331*:712–714.

Biegert J (1963) The evaluation of characteristics of the skull, hands, and feet for primate taxonomy. In SL Washburn (ed.): Classification and Human Evolution. Chicago: Aldine Publishing Company, pp. 116–145.

Cant JGH (1987) Positional behavior of female Bornean orangutans (*Pongo pygmaeus*). Am. J. Primatol. *12*:71–90.

——— (1988) Positional behavior of long-tailed macaques (*Macaca fascicularis*) in Northern Sumatra. Am. J. Phys. Anthrop. *76*:29–37.

Cartmill M (1972) Arboreal adaptations and the origin of the Order Primates. In R Tuttle (ed.): Biology of Primates. New York: Academic Press, pp. 97–122.

——— (1974) Pads and claws in arboreal locomotion. In FA Jenkins (ed.): Primate Locomotion. New York: Academic Press, pp. 45–83.

——— (1979) The volar skin of primates: Its frictional characteristics and their functional significance. Am. J. Phys. Anthrop. *50*:497–510.

Clark WEL (1936) The problem of the claw in primates. Proc. Zool. Soc. Lond. *I*:1–24.

——— (1959) The Antecedents of Man. Edinburgh: Edinburgh University Press.

Conroy GC (1976a) Hallucial tarsometatarsal joint in an Oligocene anthropoid *Aegyptopithecus zeuxis*. Nature *262*:684–686.

——— (1976b) Primate postcranial remains from the Oligocene of Egypt. Contrib. Primatol. *8*:1–134.

Conroy GC, and Rose MD (1983) The evolution of the primate foot from the earliest primates to the Miocene. Foot and Ankle *3*:342–364.

Covert HH (1988) Ankle and foot morphology of *Cantius mckennai*: Adaptations and phylogenetic implications. J. Hum. Evol. *17*:57–70

Dagosto M (1983) Postcranium of *Adapis parisiensis* and *Leptadapis magnus* (Adapiformes, Primates). Folia Primatol. *41*:49–101.

——— (1985) The distal tibia of primates with special reference to the Omomyidae. Int. J. Primatol. *6*:45–75.

——— (1986) The Joints of the Tarsus in Strepsirhine Primates: Functional, Adaptive, and Evolutionary Implications. Ph.D. dissertation, City University of New York, New York.

——— (1988) Implications of postcranial evidence for the origin of euprimates. J. Hum. Evol. *17*:35–57.

Day MH (1973) Locomotor affinities of hominoid tali from Kenya. Nature *246*:45–46.

Decker RL, and Szalay FS (1974) Origins, evolution and function of the tarsus in late Cretaceous eutherians and Paleocene primates. In FA Jenkins (ed.): Primate Locomotion. New York: Academic Press, pp. 261–292.

Delson E (1975) Evolutionary history of the Cercopithecidae. Contrib. Primatol. *5*:167–217.

Elftman H, and Manter J (1935) The evolution of the human foot, with special reference to the joints. J. Anat. *70*:56–67.

Fleagle JG (1977) Locomotor behavior and skeletal anatomy of sympatric Malaysian leaf-monkeys (*Presbytis obscura* and *Presbytis melalophos*). Yrbk. Phys. Anthrop. *20*:440–453.

——— (1983) Locomotor adaptations of Oligocene and Miocene hominoids and their phyletic implications. In RL Ciochon and RS Corruccini (eds.): New Interpretations of Ape and Human Ancestry. New York: Plenum Press, pp. 301–324.

Ford SM (1986) Systematics of the New World monkeys. In D Swindler (ed.): Comparative Primate Biology. Vol. 1, Systematics, Evolution, and Anatomy. New York: Alan R. Liss, pp. 73–135.

——— (1988) Postcranial adaptations of the earliest platyrrhine. J. Hum. Evol. *17*:155–192.

Garber PA (1980) Locomotor behavior and feeding ecology of the Panamanian tamarin (*Saguinus oedipus geoffroyi*, Callitrichidae, Primates). Int. J. Primatol. *1*:185–201.

——— (1984) Use of habitat and positional behavior in a neotropical primate, *Saguinus oedipus*. In PS Rodman and JGH Cant (eds.): Adaptations for Foraging in Nonhuman Primates. New York: Columbia University Press, pp. 112–133.

——— (1992) Vertical clinging, small body size and the evolution of feeding adaptations in the

Callitrichinae. Am. J. Phys. Anthrop. *88:*469–482.

Gebo DL (1985) The nature of the primate grasping foot. Am. J. Phys. Anthrop. *67:*269–277.

——— (1986a) The Anatomy of the Prosimian Foot and Its Application to the Primate Fossil Record. Ph.D. dissertation, Duke University, Durham.

——— (1986b) Anthropoid origins—the foot evidence. J. Hum. Evol. *15:*421–430.

——— (1986c) Miocene lorisids—the foot evidence. Folia Primatol. *47:*217–225.

——— (1987) The functional anatomy of the tarsier foot. Am. J. Phys. Anthrop. *57:*9–31.

——— (1988) Foot morphology and locomotor adaptation in Eocene primates. Folia Primatol. *50:*3–41.

——— (1989a) Locomotor and phylogenetic considerations in anthropoid evolution. J. Hum. Evol. *3:*201–234.

——— (1989b) Postcranial adaptation and evolution in Lorisidae. Primates *30:*347–367.

——— (1992) Plantigrady and foot adaptation in African apes: implications for hominid evolution. Am. J. Phys. Anthrop. *89:*29–58.

Gebo DL, and Dagosto M (1988) Foot morphology, climbing, and the evolution of the Indriidae. J. Hum. Evol. *17:*135–154.

Gebo DL, Dagosto M, and Rose KD (1991) Foot morphology and locomotor evolution in early Eocene *Cantius*. Am. J. Phys. Anthrop. *86:*51–73.

Gebo DL, Dagosto M, Rosenberger AL, and Setoguchi T (1990) New platyrrhine tali from La Venta, Colombia. J. Hum. Evol. *19:*737–746.

Gebo DL, and Simons EL (1987) Foot morphology and locomotor adaptation in early Oligocene anthropoids. Am. J. Phys. Anthrop. *74:*83–102.

Grand TI (1967) The functional anatomy of the ankle and foot of the slow loris (*Nycticebus coucang)*. Am. J. Phys. Anthrop. *28:*168–182.

Gregory WK (1920) On the structure and relations of *Notharctus,* an American Eocene primate. Mem. Amer. Mus. Nat. Hist. *3:*51–243.

Hafferl A (1932) Bau und Funktion des AffenfuBes. Ein Beitrag zur Gelenk- und Muskelmechanik, II. Die Prosimier. Z. Anat. Entwgesch. *99:*63–112.

Hall-Craggs ECB (1965a) An analysis of the jump of the lesser galago. J. Zool. Lond. *147:*20–29.

——— (1965b) An osteometric study of the hind limb of the Galagidae. J. Anat. Lond. *99:*119–125.

——— (1966) Rotational movements in the foot of *Galago senegalensis*. Anat. Rec. *154:*287–294.

Harrison T (1982) Small Bodied Apes from the Miocene of East Africa. Ph.D. dissertation, University College, London.

Hill WCO (1953) Primates: Comparative Anatomy and Taxonomy. Vol. 1, Strepsirhini. Edinburgh: Edinburgh University Press.

——— (1960) Primates: Comparative Anatomy and Taxonomy. Vol. 5, Cebidae. Edinburgh: Edinburgh University Press.

Jenkins FA (1974) Tree shrew locomotion and the origins of primate arborealism. In FA Jenkins (ed.): Primate Locomotion. New York: Academic Press, pp. 85–115.

Jenkins FA, and McClearn D (1984) Mechanisms of hind foot reversal in climbing mammals. J. Morphol. *182:*197–219.

Jolly C, and Gorton AT (1974) Proportions of the extrinsic foot muscles in some lorisid prosimians. In RD Martin, GA Doyle, and AC Walker (eds.): Prosimian Biology. London: Duckworth, pp. 801–816.

Jouffroy FK (1962) La musculature des membres chez les lemuriens de Madagascar. Mammalia *26:*1–326.

——— (1975) Osteology and myology of the lemuriform postcranial skeleton. In I Tattersall and R Sussman (eds.): Lemur Biology. New York: Plenum Press, pp. 149–192.

Jouffroy FK, Berge C, and Niemitz C (1984) Comparative study of the lower extremity in the genus *Tarsius*. In C Niemitz (ed.): Biology of Tarsiers. New York: Gustav Fischer, pp. 167–190.

Jouffroy FK, and Lessertisseur J (1979) Relationships between limb morphology and locomotor adaptations among prosimians: An osteometric study. In ME Morbeck, H Preuschoft, and N Gomberg (eds.): Environment, Behavior, and Morphology: Dynamic Interactions in Primates, New York: Gustav Fischer, pp. 143–182.

Kapandji IA (1970) The Physiology of Joints. London: Churchill Livingstone Press.

Keith A (1923) Man's posture: Its evolution and disorders. Brit. Med. J. *1:*669–672.

——— (1929) The history of the human foot and its bearing on orthopaedic practice. J. Bone Joint Surg. *11:*10–32.

Langdon JH (1984). The Miocene Hominoid Foot. Ph.D. dissertation, Yale University, New Haven.

Latimer BM, and Lovejoy CO (1989) The calcaneus of *Australopithecus afarensis* and its implications for the evolution of bipedality. Am. J. Phys. Anthrop. *82:*125–133.

Latimer BM, Lovejoy CO, Johanson DC, and Coppens Y (1982) Hominid tarsal, metatarsal and phalangeal bones recovered from the Hadar Formation: 1974–1977 collections. Am. J. Phys. Anthrop. *57:*701–719.

Latimer BM, Ohman JC, and Lovejoy CO (1987) Talocrural joint in African hominioids: Implications for *Australopithecus afarensis.* Am. J. Phys. Anthrop. *74:*155–175.

Lewis OJ (1980a) The joints of the evolving foot. Part 1, The ankle joint. J. Anat. *130:*527–543.

——— (1980b) The joints of the evolving foot. Part 2, The intrinsic joints. J. Anat. *130:*833–857.

——— (1980c) The joints of the evolving foot. Part 3, The fossil evidence. J. Anat. *131:*275–298.

——— (1981) Functional morphology of the joints of the evolving foot. Symp. Zool. Soc. Lond. *46:*169–188.

——— (1989) Functional Morphology of the Evolving Hand and Foot. Oxford: Clarendon Press.

McArdle JE (1981) The functional morphology of the hip and thigh of the lorisiformes. Contrib. Primatol., Vol. 17. Basel: Karger.

Martin RD (1968) Towards a new definition of primates. Man *3:*377–401.

——— (1986) Primates: A definition. In B Wood, L Martin, and P Andrews (eds.): Major Topics in Primate and Human Evolution. Cambridge: Cambridge Universtiy Press, pp. 1–31.

Meldrum DJ (1990). New fossil platyrrhine tali from the early Miocene of Argentina. Am. J. Phys. Anthrop. *83:*403–418.

——— (1991) Kinematics of the cercopithecine foot on arboreal and terrestrial substrates with implications for the interpretation of hominid terrestrial adaptations. Am. J. Phys. Anthrop. *84:*273–289.

Mivart St G (1873) On *Lepilemur* and *Cheirogaleus,* and on the zoological rank of the Lemuroidea. Proc. Zool. Soc. Lond. 1873, pp. 484–510.

Morbeck ME (1979) Forelimb use and positional adaptation in *Colobus guereza:* Integration of behavioral, ecological, and anatomical data. In ME Morbeck, H Preuschoft, and N Gomberg (eds.): Environment, Behavior, and Morphology. New York: Gustav Fischer, pp. 95–118.

Morton DJ (1922) Evolution of the human foot, Part 1. Am. J. Phys. Anthrop. *5:*305–336.

——— (1924a) Evolution of the human foot, Part 2. Am. J. Phys. Anthrop. *7:*1–52.

——— (1924b) Evolution of the longitudinal arch of the human foot. J. Bone Joint Surg. *6:*56–90.

——— (1927) Human origin: Correlation of previous studies of primate feet and posture with other morphologic evidence. Am. J. Phys. Anthrop. *10:*173–203.

Murie J, and Mivart St. G (1872) On the anatomy of the lemuridae. Trans. Zool. Soc. Lond. *7:*1–113.

Napier JR (1967) Evolutionary aspects of primate locomotion. Am. J. Phys. Anthrop. *27:*333–342.

Napier JR, and Napier PH (1967) Handbook of Living Primates. New York: Academic Press.

Oxnard CE (1972) Some African foot bones: A note on the interpolation of fossils into a matrix of extant species. Am. J. Phys. Anthrop. *37:*3–12.

Rasmussen DT (1990) Primate origins: Lessons from a neotropical marsupial. Int. J. Primatol. *22:*263–277.

Rollinson J, and Martin RD (1981) Comparative aspects of primate locomotion, with special reference to arboreal cercopithecines. Symp. Zool. Soc. Lond. *48:*377–421.

Rose KD, and Walker A (1985) The skeleton of early Eocene *Cantius,* oldest lemuriform primate. Am. J. Phys. Anthrop. *66:*73–90.

Rose MD (1979) Positional behavior of natural populations: Some quantitative results of a field study of *Colobus guereza* and *Cercopithecus aethiops.* In ME Morbeck, H Preuschoft, and N Gomberg (eds.): Environment, Behavior, and Morphology. New York: Gustav Fischer, pp. 75–94.

——— (1986). Further hominoid postcranial specimens from the Late Miocene Nagri Formation of Pakistan. J. Hum. Evol. *15:*333–368.

——— (1988) Functional anatomy of the cheiridia. In JH Schwartz (ed.): Orang-utan Biology. New York: Oxford University Press, pp. 299–310.

Sarmiento EE (1983) The significance of the heel process in anthropoids. Int. J. Primatol. *4:*127–152.

Schultz AH (1963) Relations between the lengths of the main parts of the foot skeleton in primates. Folia Primatol. *1:*150–171.

Simpson GG (1940) Studies on the earliest primates. Bull. Amer. Mus. Nat. Hist. *77:*185–212.

Strasser E (1988) Pedal evidence for the origin and diversification of cercopithecid clades. J. Hum. Evol. *17:*225–246.

Straus WL (1930) The foot musculature of the highland gorilla (*Gorilla beringei*). Quar. Rev. Biol. *5:*261–317.

Susman RL (1983) Evolution of the human foot: Evidence from Plio-Pleistocene hominids. Foot and Ankle *3:*365–376.

Susman RL, and Stern J (1982) Functional morphology of *Homo habilis.* Science *217:*931–934.

Sussman RW (1991) Primate origins and the evolution of angiosperms. Am. J. Primatol. *23:*209–223.

Szalay FS (1976) Systematics of the Omomyidae (Tarsiiformes, Primates): Taxonomy, phylogeny and adaptations. Bull. Amer. Mus. Nat. Hist. *156:*157–450.

——— (1977) Constructing primate phylogenies: A search for testable hypotheses with maximum empirical content. J. Hum. Evol. *6:*3–18.

Szalay FS, and Dagosto M (1980) Locomotor adaptations as reflected on the humerus of Paleogene primates. Folia Primatol. *34:*1–45.

——— (1988) Evolution of hallucial grasping in the primates. J. Hum. Evol. *17:*1–33.

Szalay FS, and Decker RL (1974) Origins, evolution and function of the tarsus in late Cretaceous eutherians and Paleocene primates. In FA Jenkins (ed.): Primate Locomotion. New York: Academic Press, pp. 223–259.

Szalay FS, and Drawhorn (1980) Evolution and diversification of the Archonta in an arboreal milieu. In WP Luckett (ed.): Comparative Biology and Evolutionary Relationships of Tree Shrews. New York: Plenum Press, pp. 133–169.

Szalay FS, and Langdon JH (1986) The foot of *Oreopithecus:* An evolutionary assessment. J. Hum. Evol. *15:*585–621.

Tuttle RH (1969) Knuckle-walking and the problem of human origins. Science *166:*953–961.

——— (1970) Postural, propulsive, and prehensile capabilities in the cheiridia of chimpanzees and other great apes. In GH Bourne (ed.): The Chimpanzee, Vol. 2. Basel: Karger, pp. 167–253.

——— (1972). Relative mass of cheiridial muscles in catarrhine primates. In RH Tuttle (ed.): In Function and Evolutionary Biology of the Primates. Chicago: Aldine-Atherton Press, pp. 262–291.

Walker A (1970) Postcranial remains of the Miocene Lorisidae of East Africa. Am. J. Phys. Anthrop. *33:*249–261.

——— (1974) Locomotor adaptations in past and present prosimian primates. In FA Jenkins (ed.): Primate Locomotion, New York: Academic Press, pp. 349–381.

Weidenreich F (1923) Evolution of the human foot. Am. J. Phys. Anthrop. *26:*473–487.

Wood BA (1974) Locomotor affinities of hominoid tali from Kenya. Nature *246:*45–46.

Wood Jones F (1916) Arboreal Man. London: E. Arnold.

Part III Extinct Primates

Postcranial Adaptations and Locomotor Behavior

C H A P T E R E I G H T

Postcranial Anatomy and Locomotor Behavior in Eocene Primates

Marian Dagosto

Primates from early Eocene deposits of Europe and North America are the first to exhibit phylogenetically derived similarities to living primates in both the skull and the postcranium. They have therefore been called the "first primates of modern aspect" (Simons, 1972). Taxonomically, these nonplesiadapiform primates (and their more recent descendants) are classified as Euprimates (Hoffstetter, 1977). The two major groups of Euprimates living in the Eocene were the Adapidae and the Omomyidae. Both groups are now extinct, but all living primates are thought to have descended from them. In many introductory texts, adapids are portrayed as similar to the living lemurs and the omomyids as similar to the living galagos and/or tarsiers. Variation within each group, especially among later members, makes it difficult to justify such broad generalizations. In fact, both groups are quite variable in their postcranial anatomy and presumed locomotor behavior. There are also significant differences between these fossil Eocene primates and their living descendants.

The following sections summarize what is known about the postcranial anatomy of each group. It would be impossible to discuss all the relevant details; the reader is referred to the original literature for more information. This review concentrates on the major differences between groups or on those anatomical features of particular functional or phylogenetic consequence.

When describing the fossil primates, comparisons are often made to the living primates commonly referred to as "prosimians." This assemblage in fact contains members of two different suborders; the Strepsirhini, which includes the lemurs and lorises, and the Haplorhini, which includes the tarsiers. (See Szalay and Delson, 1979, or Fleagle, 1988, for a classification of primates.) The living members of the Strepsirhini are grouped in the infraorder Lemuriformes (the tooth-combed lemurs) and divided into two superfamilies: Lemuroidea (including the families Lemuridae and Indriidae) and the Lorisoidea (for the families Cheirogaleidae, Lorisidae, and Galagidae). The Lemuroidea and the Cheirogaleidae inhabit the island of Madagascar and are thus referred to as the Malagasy prosimians.

Living prosimian primates exhibit a wide range of locomotor and postural behaviors. The most commonly delineated modes of behavior are called arboreal quadrupedalism (AQ) in which tree-dwelling animals move primarily by walking, running, or leaping along horizontally oriented branches; vertical clinging and leaping (VCL) in which animals progress by a series of leaps between vertical supports; and slow climbing (SC), in which the progression along branches involves varied body postures and support types, but leaping is never practiced (Napier and Napier, 1967; Napier and Walker, 1967a). Many prosimians are capable of

A. ADAPIDS

5myBP	NORTH AMERICA **Notharctinae**	EUROPE **Adapinae**	ASIA **Sivaladapinae**
Miocene			*Indraloris* *Sivaladapis* *Sinoadapis*
25			
Oligocene			
34			
Late Eocene	*Mahgarita*	*Huerzeleris* ***Adapis*** *Periconodon* ***Leptadapis*** *Microadapis* ***Caenopithecus***	
48			
Middle Eocene	***Smilodectes*** ***Notharctus***	***Pronycticebus*** ***Europolemur***	
51			
Early Eocene	*Copelemur* ***Pelycodus*** ***Cantius***	*Protoadapis* *Cantius* *Donrussellia*	
57			

FIGURE 8.1. Simplified stratigraphic range and distribution of principal genera of Adapidae (A) and Omomyidae (B). Genera for which postcranial material has been reported are in bold. *Teilhardina,* an anaptomorphine, and *Cantius,* a notharctine, are found in both Europe and North America. Although found in North America, *Mahgarita* is related to the Protoadapinae. *Ekgmowechashala* is not an omomyine but is placed in its own subfamily.

using all of these modes, but most species have a characteristic dominant mode of locomotion. With some exceptions, most lemurids use AQ; indriids, galagos, and tarsiers use VCL; and lorisines use SC.

ADAPIDAE

Adapidae is usually divided into three subfamilies: the Notharctinae, which lived primarily in western North America during the early and middle Eocene; the Adapinae, which lived primarily in Europe during most of the Eocene; and the Sivaladapinae, which lived in Asia during the Miocene (Fig. 8.1A). The limb skeleton of North American representatives of the Notharctinae is well known from nearly complete skeletons of the middle Eocene *Notharctus* and *Smilodectes,* and many elements of the early Eocene *Cantius* (see Table 8.1). There are, however, still several taxa (*Copelemur, Pelycodus,* many species of *Cantius)* that are almost completely unknown. The European adapids are less well represented. Postcranial bones of early Eocene European adapids have yet to be described. Of the middle Eocene species, only *Europolemur* and *Pronycticebus* have associated postcranial material, and this was only discovered during the last few years (Franzen, 1987; Thalmann et al., 1989). These and two other adapid hindlimb skeletons and one forelimb skeleton (von Koenigswald, 1979, 1985; Franzen, 1988) are known from fairly complete, if fragile fossils. Two late Eocene adapids, *Adapis* and *Leptadapis,* are fairly well known from

B. OMOMYIDS

		NORTH AMERICA		EUROPE
	5myBP	**Anaptomorphinae**	**Omomyinae**	**Microchoerinae**
Miocene			*[Ekgmowechashala]*	
	25			
Oligocene				
	34			
Late Eocene			*Rooneyia*, *Macrotarsius*, ***Dyseolemur***, *Ourayia*, *Chumashius*	*Pseudoloris*, ***Microchoerus***, ***Necrolemur***
	48			
Middle Eocene		*Trogolemur*, *Anaptomorphus*, *Chlorohysis*	***Washakius***, ***Hemiacodon***, ***Omomys***, ***Shoshonius***	***Nannopithex***
	51			
Early Eocene		*Steinius*, ***Absarokius***, ***Arapahovius***, ***Tetonius***, ***Anemorhysis***, *Teilhardina*	*Loveina*, *Uintanius*, *Jemezius*	***Teilhardina***
	57			

attributed elements (Dagosto, 1983; Godinot and Jouffroy, 1984; Godinot, 1991). The Miocene Asian sivaladapines are not yet represented by any postcranial remains.

The division of Adapidae into Notharctinae and Adapinae—based on dental morphology—does not describe the pattern of variation of the skeleton. The middle Eocene European adapids (the Protoadapini of Szalay and Delson, 1979) resemble the North American Notharctinae; only the late Eocene European adapinines (*Adapis, Leptadapis,* and *Caenopithecus)* depart significantly from this anatomy (Martin, 1979; Dagosto, 1983; Thalmann et al., 1989). In the following discussion, therefore, unless otherwise noted, anatomical descriptions of "notharctines" apply to all known North American and middle Eocene European adapids; the descriptions of "adapinines" apply only to the known late Eocene adapids.

The Adapidae discussed here range from approximately *Hapalemur*-sized (about 800 gm) for the smallest *Cantius* to possibly as much as 8 to 10 kg in *Leptadapis. Notharctus* and *Smilodectes* were approximately 1.5 to 4 kg (Gingerich et al., 1982; Dagosto and Terranova, 1992). Smaller adapids are known from dental remains, but these species are not yet represented by any postcranial material.

Minor differences between taxa within each group are not always noted (see Dagosto, 1983; Covert, 1985; Beard and Godinot, 1988; Gebo, 1988; Gebo et al. 1991; and Godinot, 1991, for more information).

Limb proportions

Like the VCL prosimians, *Notharctus* and *Smilodectes* have a low intermembral index (Table 8.2), indicating that the hindlimbs were markedly longer than the forelimbs (Napier and Walker, 1967a, 1967b; Stern and Oxnard, 1973).

TABLE 8.1
Eocene Primate Taxa for Which Postcranial Remains Are Known

Taxa	References
ADAPIDAE	
Notharctinae	
Cantius	Covert, 1988; Gebo, 1987; Gebo et al., 1991; Matthew, 1915; Rose and Walker, 1985
Pelycodus	Gebo et al., 1991
Notharctus	Beard and Godinot, 1988; Covert, 1985; Gebo, 1985; Gregory, 1920
Smilodectes	Beard and Godinot, 1988; Covert, 1985
Adapinae	
Europolemur	Franzen, 1987
Pronycticebus	Thalmann et al., 1989
Messel adapids	Franzen, 1988; von Koenigswald, 1979, 1985
Adapis	Beard and Godinot, 1988; Dagosto, 1983; Filhol, 1882, 1883; Godinot, 1991; Godinot and Jouffroy, 1984; Schlosser, 1887
Leptadapis	Filhol, 1883; Dagosto, 1983; Decker and Szalay, 1974; Schlosser, 1887
OMOMYIDAE	
Anaptomorphinae	
Teilhardina	Gebo, 1988; Szalay, 1976
Tetonius	Gebo, 1988; Rosenberger and Dagosto, 1992; Szalay, 1976
Arapahovius	Savage and Waters, 1978
Absarokius	Covert and Hamrick, n.d.
Omomyinae	
Hemiacodon	Dagosto, 1985; Gebo, 1988; Simpson, 1940; Szalay, 1976
Omomys	Rosenberger and Dagosto, 1992
Washakius	Szalay, 1976
Microchoerinae	
Microchoerus	Schlosser, 1907; Schmid, 1979
Necrolemur	Dagosto, 1985; Godinot and Dagosto, 1983; Schlosser, 1907; Schmid, 1979
Nannopithex	Weigelt, 1933

Pronycticebus may have had a higher intermembral index, more like that of AQ prosimians (Thalmann et al., 1989). Other similarities between *Notharctus* and VCL prosimians are the relatively long hand and long phalanges (Table 8.2; Stern and Oxnard, 1973). Compared to any leaping prosimians, however, notharctines have a low brachial index (relatively short radius compared to humerus). No associated remains can be evaluated for adapinines, but from isolated elements it appears that, unlike notharctines or any Malagasy prosimians, the forelimbs and hindlimbs were probably of nearly equal length (Dagosto, 1983).

Forelimb

Gregory (1920) and Thalmann et al. (1989) have favorably compared the shoulder blade and clavicle of notharctines to those of living lemuriforms. In *Europolemur,* Franzen (1987) noted similarities to *Galago, Indri, Propithecus,* and *Lemur* in the relative size of the infraspinous fossa compared to the supraspinous fossa. Covert (1985) found that in *Smilodectes* the infraspinous fossa projects more than in AQ lemurids, but less than in indriids, which practice orthograde climbing and arm-hanging.

As in most arboreal primates, the humeral head of notharctines faces posteriorly and is fairly spherical except in *Notharctus* and *Smilodectes,* where it is more oval (Fleagle and Simons, 1982; Gebo, 1987; Fig. 8.2), but not as mediolaterally constricted as in modern lemurids or indriids. The greater and lesser tuberosities are of average size and lie slightly below the most cranial extent of the articular surface of the humeral head. The brachioradialis crest is very broad and extends further proximally than in living prosimians (Gregory, 1920), and the deltopectoral crest extends further distally (Fig. 8.2). The humerus of the adapinines has a broader and shallower bicipital groove, a thicker deltopectoral crest, and a teres major tuberosity that is more marked and distally extensive (Filhol, 1882, 1883; Gregory, 1920; Dagosto, 1983). These features indicate more powerful forearm musculature, which probably implies more frequent climbing. The humeral head is spherical.

All adapids except *Adapis* itself share a pattern of distal humeral anatomy that is typical of most living strepsirhines, is probably primitive for euprimates,

TABLE 8.2 Comparative Data for Postcranial Indices Mentioned in the Text

	Intermembral Index	Hand Index	Phalangeal Index	Humerus/femur Humerus/tibia	Brachial Index	Trochlear Width	Trochlear Height	Lateral Condyle Ht.	Distal Calcaneus	Navicular Length
ADAPIDS										
Cantius						44	73	99	43	
Notharctus	60	[35]	[60–65]	[57/66]	94	38	97	99	41	98
Smilodectes	61			[xx/68]	89	35	90	97	40	
Europolemur					80					
Pronycticebus	(<73)			[xx/73]	[80]	[45]	[71]			
Adapis				[84/95]		38	76	80	32	
Leptadapis				[xx/90]		39	77		31	
OMOMYIDS										
Teilhardina									51	
cf. *Tetonius*						44	62			
Arapahovius										[185]
Hemiacodon						43	48	116	52	186
cf. *Washakius*									52	
Microchoerus				[59/xx]		44	72	[>99]	[>50]	
Necrolemur				[xx/53]		49	76		[>67]	
LIVING PROSIMIANS										
VCLs										
Avahi	56	34	58	48/55	117	36	106	115	41	82
Propithecus	64	33	56	53/61	109	31	99	103	44	74
Indri	64	33	55	53/61	121	35	112	100	44	67
Lepilemur	64	32	56	55/58	114	36	97	106	48	126
Galago	62	36	63	48/54	109	39	108	110	72	401
Tarsius	55	40	64	48/48	128	40	131	115	75	500
AQs										
Lemur	70	29	55	69/71	106	48	87	95	44	104
Varecia	71	30	56	70/75	97	47	89	96	41	88
Cheirogaleus	71	31	58	71/70	97	47	73	85–95	49	115–120
Microcebus	71	28	61	71/64	119	48	76	95	64	213
Phaner	67	30	57	66/63				95	59	209

Intermembral index = ([humerus length + radius length]/[femur length + tibia length] × 100); the lower the index, the longer the hindlimb is relative to the forelimb. VCLs have relatively low indices. **Hand index** = [hand length/forelimb length × 100]. The higher the index, the longer the hand. VCLs have relatively high indices. **Phalangeal index** = [length of longest manual digit/length of hand]. VCLs have high indices reflecting relatively long phalangeal portion of hand. **Humerus/femur** = [humerus length/femur length × 100]; **Humerus/tibia** = [humerus length/tibia length × 100]. Like the intermembral index, the lower indices of VCLs indicate longer hindlimb elements. **Brachial index** = [radius length/humerus length × 100]. A higher index denotes a relatively long radius or short humerus. **Trochlear width** = [humeral trochlea width/distal humeral articular surface width × 100]. VCLs have low indices reflecting a relatively narrow trochlea. **Trochlear height** = [humeral trochlea height/trochlea width × 100]. VCLs have high indices indicating high, narrow trochleas. **Lateral condyle height** = [height of lateral condyle of femur/width of distal end of femur × 100]. Leapers have high indices reflecting deep and narrow knee joints. **Distal calcaneus** = [length of calcaneus distal to the talocalcaneal joint/total length of calcaneus × 100]. Small VCLs have high indices reflecting elongated calcanei. **Navicular length** = [length of navicular/width of navicular × 100]. Tarsiers, galagos, omomyids, and some cheirogaleids have high indices denoting elongated naviculars. Values in parentheses are estimates. Data for living species and fossils from: Covert, 1985; Dagosto, 1983, 1986, 1990; Jouffroy and Lessertisseur, 1979; McArdle, 1981; Napier and Napier, 1967; Napier and Walker, 1967a; Stern and Oxnard, 1973; Szalay and Dagosto, 1980; Thalmann et al., 1989. Values for *Galago* are for *Galago senegalensis,* values for *Lemur* are for *Lemur fulvus*.

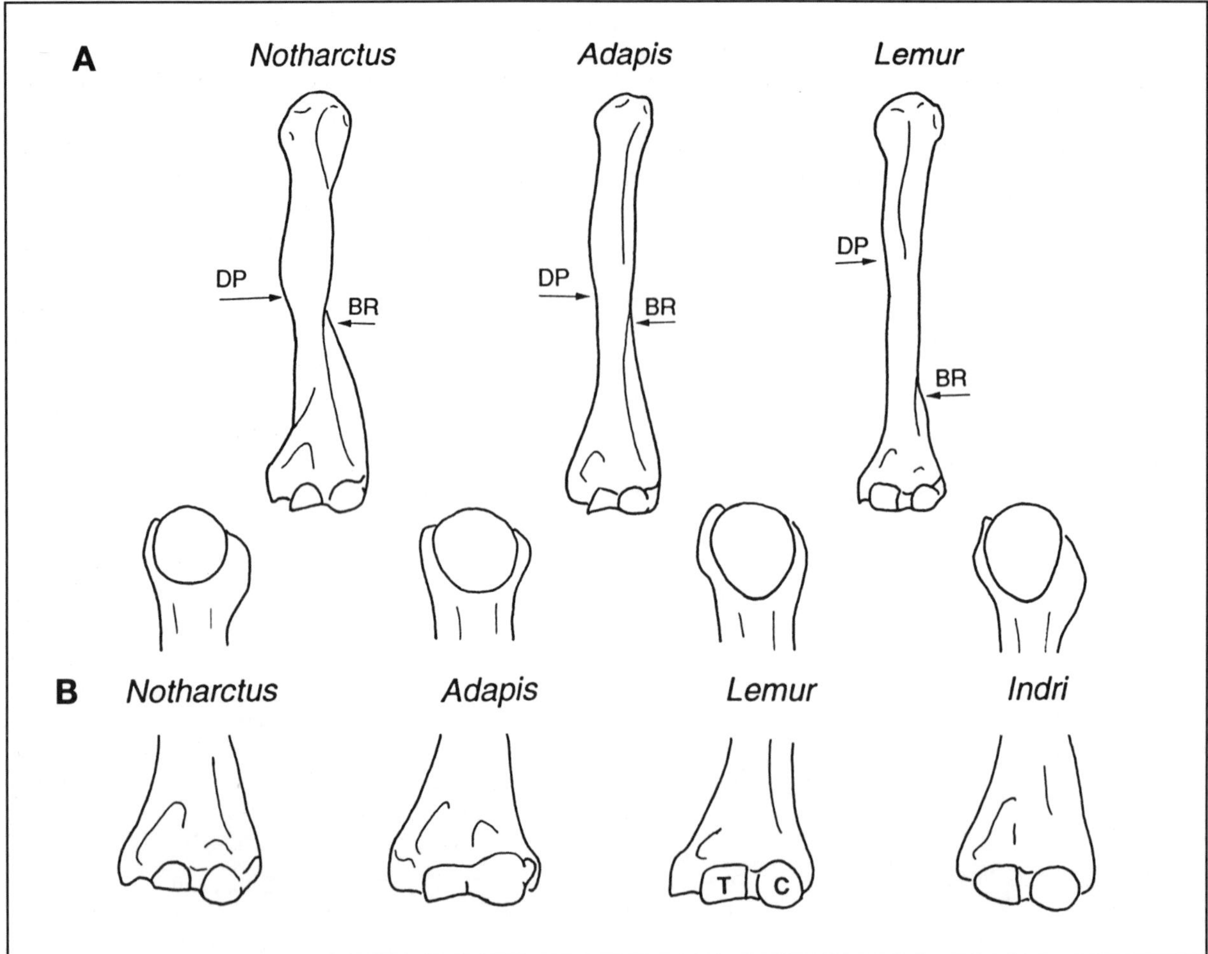

FIGURE 8.2. Anatomy of the humerus in Adapidae and Lemuriformes. A: Anterior view of left humerus drawn to the same length. Note the further proximal extent of the brachioradialis crest and the further distal extent of the deltopectoral crest in the adapids compared to *Lemur.* B: Left humerus, posterior view of proximal end (top) and anterior view of distal end (bottom), all drawn to the same width. Note the mediolateral constriction of the humeral head in *Lemur* and *Indri,* and the relatively short trochlea in *Indri* and *Notharctus.* BR = brachioradialis crest; DP = deltopectoral crest; C = capitulum; T = trochlea. A after Gregory (1920) and Dagosto (1983). B after Gebo (1987) and Szalay and Dagosto (1980).

and indicates the importance of supination of the forearm and hand (Szalay and Dagosto, 1980). The elements of this pattern are a mediolaterally broad, cylindrically shaped trochlea that is widely separated from a spherical capitulum by a deep groove (Fig. 8.2); a spherical or ovoid radial head that articulates with the rounded capitulum; and a shallow olecranon fossa that limits full extension of the forearm (Gregory, 1920). *Adapis* (but not *Leptadapis)* exhibits a different anatomy. Like lorisines, *Adapis* has a relatively short, conically shaped trochlea that is not strongly demarcated from the capitulum.

Smilodectes and *Notharctus* (but not *Cantius* or *Pronycticebus)* exhibit some similarities to the living indriids and *Lepilemur* (both VCL prosimians) in that the humeral trochlea is relatively short and high (Table 8.2), a very deep and wide groove separates the trochlea and capitulum, and sometimes a diagonal ridge runs across the distal surface of the trochlea (Szalay and Dagosto, 1980). The capitulum also extends farther distally than the trochlea, as is typical of living indriids and *Lepilemur* and this is seen in *Cantius* as well (Gebo, 1987).

The radius and ulna of notharctines differ from modern Malagasy prosimians in that the radial shaft is wider, especially at the distal end, the ulna has a smaller styloid process, a shorter and more bowed shaft, and a longer olecranon (Gregory, 1920; Rose and Walker, 1985). Radii and ulnae are reported for *Adapis* (Schlosser, 1887), but have never been studied further.

The notharctines exhibit the presumably primitive euprimate proportions and articular relationships of the carpal bones: small lunate and capitate; capitate articulates solely with the third metacarpal; triangular-shaped trapezium (Beard and Godinot, 1988). Except for *Pronycticebus,* they also have the probably primitive proportions of the hand, exhibiting relatively long digits (Jouffroy et al., 1991; Godinot, 1992). The manual distal phalanges of North American notharctines are nail bearing but are narrow and long (somewhat like those of indriids), rather than short and broad like those of lemurids (Gregory, 1920). The terminal phalanges of *Smilodectes* and *Adapis* are more dorso-volarly robust than in living lemuriforms (Godinot, 1992). The phalanges of *Europolemur,* however, are apparently more like those of extant lemuriforms (Franzen, 1987).

Adapis exhibits a more expanded radial facet on the lunate; lateral expansion of the centrale facet on the capitate; more proximodistally oriented facet for the triquetrum on the hamate; relatively flat rather than sellar-shaped articulation for the thumb and metacarpus; and broader ulna-pisiform contact (Godinot and Jouffroy, 1984; Beard and Godinot, 1988). These features indicate somewhat greater ranges of proximal carpal and midcarpal mobility in *Adapis* compared to notharctines (Beard and Godinot, 1988). Compared to living lemurids, in *Adapis* and *Leptadapis* the metacarpals are slightly shorter, with broader heads (Schlosser, 1887). The proximal articular surfaces of the metacarpals, however, are like those of lemurids rather than those of lorisines (Dagosto, 1983). Surprisingly, the pollicial-metacarpal joint is relatively flatter than in notharctines, indicating a smaller range of thumb movement (Godinot and Jouffroy, 1984). Unlike notharctines, *Adapis* had relatively short digits like many arboreal quadrupeds (Godinot, 1992).

Hindlimb

Notharctines share the "lyre-shaped" innominate of lemurids and indriids, which indicates expansion of the attachment area for the gluteal muscles (Gregory, 1920). The pelvis of notharctines differs from most living prosimians because the ilium is short and wide and the ischium is long, indicating more powerful but less speedy action of the hamstring muscles (Gregory, 1920; Stern and Oxnard, 1973; Covert, 1985; Rose and Walker, 1985). The degree of dorsal extension of the ilium is, however, similar to that of lemurids known to use vertical supports rather than the more generalized AQ forms (Fleagle and Anapol, 1992).

In notharctines, unlike in living lemurids and indriids, the greater trochanter and the head of the femur are equal in height (Gregory, 1920; von Koenigswald, 1985; Rose and Walker, 1985; Fig. 8.3). The greater trochanter does overhang the shaft anteriorly, as is usual in leaping prosimians. This gives better mechanical advantage to the vastus lateralis, which is a large and important knee extensor in prosimians (Jungers et al., 1980). Femoral head shape varies quite a bit in notharctines from nearly spherical in *Cantius* and *Smilodectes* to more cranio-caudally flattened in *Notharctus* (Covert, 1985).

In comparison to lemurids or notharctines the greater trochanter of adapinines is small, and is less laterally and anteriorly extensive. The very spherical femoral head is oriented proximomedially, rather than medially, and is situated on a relatively short neck. As in notharctines but not in living lemurids, the more distally positioned lesser trochanter is a large flat plate, oriented medially, rather than a tubercle oriented posteromedially. The third trochanter is very small in *Adapis,* but not in *Leptadapis.*

The distal femur of notharctines exhibits the anatomy typical of leaping prosimians. The articular end is deep anteroposteriorly, and narrow mediolaterally (Table 8.2; Fig. 8.3). The narrow patellar groove has a high, rounded lateral rim and is slightly raised above the level of the femoral shaft (Covert, 1985; Rose and Walker, 1985). In contrast to extant leaping prosimians, however, the inferior tibial tuberosity of the tibia (for insertion of gracilis, sartorius, and hamstrings) is more distally placed (Covert, 1985; Rose and Walker, 1985).

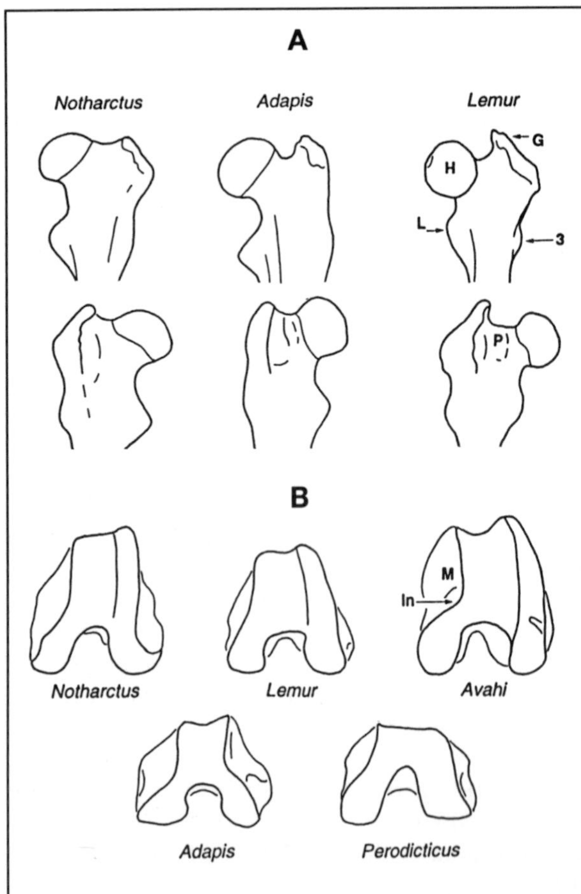

FIGURE 8.3. Anatomy of the femur in Adapidae and Lemuriformes. **A**: Left femur, anterior and posterior views of proximal end drawn to the same width. Note the high greater trochanter in *Lemur* compared to in adapids. **B**: Left femur, distal view of distal end drawn to the same width. Note the relatively deep distal femur in *Notharctus* and *Avahi,* and the relatively shallow distal femur in *Adapis* and *Perodicticus.* Also note the large medial bulge and the indentation in *Avahi,* which are not as well developed in other prosimians, except *Tarsius* (see Fig. 8.6). G = greater trochanter; H= head; L= lesser trochanter; 3 = third trochanter; P = paratrochanteric crest; M = medial bulge; In = indentation.

The knee of *Adapis* differs from notharctines and living lemuroids, and is similar to lorises or anthropoids in that the distal end of the femur is much broader than deep, the patellar surface is broad and flat, and the crests on the medial and lateral sides of the patellar grooves are not very high (Dagosto, 1983).

Notharctines exhibit typically lemuriform-like morphology of the distal tibia (Dagosto, 1985): the strongly convex articular surface of the medial malleolus is medially rotated, and the distal joint surface of the tibia, which articulates with the dorsal surface of the talus, is as long or longer (anteroposteriorly) than it is wide. The distal tibio-fibular joint is synovial and is anteriorly placed. *Leptadapis* shares these features, but exhibits a much more robust bone. *Adapis* differs from notharctines and lemuroids and is similar to lorises in its deeply excavated grooves for the tibialis posterior and flexor digitorum tibialis (Dagosto, 1983).

The tarsus of notharctines resembles most leaping prosimians in having a high, short talar body, long talar neck, large posterior trochlear shelf, elongated distal portion of the calcaneus and long tarsal bones (Figs. 8.4 and 8.5). In adapines, however, the tarsal bones are short: the talar trochlea is long (60 percent of total talus length), but the talar neck is short; the anterior portion of the calcaneus is quite short, but the heel is relatively long (Table 8.2 and Fig. 8.5) (Decker and Szalay, 1974; Martin, 1979; Dagosto, 1983; Gebo, 1988; Godinot, 1991; Martin, 1979). The medial edge of the talus curves strongly medially as it progresses distally, rather than being parallel as in notharctines. The body of the talus is dorso-ventrally low. In *Adapis* the posterior trochlear shelf is poorly developed. The non-leaping prosimians (lorises) share many of these features.

In *Notharctus,* the metatarsals and phalanges are short and broad; in *Smilodectes* and the Messel adapid, they are longer and more slender (Gregory, 1920; von Koenigswald, 1979; Covert, 1985). The terminal phalanx of the hallux is broad, flat, and nail bearing. The lateral terminal phalanges (where known) are nail bearing, but rather longer, more slender, and more robustly built than in living lemuriforms (Gebo et al., 1991). The metatarsal region of *Adapis* shows no special resemblances to lorises; the metatarsals have articular surfaces that are shaped like those of lemurids. In both groups, the first metatarsal exhibits the typically strepsirhine saddle-shaped joint and large peroneal tubercle that are the hallmarks of an opposable hallux (Szalay and Dagosto, 1988).

POSITIONAL BEHAVIOR OF NOTHARCTINES AND ADAPINES

For the most part, further study of North American notharctine postcranial material has done

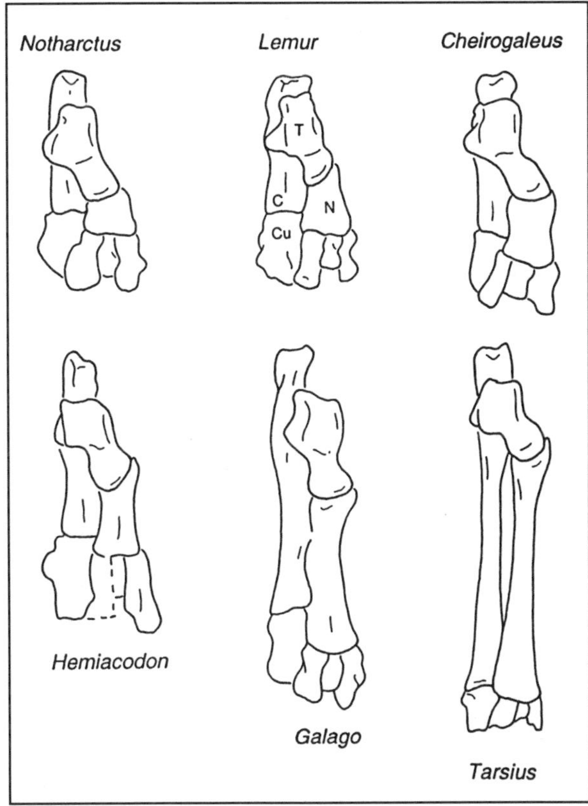

FIGURE 8.4. Tarsals of prosimian primates drawn to the same talar width. T = talus; C = calcaneus; N = navicular; Cu = cuboid. Note the extensive lengthening of the calcaneus and navicular in *Galago* and *Tarsius*. After Schultz (1963), Simpson (1940), and Dagosto (1990).

little to alter Gregory's (1920) astute assessment of *Notharctus*. He compared this genus with *Lepilemur* and *Propithecus* as well as *Lemur,* finding many similarities between *Notharctus* and the former taxa, concluding that it was an arboreal animal suited for both effective grasping and leaping. Since Gregory's monograph was written, behavioral studies of the Malagasy lemurs have led to the identification of two locomotor modes in this group: arboreal quadrupedalism (AQ) and vertical clinging and leaping (VCL) (Napier and Walker, 1967a, 1967b). Although animals in both groups leap, the vertical clingers and leapers leap more frequently, use quadrupedalism only rarely, and frequently use body postures where the trunk is positioned vertically rather than horizontally. Given Gregory's favorable comparison of *Notharctus* to *Lepilemur* and *Propithecus* (both VCLs) the question has been raised whether some or all notharctines may also have used this specialized positional behavior.

Based primarily on limb indices, Napier and Walker thought that indeed, the notharctines were VCLs. Other characters cited by Napier and Walker later proved to be unsuccessful in clearly distinguishing VCLs from AQs (Cartmill, 1972; Szalay, 1972; Stern and Oxnard, 1973; Ford, 1980). Later work led to the realization that there are at least two different postcranial anatomies associated with VCL primates, the galago-tarsier type and the lepilemur-indriid type (Cartmill, 1972; Stern and Oxnard, 1973). The notharctines most certainly do not exhibit the special features of the galago-tarsier group (extremely long foot bones, cylindrical femoral heads), but it is proving very difficult to find features that distinguish the lepilemur-indriid group from the arboreal quadrupedal lemurids (who also leap quite frequently, but do not engage in vertical postures as often as indriids or *Lepilemur).* Even when such features are identified (e.g., Gebo and Dagosto, 1988; Anemone, 1990), it is quite difficult to establish that the feature in question is functionally or adaptively related to either vertical clinging or more frequent leaping, making it difficult to infer that the reason for the presence of the feature in a fossil is the consequence of these behaviors.

Anatomical evidence in favor of the hypothesis that some or all notharctines were VCLs is derived from limb proportions and some aspects of elbow, knee, and foot morphology:

1. Limb proportions. The low, long hands (compared to arm length), and long phalanges (compared to hand length) of *Notharctus* and *Smilodectes* are all features also found primarily in VCLs (Stern and Oxnard, 1973; Napier and Walker, 1967a, 1967b). The relative length of the limbs compared to trunk length or vertebral column length, which might be more successful in separating AQs from VCLs (Napier and Walker, 1967b; Jouffroy and Lessertisseur, 1979) has not been fully investigated in notharctines because so few specimens preserve complete axial skeletons. Covert (1985) found that, compared to indriids, *Smilodectes* had short limbs relative to the length of the thoracic and lumbar vertebrae.

2. The humero-ulnar joint of *Notharctus* and *Smilodectes* (but not *Cantius* or *Pronycticebus)* exhibits a high and short trochlea, which is thought to

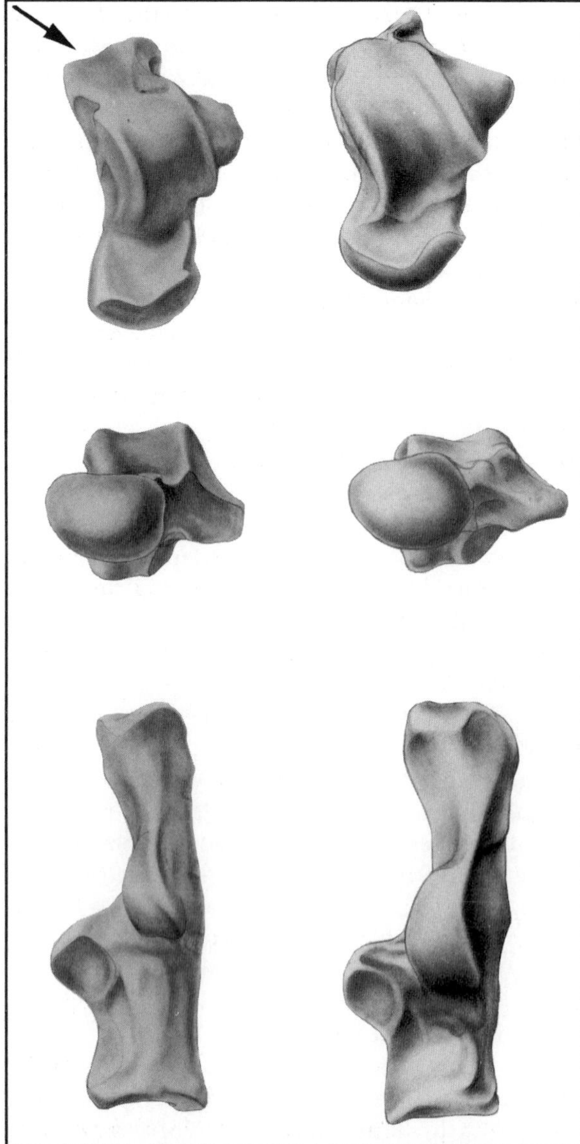

FIGURE 8.5. The talus and calcaneus of a notharctine (*Smilodectes,* right) and adapinine (*Adapis,* left). Top row: Dorsal view of the talus. Note the longer posterior trochlear shelf (arrow) and the longer neck in the notharctine. Middle row: Distal view of talus. Note the higher body in the notharctine. Bottom row: Dorsal view of calcaneus. Note the relatively larger proportion of the total length of the bone taken up by the portion anterior to the talocalcaneal joint in the notharctine. *Smilodectes* drawn by P. van Tassel; *Adapis* drawn by Anita J. Cleary. *Adapis* adapted from Szalay and Drawhorn (1980).

reflect the strongly flexed position of the forelimb in vertical clinging (Szalay and Dagosto, 1980).

3. One of the few femoral traits that apparently can distinguish between lemur-indriid VCLs and AQs is that the patellar surface is raised anteriorly beyond the level of the femoral shaft in the VCLs (Napier and Walker, 1967b; Anemone, 1990). *Notharctus* and *Smilodectes* may exhibit this trait (Covert, 1986), but in my opinion they do not differ significantly from AQ lemurids in this regard. *Cantius* exhibits a morphology intermediate between VCLs and AQs (Rose and Walker, 1985).

4. Like indriids (but not other VCLs), notharctines have long articulations of the talus and calcaneus, and a very large posterior trochlear shelf. The biomechanical relationship between these traits and vertical clinging or leaping is tenuous (Dagosto, 1986; Gebo and Dagosto, 1988).

There are also several lines of evidence that argue against the hypothesis that some or all notharctines were specialized VCLs like indriids or *Lepilemur:*

1. The brachial index of notharctines is lower than in any known VCL (and most lemurid AQs). The index is most similar to that of less frequently leaping AQs like New and Old World monkeys.

2. The carpus of *Notharctus* has several features that have been interpreted as ways of dealing with forces generated during the propulsive phase of quadrupedal walking (Beard and Godinot, 1988).

3. Unlike modern leaping prosimians (either AQs or VCLs), notharctines have a long ischium and short ilium, and a more distal insertion for the hamstrings.

4. The knee of notharctines resembles that of AQ lemurids rather than VCL indriids. The lateral condylar height index of the knee of notharctines falls between the range of AQs and VCLs (Table 8.2). In some indriids, especially *Avahi,* the medial epicondyle exhibits a distinct bulge anteriorly that is smooth and shiny like the patellar articular surface, and the junction between the patellar and condylar articular surfaces is marked by a strong indentation (Fig. 8.3). Tarsiers also exhibit this morphology, but it is not as well expressed in galagos. Notharctines do not share the medial bulge and indentation seen in many other VCLs.

5. Large prosimians that frequently use vertical supports have modified the articulation of the cuneiforms, cuboid, and metatarsals and have increased the size of the adductor hallucis muscle to

provide a strong grasp (the details are discussed in Gebo, 1985). Notharctine primates retain the primitive arrangement of these joints, and so were presumably not capable of sustained grasping of large vertical supports.

6. Unlike indriines, notharctines do not have a twisted posterior trochlear shelf of the talus, or plantar angulation of the anterior talocalcaneal facet, features that may be related to vertical clinging (Dagosto, 1986; Gebo and Dagosto, 1988).

Although notharctines (especially *Smilodectes* and to a lesser extent *Notharctus)* share several features with the VCL indriids, they lack other attributes that seem characteristic of this group and appear to be mechanically important for the performance of this behavior. Except for the intermembral index, the features associated with leaping are not as highly developed as in modern VCLs but show a level of development equivalent to (or lesser than) AQs. Features associated with vertical clinging seem to be expressed in the forelimb but not the lower limb.

Unfortunately, no living taxon provides a perfect anatomical analog for any of the extinct species (Rose and Walker, 1985). Presumably there was variation in positional behavior among the many species of notharctines, comparable to that seen in the extant lemurids and indriids of Madagascar. Thus, it is also difficult to choose a single living species to use as an anatomical or behavioral model for the notharctine group as a whole.

In contrast to the North American notharctines, vertical clinging and leaping has never been stressed in interpretations of the middle Eocene European adapids. Franzen (1987) has reconstructed *Europolemur koenigswaldi* as an arboreal quadruped with some potential for vertical clinging and leaping. *Pronycticebus* has been classified as a "grasp-leaper" (Szalay and Dagosto, 1980) by Thalmann et al. (1989).

In strong contrast to any "notharctine pattern" adapids, the available postcranial evidence strongly suggests that *Adapis* was an arboreal quadruped, from 1.5 to 2kg, in which leaping was practiced infrequently, if at all. In particular, its limb proportions, elbow joint morphology, forelimb muscle placement, femoral head morphology, knee morphology, and talar morphology all imply that it was well adapted for quadrupedalism and climbing, but only poorly adapted for leaping. The similarity of much of the *Adapis* postcranium to that of the extant lorisines (especially *Perodicticus,*) is striking, but the hands and the distal parts of the feet lack the extreme modifications characteristic of lorisines, which suggests a less specialized mode of locomotion (Dagosto, 1983). Godinot and Jouffroy (1984) and Godinot (1991) have suggested a more monkeylike model for *Adapis,* stressing the use of above-branch, horizontal quadrupedalism on the basis of hand anatomy, proportions of the calcaneus, and other aspects of joint form. *L. magnus* was a relatively large (from 6 to 10kg) short-limbed, robustly built arboreal quadruped, also well adapted for climbing and quadrupedalism but only poorly suited for leaping (Dagosto, 1983).

Recent work at new exposures of the Quercy deposits has complicated the picture by supporting earlier suggestions by Stehlin (1912) that there are in fact several species, and perhaps even genera, represented in the material attributed to *Adapis parisiensis* and *Leptadapis magnus* by Dagosto (1983) and others (Godinot, 1991; Laneque, 1992). These species exhibit variation in aspects of joint anatomy that indicates differing degrees of adaptation to quadrupedalism and climbing (Godinot, 1991).

In my opinion, the notharctine pattern of morphology (widespread among early and middle Eocene adapids from both Europe and North America) while exhibiting some unique features, closely approaches the presumed primitive euprimate and strepsirhine morphotypes; the adapinine pattern and the behaviors associated with it are more divergent and are considered phylogenetically derived (Dagosto, 1983, 1988). In contrast, other paleontologists believe that the nonleaping arboreal quadrupedalism of *Adapis* and the anatomical features associated with it are more likely to be primitive for euprimates (Godinot and Jouffroy, 1984; Godinot, 1991).

COMPARISON OF ADAPIDAE WITH LIVING STREPSIRHINI

Although the living lemurs and lorises are reasonably good models to use for understanding the behavior and morphology of Eocene Adapidae, it must also be appreciated that the fossil forms differ in many ways from their recent relatives. Some of those differences have been noted previously. The forelimb bones of the Eocene species are often more robust, more heavily muscle marked, and may have been short compared to body size. Muscles of the shoulder, hip, forearm, and leg tend to have insertions farther away from joints and appear to be set up to optimize power rather than speed. The brachial index is lower than in most living prosimians, indicating either a relatively longer humerus, or a shorter radius. In adapids where this feature is known, the third digit of the hand is equal in length to or longer than the fourth (Gregory, 1920; Thalmann, et al., 1989; but see Godinot, 1992); whereas in living lemuriforms, the fourth digit is always the longest (Jouffroy and Lessertisseur, 1979). These features point to a different pattern of forelimb use in the fossils, possibly one emphasizing climbing and/or arm-hanging (Covert, 1985).

The proximal femur is distinguished from lemurids and indriids by the low greater trochanter. The third digit of the adapid foot is equal to or longer than the fourth, in contrast to living prosimians where the fourth digit is always the longest (Lessertisseur and Jouffroy, 1973; Dagosto, 1990). Adapids do not appear to have had the toilet claw on the second digit that is characteristic of tooth-combed lemurs (Covert, 1986). Modern lemuriforms differ from extinct species in the relative size and articular relationships of the carpals and tarsals: for example, in all living tooth-combed lemurs, the large centrale contacts the hamate (Beard and Godinot, 1988); the mesocuneiform contacts the cuboid (Dagosto, 1986); and the joint surface for the first metatarsal on the entocuneiform is more laterally extensive than in adapids (Gebo, 1988). The pedal differences have been related to improved grasping abilities in the modern taxa.

Despite its overall unique and specialized anatomy, *Adapis*—unlike notharctines or middle Eocene European adapids—exhibits some shared specializations with living lemuriforms, namely the form of the ulnocarpal articulation (Beard and Godinot, 1988), the more anterior position of the peroneal tubercle on the calcaneus, and the more medial position of the calcaneo-cuboid depression on the calcaneus (Gebo, 1988). Thus, it is probably more closely related to living lemuriforms than are the notharctines (Beard and Godinot, 1988; Beard et al., 1988).

OMOMYIDAE

The postcranium of omomyids is not nearly as well known as that of the adapids. Except for the hindlimb bones attributed to *Hemiacodon* (Simpson, 1940), all known omomyid postcranial bones are isolated elements, and none are associated with teeth. Thus, attribution to taxa established on the basis of teeth is often difficult and relies primarily on locality and size. Fewer different bones are known; the forelimb is especially poorly represented. The family Omomyidae is divided into four subfamilies: an early-to-middle Eocene Anaptomorphinae with members in both Europe and North America; the Omomyinae, a North American early Eocene–early Oligocene group; the middle-to-late Eocene European Microchoerinae; and the early Miocene North American Ekgmowechashalinae (Fig. 8.1B). Postcranial remains are known for a few members of each group except the Ekgmowechashalinae. Enigmatic omomyids are also known from Asia, but as yet no postcranial material has been reported. Three morphological groups can be defined within the Omomyidae (Gebo, 1988). These groups correspond to the Omomyinae, Anaptomorphinae, and Microchoerinae as currently defined by dental evidence (Szalay and Delson, 1979). The postcranial differences between the Omomyinae and Anaptomorphinae, however, are very slight, and may not prove valid when more taxa are represented. On the other hand, the postcranial morphology of both omomyines and anaptomorphines contrasts quite markedly with that of the more derived late Eocene microchoerines.

All omomyids for which postcranial material is known are small animals. Estimates from tooth area and postcranial bones range from 40 gm (*Teilhardina*) to 500 gm (*Microchoerus*) (Gingerich, 1981; Dagosto and Terranova, 1992). No postcranial remains from larger omomyids have yet been found.

Forelimb

The proximal humerus is only known in the microchoerines *Microchoerus* and *Necrolemur* (Szalay and Dagosto, 1980). In these taxa, the humeral head is ovoid shaped, tapering distally, as in galagos and tarsiers (Fig. 8.6). The tubercles lie below the

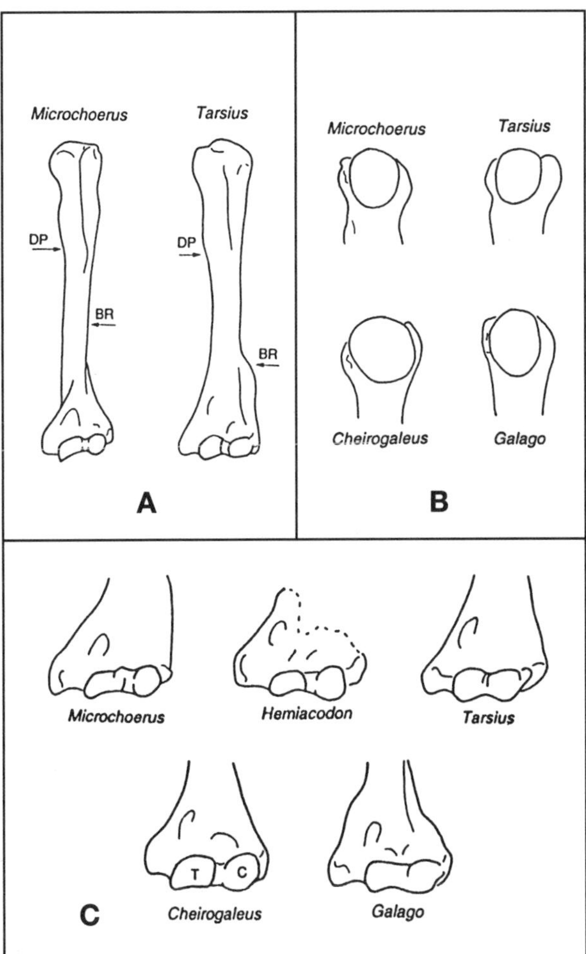

FIGURE 8.6. Anatomy of the humerus in omomyids and comparisons with *Tarsius* and lorisiformes. **A**: Anterior view of left humerus in *Microchoerus* and *Tarsius,* drawn to the same length. **B**: Posterior view of proximal left humerus drawn to the same width. Note the mediolaterally constricted humeral head in all except *Cheirogaleus.* **C**: Anterior view of distal humerus drawn to the same width. Note the relatively short trochlea in *Tarsius.* Abbreviations defined in Figure 8.2. After Szalay and Dagosto (1980).

level of the head. The muscle markings are less prominent than in adapids. In *Microchoerus,* the broad deltopectoral crest extends distally to only 40 percent of the total length of the humerus, compared to 50 percent in adapids. Like adapids, the brachioradialis crest is proximally extensive (to midshaft level) but it is not as broad.

The distal humerus is represented in all three groups (Fig. 8.6). All specimens show the typical euprimate conditions described above for adapids. Omomyines and anaptomorphines can be distinguished from microchoerines by their shorter trochleae (Table 8.2). No omomyids have the extremely short, high trochleae typical of *Tarsius* or indriids.

Hindlimb

The pelvis is known only in the omomyine *Hemiacodon.* The ilium is rod shaped like that of galagos, cheirogaleids, and *Tarsius.* Like *Notharctus* (and unlike most frequently leaping prosimians), however, the ischium is long compared to the ilium (Stern and Oxnard, 1973; Fleagle and Anapol, 1992). Like other prosimians that use vertical supports, *Hemiacodon* does exhibit some dorsal expansion of the ischium (Fleagle and Anapol, 1992).

The proximal femur is known in *Hemiacodon,* a small omomyid from the Bridger Basin, *Necrolemur*, and *Microchoerus* (Fig. 8.6) (Schlosser, 1907; Weigelt, 1933; Simpson, 1940; Dagosto and Schmid, in prep.). In all taxa, the femur is more robust than in *Tarsius* (Schlosser, 1907; Simpson, 1940). Like galagos and *Tarsius,* but unlike lemuroids or cheirogaleids, the greater trochanter and the head are at approximately the same level; the head of the femur is oriented medially and is mounted on a short neck, except in *Necrolemur* (Schlosser, 1907; Simpson, 1940). In omomyids, the greater trochanter overhangs the shaft anteriorly and continues as a thick pillar on the anterior surface of the proximal shaft.

In *Hemiacodon, Microchoerus,* and *Tarsius,* as in most prosimians, the third trochanter forms a small but laterally extended protuberance or plate, separated from the greater trochanter by an indentation. In galagos, however, the third trochanter takes the form of a ridge continuous with the greater trochanter. In *Nannopithex* and *Necrolemur,* the third trochanter is small or absent. In *Microchoerus* and *Tarsius,* the lesser trochanter is distal to the third trochanter; in lemuroids, notharctines, galagos, and *Hemiacodon,* the third trochanter is at the same level as or more distal to the lesser trochanter.

One of the most distinctive features of galagos and *Tarsius* is the cylindrical shape of the femoral

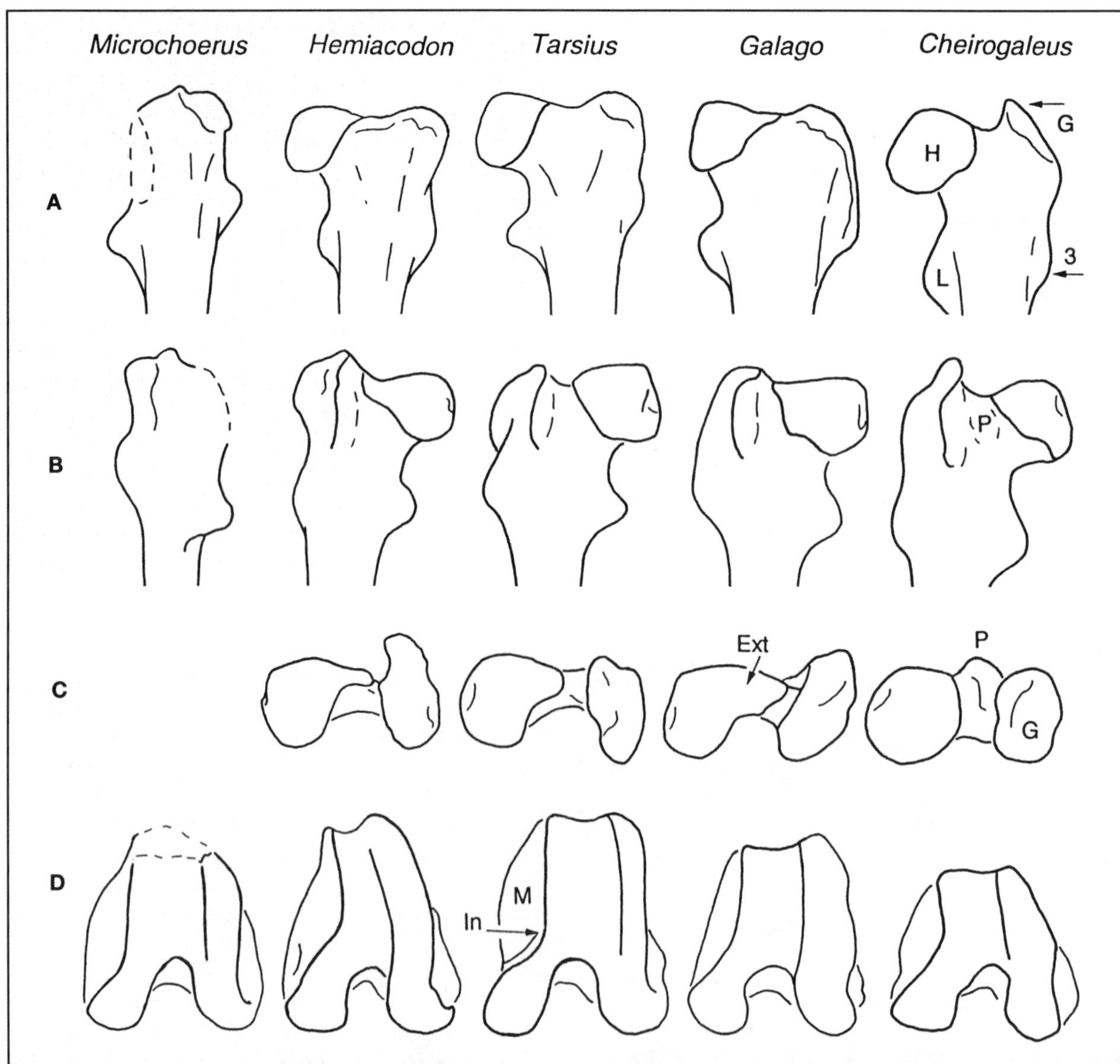

FIGURE 8.7. Anatomy of the femur in omomyids and comparisons with *Tarsius* and lorisiformes. **A**: Anterior view of proximal left femur. Note the relatively higher position of the third trochanter in *Tarsius* and *Microchoerus*. **B**: Posterior view of proximal left femur. Note the dorsolateral extension of the joint surface in all except *Cheirogaleus*. **C**: Cranial view of proximal left femur. Note the dorsolateral extension of the femoral joint surface, the presence of a paratrochanteric crest in *Cheirogaleus* only. **D**: Distal view of distal left femur. Note the relatively deep distal end in all except *Cheirogaleus*. Abbreviations defined in Figure 8.3. Ext = dorsolateral extension of joint surface.

head that is thought to be related to the flexed, abducted, and laterally rotated position of the femur during vertical clinging (Napier and Walker, 1967b; Anemone, 1990). In *Hemiacodon* and the small Bridger omomyid, this feature is moderately expressed. Like *Tarsius* and galagos, the joint surface is extended onto the posterodorsal aspect of the femoral neck (Fig. 8.7). Both *Hemiacodon* and *Tarsius* differ from galagos in that the posterior and dorsal aspects of the femoral head are not as flattened; in *Hemiacodon,* the medial side of the head is also rounder than in the extant genera. The femoral head is not preserved in *Microchoerus,* but in *Nannopithex* and *Necrolemur* the head is spherical rather than cylindrical.

The distal femur is preserved in *Hemiacodon, Microchoerus,* and *Necrolemur* (Schlosser, 1907; Simpson, 1940; Schmid, 1982). In both species, the joint surface is very deep anteroposteriorly and narrow mediolaterally (Fig. 8.7). In *Hemiacodon,* an index of relative height of the lateral condyle yields high values equal to those of tarsiers (Table 8.2). *Microchoerus* may have been more moderate in its proportions, but damage to the bone prevents an accurate measurement. Unlike *Tarsius* and indriids, neither *Hemiacodon* nor *Microchoerus* exhibits the strong medial epicondylar bulge and indentation, but like galagos, each has more moderate expression of these features.

Compared to adapids and living strepsirhine primates, known omomyids (*Absarokius, Hemiacodon,* and *Necrolemur* represented) exhibit a lesser degree of rotation of the tibial malleolus, which is convex anteriorly but flat posteriorly (in strepsirhines, including adapids, it is convex for its entire length). This is related to a less exaggerated twisting of the medial tibioastragalar facet, and a shallower tibioastragalar stop. The joint between the distal tibia and fibula is not synovial, but either probably syndesmotic as in *Hemiacodon* and *Absarokius* or synostotic (fused) in *Necrolemur* (Dagosto, 1985; Covert and Hamrick, 1993). These features show that in omomyids there is less conjunct rotation during flexion and extension of the ankle, a situation typical of many small, frequently leaping primates (Dagosto, 1985).

Omomyids exhibit many primitive euprimate features of the tarsus discussed above for notharctine adapids. All omomyids differ from adapids in possessing even more elongated foot bones: the distal calcaneus, talar neck, navicular, cuboid, and cuneiforms are longer than in adapids and in most living prosimians except some chierogaleids, galagos, and tarsiers (Fig. 8.4; compare Figs. 8.5 and 8.8). Omomyids (except *Necrolemur)* have much smaller posterior trochlear shelves than adapids, and the groove for flexor fibularis is more centrally located on the talus (Beard et al., 1988; Gebo, 1988). Flat nail-bearing terminal phalanges are known for omomyids. The pollex, hallux, and at least some lateral digits are represented (Dagosto, 1988).

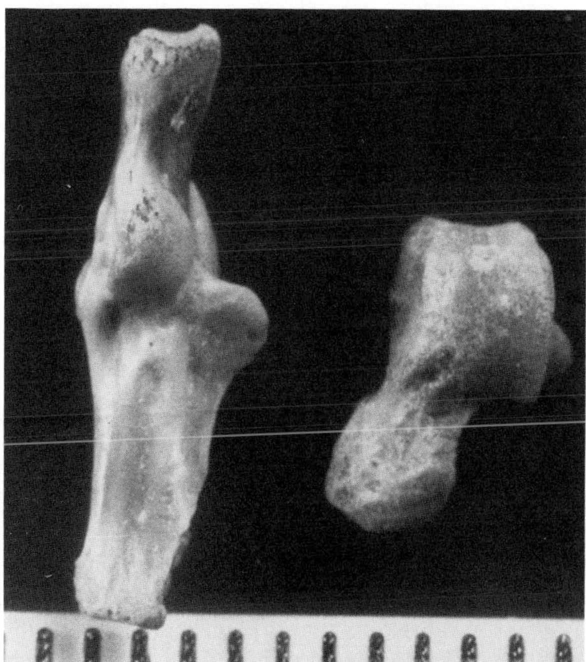

FIGURE 8.8. Tarsal bones of omomyids. Right: Dorsal view of right calcaneus of *Omomys.* Left: Dorsal view of left talus of *Omomys.* Scale in millimeters. Compare to Figure 8.5 and note the longer calcaneus and shorter posterior trochlear shelf in the omomyid compared to a notharctine. Adapted from Rosenberger and Dagosto (1992).

The talus of known omomyines (*Hemiacodon, Washakius, Omomys)* differs from that of known anaptomorphines (*Teilhardina, Tetonius)* in several minor respects. The medial and lateral rims of the talar trochlea are both quite thin and sharp; in anaptomorphines the medial rim tends to be more rounded. In omomyines both rims extend to the same point distally; in anaptomorphines the medial rim extends further distally than the lateral. Omomyines have a smaller posterior trochlear shelf (Rosenberger and Dagosto, 1992).

The Late Eocene microchoerines are very distinct compared to other omomyids. In foot proportions (relative distal calcaneal length, relative navicular length, Table 8.2), both anaptomorphines and omomyines have moderate dimensions most comparable to those of cheirogaleids. In contrast, *Necrolemur* and *Microchoerus* approach the extremes of lengthening seen in galagos or tarsiers (Schmid, 1979). *Necrolemur* is also distinguished by its short talar neck, deeply grooved trochlea, large posterior trochlear shelf, and bowing of the calcaneus (Godinot and Dagosto, 1983; Gebo, 1988).

POSITIONAL BEHAVIOR OF OMOMYIDS

Reconstructions of the locomotor behavior of omomyids are limited because evidence comes from only a small number of taxa and from a restricted range of body areas. Overgeneralizing from the few taxa for which we do have evidence to the very speciose groups of which they were members is likely to underestimate the variability within each group. Given the evidence so far available, Gebo (1988) concluded that Anaptomorphinae is the most primitive group, with its conservative talar features suggesting that these species engaged in a more varied locomotor repertoire than did omomyines, which may have been more committed to leaping. The postcranial morphology of both omomyines and anaptomorphines differs markedly from the more derived microchoerines. *Necrolemur,* the genus better represented by postcranial remains, is distinguished by a mediolaterally elongated humeral trochlea, a much more elongated calcaneus, a fused tibio-fibula, anterior-posterior compression of distal tibial shaft, a shallow medially directed groove on the malleolus for the tibialis posterior tendon, and a talus that is unusual in the great size of its posterior trochlear shelf and relative shortness of the talar neck (Schlosser, 1907; Godinot and Dagosto, 1983; Dagosto, 1985). *Microchoerus* is known from less material; however, it also exhibits an elongate calcaneus and humeral trochlea. The hindlimb features indicate the significance of saltatory leaping in these primates (Schlosser, 1907; Schmid, 1979; Godinot and Dagosto, 1983; Gebo, 1988).

As is true with adapids, some omomyids share features with living prosimians that practice vertical clinging and leaping. Thus the same question has been raised: Were some or all of the omomyids VCLs? In the absence of complete information, it is sometimes difficult to distinguish small specialized VCLs (*Tarsius,* some galagos) from small AQs (some galagos, cheirogaleids) that may employ leaping and use vertical postures with varying frequencies. The intermembral index would be most useful for making this distinction, but cannot yet be determined for any omomyid. Humerus/femur and humerus/tibia measures indicate that microchoerines potentially had relatively short forelimbs, although not as short as small VCLs (Table 8.2). Additional aspects of limb morphology exhibited by omomyids also indicate that leaping was an important part of the locomotor repertoire of all known taxa. For example, the distal knee proportions of *Hemiacodon* are only found in vertical clinging and leaping primates. The degree of calcaneal lengthening of *Necrolemur* is exhibited only by galagos and tarsiers, most of which practice frequent leaping and some degree of vertical clinging. Unlike *Tarsius* or indriid VCLs, however, the elbow joint of omomyids shows no vertical extension of the trochlea, or a particularly short trochlea (but some galagos do not exhibit these features either). The semicylindrical shape of the hip joint (known only in North American omomyids, however) is potentially indicative of use of vertical postures. *Necrolemur* and *Nannopithex,* however, have spherical femoral heads. Paradoxically, microchoerines—the omomyids with the most similarities to *Tarsius* in tibial and calcaneal anatomy—show the least similarity to *Tarsius* in the elbow, hip, and upper ankle joints.

As with adapids, the positional behavior of different omomyid species was varied, and no one living primate is a perfect analog for any of the fossil species or for the group as a whole. Although all known omomyids have anatomical features indicative of leaping, the amount of vertical clinging that may have been practiced is, again, very difficult to assess, since very few anatomical features have been unequivocally linked to this behavior.

COMPARISON OF OMOMYIDAE WITH LIVING PROSIMIANS AND ADAPIDS

As pointed out previously, there are many similarities between omomyids and living cheirogaleids, galagos, and tarsiers, but significant differences also exist. Anaptomorphines and omomyines are not typically as morphologically distinctive as galagos or tarsiers, but are more similar to the less specialized cheirogaleids. For example, they do not have foot bones that are as elongated as galagos or tarsiers, but show proportions more like those of cheirogaleids (Cartmill, 1972; Szalay, 1976; Gebo, 1988; Fig. 8.4). The only features in which

TABLE 8.3
Features That Distinguish Adapids and Omomyids

Humerus

1. Omomyids have a more ovoid-shaped glenohumeral joint (known for microchoerines only).
2. In omomyids, the insertions of the shoulder musculature are closer to the joint (microchoerines only).

Femur

3. Omomyids have a more cylindrical femoral head (*Hemiacodon* only).
4. In omomyids the lateral condyle of the knee is deeper (*Hemiacodon* only).

Tibia

5. Omomyids have less tibial malleolar rotation; flatter medial malleolar articular surface; and less curvature of the tibioastragalar-articulation (all taxa for which tali are known; see Dagosto, 1988, for list).
6. The inferior tibiofibular joint is not synvial (*Hemiacodon* and *Necrolemur* only).

Foot

7. Omomyids have more elongated tarsal bones (all known taxa).
8. In omomyids the posterior trochlear shelf is smaller (all known taxa except *Necrolemur*).
9. In omomyids, the cuboid facet on the navicular contacts only the ectocuneiform facet (all taxa for which naviculars are known; see Dagosto, 1988, for list). In adapids and lemuriforms, the cuboid facet contacts the mesocuneiform facet as well.
10. In omomyids, the talofibular facet is flat, and has a small pointed process plantarly (all taxa for which tali are known). In adapids and lemuriforms, the talofibular facet angles gradually laterally.

omomyines are more specialized than cheirogaleids are the depth of the lateral condyle of the knee and the shape of the femoral head. On the other hand, *Necrolemur* shares many highly specialized features with *Tarsius*, most notably fusion of the tibia and the fibula, and the extensive degree of calcaneal lengthening. The talus and the elbow joint of these two species differ, suggesting that the tibial and calcaneal similarities may have been attained convergently (Schlosser, 1907; Szalay and Dagosto, 1980; Godinot and Dagosto, 1983).

Adapids and omomyids share many aspects of postcranial morphology that reflect their common origin from an ancestral euprimate (Dagosto, 1988). They can also be distinguished by several characteristics, many of which are most likely related to saltatorial locomotion in small-bodied omomyids. (Table 8.3; Dagosto, 1985; Gebo, 1988). It is important to determine whether the morphology of adapids or omomyids (if either) is more likely to be primitive for euprimates, since this affects phylogenetic hypotheses and adaptive scenarios of primate evolution. There has been a tendency to accept without question the tenet that adapids are more primitive than omomyids, leading to unwarranted assumptions that postcranial features exhibited by adapids are primitive, while character states of omomyids are derived (e.g., Dagosto, 1985). More recent analyses of Eocene primate anatomy have suggested that, in fact, some features of the omomyid postcranium are probably primitive for euprimates, since many are seen in other groups of mammals as well (Features 5, 9, and 10 in Table 8.3; Beard et al., 1988; Dagosto, 1988; Gebo, 1988). Some features (3, 6, and 8) are almost certainly derived characters for particular omomyid taxa; whether these features were characteristic of all omomyids remains to be shown. The polarity of other features (1, 2, 4, and 7) still needs to be critically evaluated.

Decisions about the polarity (relative primitiveness) of features are important for determining the likelihood of different phylogenetic hypotheses, like those concerning the relationship of Eocene primates to living ones. It has been shown that adapids and living lemuriforms share several derived features of the foot (Features 9 and 10 in Table 8.3) and some features of the hand (see previous discussion of adapid forelimb) strengthening the hypothesis of a sister group relationship between them (Beard et al., 1988). The relationship of omomyids to living primates is a more contentious issue. Some primatologists argue that omomyids are the sister group of anthropoids, but others contend that it is the Adapidae that is more closely related to anthropoids. Cranial and dental evidence has been cited to support both positions (e.g., Gingerich, 1980;

Rosenberger and Szalay, 1980). No one has yet identified a derived feature of the postcranium that supports an adapid-anthropoid connection; the omomyid-anthropoid hypothesis is supported by one (the angle of the humeral trochlea relative to the shaft; Szalay and Dagosto, 1980). Anthropoids do not share the derived features of the hand and foot that link adapids and living lemuriforms; like omomyids, anthropoids exhibit the primitive euprimate character states. There are other features of the distal tibia in which omomyids are similar to anthropoids (Dagosto, 1985); however, these are also currently considered primitive euprimate features and thus not indicative of any special relationship between anthropoids and omomyids.

SUMMARY

The two groups of Eocene euprimates, the Adapidae and Omomyidae, share many features of limb anatomy that indicate their relationship with each other and with extant euprimates. These features indicate that the origin of Euprimates involved an adaptation to a regime of positional behavior emphasizing leaping as a primary method of displacement and both locomotor and postural behaviors dependent on grasping with an opposable hallux (Dagosto, 1988). Both Eocene groups show a wide range of anatomy and presumed positional behavior, and together comprise a spectrum of types comparable to that of extant prosimians. In both groups, the Late Eocene species (Adapininae and Microchoerinae) have evolved the most divergent features. In very general terms, notharctines are structurally most like the extant lemurids and indriids, which suggests that they engaged in both arboreal quadrupedalism (including leaping) and potentially adopted vertical clinging postures. The Adapininae converge on the anatomy of lorises, implying that quadrupedalism and climbing were more frequent behaviors than leaping. The poorly known Omomyinae and Anaptomorphinae are most like extant cheirogaleids; microchoerines exhibit more extreme specializations like those shown by today's vertically clinging and leaping galagos and tarsiers. There are also many substantial differences between the fossil and the living species. Thus, even though the anatomy of many of these species is quite well known, it is often difficult to provide precise depictions of their positional behavior comparable to those that can be produced for living species.

Acknowledgments: I would like to thank Dan Gebo for inviting me to contribute to this volume. The constructive comments of the editor and reviewers were greatly appreciated. I am also pleased to acknowledge the assistance of the staffs at the many museums in the United States and Europe at which I have worked. The work discussed in this paper was supported by NSF BNS-871926 and the Wenner-Gren Foundation.

LITERATURE CITED

Anemone RL (1990) The VCL hypothesis revisited: Patterns of femoral morphology among quadrupedal and saltatorial prosimian primates. Am. J. Phys. Anthrop. *83*:373–393.

Beard KC, Dagosto M, Gebo DL, and Godinot M (1988) Interrelationships among primate higher taxa. Nature *331*:712–714.

Beard KC, and Godinot M (1988) Carpal anatomy of *Smilodectes gracilis* (Adapiformes, Notharctinae) and its significance for lemuriform phylogeny. J. Hum. Evol. *17*:71–92.

Cartmill M (1972) Arboreal adaptations and the origin of the Order Primates. In R Tuttle (ed.): The Functional and Evolutionary Biology of Primates. Chicago: Aldine-Atherton, pp. 97–122.

Covert HH (1985) Adaptations and evolutionary relationships of the Eocene primate family Notharctidae. Ph.D. dissertation, Duke University.

——— (1986) Biology of early Cenozoic primates. In DR Swindler and J Erwin (eds.): Comparative Primate Biology. New York: Alan R. Liss, pp. 335–349.

——— (1988) Ankle and foot morphology of *Cantius mckennai:* Adaptations and phylogenetic implications. J. Hum. Evol. *17*:57–70.

Covert HH, and Hamrick MW (1993) Description of new skeletal remains of the Early Eocene Anaptomorphine Primate *Absarokius* (Omomyidae) and a discussion of its adaptive profile. J. Hum. Evol. :

Dagosto M (1983) Postcranium of *Adapis parisiensis* and *Leptadapis magnus* (Adapiformes, Primates). Folia Primatol. *41*:49–101.

——— (1985) The distal tibia of primates with special reference to the Omomyidae. Int. J. Primatol. *6*:45–75.

——— (1986) The joints of the Tarsus in the Strepsirhine Primates. Ph.D. dissertation, City University of New York.

——— (1988) Implications of postcranial evidence for the origin of Euprimates. J. Hum. Evol. *17*:35–56.

——— (1990) Models for the origin of the anthropoid postcranium. J. Hum. Evol. *19*:121–140.

Dagosto M, and Terranova CJ (1992) Estimating the body size of Eocene primates: A comparison of results from dental and postcranial variables. Int. J. Primatol. *13*:307–344.

Decker RL, and Szalay FS (1974) Origin and function of the pes in the Eocene Adapidae (Lemuriformes, Primates). In FA Jenkins (ed.): Primate Locomotion. New York: Academic Press, pp. 261–291.

Filhol H (1882) Mémoires sur quelques mammifères fossiles des phosphorites du Quercy. Ann. Soc. Sci. Phys. Nat., Toulouse *5*:19–156.

——— (1883) Observations relatives au mémoire de M. Copé intitulé: Relation des horizons renfermant des débris d'animaux vertébrés fossiles en Europe et en Amérique. Ann. Sci. Geol. *14*:1–51.

Fleagle JG (1988) Primate Adaptation and Evolution. New York: Academic Press.

Fleagle JG, and Anapol FC (1992) The indriid ischium and the hominid hip. J. Hum. Evol. *22*:285–306.

Fleagle JG, and Simons EL (1982) The humerus of *Aegyptopithecus zeuxis,* a primitive anthropoid. Am. J. Phys. Anthrop. *59*:175–193.

Ford SM (1980) A Systematic Revision of the Platyrrhini Based on Features of the Postcranium. Ph.D. dissertation, Univeristy of Pittsburgh, Pittsburgh.

Franzen JL (1987) Ein neuer Primate aus dem Mitteleozän der Grube Messel (Deutschland, S.-Hessen). Cour. F. Sencken. *91*:151–187.

——— (1988) Ein weiterer Primatenfund aus der Grube Messel bei Darmstadt. Cour. F. Sencken. *107*:275–289.

Gebo DL (1985) The nature of the primate grasping foot. Am. J. Phys. Anthrop. *67*:269–278.

——— (1987) Humeral morphology of *Cantius,* an early Eocene adapid. Folia Primatol. *49*:52–56.

——— (1988) Foot morphology and locomotor adaptation in Eocene Primates. Folia Primatol. *50*:3–41.

Gebo DL, and Dagosto M (1988) Foot anatomy, climbing, and the origin of the Indriidae. J. Hum. Evol. *17*:135–154.

Gebo DL, Dagosto M, and Rose KD (1991) Foot morphology and evolution in early Eocene *Cantius.* Am. J. Phys. Anthrop. *86*:51–73.

Gingerich PD (1980) Eocene Adapidae, paleobiogeography, and the origin of South American Platyrrhini. In RL Ciochan and AB Chiarelli (eds.): Evolutionary Biology of the New World Monkeys and Continental Drift. New York: Plenum Press, pp. 123–138.

——— (1981) Early Cenozoic Omomyidae and the evolutionary history of tarsiiform primates. J. Hum. Evol. *10*:345–374.

Gingerich PD, Smith H, and Rosenberg K (1982) Allometric scaling in the dentition of primates and prediction of body weight from tooth size in fossils. Am. J. Phys. Anthrop. *58*:81–100.

Godinot M (1991) Toward the locomotion of two contemporaneous *Adapis* species. Z. Morph. Anth. *78*:387–405.

——— (1992) Early euprimate hands in evolutionary perspective. J. Hum. Evol. *22*:267–284.

Godinot M, and Dagosto M (1983) The astragalus of *Necrolemur* (Primates, Microchoerinae). J. Paleont. *57*:1321–1324.

Godinot M, and Jouffroy F-K (1984) La main d'*Adapis* (Primates, Adapidae). In JM Mazin and E Salmion (eds.): Actes du symposium paléontologique G. Cuvier. France: Montbéliard, pp. 221–242.

Gregory WK (1920) On the structure and relations of *Notharctus:* An American Eocene primate. Mem. Am. Mus. Nat. Hist. *3*:51–243.

Hoffstetter R (1977) Phylogénie des Primates: Confrontation des résultats obtenus par les diverses voies d'approche de problème. Bull. et Mem. Soc. d'Anth. Paris *4*:327–346.

Jouffroy FK, Godinot M, and Nakano Y (1991) Biometrical characteristics of primate hands. J. Hum. Evol. *6*:269–306.

Jouffroy FK, and Lessertisseur J (1979) Relationships between limb morphology and locomotor adaptations among prosimians: An osteometric study. In ME Morbeck, H Preuschoft, and N Gomberg (eds.):

Environment, Behavior, and Morphology: Dynamic Interactions in Primates. New York: Fischer, pp. 143–182.

Jungers WL, Jouffroy FK, and Stern JT (1980) Gross structure and function of the quadriceps femoris in *Lemur fulvus:* An analysis based on telemetered electromyography. J. Morph. *164:*287–299.

von Koenigswald W (1979) Ein Lemurenreste aus dem eozänen Ölschiefer der Grube Messel bei Darmstadt. Palaont. Z. *53:*63–76.

——— (1985) Der dritte Lemurenrest aus dem mitteleozänen Ölschiefer der Grube Messel bei Darmstadt. Carolinea *42:*145–148.

Lanèque L (1992) Analyse de matrice de distance euclidienne de la région du museau chez *Adapis* (Adapiforme, Eocène). C. R. Acad. Sci. Paris *314:*1387–1393.

Lessertisseur J, and Jouffroy FK (1973) Tendances locomotrices des primates traduites par les proportions du pied. Folia Primatol. *20:*125–160.

McArdle JE (1981) Functional Morphology of the Hip and Thigh of the Lorisiformes. Contrib. Primatol. *17:*1–132.

Martin RD (1979) Phylogenetic aspects of prosimian behavior. In GA Doyle and RD Martin (eds.): The Study of Prosimian Behavior. New York: Academic Press, pp. 45–77.

Matthew WD (1915) A revision of the lower Eocene Wasatch and Wind River faunas. Vol. 4, Entelonychia, Primates, Insectivora (part). Bull. Am. Mus. Nat. Hist. *34:*429–483.

Napier JH, and Napier PR (1967) Handbook of Living Primates. New York: Academic Press.

Napier JH, and Walker A (1967a) Vertical clinging and leaping—a newly recognised category of locomotor behaviour of primates. Folia Primatol. *6:*204–219.

——— (1967b) Vertical clinging and leaping in living and fossil primates. In D Starck, R Schneider, and H-J Kuhn (eds.): Neue Ergebnisse der Primatologie (Progress in Primatology). Stuttgart: Fischer, pp. 66–69.

Rose KD, and Walker AC (1985) The skeleton of early Eocene *Cantius,* oldest lemuriform primate. Am. J. Phys. Anthrop. *66:*73–90.

Rosenberger AL, and Dagosto M (1992) New craniodental and postcranial evidence of fossil tarsiiforms. In S Matano, RH Tuttle, H Ishida, and M Goodman (eds.): Topics in Primatology, Vol. 3. Kyoto: University of Kyoto Press, pp. 37–51.

Rosenberger AL, and Szalay FS (1980) On the tarsiiform origins of the anthropoidea. In RL Ciochan and AB Chiarelli (eds.): Evolutionary Biology of New World Monkeys and Continental Drift. New York: Plenum Press, pp. 139–157.

Savage DE, and Waters BT (1978) A new omomyid primate from the Wasatch formation of Southern Wyoming. Folia Primatol. *30:*1–29.

Schlosser M (1887) Die Affen, Lemuren, Chiropteren, Insectivoren, Marsupialier, Creodonten, und Carnivoren des europäischen Tertiärs. Bietr. Palaont. Ost-Ung. *6:*1–162.

——— (1907) Beitrag zur Osteologie und systematischen Stellung der Gattung *Necrolemur,* sowie zur Stammesgeschichte der Primaten überhaupt. Neues Jb. Miner. Geol. Palaont. Mh. *1907:*199–226.

Schmid P (1979) Evidence of microchoerine evolution from Dielsdorf (Zurich Region, Switzerland)—a preliminary report. Folia Primatol. *31:*301–311.

——— (1982) Die systematische Revision der Europäischen Microcheridae Lydekker, 1887 (Omomyiformes, Primates). Universität Zürich.

Schultz AH (1963) Relations between the lengths of the main parts of the foot skeleton in primates. Folia Primatol. *1:*150–171.

Simons EL (1972) Primate Evolution. New York: Macmillan.

Simpson GG (1940) Studies on the earliest primates. Bull. Am. Mus. Nat. Hist. *77:*185–212.

Stehlin HG (1912) Die Säugetiere des schweizerischen Eocaens. Abh. schweiz. paläont. Ges. *38:*1163–1298.

Stern JT, and Oxnard CE (1973) Primate locomotion: Some links with evolution and morphology. Primatologia *4:*1–93.

Szalay FS (1972) Paleobiology of the earliest primates. In R Tuttle (ed.): The Functional and Evolutionary Biology of the Primates. Chicago: Aldine-Atherton, pp. 3–35.

——— (1976) Systematics of the Omomyidae (Tarsiiformes, Primates): Taxonomy, phylogeny, and adaptations. Bull. Am. Mus. Nat.

Hist. *156*:157–450.

Szalay FS, and Dagosto M (1980) Locomotor adaptations as reflected on the humerus of Paleogene primates. Folia Primatol. *34*:1–45.

——— (1988) Evolution of hallucal grasping in the primates. J. Hum. Evol. *17*:1–33.

Szalay FS, and Delson E (1979) Evolutionary History of the Primates. New York: Academic Press.

Szalay FS, and Drawhorn G (1980) Evolution and diversification of the Archonta in an arboreal milieu. In Luckett WP (ed.): Comparative Biology and Evolutionary Relationships of Tree Shrews. New York: Plenum Press, pp. 133–169.

Thalmann U, Haubold H, and Martin RD (1989) *Pronycticebus neglectus*—an almost complete adapid primate specimen from the Geiseltal (GDR). Palaeovertebrata *19*:115–130.

Weigelt J (1933) Neue Primaten aus mitteleozänen (oberlutetischen) Braunkhole des Geiseltals. Nova acta Leopolda *1*:97–156.

CHAPTER NINE

Postcranial Anatomy and Locomotor Adaptation in Early African Anthropoids

Daniel L. Gebo

Extinct early African anthropoids have been found in Algeria, Egypt, and nearby in Oman (Simons, 1972; de Bonis et al., 1988; Thomas et al., 1988; Godinot and Mahboubi, 1992). Unfortunately, only the fossil material from Egypt has yielded any substantial postcranial material (but see Senut and Thomas, 1992) and thus this locality, the Fayum, provides our first clues concerning postcranial adaptations in early African anthropoids (see Simons, 1972; Bown and Kraus, 1988, for historical and geological overviews of this important locality). The Fayum fossils date from 31 to 37 million years ago (Fleagle et al., 1986a, 1986b; Fleagle, 1988; Simons, 1990; Kappelman, 1992; Rasmussen et al., 1992; and Rasmussen and Simons, 1992; but see Van Couvering and Harris, 1991). The first fossil anthropoid described from the Fayum was by Osborn in 1908 (see Simons, 1972, for a historical review) and today, seventeen species of extinct anthropoids have been named (Tables 9.1 and 9.2). These extinct anthropoids are small to medium sized and appear to be adaptively similar to South American squirrel and howling monkeys. All of the postcranial elements are unassociated with dentitions and thus are attributed to the dentally recognized species on the basis of size, commonality at a specific quarry, and morphological similarity.

Two primate families have been recognized at the Fayum, the Parapithecidae (*Apidium, Parapithecus, Serapia,* and *Qatrania)* and the Propliopithecidae (*Aegyptopithecus, Propliopithecus, Oligopithecus,* and *Catopithecus). Oligopithecus* and *Catopithecus* are placed in a separate subfamily from the other propliopithecids (Table 9.2) (Simons, 1989; Rasmussen and Simons, 1992). There are still some questions concerning the allocation of *Proteopithecus, Plesiopithecus,* and *Arsinoea.* Dr. E. L. Simons is currently working on this new material from the lowest and oldest locality (L-41) at the Fayum (see Simons, 1992).

The phyletic relationships of these two families to living and other extinct anthropoids has been controversial and is still an active area of ongoing debate within the field. Simons (1970, 1972) and Kay (1977) originally considered parapithecids to be ancestral to Old World monkeys on the basis of some similarities in the dentition of *Parapithecus grangeri,* a form that would later be shown to lack lower incisors (Simons, 1986). Later, Szalay and Langdon (1986) and Gebo and Simons (1987) noted similarities in foot anatomy to Old World monkeys and Gebo (1989) discussed the possibility that parapithecids could represent the sister taxon to Old World monkeys on the basis of these foot characteristics.

TABLE 9.1
Fayum Anthropoids[1]

Upper Level	
Aegyptopithecus zeuxis	Simons, 1965
Propliopithecus chirobates	Simons, 1965
Apidium phiomense	Osborn, 1908
Parapithecus grangeri	Simons, 1974
Parapithecus frassi	Schlosser, 1911
Qatrania fleaglei	Simons and Kay, 1988
Middle Level	
Propliopithecus ankeli	Simons et al., 1987
Apidium moustafai	Simons, 1962
Lower Level	
Oligopithecus savagei	Simons, 1962
Qatrania wingi	Simons and Kay, 1983
Catopithecus browni	Simons, 1989
Proteopithecus sylviae	Simons, 1989
Serapia eocaena	Simons, 1992
Arsinoea kallimos	Simons, 1992
Plesiopitherus teras	Simons, 1992
Unknown Province	
Propliopithecus markgrafi	Schlosser, 1910
Propliopithecus haeckeli	Schlosser, 1910

[1] Anthropoid species by level and including the original descriptive references.

In contrast, Delson (1975), Delson and Andrews (1975), and Szalay and Delson (1979) have considered parapithecids to represent primitive catarrhines, not specifically related to Old World monkeys. Hoffstetter (1977) linked parapithecids with platyrrhines (New World monkeys), but more recently, Harrison (1987), Fleagle and Kay (1987), and Ford (1990) have proposed that parapithecids actually evolved before the platyrrhine radiation and thus represent very primitive anthropoids indeed (see also Simons and Kay, 1988).

Like the parapithecids, the phyletic relationships of the Propliopithecidae have been interpreted in a number of ways. Propliopithecids either represent primitive hominoids and thus "dawn apes" (Simons, 1972, 1985; Simons and Pilbeam, 1972; Szalay and Delson, 1979) or they are more primitive and represent early catarrhines that are ancestral to both the Old World monkey lineage and to later hominoids (apes) (Kay et al., 1981; Fleagle and Simons, 1982a; Fleagle and Kay, 1983; Andrews, 1985; Harrison, 1987). The phylogenetic debates continue over these two very interesting and enigmatic anthropoid families known from the Fayum and more recently from Oman and Algeria on the basis of dentitions (Thomas et al., 1988; Thomas et al., 1989; Godinot and Mahboubi, 1992). Although the precise phylogenetic position of these two families is still uncertain, reconstructive analyses of their skeletal anatomy and the inferences concerning postcranial adaptation and locomotor behavior of these extinct species are far less controversial.

PARAPITHECIDAE

Conroy (1976b) was the first to describe the limb elements of Fayum primates. These unassociated elements would later be allocated to *Apidium*. It was clear from the very first analysis that these fossils were not exactly similar to any one species or group of living primates. *Apidium*'s postcranium is a mosaic of unusual features, which has in part led researchers to widely different phylogenetic interpretations (see Fleagle and Kay, 1987, for a review). In terms of locomotor adaptation, however, later views based on new discoveries have differed very little from Conroy's initial assessment that *Apidium phiomense* is an arboreal quadrupedal leaping primate similar to *Saimiri* or *Cebus* in its overall locomotor profile (Conroy, 1976b; Fleagle, 1980). Field studies on living species of cebines since 1980 (e.g., Fleagle and Mittermeier, 1980, 1981; Terborgh, 1983; Gebo, 1992) suggest that *Saimiri*, a more frequent leaper than *Cebus*, may be a slightly better model of the two cebines for *Apidium*. The following sections of this chapter will discuss several of the important morphological features known for *Apidium phiomense* and will summarize the latest functional interpretations.

Apidium phiomense

Scapula Anapol (1983) described the scapula of *Apidium*. Using an analysis of scapular angles and the surface of the glenoid fossa, he found that this specimen, DPC 1007, was most similar to many

TABLE 9.2
Fayum Species[1]

	Quarry	ML	Body Weight(g)	PE
Propliopithecidae				
Propliopithecinae				
Aegyptopithecus zeuxis	I,M	249	6700	+
Propliopithecus chirobates	I,M	249	4200	+
Propliopithecus ankeli	V	165	5700	-
Propliopithecus haeckeli	?	?	4000	-
Propliopithecus markgrafi	?	?	4000	-
Oligopithecinae				
Oligopithecus savagei	E	92	1500	-
Catopithecus browni	L-41	47	750	?
Parapithecidae				
Apidium moustafai	V	165	850	-
Apidium phiomense	I,M	249	1600	+
Parapithecus grangeri	I,M	249	3000	?
Qatrania fleaglei	I,M	249	600	?
Qatrania wingi	E	92	300	-
Parapithecus frassi	?M	?249	1700	-
Serapia eocana	L-41	47	1000	-
Family Uncertain				
Plesiopithecus teras	L-41	47	600	-
Arsinoea kallimos	L-41	47	500	-
Proteopithecus sylviae	L-41	47	500	?

[1] Taxonomic allocations, locality, and meter level (ML) information with body weights from Fleagle (1988) and DT Rasmussen (pers. com.). The presence of postcranial elements (PE) attributed to these species is noted by a plus sign.

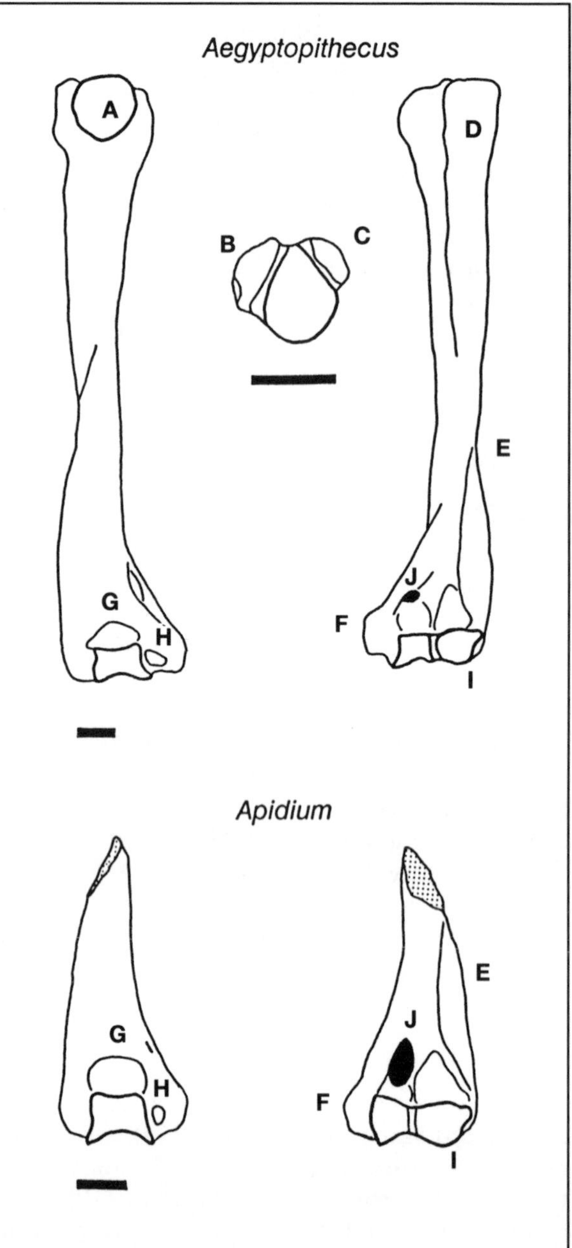

FIGURE 9.1. Posterior (left) and anterior (right) views of *Apidium* (DPC 1008, bottom) and *Aegyptopithecus* (DPC 1275, top) humeri, plus a proximal view of the humeral head for *Aegyptopithecus* (top middle): a = posteriorly facing humeral head; b = prominent greater tubercle; c = lesser tubercle; d = prominent deltopectoral crest; e = wide brachial flange; f = wide medial epicondyle; g = olecranon fossa (circle below g); h = prominent dorsoepitrochlear fossa (circle below h); i = elongate capitulum; and j = entepicondylar foramen (blackened circles). The dotted area represents a broken surface. Black bars equal 1 cm.

arboreal quadrupedal primates, particularly colobines. The metric analysis of the axilloglenoid angle (orientation of the glenoid fossa to the axillary border) shows *Apidium* to be similar to a variety of leaping primates, while the spinoglenoid angle (angle of the scapular spine to the glenoid fossa) indicates similarities to arboreal quadrupeds. The shape of the glenoid fossa is similar to *Saimiri*, an arboreal quadruped and frequent leaper (Fleagle and Mittermeier, 1980; but see Boinski, 1989).

TABLE 9.3
The Entepicondylar Foramen in Numbers of Platyrrhine Humeri Housed at the Field Museum of Natural History (Chicago)[1]

	Present	Absent
Callitrichines		
Callimico goeldii	3	0
Cebuella pygmaea	0	7
Callithrix jacchus	0	4
Callithrix argentata	0	2
Saguinus mystax	7	0
Saguinus labiatus	4	0
Saguinus fuscicollis	5	0
Saguinus leucopus	6	0
Saguinus imperator	1	4
Saguinus oedipus	8	4
Leontopithecus rosalia	0	10
Cebines		
Saimiri sciureus	12	0
Cebus capucinus	7	0
Cebus apella	18	0
Cebus nigrivattus	2	0
Cebus albifrons	5	0
Aotines		
Aotus trivirgatus	18	3
Callicebus torquatus	2	0
Callicebus cupreus	0	4
Callicebus donacophilus	0	5
Pithecines		
Cacajao calvus	3	0
Pithecia pithecia	11	0
Pithecia hirsuta	2	0
Pithecia monachus	2	0
Chiropotes sp.	4	0
Atelines		
Alouatta seniculus	0	4
Ateles geoffroyi	0	3
Ateles belzebuth	0	2
Brachyteles arachnoides	0	2
Lagothrix lagotricha	0	2

[1] As shown, this foramen is lost in the atelines; in the callitrichines by *Cebuella, Callithrix,* and *Leontopithecus;* in the aotines by *Callicebus cupreus* and *Callicebus donaphilus;* and is variably present in *Saguinus imperator, Saguinus oedipus,* and *Aotus trivirgatus.* This foramen has been lost a minimum of three times in platyrrhine evolution.

Humerus The humerus of *Apidium* is comparable in morphology to a wide variety of primates including the callitrichines in its shortness and robusticity (Fleagle and Kay, 1987). The proximal humerus possesses a prominent deltopectoral crest, a broad and shallow bicipital groove, and a well-delineated deltoid plane on the lateral side (Fleagle and Kay, 1987). The shape of the humeral head and the placement of the tubercles is similar to that of *Microchoerus,* an European omomyid (Fleagle, pers. com.).

The proximal part of the humeral shaft is compressed laterally and is deep dorso-ventrally, while the distal part is broad and flattened with a moderate to small-sized brachial flange (attachment of the brachialis muscle) (Fig. 9.1; Conroy, 1976b). The supinator crest extends well up the posterior part of the humeral shaft where the brachialis, brachioradialis, and extensor carpi radialis musculature originate.

The distal humerus possesses a primitive euprimate feature in retaining an entepicondylar foramen (Fig. 9.1). The median nerve and brachial artery pass through this hole (Conroy, 1976b). This foramen is absent in living apes and Old World monkeys, in atelines (the spider monkey group), and in some other species and genera of New World monkeys (Table 9.3). The distal humerus also has a large and medially projecting medial epicondyle (attachment area for the carpal and digital flexors—pronator teres, flexor carpi radialis, flexor digitorum profundus, palmaris longus, and flexor carpi ulnaris) with a deep dorsoepitrochlear fossa for the ulnar collateral ligament (Fig. 9.1; Conroy, 1976b). A faint ridge separates the trochlea and the elongate capitulum. The wide anteroposterior width of the ventral surface of the trochlea (Fig. 9.1) and the sharp ventral extension of the trochlea medially suggests mechanical stability during quadrupedalism (Szalay and Dagosto, 1980). The coronoid fossa and the posteriorly located olecranon fossa, are both shallow in depth (Conroy, 1976b) while the radial fossa is deeper and more excavated.

Ulna The shaft is moderately robust with a very slight dorsal convexity proximally (Fig. 9.2). Distally, there is a prominent pronator crest (Fig. 9.2). The olecranon process is moderate in length (Fig. 9.2) and is not bent medially or posteriorly as in cercopithecoids. The coronoid process is higher

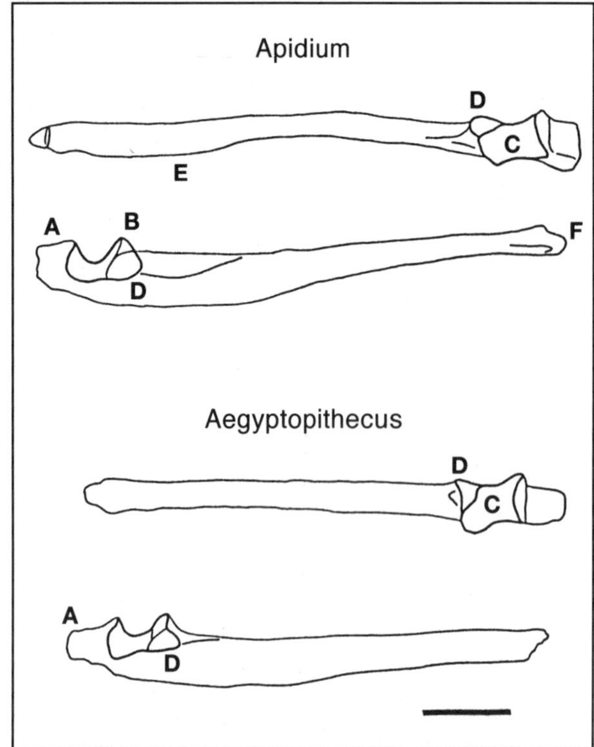

FIGURE 9.2. Dorsal (top of each) and lateral views (bottom of each) of *Apidium* (DPC 1295) and *Aegyptopithecus* (YPM 23940) ulnae: a = moderately long olecranon process; b = high coronoid process; c = broad sigmoid notch; d = radial facet; e = pronator crest; f = styloid process. The ulna of *Aegyptopithecus* is broken distally. Black bar equals 1 cm.

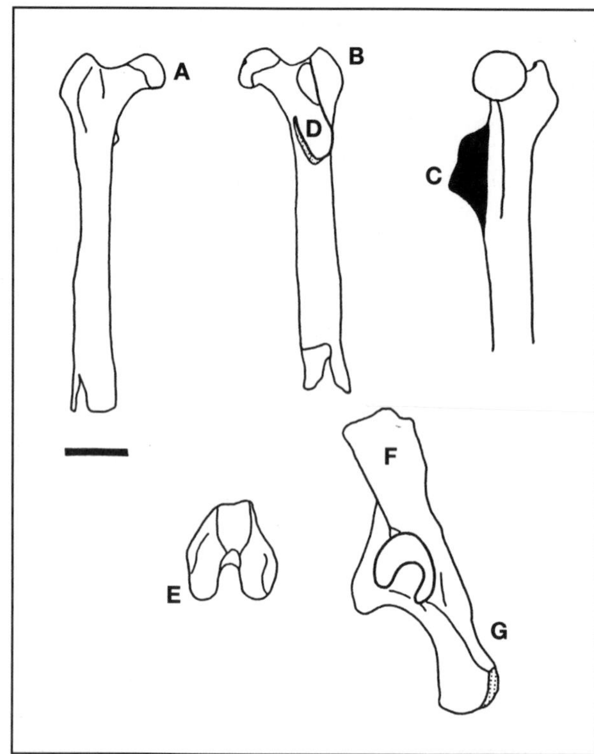

FIGURE 9.3. Anterior (left), posterior (middle), and oblique (right) views of a proximal femur (DPC 1081, top); distal femur (bottom left, drawn from Fleagle and Kay, 1987: Fig. 9b); and lateral view of a pelvis (DPC 1036, bottom right) of *Apidium*: a = oval femoral head; b = greater trochanter; c = lesser trochanter (blackened in oblique view; note its great size); d = walled-off posterior femoral fossa; e = anteroposteriorly high condyles; f = broad ilium; and g = long ischium. Dotted areas represent broken or eroded surfaces. Black bar equals 1 cm.

than the condyloid process as in Old World monkeys but is not well buttressed (Conroy, 1976b). The sigmoid notch is wide and lacks any waisting of the facet as in Old World monkeys (Conroy, 1976b). The anconeal process is relatively broad with a prominent lateral process (Conroy, 1976b). The radial notch is more anteriorly located, and this likely restricts forearm pronation and supination (Conroy, 1976b). A long styloid process is present distally (Fig. 9.2).

Radius The slope of the radial head is like most primates (other than hominoids), having a downward slope toward the lateral side. It lacks a projection of the radial rim onto the surface of the radial head as in Old World monkeys. The oblique line is well developed for the attachment of the superficial flexors (Conroy, 1976b).

Innominate There are three clear surfaces or planes (the gluteal, sacral, and iliac) to the ilium of *Apidium*. The gluteal plane is broad and quadrangular in shape (Fig. 9.3) and may represent a shared-derived feature of anthropoids (Fleagle and Kay, 1987). The sacral plane shows a short length between the acetabulum and the auricular articular surface. In *Apidium*, the iliac plane is especially broad near the acetabulum (see Fleagle and Kay, 1987). As the iliac plane extends cranially away from the acetabular region, it becomes smaller in prosimians, most nonateline platyrrhines, and in *Apidium*. In contrast, living catarrhines and atelines show this plane to increase in width (see Gebo et al., in press). The ratio describing the acetabular region (ventral/dorsal breadth of the acetabular rim from Schultz, 1969) shows *Apidium* to have dorsal thickening and thus to be similar to quadrupedal primates. The pubis is stout and is oriented posteroventrally.

A ratio by Leutenegger (1970) to estimate the lever arm lengths for the hamstrings about the hip joint to the load arm of body weight shows a very low value of 0.84 for *Apidium* (Fleagle and Simons, 1979; but see McCrossin and Benefit, 1992), indicating that the ischium is very long in *Apidium* (Fig. 9.3). A long ischium has been interpreted to be indicative of frequent leaping abilities (Fleagle, 1977; but see McCrossin and Benefit, 1992, and Fleagle and Anapol, 1992). The ischium lacks the ischial callosity seen in Old World monkeys and some living apes (Fleagle and Simons, 1979).

Femur Proximally, the greater trochanter projects over the femoral shaft and has a flattened lateral surface for gluteus superficialis (Fleagle and Kay, 1987). There is a very large lesser trochanter, which meets with the greater trochanter and walls off this region of the posterior femur (Fig. 9.3; Fleagle and Kay, 1987). The femoral neck is long and the femoral neck and head are approximately right-angled relative to the shaft (Fig. 9.3). The shape of the femoral head is ovoid rather than very round and ball-like. The distal articular edge of the femoral head is oriented obliquely as the edge continues mediolaterally across the anterior and posterior surfaces. The femoral shaft is flattened and broad mediolaterally as one moves distally down the shaft. Distally, the condyles are very high (anteroposteriorly) (Fig. 9.3) with a well-developed and rounded lateral rim (Fleagle and Kay, 1987). Both features are similar to a variety of living and extinct prosimians and are suggestive of habitually leaping primates.

Tibia and Fibula The tibia is very long (Fig. 9.4). It is unique among anthropoids in its transverse narrowness in the proximal part of the shaft (Fleagle and Simons, 1983). The insertion of semitendinosus and gracilis muscles are likewise very proximally positioned in *Apidium*. The fibula is closely apposed to the tibia—40 percent of the length of the tibia (Fleagle and Simons, 1983) (Fig. 9.4). "The extensive tibio-fibular syndesmosis in *Apidium phiomense* is unique among higher primates" (Fleagle and Simons, 1983:239). There is an anteroposteriorly long distal facet on the distal tibia for the fibula that is similar to that of New World monkeys (Ford, 1988).

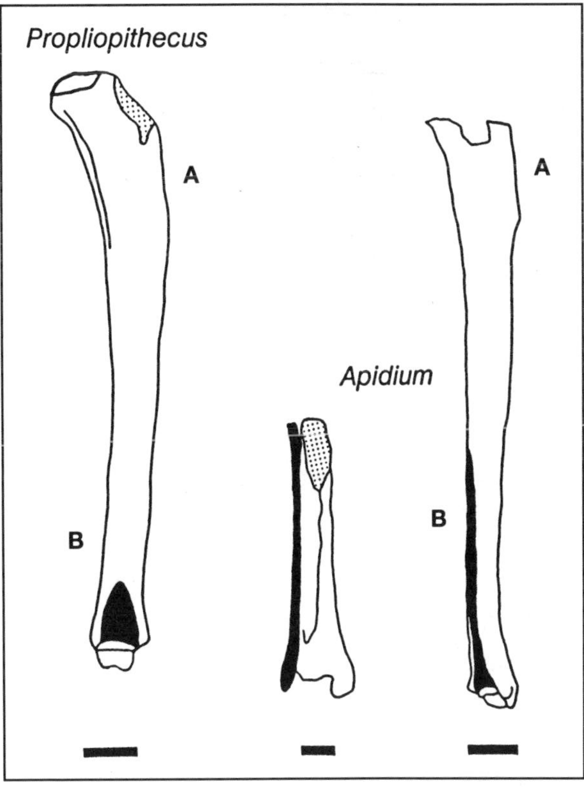

FIGURE 9.4. Lateral view of right tibia of *Propliopithecus chirobates* (DPC 2384, left, drawn reversed) and lateral view (DPC 1149, right) and an anterior view (middle, drawn from Fleagle and Kay, 1987: Fig. 8C) of *Apidium phiomense*. Note the extensive fibular apposition in *Apidium*: A = tibial tuberosity; B = fibular facet (blackened). The distal fibula is blackened in the middle illustration. Dotted areas represent broken or eroded surfaces. Black bar equals 1 cm.

Talus At the upper ankle joint, *Apidium* shows a parallel-sided trochlear surface with rims of equal height and a moderately deep trochlear groove (Conroy, 1976b; Gebo and Simons, 1987; Gebo,

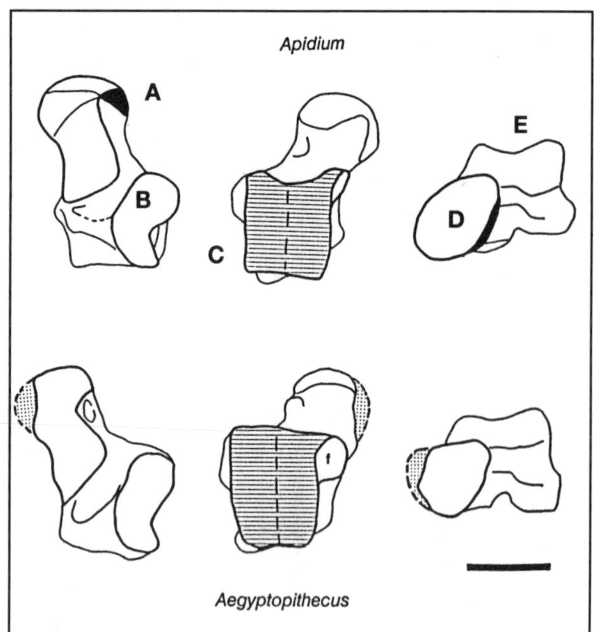

FIGURE 9.5. Plantar (left), dorsal (middle), and anterior views (right) of *Apidium* (YPM 25801, top) and *Aegyptopithecus* (DPC 1301, bottom) tali: A = corresponding facet for the anterior or most distal calcaneal facet (blackened, left and right views); B = asymmetrical posterior facet; C = parallel trochlear rims (horizontal hatching); D = twisted talar head; E = deep trochlear groove; and F = cup-shaped medial malleolus. Dotted areas represent broken or eroded surfaces. Black bar equals 1 cm.

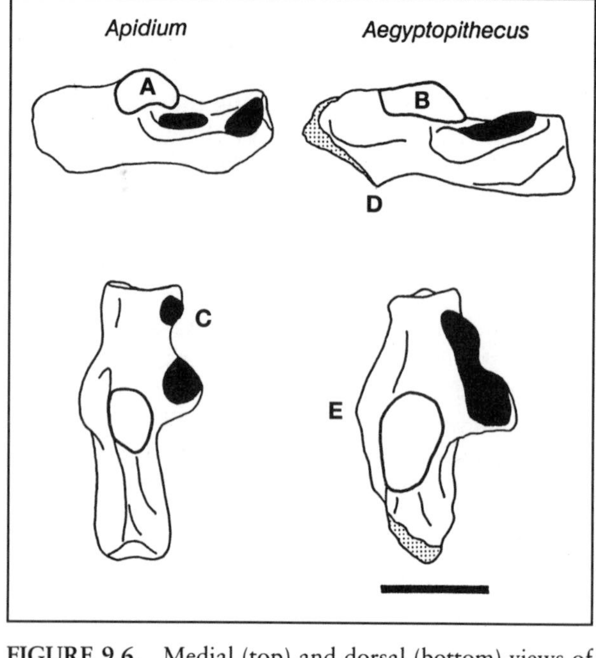

FIGURE 9.6. Medial (top) and dorsal (bottom) views of *Apidium* (YPM 25804, left) and *Aegyptopithecus* (DPC 3051, right) calcanei: A = shortened and downward bending posterior calcaneal facet; B = long posterior calcaneal facet; C = separate anterior or distal calcaneal facets; D = plantar curvature; and E = mediolaterally wide calcaneal shape. Anterior calcaneal facets are blackened. Dotted areas represent broken or eroded surfaces. Black bar equals 1 cm.

1989)(Fig. 9.5). The talar body is high and circular in outline (Conroy, 1976b). These features are characteristically associated with leaping primates and are likely euprimate features (Dagosto, 1988) and primitive for anthropoids. The talar head is dorsolateral in its orientation to the talar body and flattened plantolaterally to articulate with the most distal and medially facing calcaneal facet (Fig. 9.5). The posterior talar facet is short and asymmetrical. Features of the talar head and posterior facets are very similar to the same joint surfaces in Old World monkeys. The talar neck is relatively short and is oriented proximodistally rather than being long and widely angled medially as in most primates (Fig. 9.5). The posterior surface of the trochlea lacks a lateral tubercle for the flexor groove.

Calcaneus The calcaneus is shortened distally and broadened proximally (Fig. 9.6). At the subtalar joint, *Apidium* possesses two distal facets on the calcaneus and a proximodistally short and downward bending proximal facet (Fig. 9.6). The subtalar joint of *Apidium* mirrors that of Old World monkeys (Gebo and Simons, 1987). The calcaneocuboid joint is very shallow and is also similar to the same joint in Old World monkeys (Szalay and Langdon, 1986; Gebo, 1989).

Navicular The navicular is moderate in length, the naviculocuboid facet is small, and the naviculoentocuneiform facet is large. All of these features are most similar to New World monkeys (Szalay and Langdon, 1986).

Cuboid In *Apidium,* the cuboid is moderately long with no sesamoid facet along the peroneal groove laterally (Strasser, 1988). Medially, the two facets for the ectocuneiform are similar in shape to that in New World monkeys (Strasser, 1988).

Phalanges There are several unstudied specimens that belong to either the hand or the foot of *Apidium*. The distal elements suggest flattened nails rather than claws (Fleagle and Kay, 1987).

Parapithecid Summary

Although found in Africa, *Apidium phiomense* clearly lacks the specialized limb anatomy of living Old World monkeys and apes. Its greatest morphological similarity appears to be with the smaller New World monkeys of South America, and many of these shared similarities are likely to be primitive features for anthropoids.

Functionally, three groups of morphological features appear to be commonly observed in the postcranial anatomy of *Apidium*. The first group of morphological features is associated with arboreal quadrupedal primates, the second set is found in frequently leaping primates, and the third set of features is more commonly observed in frequent climbing primates. The arboreal quadrupedal features include the pear shape of the glenoid fossa (scapula), a shallow olecranon fossa (humerus), an expanded anteroposterior width of the ventral trochlea (humerus), the length of the olecranon process (ulna), the shape of the radial head, the wide gluteal plane (innominate), and the dorsal thickening of the acetabular rim (innominate).

Features associated with leaping include the low intermembral index of 70 (Fleagle, 1988), the axilloglenoid angle of the scapula, the long ischium (innominate), the perpendicular angle of the femoral neck and head to the shaft (femur), the anteroposteriorly high distal condyles (femur), the long and proximally compressed tibia, the proximal insertions of semitendinosus and gracilis muscles (tibia), the closely apposed distal fibula, the shape of the upper ankle joint with its parallel-sided trochlear rims (talus), and the anteroposteriorly high talar body. The morphology of the distal femur, fibular apposition, and the shape of the upper ankle joint in *Apidium* are highly specialized features relative to other leaping anthropoids.

Features such as the robust humerus and ulna and the prominent muscle attachment areas (deltopectoral crest, flat deltoid plane, medial epicondyle, brachial flange, oblique line, and supinator crest) suggest powerful and complex arm motions that are used in climbing activities. This is also true of the very large and flattened lesser trochanter of the femur, which is associated with the iliopsoas musculature and is important in powerful flexion and rotation of the thigh during climbing (Taylor, 1976).

The unusual shape of the subtalar and transverse tarsal joints in the foot of *Apidium* as well as the robust (especially proximally) and short distal calcaneal region are problematical in their mechanical interpretation relative to other postcranial features. The subtalar and transverse tarsal joints indicate a reduction in the potential for inversion of the foot, a movement commonly associated with grasping and climbing activities. The shape of these two joints in *Apidium* mirrors that of living Old World monkeys, animals that still retain feet capable of climbing and grasping branches in the arboreal environment, but that show many adaptations for terrestriality. Of all of the known living or extinct primates, only Old World monkeys and *Apidium* have this variety of reduced foot mobility. The modified foot joints, as well as other parts of the skeleton of Old World monkeys, have largely been interpreted as adaptations for terrestrial cursoriality (Langdon, 1984; Strasser and Delson, 1987; Strasser, 1988).

Apidium also has a relatively short distal calcaneus. Frequently leaping arboreal primates, including anthropoid leapers, generally lengthen the distal calcaneus. If *Apidium* was a frequent leaper and arboreal quadruped similar to *Saimiri,* why did it evolve reduced foot mobility and a shortened distal calcaneal length? One would expect the foot bones of *Apidium* to look and work like those of cebine monkeys (e.g., *Saimiri* or *Cebus)* because so many of the other limb elements do. The modified talocalcaneal joint and the relatively short distal calcaneus indicate that *Apidium* may well have had a terrestrial adaptive past like that of Old World monkeys. The evolutionary implication of this interpretation is that the long tibia, the fibular apposition, the high distal condyles of the femur, and the shape of the upper ankle joint represent leaping adaptations that are secondarily imposed upon an earlier terrestrially adapted lifestyle. This interpretation contrasts with the explanation given by Fleagle and Kay (1987), who interpret parapithecid foot anatomy as being adaptive for leaping; and thus, many if not all of the leaping features noted in *Apidium* are thought to be primitive features inherited from an ancestral arboreal leaping haplorhine.

PROPLIOPITHECIDAE

Fleagle et al. (1975) and Conroy (1976b) described the first known postcranial element of a propliopithecid, the YPM 23940 ulna attributed to *Aegyptopithecus zeuxis*. This ulna was interpreted to suggest that *Aegyptopithecus* was an arboreal quadruped like *Alouatta* and preserved a primitive adaptive pattern unlike any of the living hominoids (apes). This interpretation has not altered very much with the additional recovery of other limb elements.

Aegyptopithecus zeuxis and *Propliopithecus chirobates*

Humerus Fleagle and Simons (1978) initially described the humeri of *Aegyptopithecus* and *Propliopithecus* as being very primitive in their morphology and most similar to larger arboreal platyrrhine quadrupeds, particularly *Alouatta*. These humeri lack the specialized features characteristic of either of the two living catarrhine lineages, the Old World monkeys (cercopithecoids) and apes (hominoids) (Fleagle and Simons, 1978, 1982a, 1982b; Fleagle, 1983; Fleagle and Kay, 1983; Ford, 1988; Gebo et al., 1988; Rose, 1988, 1989; Senut, 1989).

The humeri attributed to *Aegyptopithecus* are more complete and thus better studied than those of *Propliopithecus* (but see Fleagle and Simons, 1982b. The following section will largely draw on this work (e.g., Fleagle and Simons, 1978; Fleagle and Simons, 1982a; Fleagle, 1983; Rose, 1989). The humerus of *Aegyptopithecus* is very robust with large muscle crests (Fig. 9.1). It is not elongated as in living apes. The greater and lesser tubercles are prominent and are laterally placed relative to the humeral head as in arboreal quadrupeds (Fleagle and Simons, 1982a). The humeral head faces posteriorly and is not broad or ball-like in shape (Fig. 9.1), a distinct difference compared to living apes. There is some slight flattening on the top of the humeral head, which suggests a stable position for the humerus in full flexion, a position emphasized in arboreal quadrupedalism (Rose, 1989). A pronounced deltopectoral crest is prominent and very long (about 40 percent of the length of the shaft) (Fleagle and Simons, 1982a). The bicipital groove is broad and shallow. There is a very distal crest for the insertion of the teres major muscle (extensor and medial rotator of the humerus), which is characteristic of arboreal quadrupeds and climbers (Fleagle and Simons, 1982a).

The distal humerus is broad with a prominent medial epicondyle (for digital and carpal flexors and pronators) and brachial flange (for brachial flexors and supinator) (Fig. 9.1). The lateral development of the brachialis flange exceeds that of any living anthropoid (Fleagle and Simons, 1982a). The distal humerus shows an emphasis on flexor rather than extensor musculature, as is found in more climbing and quadrupedal primates like prosimians and New World monkeys (Fleagle and Simons, 1982a). The trochlea is relatively broad, not restricted as in Old World monkeys, and the capitulum is round, not spherical as in living apes, and has a small proximolateral tail (Rose, 1988). The space between the trochlea and the capitulum, the zona conoidea, is wide and shallow (Rose, 1988). The olecranon fossa is similarly wide and shallow and compares best with a variety of living prosimians and New World monkeys rather than being deeply excavated as in Old World monkeys and apes. An entepicondylar foramen and a dorsoepitrochlear fossa are also present on the distal humerus. Despite the presence of a variety of primitive anthropoid features in the humerus, the elbow is of the nontranslatory type as in living and extinct catarrhines (see Rose, 1988, for details).

Ulna The ulnae of *Aegyptopithecus zeuxis* and *Propliopithecus chirobates* are robust bones with low coronoid processes (Fig. 9.2), a feature that is similar to arboreal quadrupedal primates. The ulna has a convex curvature proximally, and this is also similar to arboreal primates (Conroy, 1976b). The sigmoid notch is broad and shallow (Fig. 9.2) as in most New World monkeys; it is not reduced as in Old World monkeys. The flexion-extension axis of the elbow through the sigmoid notch is oriented obliquely, and this places the ulna in a laterally deviated position when the ulna is extended and a medial position when flexed. This oblique axis is characteristic of arboreal forms (Conroy, 1976b). The triangular-shaped radial facet is small like many prosimians and New World monkeys. A relatively long olecranon process is present (Fig. 9.2), and this characterizes arboreal quadrupeds (Oxnard, 1963; Jolly, 1967). A prominent pronator crest is present and is similar to arboreal climbing primates. The ulna lacks the sharp interosseous ridge observed in most Old World monkeys (Fleagle et al., 1975). Other nonmetric traits liken this ulna to that of *Alouatta* (Schön Ybarra and Conroy, 1978).

Tibia The tibia of *Propliopithecus chirobates* is relatively short and robust (Fig. 9.4) and is best interpreted as an adaptation to either arboreal quadrupedalism or climbing (Fleagle and Simons, 1982). The tibial condyles are slightly asymmetrical as seen in arboreal quadrupeds. This asymmetry is related to the use of abducted thighs during arboreal quadrupedalism (Grand, 1968; Jenkins and Camazine, 1977). The distal insertion of the semitendinosus muscle is comparable with that of arboreal quadrupeds (Fleagle and Simons, 1982b) rather than to leaping primates. The distal articular surface is extremely similar to known isolated hominoid tibia from the early Miocene of Africa (Fleagle and Simons, 1982b).

Talus Two tali are known for *Aegyptopithecus zeuxis* (Fleagle, 1983; Gebo and Simons, 1987), and both are moderate in their dorsoplantar height and depth of trochlear grooving (Fig. 9.5). The trochlear rims are parallel and the medial rim curves distally to form a deep medial malleolar cup for the tibial malleolus (Fig. 9.5). This curvature allows the foot to abduct during dorsiflexion of the foot and to adduct during plantarflexion. The talar neck is angled medially rather than being oriented anteroposteriorly. The plantar facets are indicative of extensive joint mobility and similar to most primates with the exception of Old World monkeys. The parallel trochlear rims, the moderate talar body height, and trochlear grooving are reminiscent of primates that characteristically leap, while the curvature of the medial trochlear rim is associated with the increased foot mobility seen in climbing primates.

Calcaneus Several calcanei of *Propliopithecus chirobates* (Fleagle and Simons, 1982b) and of *Aegyptopithecus zeuxis* (Gebo and Simons, 1987) have been recovered from the Fayum. All calcanei are very broad mediolaterally and are moderately long distally (approximately 38 percent of the length of the calcaneus) (Fig. 9.6). This distal length is comparable to a wide variety of primates (including *Notharctus)* but is generally shorter than in most frequently leaping primates. The plantar surface of these calcanei is curved anteroposteriorly, and the posterior heel region bends medially although the heel tubercle is broken off (Fig. 9.6). The shape of the plantar calcaneal surface suggests that a plantar tubercle is present (Szalay and Langdon, 1986), a feature that Sarmiento (1983) has linked to frequent climbing or hindlimb suspension. The plantar surface of the sustentaculum tali shows a broad and deep groove for the flexor hallucis longus muscle (an important muscle for toe flexion). The calcaneocuboid joint surface shows a deep pivot that is indicative of a wide degree of rotational capabilities at the transverse tarsal joint during foot inversion and grasping and climbing activities. This feature is observed in atelines and living and extinct hominoids. Similarly, the posterior calcaneal facet is very long (Fig. 9.6) and would provide an extensive surface for the talus to slide back along while inverting the foot during grasping and climbing activities. Overall, these calcanei are morphologically very similar to those calcanei attributed to the early Miocene hominoids and to the living atelines. Atelines are known for their frequent climbing activities and their foot suspensory feeding postures (Richard, 1970; Mendel, 1976; Mittermeier, 1978; Schön Ybarra, 1984; Cant, 1986; Schön Ybarra and Schön, 1987; Gebo, 1992).

Metatarsals and Phalanges A first metatarsal was described by Conroy (1976a) and is very anthropoid-like in having shortened the peroneal tubercle (the peroneus longus muscle pulls on this tubercle to adduct the big toe) and a saddle-shaped proximal facet. This metatarsal has a facet for the prehallux, a bone commonly oberved in New World monkeys and gibbons but not Old World monkeys or the great apes (Conroy, 1976a). The first metatarsal is also relatively long (Gebo, 1987). It is curved longitudinally and this suggests strong flexor tendons (Preuschoft, 1974).

A second and a fourth metatarsal have been attributed to *Aegyptopithecus* (Gebo and Simons, 1987). Both appear to show several resemblances to prosimians, especially in their equal heights at the proximal and distal ends and in the continuous medial facet on the fourth metatarsal.

The proximal phalanx is long and slender with moderate longitudinal curvature as in climbers (Preuschoft, 1974). The joint surface is tilted distally and is proximally deep as in more terrestrial forms (Preuschoft, 1974). The curved phalanges are not long like living apes nor are they robust like terrestrial monkeys (Fleagle, 1983). They suggest arboreal habits (Preuschoft, 1974).

Propliopithecid Summary

The postcranial anatomy of propliopithecids clearly shows that these species are decidedly unlike either of the two living groups of catarrhines, the modern apes (hominoids), and the Old World monkeys (cercopithecoids). Given this evidence, propliopithecids are increasingly thought of as a group that evolved before cercopithecoids and hominoids separated off from the catarrhine stock and that propliopithecids are structurally suitable to be ancestral to both groups.

Like *Apidium, Aegyptopithecus* and *Propliopithecus* share a number of anatomical features, which suggests a locomotor adaptation of arboreal quadrupedalism. Propliopithecids differ, however, in their emphasis of climbing over leaping features. Features such as the robust humerus and ulna with prominent muscle attachment areas (deltopectoral crest, medial epicondyle, and brachial flange) suggest powerful and complex arm motions in the propliopithecids as they do in *Apidium.* In the foot, the wide calcaneal shape, the anteroposterior curvature of the plantar calcaneal surface, the elongated posterior calcaneal facet, the deep calcaneocuboid pivot, the distal curvature of the medial trochlear rim, the angle of the talar neck, and the long and curved first metatarsal are all important features that suggest frequent climbing abilities and that are entirely lacking in the foot anatomy of *Apidium.*

Other features, which support arboreal quadrupedalism in the propliopithecids and which are similar to those noted for *Apidium,* would include the flattened humeral head, the location of the humeral tuberosities, the shallow olecranon fossa (humerus), the expanded anteroposterior width of the ventral trochlea (humerus), the length of the olecranon process (ulna), and the asymmetrical condyles of the tibia.

The parallel trochlear rims (talus), the moderate talar body height, and trochlear grooving (talus) suggest some leaping capabilities in *Aegyptopithecus* but far less than those noted in *Apidium.*

The overall interpretation of the known propliopithecid postcranial elements suggests that these forms resemble larger arboreal quadrupedal platyrrhines in their locomotor habits, especially *Alouatta,* a frequent climber and occasional leaper, and a quadruped that often feeds by hindlimb suspension (Richard, 1970; Mendel, 1976; Cant, 1986; Schön Ybarra, 1984; Schön Ybarra and Schön, 1987; Cant, 1986; Gebo, 1992).

NEW L-41 PRIMATES

Recently, five limb bones (two humeri, two femora, and an innominate) were discovered from the stratigraphically lowest and thus oldest locality at the Fayum (L-41). These elements represent the first known postcranial remains of Eocene anthropoids and are most likely allocated to *Catopithecus* and *Proteopithecus.* These bones are more primitive in appearance (i.e., in their possession of large femoral third trochanters and longer humeral capitular tails) than those of middle- or upper-level parapithecids and propliopithecids from the Fayum (see Gebo et al., 1994, for more details), and this new material certainly helps in our efforts to reconstruct the behavioral adaptations of early anthropoids.

CONCLUSION

Despite the fact that only a few out of the seventeen early African species known from the Fayum have postcranial remains unequivocally attributed to them, the two families the Parapithecidae and the Propliopithecidae appear to show a skeletal anatomy that is more similar phenetically to some South American monkeys (platyrrhines) than to living Old World catarrhines. Although the phylogenetic placement of these two families is not yet certain, the interpretation of the postcranial remains in terms of locomotor adaptation is fairly consistent across a variety of studies. *Apidium phiomense* is inferred to move in a manner similar to that of the living squirrel monkey, an arboreal quadruped and frequent leaper, but with more robust arms for enhanced climbing abilities, while *Aegyptopithecus zeuxis* and *Propliopithecus chirobates* are thought to mirror the locomotor abilities of howling monkeys.

LITERATURE CITED

Anapol F (1983) Scapula of *Apidium phiomense:* A small anthropoid from the Oligocene of Egypt. Folia Primatol. *40:*11–31.

Andrews P (1985) Family group systematics and evolution among catarrhine primates. In E Delson (ed.): Ancestors: The Hard Evidence. New York: Alan R. Liss, pp. 14–22.

Boinski S (1989) The positional behavior and substrate use of squirrel monkeys: Ecological implications. J. Hum. Evol. *18:*659–677.

Bonis L de, Jaeger JJ, Coiffat B, and Coiffat PE (1988) Découverte du plus ancien primate catarrhinien connu dans l'Eocène supérieur d'Afrique du Nord. C.R. Séanc. Acad. Sci. Paris, II *306:*929–938.

Bown TM, and Kraus MD (1988) Geology and paleoenvironment of the Oligocene Gabal El Qatrani Formation, Fayum Depression, Egypt. Geol. Surv. Egypt, Ann. *4:*115–138.

Cant JGH (1986) Locomotion and feeding postures of spider and howling monkeys: Field study and evolutionary interpretation. Folia Primatol. *46:*1–14.

Conroy GC (1976a) Hallucial tarsometatarsal joint in an Oligocene anthropoid *Aegyptopithecus zeuxis.* Nature *262:*684–686.

——— (1976b) Primate postcranial remains from the Oligocene of Egypt. In FS Szalay (ed.): Contributions to Primatology, Vol. 8. New York: S. Karger.

Dagosto M (1988) Implications of postcranial evidence for the origin of euprimates. J. Hum. Evol. *17:*35–56.

Delson E (1975) Toward the origin of the Old World monkeys. Evolution des vertèbres: Problèmes actuels de palaéontologie. Actes CNRS Coll. Int. *218:*839–850.

Delson E, and Andrews P (1975) Evolution and interrelationships of the catarrhine primates. In WC Luckett and FS Szalay (eds.): Phylogeny of the Primates: A Multidisciplinary Approach. New York: Plenum Press, pp. 405–446.

Fleagle JG (1977) Locomotor behavior and skeletal anatomy of sympatric Mayssian leaf-monkeys (*Presbytis obscura* and *Presbytis melalophos).* Yrbk. Phys. Anthrop. *20:*440–453.

——— (1980) Locomotor behavior of the earliest anthropoids: A review of the current evidence. Z. Morph. Anthrop. *71:*149–156.

——— (1983) Locomotor adaptations of Oligocene and Miocene hominoids and their phyletic implications. In RL Ciochon and RS Corruccini (eds.): New Interpretations of Ape and Human Ancestry. New York: Plenum Press, pp. 301–324.

——— (1988) Primate Adaptation and Evolution. New York: Academic Press.

Fleagle JG, and Anapol FC (1992) The indriid ischium and the hominid hip. J. Hum. Evol. *22:*285–306.

Fleagle JG, Bown TM, Obradovich JD, and Simons EL (1986a) Age of the earliest African anthropoids. Science *234:*1247–1249.

——— (1986b) How old are the Fayum primates? In J Else and P Lee (eds.): Primate Evolution, Proceedings of the Xth Congress of the International Primatological Society, Nairobi, 1985. Cambridge: Cambridge University Press, pp. 3–17.

Fleagle JG, and Kay RF (1983) New interpretations of the phyletic position of Oligocene hominoids. In RL Ciochon and RS Corruccini (eds.): New Interpretations of Ape and Human Ancestry. New York: Plenum Press, pp. 181–210.

——— (1987) The phyletic position of Parapithecidae. J. Hum. Evol. *16:*483–532.

Fleagle JG, and Mittermeier RA (1980) Locomotor behavior, body size, and comparative ecology of seven Surinam monkeys. Am. J. Phys. Anthrop. *52:*301–314.

——— (1981) Differential habitat use by *Cebus apella* and *Saimiri sciureus* in Central Surinam. Primates *22:*361–367.

Fleagle JG, and Simons EL (1978) Humeral morphology of the earliest apes. Nature *276:*705–707.

——— (1979) Anatomy of the bony pelvis in Parapithecid primates. Folia Primatol. *31:*176–186.

——— (1982a) The humerus of *Aegyptopithecus zeuxis:* A primitive anthropoid. Am. J. Phys. Anthrop. *59:*175–193.

——— (1982b) Skeletal remains of *Propliopithecus chirobates* from the Egyptian Oligocene. Folia Primatol. *39:*161–177.

——— (1983) The tibio-fibular articulation in

Apidium phiomense, an Oligocene anthropoid. Nature *301:*238–239.

Fleagle JG, Simons EL, and Conroy GC (1975) Ape limb bone from the Oligocene of Egypt. Science *189:*135–137.

Ford SM (1988) Postcranial adaptations of the earliest platyrrhine. J. Hum. Evol. *17:*155–192.

——— (1990) Locomotor adaptations of fossil platyrrhines. J. Hum. Evol. *19:*141–174.

Gebo DL (1989) Locomotor and phylogenetic considerations in anthropoid evolution. J. Hum. Evol. *18:*201–233.

——— (1992) Locomotor and postural behavior in *Alouatta palliata* and *Cebus capucinus.* Am. J. Primatol. *26:*277–290.

Gebo DL, Beard KC, Teaford MF, Walker A, Larson SG, Jungers WL, and Fleagle JG (1988) A hominoid proximal humerus from the early Miocene of Rusinga Island, Kenya. J. Hum. Evol. *17:*393–401.

Gebo DL, and Simons EL (1987) Morphology and locomotor adaptations of the foot in early Oligocene anthropoids. Am. J. Phys. Anthrop. *74:*83–101.

Gebo DL, Simons EL, Rasmussen DT, and Dagosto M (1994) Eocene Anthropoid Postcrania from the Fayum, Egypt. In JG Fleagle and RF Kay (eds.): Anthropoid Origins, Advances in Primatology. New York: Plenum Press.

Godinot M, and Mahboubi M (1992) Earliest known simian primate found in Algeria. Nature *357:*324–326.

Grand TI (1968) The functional anatomy of the lower limb of the howler monkey (*Alouatta caraya).* Am. J. Phys. Anthrop. *28:*163–182.

Harrison T (1987) The phyletic relationships of the early catarrhine primates: A review of the current evidence. J. Hum. Evol. *16:*41–80.

Hoffstetter R (1977) Phylogénie des primates. Bulletins et Mémoires de la Société d'Anthropologie de Paris *4*(Série XIII):327–346.

Jenkins FA, and Camazine SM (1977) Hip structure and locomotion in ambulatory and cursorial carnivores. J. Zool. Lond. *181:*281–298.

Jolly C (1967) The evolution of the baboons. In H Vagtborg (ed.): The Baboon in Medical Research. Austin: University of Texas Press, pp. 23–50.

Kappelman J (1992) The age of the Fayum primates as determined by paleomagnetic reversal stratigraphy. J. Hum. Evol. *22:*495–504.

Kay RF (1977) The evolution of molar occlusion in the Cercopithecoidea and early catarrhines. Am. J. Phys. Anthrop. *46:*327–352.

Kay RF, Fleagle JG, and Simons EL (1981) A revision of the Oligocene apes of the Fayum Province, Egypt. Am. J. Phys. Anthrop. *55:*293–322.

Langdon J (1984) A Comparative Functional Study of the Miocene Hominoid Foot Remains. Ph.D. dissertation, Yale University, New Haven.

Leutenegger W (1970) Das Becken der rezenten Primaten. Morph. Jb. *115:*1–101.

McCrossin ML, and Benefit BR (1992) Comparative assessment of the ischial morphology of *Victoriapithecus macinnesi.* Am. J. Phys. Anthrop. *87:*277–290.

Mendel F (1976) Postural and locomotor behavior of *Alouatta palliata* on various substrates. Folia Primatol. *26:*36–53.

Mittermeier RA (1978) Locomotion and postural repertoires of *Ateles geoffroyi* and *Ateles paniscus.* Folia Primatol. *30:*161–193.

Osborn HF (1908) New fossil mammals from the Fayum Oligocene, Egypt. Bull. Am. Mus. Nat. Hist. *24:*265–272.

Oxnard C (1963) Locomotor adaptations of the primate fore-limb. Symp. Zool. Soc. Lond. *10:*165–182.

Preuschoft H (1974) Body posture and mode of locomotion in fossil primates—method and example: *Aegyptopithecus zeuxis.* In Proceedings of the Fifth Symposium of the Congress of the International Primatology Society. Tokyo: Japan Science Press, pp. 346–359.

Rasmussen DT, Bown T, and Simons EL (1992) The Eocene-Oligocene transition in continental Africa. In DR Prothero and WA Berggren (eds.): Eocene-Oligocene Climatic and Biotic Evolution. Princeton: Princeton University Press, pp. 548–566.

Rasmussen DT, and Simons EL (1992) Paleobiology of the Oligopithecines, the earliest known anthropoid primates. Int. J. Primatol. *13:*1–32.

Richard A (1970) A comparative study of the activity patterns and behavior of *Alouatta villosa* and *Ateles geoffroyi.* Folia Primatol. *12:*241–263.

Rose MD (1988) Another look at the anthropoid elbow. J. Hum. Evol. *17:*193–224.

——— (1989) New postcranial specimens of catarrhines from the Middle Miocene Chinji Formation, Pakistan: Descriptions and a discussion of proximal humeral functional morphology in anthropoids. J. Hum. Evol. *18:*131–162.

Sarmiento E (1983) The significance of the heel process in anthropoids. Int. J. Primatol. *4:*127–152.

Schlosser M (1910) Uber einige fossile Säugetiere aus dem oligocan von Äegypten. Zoo. Anz. *34:*500–508.

——— (1911) Beiträge zur Kenntnis der Oligozanen Landsäugetiere aus dem Fayum, Äegypten, Beitr. Palaeontol. Oesterreich-Ungarns Orients *6:*1–227.

Schön Ybarra MA (1984) Locomotion and postures of red howlers in a deciduous forest-savanna interface. Am. J. Phys. Anthrop. *63:*65–76.

Schön Ybarra MA, and Conroy GC (1978) Nonmetric features of the ulna of *Aegyptopithecus, Alouatta, Ateles* and *Lagothrix.* Folia Primatol. *29:*178–195.

Schön Ybarra MA, and Schön MA (1987) Positional behavior and limb bone adaptations in red howling monkeys (*Alouatta seniculus).* Folia Primatol. *49:*70–89.

Schultz AH (1969) Observations of the acetabulum of primates. Folia Primatol. *11:*181–199.

Senut B (1989) Le coude des primates hominoïdes. Paris: CNRS.

Senut B, and Thomas H (1992) First discoveries of anthropoid postcranial remains from Taqah (Early Oligocene, Sultanate of OMAN). 14th Congress of the International Primatological Society, abstract number 0502, p. 258.

Simons EL (1962) Two new primate species from the African Oligocene. Postilla *64:*1–12.

——— (1965) New fossil apes from Egypt and the initial differentiation of Hominoidea. Nature *205:*135–139.

——— (1970) The deployment and history of Old World monkey (Cercopithecidae, Primates). In JR Napier and PH Napier (eds.): Old World Monkeys. New York: Academic Press, pp. 97–137.

——— (1972) Primate Evolution: An Introduction to Man's Place in Nature. New York: Macmillan Press.

——— (1974) *Parapithecus grangeri* (Parapithecidae, Old World Higher Primate): New species from the African Oligocene of Egypt and the initial differentiation of Cercopithecoidea. Postilla *64:*1–12.

——— (1985) Origins and characteristics of the first hominoids. In E Delson (ed.): Ancestors: The Hard Evidence. New York: Alan R. Liss, pp. 37–41.

——— (1986) *Parapithecus grangeri* of the African Oligocene: An archaic catarrhine without lower incisors. J. Hum. Evol. *15:*205–213.

——— (1989) Description of two genera and species of Late Eocene Anthropoidea from Egypt. Proc. Natl. Acad. Sci. *86:*9956–9960.

——— (1990) Discovery of the oldest known anthropoidean skull from the Paleogene of Egypt. Science *247:*1567–1569.

——— (1992) Diversity in the early Tertiary anthropoidean radiation in Africa. Proc. Nat. Acad. Sci. *89:*10743–10747.

Simons EL, and Kay RF (1983) *Qatrania,* a new basal anthropoid primate from the Fayum, Oligocene of Egypt. Nature *304:*624–626.

——— (1988) New material of *Qatrania* from Egypt with comments on the phylogenetic position of the Parapithecidae (Primates, Anthropoidea). Am. J. Primatol. *15:*337–347.

Simons EL, and Pilbeam D (1972) Hominoid paleoprimatology. In R Tuttle (ed.): The Functional and Evolutionary Biology of Primates. Chicago: Aldine-Atherton Press, pp. 36–62.

Simons EL, Rasmussen DT, and Gebo DL (1987) A new species of *Propliopithecus* from the Fayum, Egypt. Am. J. Phys. Anthrop. *73:*139–147.

Strasser E (1988) Pedal evidence for the origin and diversification of cercopithecid clades. J. Hum. Evol. *17:*225–246.

Strasser E, and Delson E (1986) Cladistic analysis of cercopithecoid relationships. J. Hum. Evol. *16:*81–99.

Szalay FS, and Dagosto M (1980) Locomotor adaptations as reflected on the humerus of Paleogene primates. Folia Primatol. *34:*1–45.

Szalay FS, and Delson E (1979) Evolutionary History of the Primates. New York: Academic Press.

Szalay FS, and Langdon JH (1986) The foot of

Oreopithecus: An evolutionary assessment. J. Hum. Evol. *15*:585–621.

Taylor ME (1976) The functional anatomy of the hindlimb of some African Viverridae (Carnivora). J. Morph. *143*:307–336.

Terborgh J (1983) Five New World Primates: A Study in Comparative Ecology. Princeton: Princeton University Press.

Thomas H, Roget J, Sen S, and Al-Sulaimani Z (1988) Découverte des plus anciens "Anthropoïdes" du continent arabo-africain et d'un Primate tarsiiforme dans l'Oligocène du Sultanat d'Oman. C.R. Séanc. Acad. Sci. Paris Ser. II, *306*:823–829.

Thomas H, Roget J, Sen S, Bourdillon-de-Grissac C, and Al-Sulaimani Z (1989) Découverte de vertébrés fossiles dans l'Oligocène inférieur de Dhofar (Sultanat d'Oman). Geobios *22*:101–120.

Van Couvering JA, and Harris JA (1991) Late Eocene age of Fayum mammal faunas. J. Hum. Evol. *21*:241–260.

CHAPTER TEN

Postcranial Adaptations and Positional Behavior in Fossil Platyrrhines

Jeff Meldrum

The living platyrrhine monkeys of South America span a considerable range of body size, from the 100 g pygmy marmoset *Cebuella pygmaea,* to the 10 kg woolly spider monkey *Brachyteles arachnoides.* The evolution of body size as an adaptive character related to food resource exploitation is recognized as a principal factor in the diversification of platyrrhine locomotor and postural adaptations. Consequently, many features of the craniodental and postcranial skeleton are correlated with increasing and decreasing body size within the context of an arboreal habitat. Among the diverse positional repertoires and associated postcranial morphologies represented by the living New World monkeys are a number of striking convergences with behaviors and morphologies of some Old World primates. For example, the suspensory and climbing behaviors of the large-bodied atelines bear many similarities to those behaviors of the Old World hominoids. The vertical clinging postures adopted by species in such genera as *Pithecia, Callicebus,* and *Cebuella* are also seen in many prosimian primates. On the other hand, some adaptations are quite distinctive of the platyrrhines, such as the prehensile tail of the atelines or the nonopposable hallux of the callitrichines. Novel locomotor strategies have played a prominent role in the adaptive radiation of primate taxa at all levels (e.g., Dagosto, 1988).

The radiation of the platyrrhine primates can also be defined in terms of locomotor diversification associated with ecological adaptation and niche partitioning. Reconstructing the evolutionary history of this diversification is one of the principal aims of paleoprimatology. Unfortunately, the postcranial fossil record for platyrrhine primates is very sparse (see Tables 10.1 and 10.2), somewhat limiting the potential for reconstructing positional behavior diversity in extinct platyrrhine communities. Fieldwork continues to produce new finds, however, adding further elements to the picture of fossil platyrrhine postcranial morphology and diversity. In this chapter, I review the Patagonian platyrrhine postcranial fossils from the early Miocene of Argentina and the La Venta platyrrhines from the middle Miocene of Colombia, contrast these two platyrrhine radiations, offer preliminary observations regarding several new specimens, and discuss the establishment of the distinctive morphotypes of the modern platyrrhine subfamilies.

DISTINCTIVENESS OF THE PATAGONIAN AND LA VENTA PRIMATES

The fossil primates from the Miocene of Argentina—as with the entire mammalian paleofaunas—are quite distinct from those of the Miocene of

TABLE 10.1
Platyrrhine Primate Postcranial Fossils from the Miocene of Argentina

Specimen Number	Description	Taxon	Stratigraphic Position	Locality
MACN-CH 362	L. talus	*Dolichocebus gaimanensis* (Reeser, 1984; Gebo & Simons, 1987; Meldrum, 1990)	Gaiman Formation	Gaiman
MACN-A 635	R. femur, radius, distal L. humerus	*Homunculus patagonicus* (Bluntschli, 1931)	Santa Cruz Formation	Corriguen Aike
MACN-SC 15	L. distal humerus	cf. *Carlocebus carmenensis* (Ford, 1990)	Pinturas Formation	Rio Pinturas
MACN-SC 101	R. scapula, L. ulna	*Carlocebus carmenensis* (Anapol & Fleagle, 1988)	Pinturas Formation, Middle Sequence	Cerro de los Monos
MACN-SC 304	L. talus	*Carlocebus* cf. *carmenensis* (Meldrum, 1990)	Pinturas Formation, Middle Sequence	Cauce Seca
MACN-SC 368	L. talus	*Carlocebus* cf. *carmenensis* (Meldrum, 1990)	Pinturas Formation, Middle Sequence	Loma de las Ranas
MACN-SC 271	R. talus	*Carlocebus* cf. *carmenensis* (Meldrum, 1990)	Pinturas Formation, Middle Sequence	Loma de la Lluvia
MACN-SC 396	R. talus	*Carlocebus* cf. *carmenensis* (Meldrum, 1990)	Pinturas Formation, Middle Sequence	Cerro de los Monos
MACN-SC 397	R. talus	*Soriacebus ameghinorum* (Meldrum, 1990)	Pinturas Formation, Lower Sequence	Pt. Sumich Norte
MACN-SC 350	R. tistal tibia	*Carlocebus* cf. *carmenensis*	Pinturas Formation	Pt. Sumich Sur
MACN-SC 367	L. calcaneus	*Carlocebus* cf. *carmenensis*	Pinturas Formation	Loma de las Ranas
MACN-SC 377	L. calcaneus	cf. *Soriacebus adrianae*	Pinturas Formation	Pt. Sumich Sur

MACN = Museo Argentino de Ciencias Naturales; CH = Chubut Province; A = Aike; SC = Santa Cruz Province.

Colombia, suggesting a rather high level of endemism. Cladistic analysis of dental characters demonstrates that the Patagonian primates form a single cluster separate from the Colombian species (Kay, personal communication). Their display of numerous unique features suggests they are more closely related to one another than to the more diverse Colombian primates (Fleagle, 1990; Fleagle et al., in press). Furthermore, the Patagonian primates are not as closely allied with modern platyrrhine subfamilies as are many of the middle Miocene primates from Colombia. The distinction between the two primate faunas is the result of a combination of several likely factors. First, they are not contemporaneous. The Colombian primates are slightly younger, ranging in age between about 11.0 and 13.5 million years (Guerrero, in press). The Patagonian primates range in age between about 16 and 19 mya (Marshall et al., 1986; MacFadden, 1990). Second, geographic barriers separated the ranges of the respective primate assemblages during much of the early and middle Miocene, probably resulting in separate and distinct radiations. An extensive marine trangression just north of Patagonia, in the Paraná-Paraguay basin, nearly divided the continent at 25° south latitude (Pascual et al., 1985) when sea levels were maximal near the onset of the middle Miocene, 16 mya (Haq et al., 1987). Third,

TABLE 10.2
Platyrrhine Primate Postcranial Fossils from the Miocene of Colombia

Specimen Number	Description	Taxon	Stratigraphic Position	Locality
UCMP 38762	partial skeleton ~70% complete	Holotype of *Cebupithecia sarmientoi* (Stirton & Savage, 1951)	Villavieja Formation, Baraya Member, Monkey Unit	UC locality V4517
IGM 184667	partial pelvis and pelvic limbs	*Cebupithecia sarmientoi* (Meldrum & Kay, 1990)	La Victoria Formation, Perico Member, below Chunchullo Sandstone	Duke locality 79
IGM 183420	L. distal humerus	*Cebupithecia sarmientoi* (Meldrum et al., 1990)	La Victoria Formation, Perico Member, below Chunchullo Sandstone	Duke locality 21
IGM 182858	L. proximal humerus	*Cebupithecia sarmientoi* (Meldrum & Kay, n.d.)	La Victoria Formation, Perico Member, below Chunchullo Sandstone	Duke locality 21
IGM 184074	L. talus	Genus & species indet. (Ford et al., 1991)	Villavieja Formation, Cerro Colorado Member, El Cardon Red Beds	Duke locality 43
IGM 183512	L. distal humerus	cf. *Neosaimiri fieldsi* (Meldrum et al., 1990)	Villavieja Formation, Baraya Member, Monkey Unit	Duke locality 54
IGM 250436	L. distal tibia	*Laventiana annectens* (Meldrum & Kay, n.d.)	Villavieja Formation, Cerro Colorado Member, El Cardon Red Beds	Duke locality 32
IGM-KU 8803	R. talus	*Laventiana annectens* (Gebo et al., 1990; Rosenberger et al., 1991)	Villavieja Formation, Cerro Colorado Member, El Cardon Red Beds	Kyoto locality Masato site
IGM-KU 8802	R. talus	cf. *Aotus dindensis* (Gebo et al., 1990)	Villavieja Formation, Baraya Member, Monkey Unit	Kyoto locality 9–86A

IGM = INGEOMINAS Geological Museum, Bogota; UCMP = University of California, Museum of Paleontology, Berkeley, CA; KU = Kyoto University

regional climatic deterioration in Patagonia during the middle Miocene led to the extinction of many climate-sensitive taxa, including platyrrhine primates (Pascual and Jaureguizar, 1990), with the result that the Patagonian primates may have contributed very little, if at all, to subsequent platyrrhine evolution (cf. Fleagle et al., 1987). In contrast, the La Venta primates had achieved a considerable taxonomic diversity by the middle Miocene and attained a wide range of body size. Most of the La Venta primates can be readily allied with living platyrrhine subfamilies.

PATAGONIAN PRIMATES

Dolichocebus gaimanensis

This species is known from a nearly complete, but badly crushed skull and a number of isolated teeth. Its dentition bears resemblances to *Saimiri* and *Aotus,* but also shares a number of primitive dental traits with some of the Fayum anthropoids from the Oligocene of Egypt. It has a body weight estimate of 3.0 kg (Fleagle, 1988). A single postcranial element (MACN-CH 362) is known for this early Miocene platyrrhine, a left talus with a badly eroded head (Fig. 10.1). It is rather distinctive from any extant platyrrhine genus in morphology, but

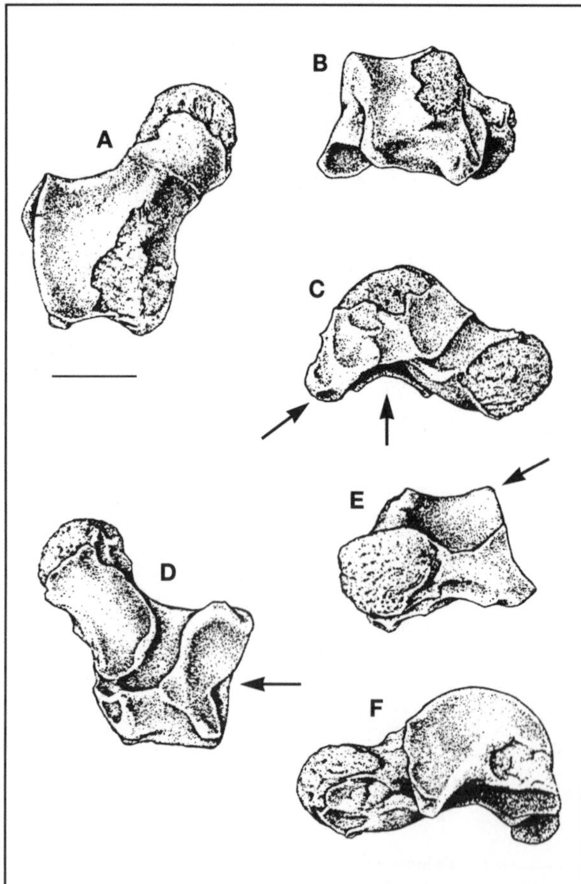

FIGURE 10.1. Left talus (MACN-CH 362) attributed to *Dolichocebus gaimanensis*, as seen in dorsal (A), proximal (B), medial (C), plantar (D), distal (E), and lateral views (F). Scale bar equals 0.5 cm.

bears the greatest similarity to the small-bodied cebids (Reeser, 1984; Gebo and Simons, 1987; Meldrum, 1990). The talar trochlea is moderately high, nearly parallel-sided, and the lateral margin is very sharp (Fig. 10.1E). The posterior calcaneal facet is very narrow and has a small radius of curvature (Fig. 10.1C and D). There is a well-developed posterior medial tubercle (Fig. 10.1C). These features indicate a restricted range of motion at the subtalar joint and a propensity for quadrupedal leaping (Meldrum, 1990). Ford (1990) has suggested that *Dolichocebus* most closely approximates the morphology for the hypothetical ancestral platyrrhine, with the exception of derived features associated with increased frequencies of leaping.

Homunculus patagonicus

With an estimated body weight of 2.7 kg (Fleagle, 1988), *Homunculus* is known from a partial cranial and facial fragment, a nearly complete lower mandible, and assorted isolated teeth. The few postcranial remains of *Homunculus* (MACN 5760) are perhaps the best preserved among the Patagonian primates. They consist of a complete right femur, a distal fragment of the right humerus, a complete left radius, and a nondescript fragment of the ulna diaphysis, associated with a nearly complete mandible (Fig. 10.2). Bluntschli (1913, 1931) concludes that the limb bones of *Homunculus* most resemble those of the small cebids, particularly *Aotus* and *Callicebus,* with some notable exceptions. The bones are larger and more robust than those of modern small cebids, and the radius is much longer relative to the femur than *Aotus,* approaching the proportions of the alouattines (*Homunculus* 84.5%, *Aotus* 72.3%, and *Alouatta* 89.9%). *Homunculus* appears to have had more powerful forelimbs than the living small cebids and may have employed more forelimb-assisted climbing such as practiced by *Lagothrix* or *Alouatta.*

A morphometric analysis of the femur by Ciochon and Corruccini (1975) has generally been taken to indicate a close resemblance to the callitrichines. This assertion is, however, based on the first two principal components axes, which account for only 39.5 percent of the variation. Ciochon and Corruccini (1975) go on to report the results of an analysis of species mean values that show *Homunculus* to be quite distinct from the callitrichines, and in fact from all platyrrhines, in the second and subsequent axes. Its distinctiveness resides primarily in the proportionately larger dimensions of the trochanters, condyles, and robust shaft.

A number of features of the *Homunculus* femur are associated with adaptations for leaping. The articular surface of the femoral head extends onto the posterosuperior aspect of the neck and the fovea capitus is quite distally located. Both features are characteristic of habitually adducted hindlimbs, moving primarily in a sagittal plane. The greater trochanter is very broad and rugous, and overhangs the femoral shaft anteriorly, providing extensive attachment for the m. vastus lateralis, a powerful extensor of the knee. The intertrochanteric line is very prominent, marking the attachment of the il-

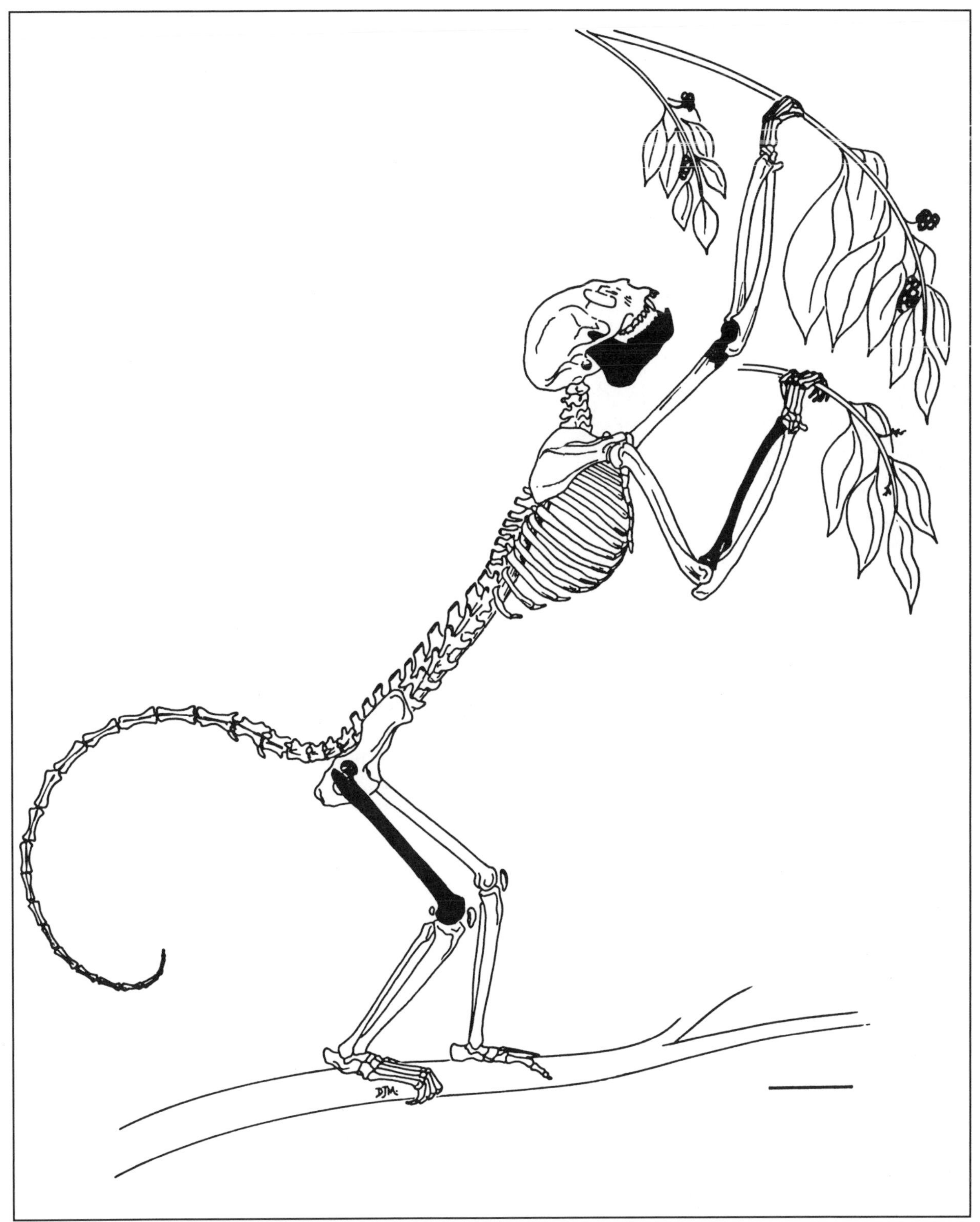

FIGURE 10.2. Speculative reconstruction of the skeleton of *Homunculus patagonicus*. Darkened areas indicate the associated elements of specimen MACN-A 635. Scale bar equals 5.0 cm.

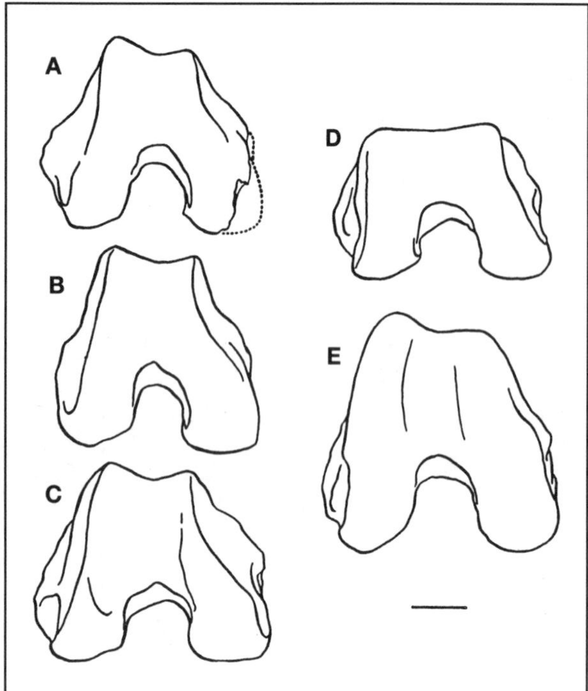

FIGURE 10.3. Distal views of the right femoral outlines of A: *Cebupithecia sarmientoi* (UCMP 38762, reconstructed); B: *Cebupithecia sarmientoi* (IGM 184667, reconstructed); C: *Homunculus patagonicus* (MACN-A 635); D: *Pithecia pithecia* (USNM 339660); and E: *Lemur macaco* (ANSP 1422). Scale bar equals 0.5 cm.

iofemoral and ischiofemoral ligaments that check extension of the hip. The rims of the patellar groove are raised, especially the lateral rim. The condyles are rather deep and the biepicondylar breadth is moderate (Fig. 10.3C), giving a ratio of 0.82, comparable to *Aotus* (0.82) or *Saguinus* (0.83).

The humerus of *Homunculus* displays a number of features that further suggest climbing, in agreement with the proportionately long robust radius. The capitulum is quite spherical and "unrolled" relative to the long axis of the humeral shaft, providing a greater range of flexion-extension (Napier and Davis, 1959; Meldrum et al., 1990). The lateral epicondyle is relatively large (Ford, 1990), the trochlea is cylindrical, and the medial epicondyle has little dorsal angle (Fleagle and Meldrum, 1988).

Carlocebus carmenensis

The dentition of *Carlocebus carmenensis* is very similar to that of *Homunculus patagonicus* and these two species are most likely sister taxa (Fleagle, 1990). *Carlocebus* differs in its slightly greater absolute body size (ca. 3.5-4.0 kg) and less prominent molar cusps and crests. It shows the greatest similarity, in features of the postcanine dentition, to the extant genus *Callicebus,* but has much smaller incisors.

Four fossil tali have been referred to *Carlocebus* cf. *carmenensis* (Meldrum, 1990; Fig. 10.4B–E). The talar trochlea are parallel-sided and moderately tall. The necks are moderately long and the heads rounded. The proportions of these tali fall intermediate to those of the callitrichines and small-bodied cebids. Despite their overall similarity to one another, subtle but consistent differences among these four tali suggest that two species may be represented. MACN-SC 271 and 368 are distinguished from MACN-SC 304 and 396 in that the trochlea is somewhat higher and narrower, the neck is slightly longer and has a greater medial angulation, the head is rounder, and the posterior calcaneal facet is narrower. Both talar morphs suggest basically above-branch arboreal quadrupedalism, each with varying frequencies of leaping and perhaps clinging.

An isolated distal fragment of a right tibia (MACN-SC 350)—also referred to *Carlocebus* cf. *carmenensis*—articulates conformably with the tali previously discussed. The shaft is very robust, similar to the condition characterizing the long bones of *Homunculus,* and it broadens due to an expansion of the distolateral border. The broad fibular incisure is bounded by well-developed crests that extend and converge proximally. There was apparently a strong distal tibiofibular syndesmosis, with the two leg bones tightly appressed for roughly 15 mm. This indicates a more stable talocrural joint with movement restricted to flexion/extension, suggesting rapid quadrupedalism and/or leaping. A partial calcaneus (MACN-SC 367) is also referred to *Carlocebus* cf. *carmenensis.* The posterior articular facet of the calcaneus is quite long, with a relatively low angle of orientation to the long axis of the calcaneus. The anterior facet is very broad and extends to the distal end of the calcaneus. These features have been associated with either leaping or suspen-

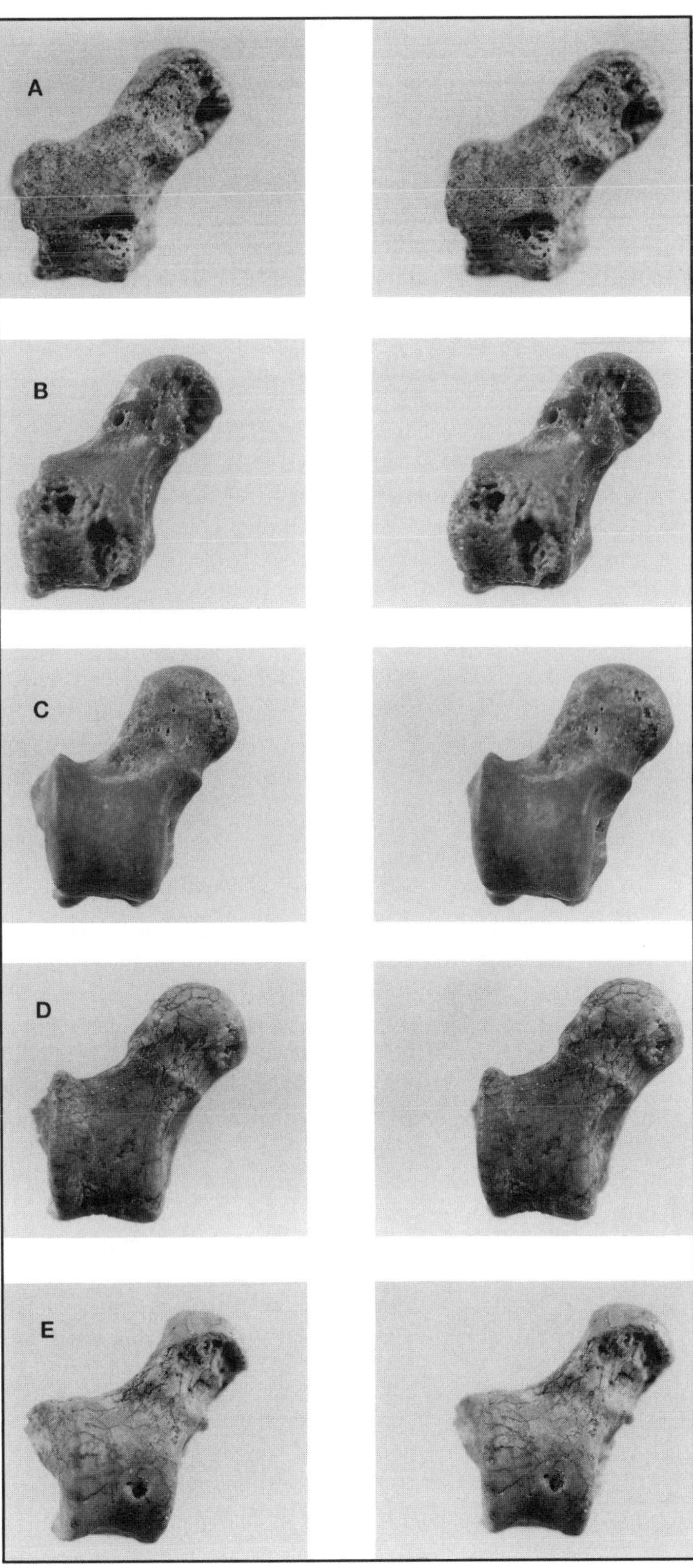

FIGURE 10.4. Stereophotographs of five tali from Pinturas primates. **A**: MACN-SC 397 referred to *Soriacebus ameghinorum* (photographically reversed); **B**: MACN-SC 271 (photographically reversed); **C**: MACN-SC 368; **D**: MACN-SC 304; and **E**: MACN-SC 396 (photographically reversed). All referred to *Carlocebus* cf. *carmenensis*.

sory/climbing behaviors, or a combination of the two (Ford, 1990). A more extensive analysis of the tibia fragment and partial calcaneus will be considered elsewhere (Meldrum and Fleagle, in prep.).

A fragmentary scapula and ulna (MACN-SC 101), recovered in association with a facial fragment and a worn mandibular specimen, are also referred to here as *Carlocebus carmenensis.* They are of the appropriate size, with body weight estimated at 3.6 kg (Anapol and Fleagle, 1988). The pear-shaped glenoid outline, the intersection of the root of the scapular spine relatively low on the glenoid, and the very robust axillary border are all conditions found in platyrrhine species characterized by increased use of the forelimbs in climbing, such as *Lagothrix* and *Alouatta.* The relative length of the olecranon is comparable to values for *Alouatta,* as are dimensions of the sigmoid notch (Anapol and Fleagle, 1988). The ulnar shaft was oval in cross section, in contrast to the inverted T-shape characterizing clinging primates (Fleagle and Meldrum, 1988). These features suggest a mode of locomotion similar to that of extant *Lagothrix* or *Alouatta,* consisting of arboreal quadrupedalism, with increased emphasis on the use of flexed forelimbs for climbing. This observation is very interesting in light of the foregoing discussion of the forelimb anatomy of *Homunculus,* which indicated relatively long forearms (with proportions approaching those of *Alouatta)* and features of the elbow suggesting climbing behavior. Furthermore, a partial distal humerus (MACN-SC 15) that may also be referred to *Carlocebus* cf. *carmenensis,* is very similar in morphology to the Corriguen Aike humerus, and is approximately 10 percent larger (Ford, 1990), in agreement with differences in molar linear dimensions. Ford (1990) has noted features of this humerus, including cylindrical trochlea, clinging facet, and rounded trochleocapitular ridge, suggesting greater frequency of climbing/suspensory postures.

Soriacebus ameghinorum

Soriacebus ameghinorum is a saki-sized monkey (ca. 2.0 kg) with unusual dental and mandibular proportions. It shares individually various dental similarities with the living callitrichines, pitheciines, and *Callicebus* (Fleagle et al., 1987). Rosenberger et al. (1990) have proposed placing *Soriacebus* in the Pitheciinae on the basis of several phenetic similarities of the dentition. On the other hand, Kay (1990) concludes that *Soriacebus* is a sister taxon to all other platyrrhines, and that limited similarities to pitheciines evolved independently.

A single talus (MACN-SC 397) has been referred to this Pinturas primate (Fig. 10.4A; Meldrum, 1990). The talar trochlea is relatively low and broad, with a high degree of wedging. The neck is short and moderately angled (40° relative to the fibular facet), and the head is relatively dorsoventrally compressed. These features distinguish the living platyrrhine taxa that are characterized by frequent climbing behaviors, as opposed to simple above-branch quadrupedalism. These include the atelines, *Alouatta,* and *Pithecia.* The talus of *Soriacebus ameghinorum* bares the greatest similarity to *Pithecia.* The locomotor behavior of *Pithecia pithecia* is dominated by leaping, followed by quadrupedal walking and running, and finally climbing (Fleagle and Meldrum, 1988). The other species of *Pithecia* for which locomotor observations are available, *P. hirsuta* (= *monachus),* leap less often and have somewhat higher frequencies of quadrupedalism and climbing (Happel, 1982). Further details of pitheciine positional behavior are discussed under *Cebupithecia.*

Soriacebus adrianae

This species is considerably smaller than *S. ameghinorum,* with an estimated body weight of from 800 to 900 g (Fleagle, 1990). *S. adrianae* is found in younger strata than *S. ameghinorum* (Meldrum, 1990). It differs only in a few dental traits, such as a relatively smaller M2 trigonid, much smaller M3, and perhaps a taller P2. A small partial calcaneus (MACN-SC 377) was recovered from a locality rich in remains of *S. adrianae* and is of the appropriate size for allocation to this species. The fossil preserves the distal two-thirds of the calcaneus. The anterior talocalcaneal facet is very narrow and does not appear to reach the distal end of the calcaneus. Dagosto (1986) states this condition is associated with an alternating tarsus with limited mobility, especially in eversion and plantarflexion.

ADAPTIVE DIVERSITY OF THE PATAGONIAN PRIMATES

It has been proposed that the ancestral platyrrhine was a fairly small, *Aotus*-like monkey, weighing approximately 1.0 kg, characterized by arboreal quadrupedalism on horizontal supports, with limited leaping or climbing (Ford, 1988). Assuming that this model is accurate, to what extent have the primates of the Patagonian radiation diverged from the ancestral condition and how diverse have their positional adaptations become?

Dolichocebus is the oldest species for which postcranial remains are represented. It is similar to the ancestral morphotype only in that the incomplete talus is similar in some ways to tali of the small-bodied cebids, including *Aotus* and *Callicebus*. It differs from the ancestral condition in that *Dolichocebus* is considerably larger and appears to have a number of derived features of the talus associated with frequent leaping behavior (cf. Ford, 1990). *Homunculus* and its closely related, and perhaps morphologically similar, sister taxon *Carlocebus* are also considerably larger than the hypothetical ancestor and have diverged from its basic quadrupedal pattern. *Homunculus* and *Carlocebus* display limb proportions and a combination of hindlimb and forelimb morphologies that suggest frequent forelimb dominated climbing interspersed with leaping. *Soriacebus ameghinorum* is also more derived than the hypothetical ancestor, in that similarities of the talus to the extant *Alouatta* and *Pithecia* indicate greater commitment to climbing behaviors. On very limited and preliminary evidence, it can be said that *S. adrianae* was slightly smaller than the hypothetical ancestor and may have employed a similar generalized above-branch arboreal quadrupedalism, perhaps with reduced tarsal mobility. This similarity to the hypothetical ancestor is likely the result of convergence, as *S. adrianae* is almost certainly derived from the larger *S. ameghinorum*.

The radiation of primates in the early Miocene of Argentina is potentially unique. Despite the functional analogies drawn to extant platyrrhine taxa, these Argentine fossils cannot be readily associated with modern platyrrhine subfamilies, and they may have had little, if any, connection to later platyrrhine evolution. Inferences made from adaptive characters of the postcranial fossils indicate a somewhat limited range of positional behaviors by comparison with the extant platyrrhines. Neither the specialized suspensory behavior of the atelines nor the scansorial behavior of the callitrichines had their counterparts among the known Patagonian primates.

LA VENTA PRIMATES

Cebupithecia sarmientoi

Postcranially, this is by far the most extensively represented platyrrhine primate; the holotype skeleton (UCMP 38762) is nearly 70 percent complete (Fig. 10.5; Stirton, 1951; Stirton and Savage, 1951). Additional finds, including a partial pelvis and hindlimbs (IGM 184667; Meldrum and Kay, 1990; in press), and a fragmentary humerus (IGM 183420 and IGM 182858; Meldrum et al., 1990) have added to the representation of *Cebupithecia*'s postcranial morphology. As with the craniodental remains, the postcranium shares a number of particular similarities with the extant genus *Pithecia* (Fleagle and Meldrum, 1988; Meldrum and Fleagle, 1988), although notable differences are also present (Ford, 1986, 1990; Davis, 1987). The distal humerus (Fig. 10.6A) displays a number of morphological features that suggest clinging and suspensory postures. These include a medial epicondyle with very little dorsal angulation, a cylindrical trochlea, and a contact facet for the coronoid process of the ulna. The ulnar shaft is T-shaped in crosssection, similar to that of *Pithecia pithecia* and other clinging primate species. The morphology of the distal tibia and talar trochlea also indicate adaptation for clinging postures. The distal edge of the talar trochlea extends onto the dorsal surface of the talar neck forming a small tibial stop, that in turn articulates with a proximally extended tibial trochlear surface during extreme dorsiflexion.

Features of the vertebral column, pelvis, and hindlimb indicate that *Cebupithecia* was an adept leaper. The lumbar region is very long, and the vertebrae are robust with marked muscle attachments (Meldrum and Lemelin, 1991). The ischium is relatively long, which increases leverage of the hip extensors. The femoral neck is short and nearly perpendicular to the femoral shaft. The articular surface of the head extends onto the posterior surface of the neck. On the dorsal surface of the neck the intertrochanteric line is evident, marking the insertion of the iliofemoral and ischiofemoral ligaments. These ligaments function to check the extreme extension of the hip. The femoral condyles are very deep and the biepicondylar breadth is quite narrow (Fig. 10.3), giving a ratio, 0.90, that is slightly outside the range for a sample of platyrrhines, but is closely approached by *Aotus* and *Saguinus* (Meldrum and Kay, 1990).

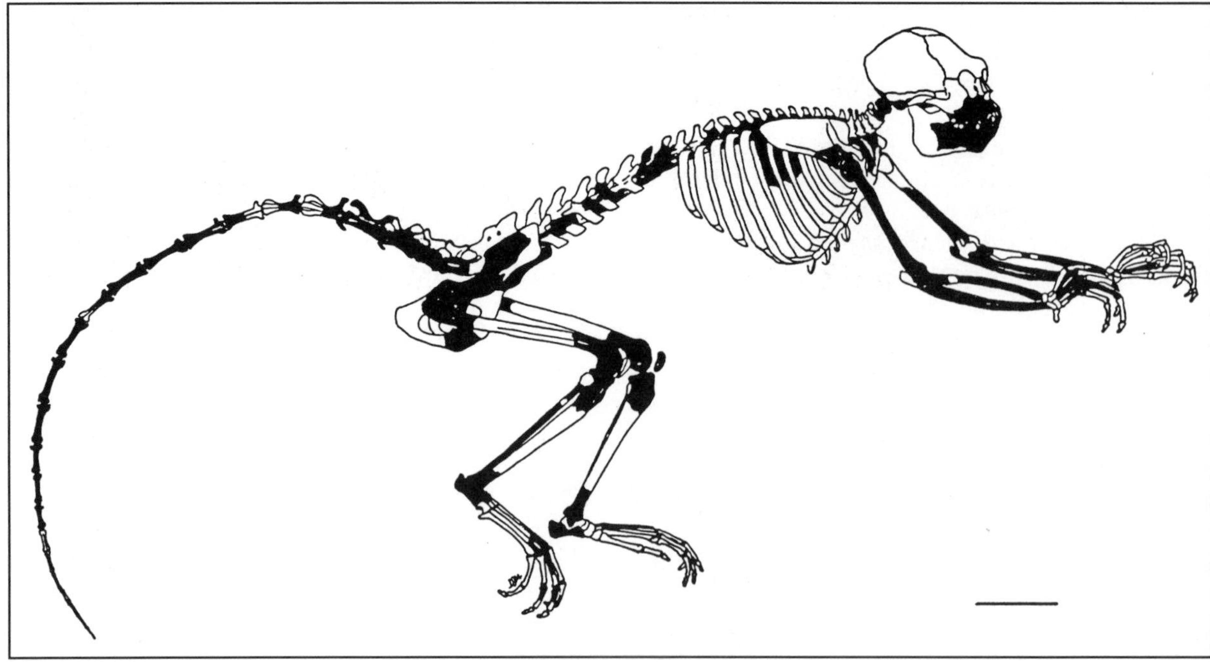

FIGURE 10.5. Reconstruction of the skeletal anatomy of *Cebupithecia sarmientoi*. Darkened areas indicate preserved portions of the skeleton. Scale bar equals 5.0 cm.

Comparable values are common among prosimian leapers. The lateral lip of the patellar groove is very prominent. The tibial shaft is very mediolaterally compressed. The distal tibial shaft is broadened and the lateral border marked by a distinct ridge indicating a tightly appressed fibula and a well-developed distal tibiofibular syndesmosis. The talar trochlear crests are relatively parallel and sharp. These and other features of the hindlimb all indicate a marked adaptation for powerful leaping in *Cebupithecia*.

Neosaimiri fieldsi

Rosenberger, Hartwig, Takai, Setoguchi, and Shighaka (1991) have recently proposed synonymizing *Neosaimiri* with the genus *Saimiri*. *Neosaimiri* differs by having proportionately smaller incisors, a complete incisor lingual cingulum, a small canine heel, narrower premolars with reduced metaconids and prominent buccal cingula, narrower molar trigonids, and a larger alveolus for M_3. The mandibular corpus is also more robust and is more convex along its inferior border. I do not think the proposed synonymy of Neosaimiri and *Saimiri* has been satisfactorily demonstrated, particularly in view of the minor distinctions cited by the same workers to differentiate the closely related *Laventiana annectens* from *Neosaimiri* (discussed next). Therefore, the generic distinction of *Neosaimiri* will continue to be made.

A single postcranial fossil has been attributed to *Neosaimiri* (Meldrum et al., 1990). IGM 183512 is a small left distal humerus (Fig. 10.6B). This specimen was recovered from the same stratigraphic horizon as the type specimen of *Neosaimiri*. It is very similar in size and morphology to the extant genus *Saimiri*. The medial epicondyle is more dorsally angled, the medial lip of the trochlea is more prominent; the capitulum is less spherical and is "rolled" ventrally by comparison with *Cebupithecia*. To the extent of these parallels in the elbow region, *Neosaimiri* appears to have utilized its forelimb in arboreal quadrupedalism much as the extant squirrel monkey, *Saimiri*.

Laventiana annectens

Based on a nearly complete mandible, *Laventiana* is very similar to *Neosaimiri*, differing only in the presence of a distinct entoconid sulcus on relatively

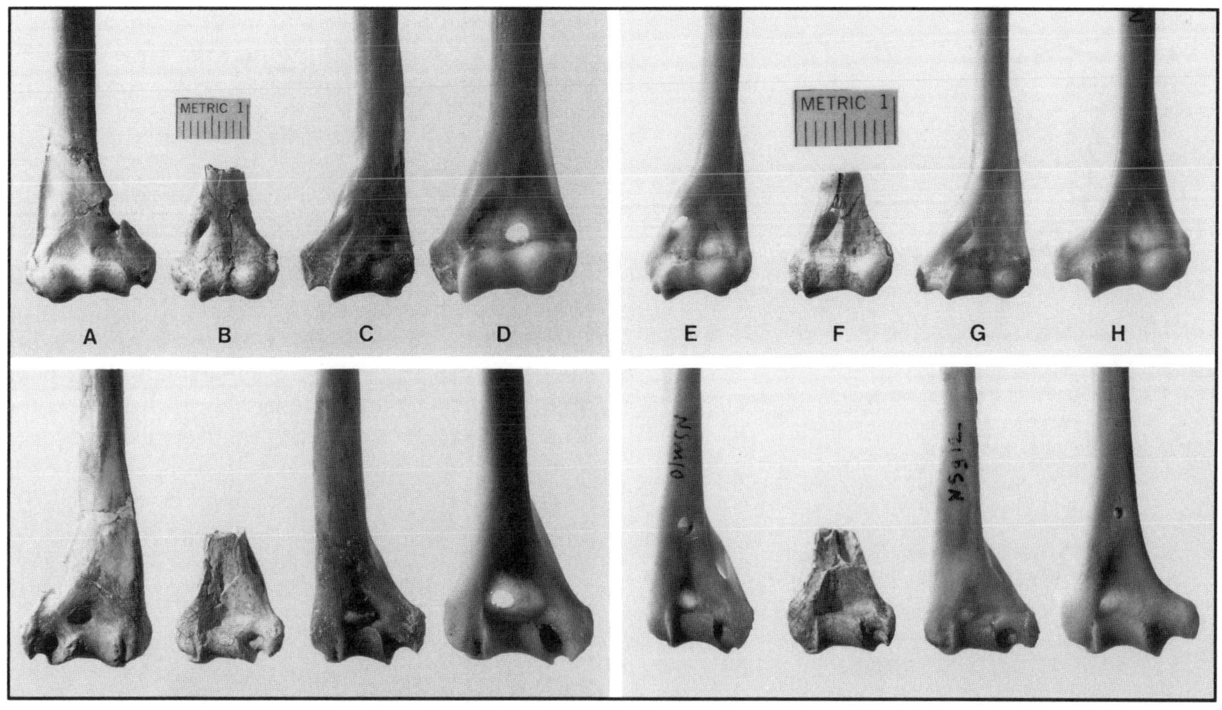

FIGURE 10.6. Ventral view (above) and dorsal view (below) of distal humeri. **A**: *Cebupithecia sarmientoi* (UCMP 38762); **B**: *Cebupithecia sarmientoi* (IGM83420); **C**: *Pithecia pithecia;* **D**: *Cebus apella;* **E**: *Saimiri sciureus;* **F**: *Neosaimiri fieldsi;* **G**: *Saguinus* sp.; and **H**: *Aotus trivirgatus.* Scale bar equals 1.0 cm.

larger M_{1-2} (Setoguchi et al., 1990; Rosenberger, Setoguchi, and Hartwig, 1991), and is certainly very closely related. A well-preserved right talus (IGM-KU 8803) was recovered in association with the type mandible (Gebo et al., 1990). The talus falls well within the range of 12 linear measurements of 30 adult *Saimiri* spp. (Meldrum, 1990; contra Gebo et al., 1990). It is most similar to *Saimiri* among the extant platyrrhines in having a relatively long narrow talar neck, narrow round head, relatively narrow trochlea with a sharp curved lateral rim, and a small nonfaceted medial protuberance (Gebo et al., 1990). It differs from *Saimiri* in the lack of a dorsal tibial stop at the distal border of the trochlea, a feature usually present in *Saimiri* (Fig. 10.7B). This morphological profile suggests a small arboreal quadrupedal runner-leaper, similar in positional behavior to the living squirrel monkey.

An isolated distal tibia (IGM 250406) was recovered from the same locality as the talus and type mandible, that is also extremely similar in size and morphology to that of *Saimiri* (Fig. 10.7A; Meldrum and Kay, in press). Of particular note is the indication of a well-developed syndesmosis of the distal tibiofibular joint. The posterior surface of the distal tibia is quite flat and the lateral border is marked by a pronounced ridge, along which the fibula would be appressed. This close appression of the distal tibia and fibula commonly occurs in *Saimiri* and several other platyrrhine genera, including *Aotus, Callithrix, Cebuella,* and *Pithecia* (see also Fleagle and Simons, 1983; Fleagle and Meldrum, 1988). Only *Saimiri* commonly displays the distinctive combination of mediolateral widening and posterior flattening of the distal tibial shaft, also evident in the fossil tibia. In contrast, the tibial shafts of *Callithrix* and *Cebuella* remain narrow and the fibula simply converges onto the distal tibial shaft. In *Aotus* the syndesmosis occurs much less frequently and is quite restricted in length along the lateral border of the distal tibial shaft. The extensive syndesmosis of the distal tibiofibular joint serves to limit lateral displacement of the fibula and rotational movement of the tibiotalar joint, by restricting

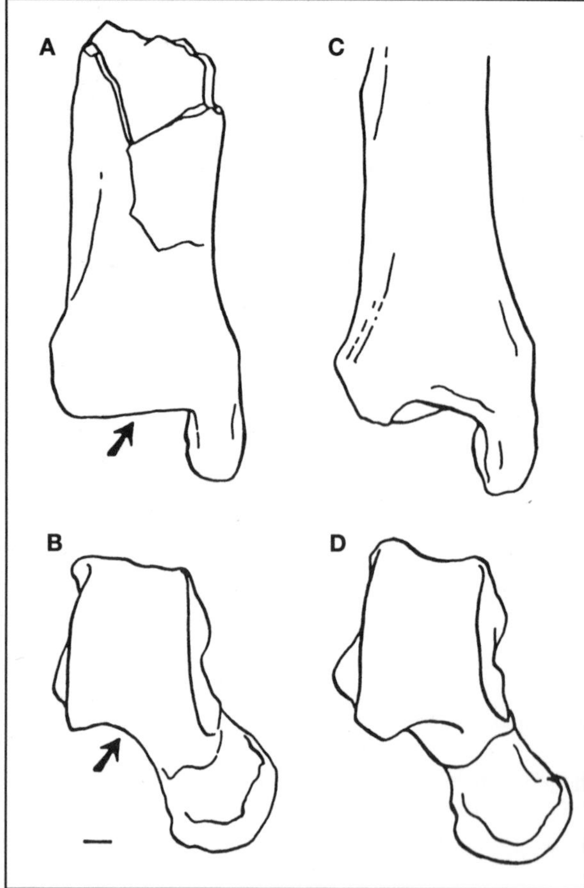

FIGURE 10.7. Distal tibia and talus of IGM 250436 (**A**) and IGM-KU 8803 (**B**), specimens referred to *Laventiana annectens,* compared with same elements (**C, D**) of *Saimiri sciureus* (USNM 397810). Scale bar equals 1.0 mm.

movement of the talus to flexion-extension in the sagittal plane. This morphology is correlated with frequent leaping. Its presence in *Saimiri* agrees with reports of frequent leaping in *Saimiri sciureus* (Fleagle et al., 1981), although some species of *Saimiri* (i.e., *S. oerstedii*) reportedly leap less frequently (Boinski, 1989). The presence of a well-developed syndesmosis in IGM 250436 suggests that this small fossil platyrrhine was a small arboreal quadrupedal leaper.

Just as with the talus, the tibia lacks an extension of the trochlear surface onto the anterior surface of the shaft, corresponding to the tibial stop (Fig. 10.7A). The correlated lack of this feature, the overall concordance in size and articular morphology, and their recovery from precisely the same location, combine to suggest that these fossils represent the same taxon.

"Aotus dindensis" (Mohanamico hershkovitzi)

A small isolated talus (IGM-KU 8802) was recovered within the Monkey Unit (Gebo et al., 1990). It falls within the size range of extant *Aotus* and was referred to cf. *Aotus dindensis* on the basis of a suite of characters, consisting of a moderately short talar neck, wide talar head, moderately high talar body, long and narrow talar body, and medium-to-large medial protuberance, sometimes faceted, which characterize the Aotini (Gebo et al., 1990). Particular emphasis is placed by Gebo et al. (1990) on the similar development of the medial protuberance in IGM-KU 8802 and *Aotus,* which provides attachment for the posterior talotibial ligament. The condition of this feature is variable, and a moderate tubercle is also present in *Pithecia* and some of the callitrichines. The presence of the facet on the medial tubercle is also quite variable, present in only one third of the specimens of Aotini, and also present in a few specimens of *Cebus,* and the fossil talus of *Cebupithecia* (UCMP 38762) (see Table 2 in Gebo et al., 1990).

Controversy surrounds the validity of this taxon, which affects the interpretation of the affinities of the talus. Kay (1990) has persuasively demonstrated that the type of "*Aotus dindensis*" does not differ significantly from the type specimen of *Mohanamico hershkovitzi* (Luchterhand et al., 1986) and the former should be considered a junior synonym of *Mohanamico hershkovitzi.* There is no compelling evidence to suggest a close affinity of *Mohanamico* to the Aotini; in fact, Rosenberger et al. (1990) have suggested that *Mohanamico* shares many similarities with the callitrichines and *Callimico,* while Kay (1990) and Meldrum and Kay (1992) have suggested affinities to the pitheciines. This raises the question of the soundness of referring the talus (IGM-KU 8802) to "*Aotus dindensis*" *(Mohanamico hershkovotzi)* on the basis of ambiguous similarities to *Aotus* or *Callicebus.* On the other hand, Gebo et al. (1990) note that IGM-KU 8802 clearly differs from extant *Aotus* in having a more square-shaped (relatively wide and short) talar body. The proportions and configuration of the talar head and trochlea are very similar to values reported for callitrichines (Meldrum, 1990). A satisfactory resolution of this issue must await the discovery of associated craniodental and postcranial materials for *"Aotus dindensis"* and/or *Mohanamico hershkovitzi.*

Specimens of Uncertain Allocation

Another talus (IGM 184074) was recovered from higher in the stratigraphic section (Ford et al., 1991). It is a left specimen, reasonably well preserved, with superficial erosion of the head, posterior tubercles, and calcaneal facets. It measures approximately 14.0 mm in length, suggesting a monkey intermediate in size to *Callicebus* and *Pithecia* (Meldrum, 1990). It shares a number of similarities with the talus of *Cebupithecia,* but has a somewhat narrower head and neck, and a slightly more rounded lateral trochlear crest. The mandible of a new species of primitive pitheciine (DING 90-23) was recovered from virtually the same locality as the talus (Meldrum and Kay, 1992). The correlation of talar length and M_1 length suggests that the talus may potentially pertain to this new species. Any allocation at this time would be premature.

ADAPTIVE DIVERSITY AND THE ESTABLISHMENT OF MODERN SUBFAMILIAL MORPHOTYPES

On the basis of dental morphology, the platyrrhine fossil record suggests that by the Miocene of Colombia, several of the modern platyrrhine subfamilies (sensu Thorington and Anderson, 1984) were represented among the primate paleofauna: the Alouattinae represented by *Stirtonia tatacoensis* and *S. victoriae;* the Pitheciinae by *Cebupithecia sarmientoi* and DING 90-23, genus and species nov.; the Saimirinae by *Neosaimiri fieldsi* and *Laventiana annectens;* and the Callitrichinae (or Callimiconinae) by IGM 184332 (DING 88-275), IGM 184531, and perhaps IGM-KU 8402 (an isolated incisor, genus and species indeterminant). Had these Miocene fossil taxa also achieved the level of diversity in postcranial adaptation that characterizes the extant platyrrhine subfamilies? The postcranial fossils of the La Venta primates provide a limited answer to this question. Only two subfamilies are well represented by fossil postcrania, the Pitheciinae and the Saimirinae. Once again, assuming the *Aotus*-like model of a hypothetical ancestral platyrrhine is accurate, to what extent had the modern subfamilial morphotypes differentiated by the middle Miocene?

The extant pitheciines are larger than the hypothetical ancestor, weighing on average between 2.0 and 3.5 kg (Fleagle, 1988). *Cebupithecia* was within the lower limits of this range, with an estimated body weight of 2.2 kg. The extant pitheciines display more climbing and limited suspensory behaviors than the hypothetical ancestor, as reflected in their proportionately longer limbs. The relative length of the forelimb (humerus + radius / trunk) ranges between 70 and 96 for species of *Chiropotes* and *Pithecia* (Hershkovitz, 1985, 1987). Using an estimated trunk length of 250 mm, *Cebupithecia* had forelimbs relatively shorter than the extant pitheciines, with an estimated value of 65. This falls within the range of values for *Aotus, Callicebus,* and *Saimiri* (52-73; Hershkovitz, 1990). The proclivity for climbing and suspensory behaviors in the extant pitheciines is also reflected in the configuration of their limb joints. For example, the distal femur of the extant pitheciines is relatively broad and shallow, with a wide, shallow patellar groove, very similar to the condition in the extant atelines. Likewise, the talar trochlea is low, with a rounded lateral margin, approaching the condition found in the ateline talus. These and other features of the pitheciine limbs, which resemble the ateline condition and are best explained as skeletal correlates of climbing and suspensory behaviors, are noticeably lacking in *Cebupithecia.* For example, the distal femur of *Cebupithecia* is very narrow and deep; and the talar trochlea is moderately high with a sharply crested lateral margin and more closely resembles such forms as *Aotus, Callicebus,* or some callitrichines (Ciochon and Corruccini, 1975; Davis, 1987, 1988; Ford, 1990). Therefore, while *Cebupithecia* had evolved the body size and dental morphology characteristic of modern pitheciines, it still lacked many of the derived postcranial features that characterize the extant pitheciines.

By comparison, much less is known concerning the postcranial anatomy of *Neosaimiri* and *Laventiana.* The known distal humerus, distal tibia, and talus are nearly identical in size and morphology to those of the extant squirrel monkey, *Saimiri. Saimiri* differs from the hypothetical ancestral morphotype in several ways. Squirrel monkeys are moderately smaller, with body weights ranging between 365 and 1135 g (Fleagle, 1988). *Neosaimiri* had an estimated body weight of 840 g. *Saimiri* has relatively longer hindlimbs than predicted for the hypothetical ancestor and leaps with greater frequency. Although nothing can be said about the limb proportions of *Neosaimiri* or *Laventiana,* the morphology of the distal tibia and talus of *Laventiana* are remarkably similar to those of the extant squirrel monkey and likewise indicate adaptations for frequent leaping. This is particularly evident in the extensive syndesmosis of the distal tibiofibular joint. Recent analysis of dental characteristics suggest that *Saimiri* and *Neosaimiri/Laventiana* are the sister taxa to the Callitrichinae (Kay and Meldrum, n.d.). Therefore, the saimirines may represent part of a radiation of platyrrhines, including the Callitrichinae and Callimiconinae, that underwent reduction in body size to exploit a primarily insectivorous-frugivorous diet, while adopting a more quadrupedal-scansorial-leaping positional behavior.

By comparing corresponding elements of the postcranium of *Cebupithecia* and *Neosaimiri/Laventiana,* some aspects of the ecological adaptive diversity present in the primate community of the Colombian Miocene can be inferred. Their modern counterparts, *Pithecia* and *Saimiri,* occur sympatrically over much of their respective ranges. *Pithecia* has a rather broad range of forest habitat tolerance. It frequents the understory and lower canopy strata and is frequently encountered on vertical supports. Locomotion is primarily by leaping and bounding. Few observations are available for postures during feeding. It is almost exclusively a frugivore seed-predator. The morphology of the distal humerus of *Cebupithecia* suggests it too adopted frequent clinging postures on vertical supports, while the morphology of the hindlimb also suggests frequent leaping (Meldrum et al., 1990), although perhaps after a different fashion, given the less derived hindlimb morphology present in *Cebupithecia.* The dentition of both taxa display specializations for seed predation.

Saimiri is also environmentally tolerant and also occupies the understory and lower canopy strata. It moves primarily by quadrupedal walking and running, interspersed with frequent leaps. Its diet differs from *Pithecia* in that it consists largely of insects and some fruit. The distal humerus of *Neosaimiri* suggests arboreal quadrupedalism, and the extensive syndesmosis of the distal tibiofibular joint of the closely related *Laventiana* suggests frequent leaping. The dentitions of *Neosaimiri* and *Laventiana* indicate perhaps more emphasis on frugivory than insectivory (Kay et al., in press). By analogy, it seems reasonable to assume that *Cebupithecia* and *Neosaimiri/Laventiana* had partitioned the forest understory in a manner comparable to living *Pithecia* and *Saimiri.*

The primitiveness of many aspects of the postcranial skeleton of *Cebupithecia* demonstrates that the distinctive features of the dentition and postcranium characterizing modern platyrrhine subfamilies may not have evolved in concert (see also Ford, 1990). Therefore, it is impossible to predict with certainty the postcranial adaptations of fossil platyrrhines known only from dental remains. For example, the platyrrhine subfamily Alouattinae is represented in the Miocene of Colombia by *Stirtonia,* which displays many dental similarities to the modern genus *Alouatta* and which had a comparably large body size (Kay et al., 1987). But only the discovery of postcranial fossils of *Stirtonia* will determine whether the distinctive climbing and suspensory adaptations associated with the foraging strategy of the large-bodied alouattines (and even more specialized atelines) had emerged at this point in their evolutionary history. Similarly, in the smaller-body-size range, a number of putative callitrichines are known from only craniodental remains. Until postcranial remains surface, it will be impossible to determine whether the scansorial and vertical clinging positional behaviors of this subfamily had evolved. Consequently, additional postcranial fossils are needed to continue the reconstruction of the behavioral ecology of fossil platyrrhine communities during the Miocene of South America.

Acknowledgments: I would like to thank John G. Fleagle and Richard F. Kay for permitting me to study fossils in their care, and for discussion and comments on this manuscript. Susan Ford also provided comments. Thanks to Daniel L. Gebo for inviting me to contribute to this volume. This work was supported in part by a grant from NSF (BNS 8719448) and a short-term fellowship from the Smithsonian Institution.

LITERATURE CITED

Anapol F, and Fleagle JG (1988) Fossil platyrrhine forelimb bones from the early Miocene of Argentina. Am. J. Phys. Anthrop. *76:*417–428.

Bluntschli H (1913) Die fossilen Affen Patagoniens und der Ursprung der Platyrrhinen Affen. Verhandl. Anat. Ges., p. 33.

——— (1931) *Homunculus patagonicus* und die ihm zugereihten Fossilfunde aus den Santa-Cruz-Schichten Patagoniens. Gegenbaurs Morphologisches Jahrbuch *67:*811–892.

Boinski S (1989) The positional behavior and substrate use of squirrel monkeys: Ecological implications. J. Hum. Evol. *18:*659–677.

Ciochon RL, and Corruccini RS (1975) Morphometric analysis of platyrrhine femora with taxonomic implications and notes on two fossil forms. J. Hum. Evol. *4:*193–217.

Dagosto M (1986) The Joints of the Tarsus in the Strepsirhine Primates: Functional, Adaptive, and Evolutionary Implications. Ph.D. dissertation, City University of New York, New York.

——— (1988) Implications of postcranial evidence for the origin of euprimates. J. Hum. Evol. *17:*35–56.

Davis LC (1987) Morphological evidence of positional behavior in the hindlimb of *Cebupithecia sarmientoi* (Primates: Platyrrhini). M.A. thesis, Arizona State University.

——— (1988) Morphological evidence of locomotor behavior in a fossil platyrrhine. Am. J. Phys. Anthrop. *75:*202.

Fleagle JG (1988) Primate Adaptation and Evolution. New York: Academic Press.

——— (1990) New fossil platyrrhines from the Pinturas Formation, southern Argentina. J. Hum. Evol. *19:*61–85.

Fleagle JG, Kay RF, and Anthony MRL (in press) Fossil New World monkeys. In RF Kay, RH Madden, R Cifelli, and J Flynn (eds.): A History of Neotropical Fauna: Vertebrate Paleobiology of the Miocene of Tropical South America. Washington DC: Smithsonian Institution Press.

Fleagle JG, and Meldrum DJ (1988) Locomotor behavior and skeletal morphology of two sympatric pitheciine monkeys, *Pithecia pithecia* and *Chiropotes satanas.* Am. J. Primatol. *16:*227–249.

Fleagle JG, Mittermeier RA, and Skopec AL (1981) Differential habitat use by *Cebus apella* and *Saimiri sciureus* in Central Surinam. Primates *22:*361–367.

Fleagle JG, Powers DW, Conroy GC, and Watters JP (1987) New fossil platyrrhines from Santa Cruz Province, Argentina. Folia Primatol. *48:*65–77.

Fleagle JG, and Simons EL (1983) The tibio-fibular articulation in *Apidium phiomense,* an Oligocene anthropoid. Nature *301:*238–239.

Ford SM (1986) Systematics of the New World monkeys. In D Swindler (ed.): Comparative Primate Biology. Vol. 1, Systematics, Evolution and Anatomy. New York: Alan R Liss, pp. 77–135.

——— (1988) Postcranial adaptations of the earliest platyrrhine. J. Hum. Evol. *17:*155–192.

——— (1990) Locomotor adaptations of fossil platyrrhines. J. Hum. Evol. *19:*141–173.

Ford SM, Davis LC, and Kay RF (1991) New platyrrhine astragalus from the Miocene of Colombia. Am. J. Phys. Anthrop. Suppl. *12:*73–74.

Gebo DL, Dagosto M, Rosenberger AL, and Setoguchi T (1990) New platyrrhine tali from La Venta, Colombia. J. Hum. Evol. *19:*737–746.

Gebo DL, and Simons EL (1987) Morphology and locomotor adaptations of the foot in early Oligocene anthropoids. Am. J. Phys. Anthrop. *74:*83–101.

Guerrero J (in press) Stratigraphy and sedimentary environments of the mammalbearing middle Miocene Honda Group of Colombia. In RF Kay, RH Madden, R Cifelli, and J Flynn (eds.): A History of Neotropical Fauna: Vertebrate Paleobiology of the Miocene of Tropical South America. Washington DC: Smithsonian Institution Press.

Happel RE (1982) Ecology of *Pithecia hirsuta* in Peru. J. Hum. Evol. *11:*581–590.

Haq BU, Hardenbol J, and Vail PR (1987) Chronology of fluctuating sea levels since the Triassic. Science *235:*1156–1167.

Hershkovitz P (1985) A preliminary taxonomic review of the South American bearded saki monkeys genus *Chiropotes* (Cebidae, Platyrrhini), with the description of a new species. Fieldiana: Zoology (n.s.) *27:*1–46.

——— (1987) The taxonomy of South American sakis, genus *Pithecia* (Cebidae, Platyrrhini): A preliminary report and critical review with the description of a new species and a new subspecies. Am. J. Primatol. *12:*387–468.

——— (1990) Titis, New World Monkeys of the genus *Callicebus* (Cebidae, Platyrrhini): A preliminary taxonomic review. Fieldiana: Zoology (n.s.) *55:*1–109.

Kay RF (1990) The phyletic relationships of extant and fossil Pitheciinae (Platyrrhini, Anthropoidea). J. Hum. Evol. *19:*175–208.

Kay RF, Madden RH, Plavcan JM, Cifelli RL, and Guerrero Diaz J (1987) *Stirtonia victoria,* a new species of Miocene Colombian primate. J. Hum. Evol. *16:*173–196.

Kay RF, and Meldrum DJ (in press) A new small platyrrhine from the Miocene of Colombia and the phyletic position of the callitrichines. In RF Kay, RH Madden, R Cifelli, and J Flynn (eds.): A History of Neotropical Fauna: Vertebrate Paleobiology of the Miocene of Tropical South America. Washington DC: Smithsonion Institution Press.

Luchterhand K, Kay RF, and Madden RH (1986) *Mohanamico hershkovitzi,* gen. et sp. nov., un primate du Miocène moyen d'Amérique du Sud. C.R. Acad. Sc. Paris *303,* Série II, 1753–1758.

MacFadden BJ (1990) Chronology of Cenozoic primate localities in South America. J. Hum. Evol. *19:*7–21.

Marshall LG, Drake RE, Curtis GH, Butler RH, Flanagan KM, and Naeser VE (1986) Geochronology of type Santacrucian (Middle Tertiary) land mammal age, Patagonia, Argentina. J. Geol. *94:*449–457.

Meldrum DJ (1990) New fossil platyrrhine tali from the early Miocene of Argentina. Am. J. Phys. Anthrop. *83:*403–418.

Meldrum DJ, and Fleagle JG (1988) Morphological affinities of the postcranial skeleton of *Cebupithecia sarmientoi.* Am. J. Phys. Anthrop. *75:*249–250.

——— (n.d.) Postcranial morphology of *Carlocebus* and *Homunculus,* in prep.

Meldrum DJ, Fleagle JG, and Kay RF (1990) Partial humeri of two Miocene Colombian primates. Am. J. Phys. Anthrop. *81:*413–422.

Meldrum DJ, and Kay RF (1990) A new partial skeleton of *Cebupithecia sarmientoi* from the Miocene of Colombia. Am. J. Phys. Anthrop. *81:*267.

——— (1992) A new specimen of pitheciine primate from the Miocene of Colombia. Am. J. Phys. Anthrop. Suppl. *14:*121.

——— (in press) Postcranial skeletal morphology of the fossil platyrrhines from the Miocene of Colombia. In R Kay, RH Madden, R Cifelli, and J Flynn (eds.): A History of Neotropical Fauna: Vertebrate Paleobiology of the Miocene of Tropical South America. Washington DC: Smithsonion Institution Press.

Meldrum DJ, and Lemelin P (1991) The axial skeleton of *Cebupithecia sarmientoi* (Pitheciinae, Platyrrhini) from the middle Miocene of La Venta, Colombia. Am. J. Primatol. *25:*69–89.

Napier JR, and Davis PR (1959) The forelimb skeleton and associated remains of *Procounsul africanus.* In Fossil Mammals of Africa, No. 16. London: British Museum (NH), pp 1–69.

Pascual R, and Jaureguizar EO (1990) Evolving climates and mammal faunas in Cenozoic South America. J. Hum. Evol. *19:*23–60.

Pascual R, Vucetich MG, Scillato-Yane GJ, and Bond M (1985) Main pathways of mammalian diversification in South America. In F Stehli and SD Webb (eds.): The Great American Biotic Interchange. New York: Plenum Press, pp. 219–247.

Reeser LA (1984) Morphological affinities of new fossil talus of *Dolichocebus gaimenensis.* Am. J. Phys. Anthrop. *63:*206–207.

Rosenberger AL, Hartwig WC, Takai M, Setoguchi T, and Shigahara N (1991) Dental variability in *Saimiri* and the taxonomic status of *Neosaimiri fieldsi,* an early squirrel monkey from La Venta, Colombia. Int. J. Primatol. *12:*291–301.

Rosenberger AL, Setoguchi T, and Hartwig WC (1991) *Laventiana annectens,* new genus and species: Fossil evidence for the origin of callitrichine New World monkeys. Proc. Natl. Acad. Sci. *88:*2137–2140.

Rosenberger AL, Setoguchi T, and Shigehara N (1990) The fossil record of calltrichine primates. J. Hum. Evol. *19:*209–236.

Setoguchi T, Takai M, and Shigehara N (1990) A

new ceboid monkey, closely related to *Neosaimiri,* found in the Upper Red Bed in the La Venta badlands, Middle Miocene of Colombia, South America. Kyoto University Overseas Research Reports of New World Monkeys, *7:*9–13.

Stirton RA (1951) Ceboid monkeys from the Miocene of Colombia. Bull. Univ. Calif. Publ. Geol. Sci. *28:*315–356.

Stirton RA, and Savage DE (1951) A new monkey from the La Venta Miocene of Colombia. Compilacion de los estudios geologicos oficiales en Colombia, Servicio Geologico Nacional Bogota *7:*345–356.

Thorington RW, Jr., and Anderson S (1984) Primates. In S Anderson, JK Jones, Jr. (eds.): Orders and Families of Recent Mammals of the World. New York: John Wiley and Sons, pp. 187–217.

C H A P T E R   E L E V E N

Locomotor Anatomy of Miocene Hominoids

Michael D. Rose

A relatively large number of noncercopithecoid fossil taxa are known from the Oligocene and Miocene epochs. There are, however, considerable differences of opinion as to which of these taxa belong with the hominids and extant great and lesser apes in the superfamily Hominoidea. Some (e.g., Szalay and Delson, 1979; Gebo, 1989) consider the Oligocene propliopithecids, together with the noncercopithecoid Miocene catarrhines, to be hominoids. By contrast Harrison (e.g., 1987, 1988) suggests that only a few of the known Miocene taxa are hominoids, and that the propliopithecids plus numerous Miocene taxa are noncercopithecoid, nonhominoid, primitive catarrhines. Whatever the final consensus may be, it is clear that the majority of noncercopithecoid Miocene catarrhines are quite different from living hominoids in most aspects of their postcranial morphology and inferred positional behavior. This chapter reviews the functional morphology of a number of Miocene taxa and leaves aside the question as to whether the individual taxa discussed are hominoids or primitive catarrhines. For the sake of convenience, they will all be referred to here as hominoids.

Although Miocene hominoids usually share some functional morphological features with some living primates, there are no close overall similarities between them and particular living taxa. This leads to interpretive problems. At best, interpretations based on similarities with living primates result in Miocene hominoids being seen as rather puzzling chimeras. This lack of overall similarity to living primates, coupled with the lack of detailed information concerning the links between morphology and positional behavior in many extant species, makes inferences concerning the positional behavior of Miocene hominoids extremely vague and tentative. In many respects it is easier to say what Miocene hominoids did not do. Thus it is fairly certain that there are no specialized leapers, no hylobatid-like brachiators, and no baboon-like or knuckle-walking ape-like terrestrialists among known Miocene hominoids. As will be seen in this chapter, through a combination of comparisons with living primates and direct deductions from the application of simple biomechanical principles, almost all Miocene hominoids seem to have been predominantly arboreal animals using some combination of quadrupedalism, climbing, and suspensory activities. Difficulties arise in determining the extent that each of these activities was used by a particular species and, more importantly, in determining which particular types of these activities were used. Thus, among extant primates there are many different types of quadrupedal, climbing, and suspensory behavior. Unfortunately, many of these have not yet been linked to specific morphological markers. These considerations, together with the fact that many functionally important regions of the postcranium remain unsampled for many Miocene hominoid species, explains the lack of specificity in most of the interpretations given here. The following account concentrates on the Miocene hominoids for which there are enough specimens to gain at least some idea of the overall functional organization of the postcranium. The postcranium of other hominoids, such as *Dryopithecus* and *Kenyapithecus*, are only currently known from a smattering of fragmentary or poorly attributed specimens.

PLIOPITHECUS

Pliopithecids are known from a variety of sites from the Middle and Late Miocene of Eurasia. The most extensive postcranial remains are of *Pliopithecus vindobonensis* from Europe, although some postcranials of other pliopithecids have been described (Begun 1987, 1989; Meldrum and Pan, 1988). The following account concentrates on *Pliopithecus*, and supplements the accounts given by Zapfe (1960) and Fleagle (1983).

Pliopithecus vindobonensis is approximately *Alouatta*-sized, with an estimated body weight of approximately 7 kg (Fleagle, 1988). It resembles a number of quadrupedal anthropoids in having an intermembral index of 94. However, for an animal of its size, it has somewhat elongated forelimbs, more so than in *Alouatta* but less than in *Ateles*, and hindlimbs that are as elongated as those of *Ateles* and *Hylobates* (Jungers, 1984, and pers. comm.; Ruff et al., 1989).

The axial skeleton is known from the sternum, some lumbar vertebrae, and the sacrum. The overall morphology of the trunk is probably similar to that found in nonhominoid anthropoids, although both the sternum and lumbar vertebrae are moderately broad. The lumbar vertebral column is relatively long, with six or seven vertebrae. The sacrum has three segments, as in most nonhominoids, and there was probably a moderately long tail (Ankel, 1965, 1972).

The various bones of the forelimb show mixed resemblances to extant primates. The clavicles are relatively straight, as in hylobatids. The rest of the shoulder region most resembles those of quadrupedal climbing platyrrhines such as *Cebus* or *Cacajao* and, in some respects, colobines (Corruccini et al., 1976). Thus, on a scapular fragment the glenoid fossa is moderately deep superoinferiorly, and the trough on the lateral border for the

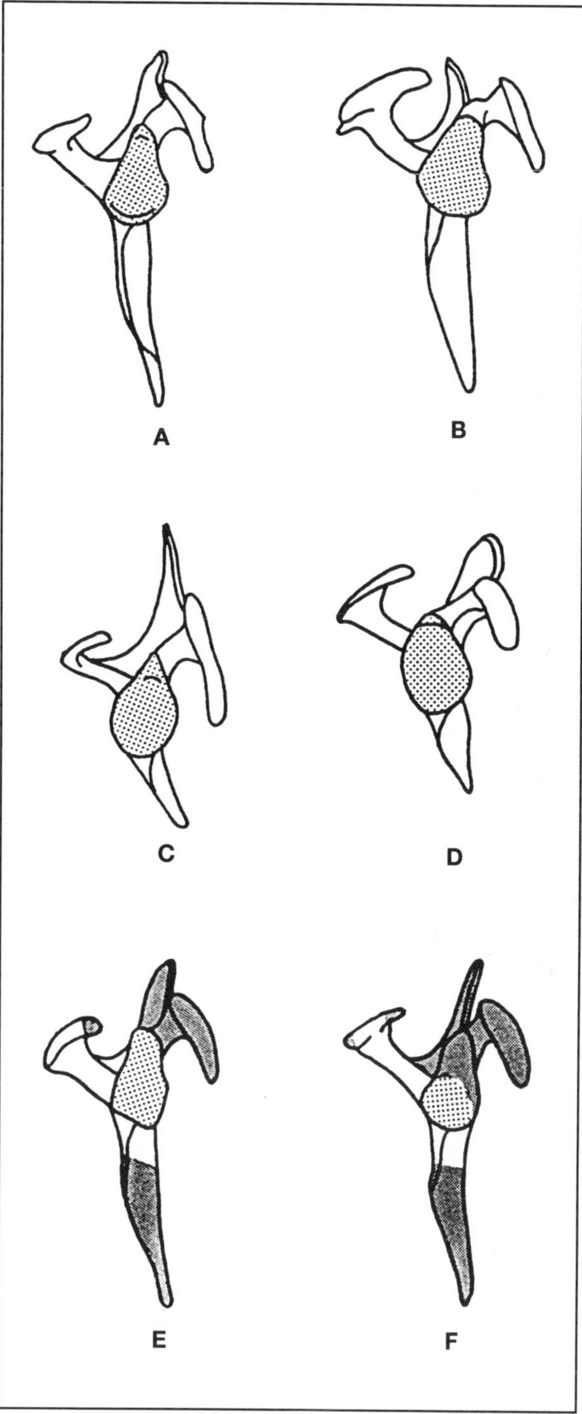

FIGURE 11.1. Left scapulae, lateral view, drawn to the same superoinferior length of the glenoid cavity. **A:** *Cebus;* **B:** *Colobus;* **C:** *Ateles;* **D:** *Hylobates;* **E:** *Pliopithecus;* **F:** *Proconsul.* The glenoid articular surface is stippled. Speculative reconstructions of missing parts of E and F are indicated with heavy stipple.

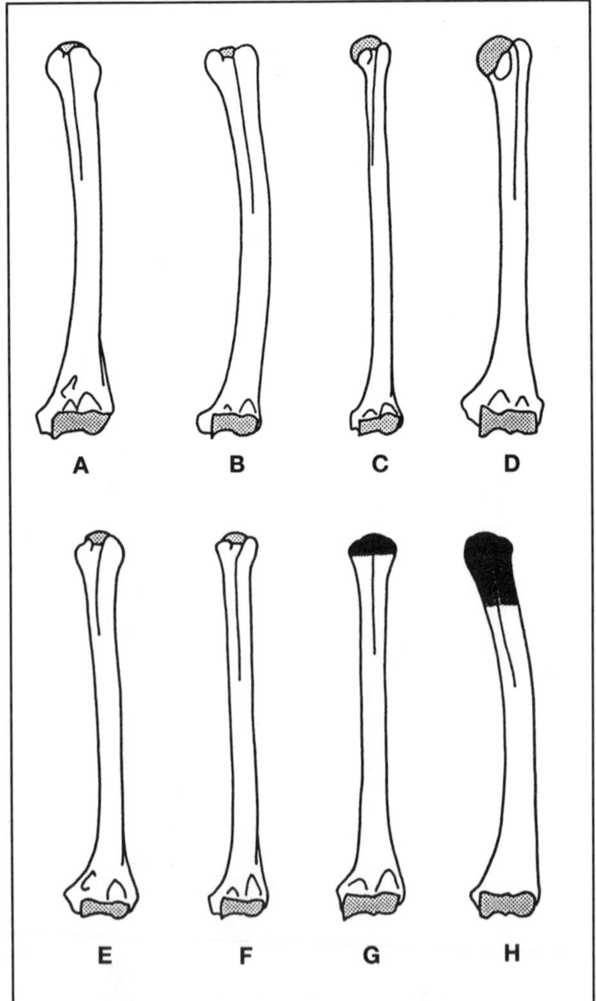

FIGURE 11.2. Left humeri, anterior view, drawn to the same proximodistal length. A: *Cebus;* B: *Cercopithecus;* C: *Ateles;* D: *Pan;* E: *Pliopithecus;* F: small ape; G: *Proconsul;* H: *Sivapithecus.* The proximal and distal articular surfaces are stippled. F, G, and H are each reconstructed from a number of partial specimens. Speculative reconstructions of missing parts of G and H are indicated with heavy stipple.

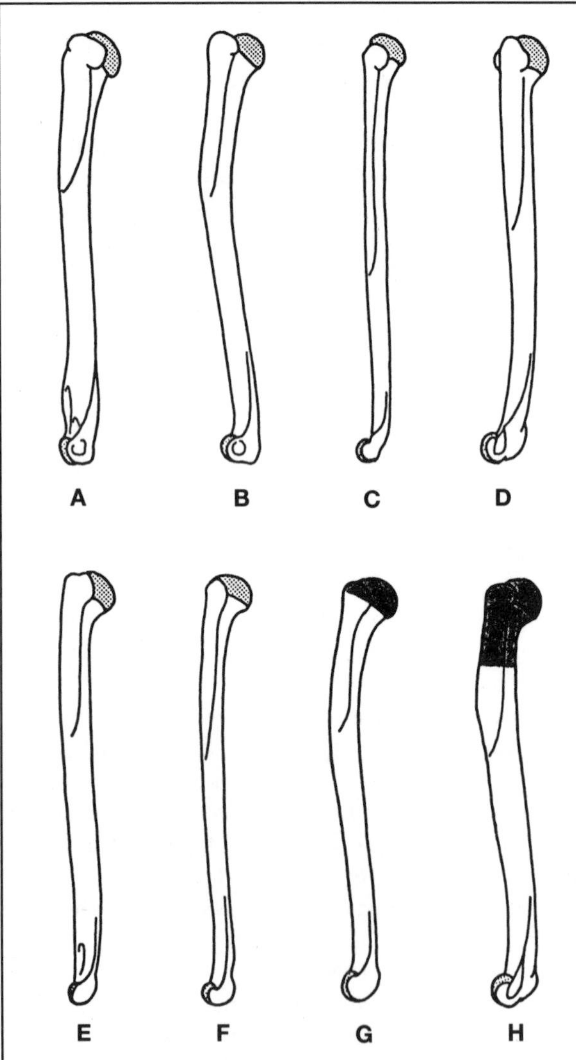

FIGURE 11.3. Left humeri, lateral view, drawn to the same proximodistal length. A: *Cebus;* B: *Cercopithecus;* C: *Ateles;* D: *Pan;* E: *Pliopithecus;* F: small ape; G: *Proconsul;* H: *Sivapithecus.* The proximal and distal articular surfaces are stippled. F, G, and H are each reconstructed from a number of partial specimens. Speculative reconstructions of missing parts of G and H are indicated with heavy stipple.

origin of m. teres minor is only modestly excavated (Fig. 11.1E). The root of the spine is relatively robust, not like the thin root found in suspensory platyrrhines (Fig. 11.1C). The acromion extends quite far laterally and bears relatively broad areas for m. trapezius and m. deltoid, as in cercopithecids. The humeral head is moderately expanded (Figs. 11.2E and 11.3E), more so than in cercopithecids (Fig. 11.3B), but not like the globular head of suspensory platyrrhines (such as *Ateles)* and hominoids (Fig. 11.3C and D). Resemblances between the shoulder region of *Pliopithecus* and generalized anthropoids were also found by Ciochon and Corruccini (1977). These features imply that movement at the glenohumeral joint is not as restricted to predominantly parasagittal movement as is typical

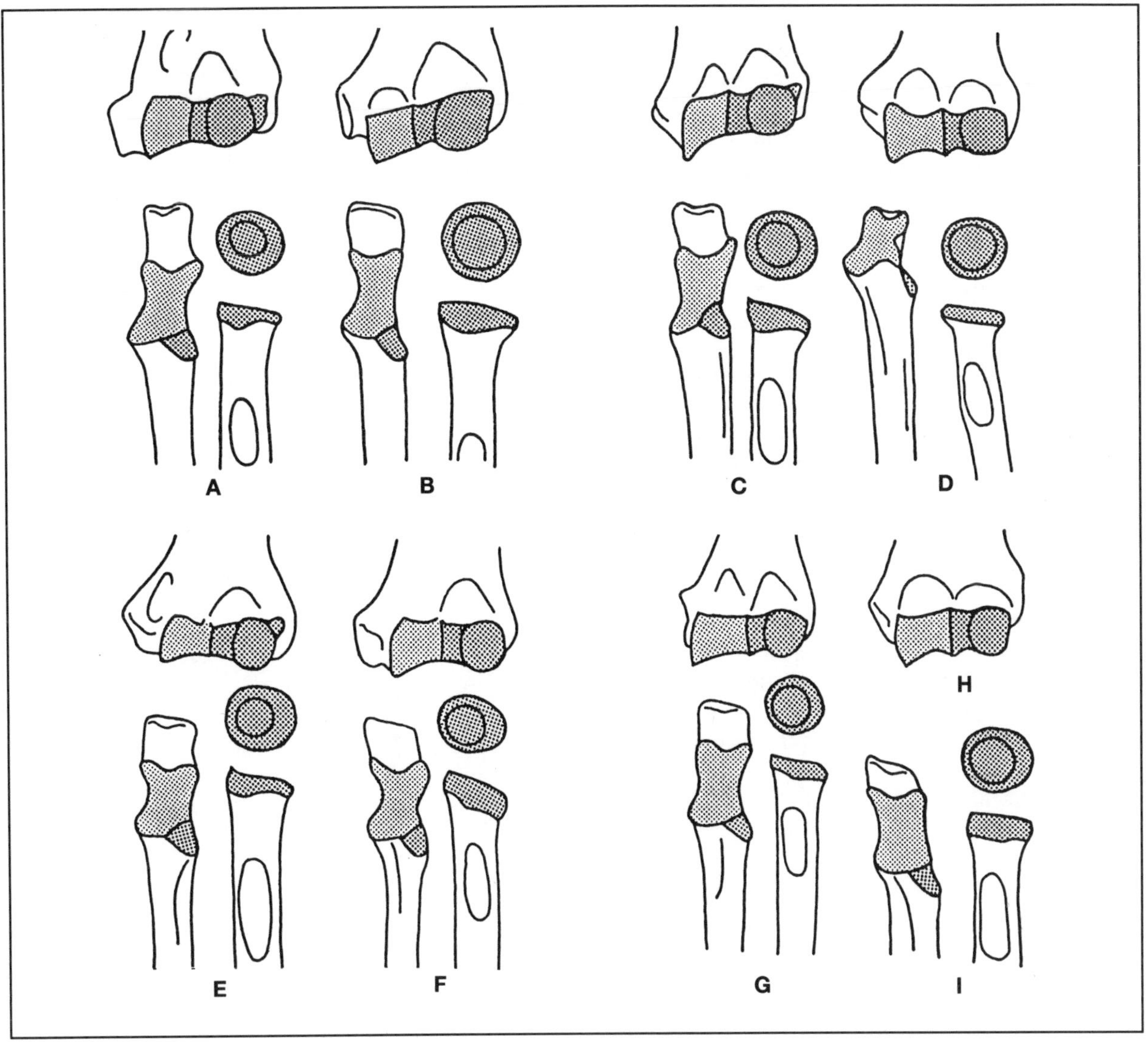

FIGURE 11.4. Anterior view of left distal humeri and proximal radii and ulnae, together with proximal view of radial heads, drawn to the same mediolateral width of the distal humeral articular surface. A: *Cebus;* B: *Ateles;* C: *Cercopithecus;* D: *Pan;* E: *Pliopithecus;* F: small ape; G: *Proconsul;* H: *Sivapithecus* (humerus only); and I: *Turkanapithecus* (radius and ulna only). The humeroulnar joint articular surfaces are lightly stippled, and the humeroradial and proximal radioulnar joint surfaces are indicated with coarse stipple.

of cercopithecids, but not as free as in suspensory platyrrhines, or in climbing and/or suspensory hominoids.

Similarities to quadrupedal-climbing platyrrhines are found in the rest of the humerus and in the joints of the elbow region. Thus the humerus is relatively straight-shafted, with a relatively anteriorly facing deltoid plane bounded by a distinct deltopectoral crest but without a distinct deltotriceps crest (Figs. 11.2E and 11.3E). This is associated with a relatively anteriorly facing greater tuberosity separated from a relatively large lesser tuberosity by a broad bicipital groove.

Distally the shaft bears a well-developed lateral supracondylar crest (Figs. 11.2E and 11.4E). Medially, the supracondylar region is pierced by an entepicondylar foramen for the passage of the ulnar nerve, as in strepsirhines and many platyrrhines (Figs. 11.2A and 11.4A). There is a relatively shallow olecranon fossa. The medial epicondyle resembles those of *Alouatta* and *Ateles* in being directed medially. The distal articular surface bears a relatively broad area for articulation with the radius (Fig. 11.4E). This includes a capitulum that is best developed anteriorly in terms of its breadth and curvature and a broad, unrecessed zona conoidea between the capitulum and the trochlea. These features are matched by radial head features that include a relatively small area for articulation with the capitulum and a well-developed lateral lip for articulation with the zona conoidea in pronation. These features are associated with the stabilization of the elbow region in a semiflexed and fully pronated position.

Anteriorly, the trochlea is relatively narrow mediolaterally, and has a truncated cone shape. There is a pronounced medial keel, but laterally there is only a minimally developed lateral keel separating the trochlea from the zona conoidea. The trochlear surface broadens posteriorly, where it becomes more spool-shaped, but does not encroach into the olecranon fossa. Corresponding features of the trochlear notch of the ulna include a distolateral lip that matches the conical shape to the anterior part of the trochlea, and a relatively broad, convex proximal part that matches the spool-shaped posterior part of the trochlea. This morphology most resembles that of generalized platyrrhines, and differs in numerous respects from the humeroulnar region of suspensory platyrrhines, cercopithecids, and hominoids. The radial notch on the ulna, for the proximal radioulnar joint, is not markedly excavated and faces anterolaterally, rather than anteriorly as in cercopithecids, or laterally as in hominoids and suspensory platyrrhines. The corresponding surface on the side of the radial head only extends part of the way round the head, and is most extensive anteromedially. There is a correspondingly emphasized flattened border posterolaterally, where articular cartilage is absent (Rose, 1988). At the distal radioulnar joint the articular surface on the ulna is limited, while the corresponding surface on the distal radius is quite flat. These features are comparable with a pronation-supination range of about 110° in the forearm. This is similar to the ranges found in *Cebus* and *Lagothrix* by Sarmiento (1985), which are in the middle part of the range of excursions found among anthropoids. Robertson (1985), using different criteria, reached similar conclusions concerning rotatory capabilities in the *Pliopithecus* forearm. These features of the radioulnar joints, together with a mediolateral compression of the radial neck are part of the same complex, mentioned above for the humeroradial joint, that provides maximal stability of the elbow and forearm in semiflexion and full pronation, as occurs during the stance phase of quadrupedal progression. In this respect *Pliopithecus* resembles predominantly quadrupedal anthropoids (Rose, 1988). In other

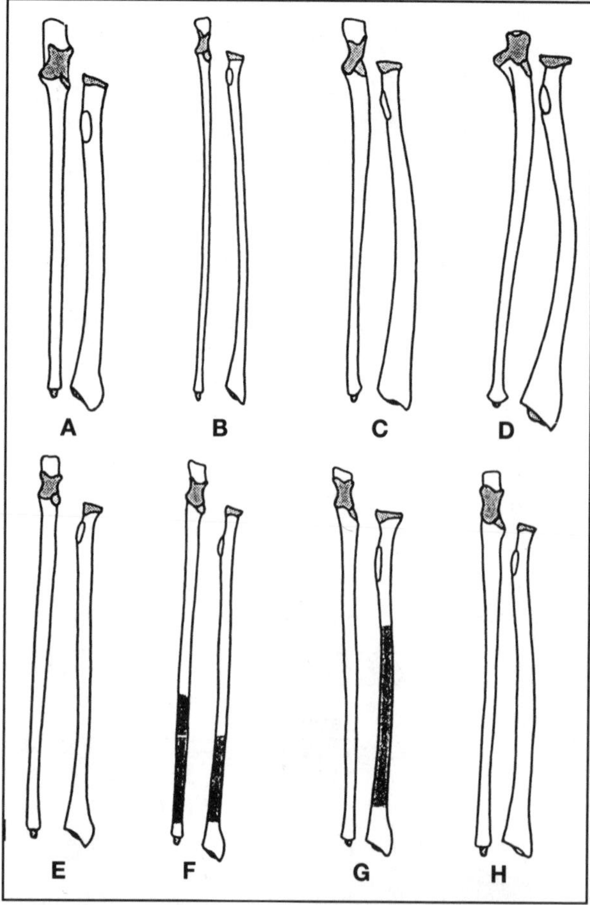

FIGURE 11.5. Left ulnae and radii, anterior view, drawn to the same proximodistal length of the ulna. A: *Cebus;* B: *Ateles;* C: *Cercopithecus;* D: *Pan;* E: *Pliopithecus;* F: small ape; G: *Turkanapithecus;* H: *Proconsul.* The articular surfaces are stippled. F and H are each reconstructed from a number of partial specimens. Speculative reconstructions of missing parts of F and G are indicated with heavy stipple.

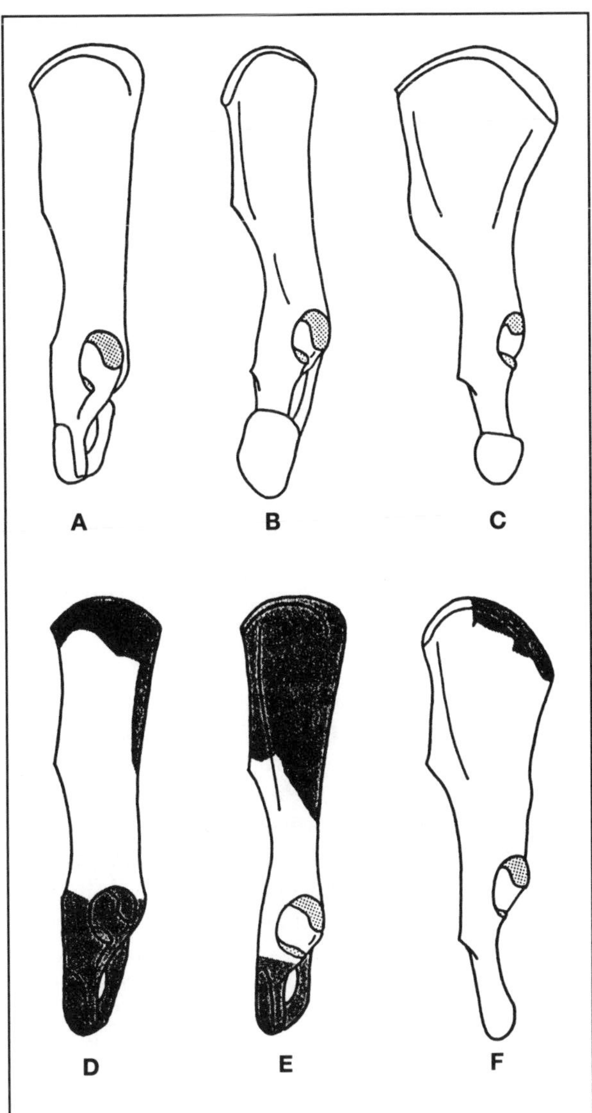

FIGURE 11.6. Right hip bones, posterolateral view to show full breadth of iliac blade, drawn to the same superoinferior length. A: *Alouatta;* B: *Cercopithecus;* C: *Pan;* D: *Pliopithecus;* E: small ape; F: *Proconsul.* Visible parts of the acetabular articular surface are stippled. Speculative reconstructions of missing parts of D, E, and F are indicated with heavy stipple.

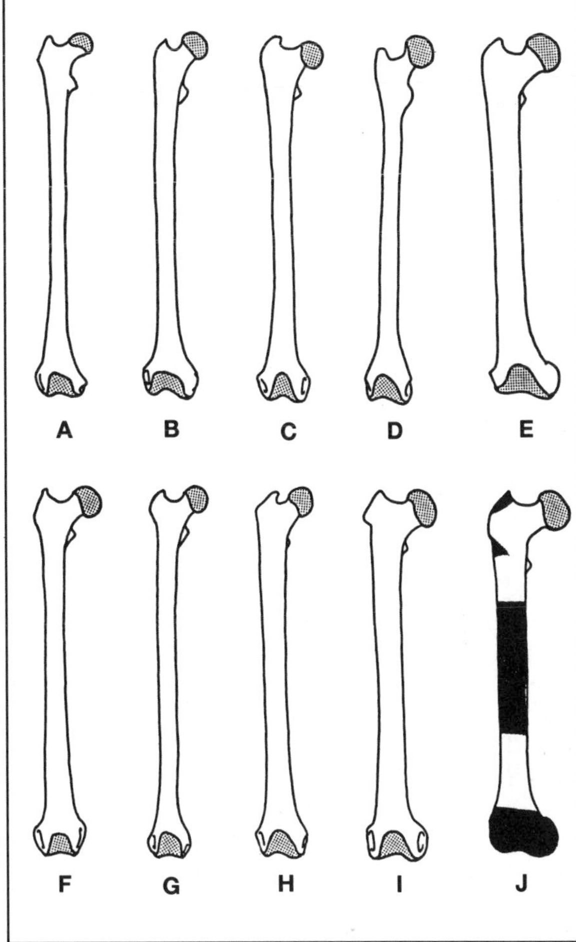

FIGURE 11.7. Right femora, anterior view, drawn to the same proximodistal length. A: *Cacajao;* B: *Ateles;* C: *Cercopithecus;* D: *Hylobates;* E: *Pan;* F: *Pliopithecus;* G: small ape; H: *Turkanapithecus;* I: *Proconsul;* and J: *Sivapithecus.* The proximal and distal articular surfaces are stippled. G, I, and J are each reconstructed from a number of partial specimens. Speculative reconstructions of missing parts of J are indicated with heavy stipple.

features of the radius and ulna (Fig. 11.5E), including the intermediate length of the neck and the anteromedial placement of the bicipital tuberosity, *Pliopithecus* most resembles generalized platyrrhines (Fig. 11.5A), particularly *Alouatta.* They indicate relatively powerful flexion at the elbow joint and supination of the forearm associated with reaching and hoisting during climbing. A well-developed crest on the lateral side of the distal radius, for the insertion of m. brachioradialis, occurs in *Ateles* (Fig. 11.5B) and, as it indicates powerful flexion of the semipronated forearm, is another feature linked to climbing. A number of bones are known from the wrist, palm, and fingers. They

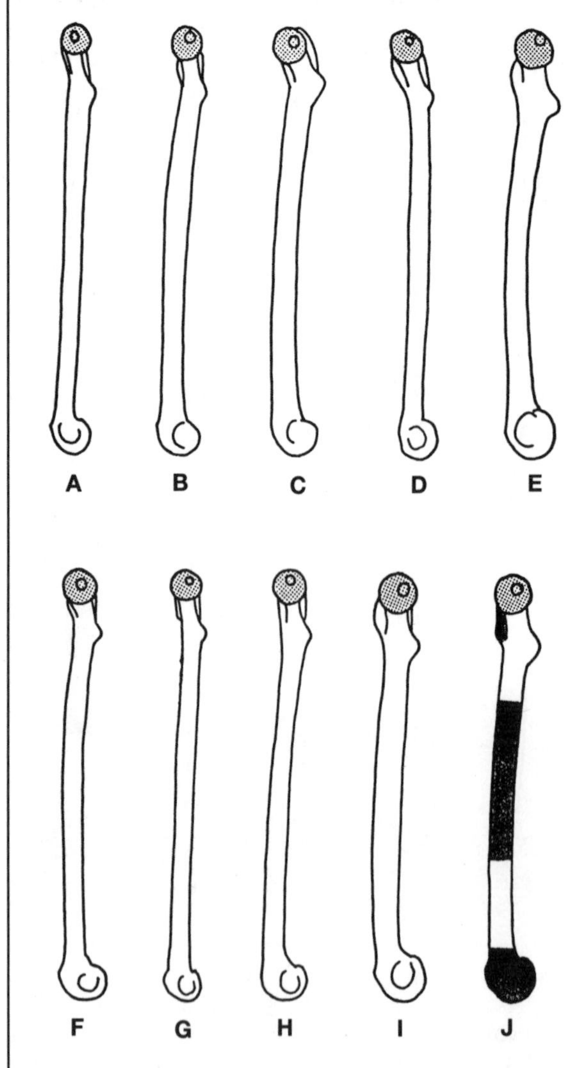

FIGURE 11.8. Right femora, medial view, drawn to the same proximodistal length. A: *Cacajao;* B: *Ateles;* C: *Cercopithecus;* D: *Hylobates;* E: *Pan;* F: *Pliopithecus;* G: small ape; H: *Turkanapithecus;* I: *Proconsul;* J: *Sivapithecus.* The proximal articular surfaces are stippled. G, I, and J are each reconstructed from a number of partial specimens. Speculative reconstructions of missing parts of J are indicated with heavy stipple.

are generally similar to those of arboreal quadrupeds (Zapfe, 1960; Corruccini et al., 1975; Fleagle, 1983; Robertson, 1985).

In the hindlimb a fragment of ilium (Fig. 11.6D) most resembles the platyrrhine condition (Fig. 11.6A). It lacks the expansion of the iliac blade and other features found in hominoids (Fig. 11.6C). The *Pliopithecus* femur (Figs. 11.7F and 11.8F) resembles those of anthropoids that practice hindlimb suspension in being relatively long and gracile (Figs. 11.7A and B, and 11.8A and B). However, the rest of the femoral morphology shows less specific resemblances. As in noncercopithecid anthropoids, the neck is angulated so that it extends proximally, and the globular head is situated proximal to the greater trochanter to the extent found in hominoids and large platyrrhines. These features relate to the thigh abduction characteristic of climbing and suspensory species (Fleagle, 1983). The neck and trochanteric regions resemble platyrrhines such as *Cebus* and *Cacajao* in lacking a tubercle of the neck and in having a laterally flaring greater trochanter and a posterolaterally directed lesser trochanter (Fig. 11.7A and F).

There are few features of the distal articular surfaces that distinguish *Pliopithecus* from other anthropoids. The medial condyle is relatively narrow mediolaterally, as in colobines. The lateral lip of the surface for the patella projects anteriorly as in all anthropoids except large hominoids. The patella itself is almost circular and anteroposteriorly shallow, as in noncercopithecid anthropoids.

The straight-shafted, relatively robust tibia (Fig. 11.9F) has a proximal surface that is relatively deep anteroposteriorly. The distal surface (Fig. 11.10E), for articulation with the talus, is relatively narrow mediolaterally and the maleolus extends distally without any marked hooking (Fig. 11.9C), as in generalized platyrrhines (Fig. 11.9A). *Pliopithecus* resembles cercopithecids (Fig. 11.10C) in lacking an extension of the articular surface proximally, onto the lateral surface of the shaft adjacent to the surface for the tibiofibular syndesmosis. It is also cercopithecid-like in having a dorsiflexion stop-facet on the junction of the anterior and distal surfaces.

The *Pliopithecus* talus (Fig. 11.11F) resembles hylobatids and some cercopithecines (Fig. 11.11C and D) in having a mediolaterally narrow trochlear surface for the tibiotalar joint. It more specifically resembles cercopithecids in having a dorsiflexion stop dorsolaterally on the neck, that contacts the above-mentioned facet on the distal tibia (Le Gros Clark and Thomas, 1951; Harrison, 1982). A distinct cup-shaped depression medially, at the junction of the trochlea and neck receives the tibial maleolus when the leg is vertically above the foot. This feature (found on cercopithecid, hylobatid, and some platyrrhine tali) is related to stabilized load bearing during quadrupedal progression. The articular sur-

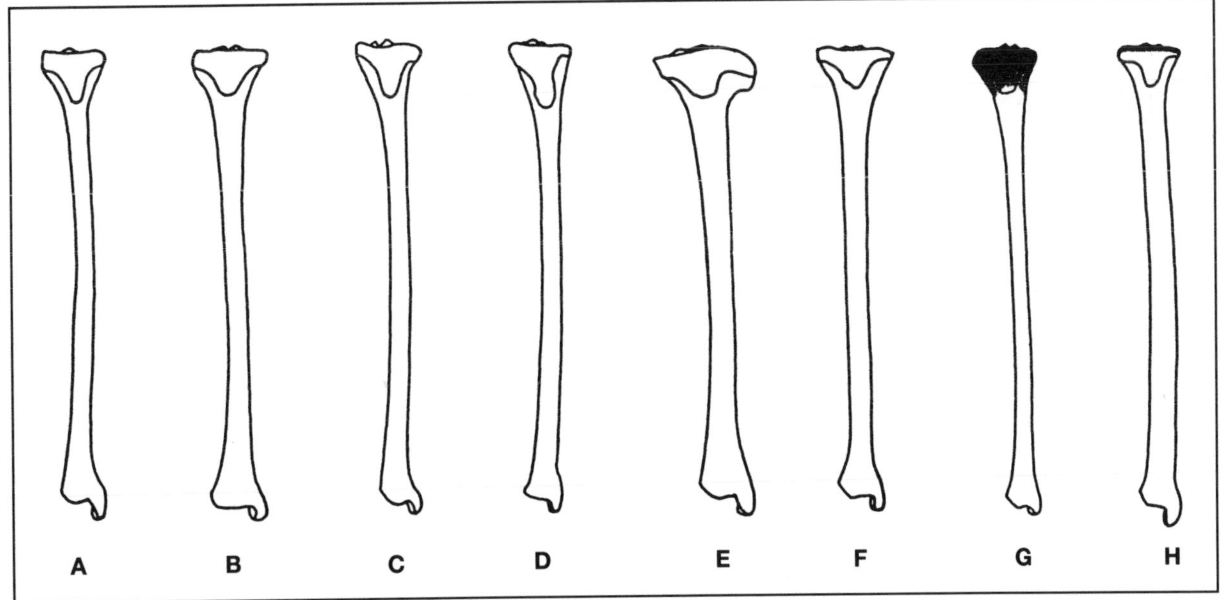

FIGURE 11.9. Right tibiae, anterior view, drawn to the same proximodistal length. A: *Cacajao;* B: *Ateles;* C: *Cercopithecus;* D: *Hylobates;* E: *Pan;* F: *Pliopithecus;* G: small ape; H: *Proconsul.* G and H are each reconstructed from a number of partial specimens. Speculative reconstructions of missing parts of G and H are indicated with stipple.

faces on both the talus and calcaneus for the distal talocalcaneal joint are also cercopithecid-like (Zapfe, 1960; Langdon, 1986). The talar head resembles that of *Alouatta* (Fig. 11.11B) in being relatively broad mediolaterally, and does not exhibit a great deal of torsion.

In its general shape and in the morphology of the proximal surface for the talus, the *Pliopithecus* calcaneus most resembles that of *Ateles.* It resembles cercopithecids in the form of sustentacular articular surfaces in the shallow, conical articular surface for the cuboid. It differs from all hominoids (Fig. 11.10D and E) in having a relatively long distal segment, as well as in the details of its articular surfaces. It differs from hylobatids in having a well-developed medial tubercle on the plantar surface of the short, deep, calcaneal tuberosity, a feature associated with the ability to grasp with the lateral four toes, especially during climbing (Sarmiento, 1983). In addition, the presence of a well-developed groove on the plantar surface of the sustentaculum, for the tendon of m. flexor hallucis longus is an indicator of powerful hallucial grasping. A medial cuneiform resembles those of nonhominoids. The metatarsals and phalanges are relatively long and resemble those of arboreal nonhominoid anthropoids, especially in terms of their contribution to the overall proportions of the foot (Schultz, 1963).

The *Pliopithecus* postcranium presents all the problems mentioned in the introduction concerning the interpretation of probable functional and behavioral capabilities. There are no consistent patterns of resemblances to extant taxa, and particular points of similarity are frequently difficult to evaluate due to a basic lack of information concerning the functional significance of the features concerned.

In its overall proportions *Pliopithecus* can be thought of as being like an *Alouatta* monkey with rather long hindlimbs, or as an *Ateles* monkey with rather short forelimbs. In the forelimb the shoulder, elbow, and forearm joints lack features associated with specialized suspensory capabilities as found in hylobatids, *Pongo,* and ateline platyrrhines, with specialized climbing capabilities as exhibited by African apes, or with committed quadrupedalism of the cercopithecid type. There are numerous indicators, especially in the elbow region, that *Pliopithecus* had a well-developed capability for quadrupedalism, but the modest (not

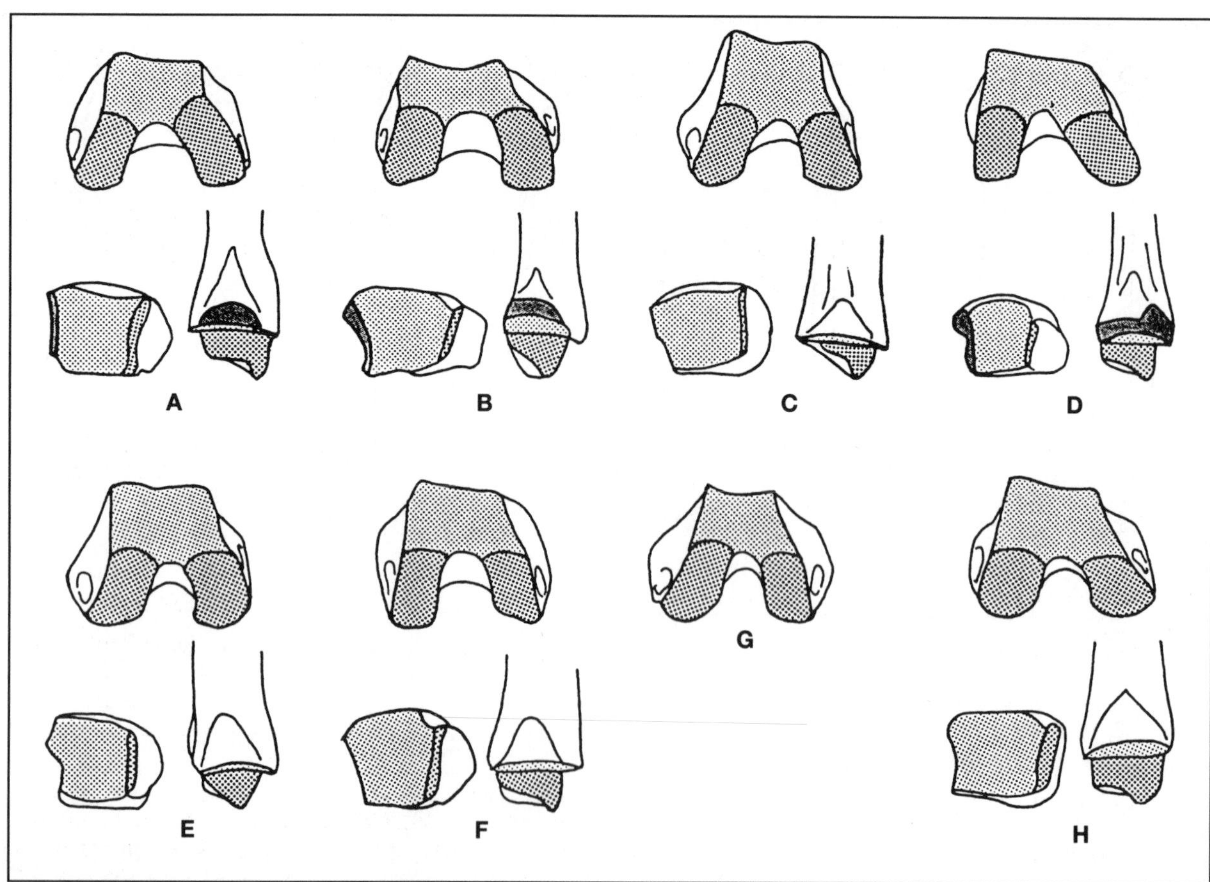

FIGURE 11.10. Right femora (distal view) and right tibiae (distal and lateral views) drawn to the same mediolateral width of the distal femoral articular surface. **A**: *Cacajao;* **B**: *Ateles;* C: *Cercopithecus;* **D**: *Hylobates;* **E**: *Pliopithecus;* **F**: small ape; **G**: *Turkanapithecus;* and **H**: *Proconsul.* The femeropatellar and the distally facing part of the tibiotalar articular surfaces are indicated with light stipple. The femerotibial (condylar) and maleolar part of the tibiotalar articular surfaces are indicated with coarse stipple. Encroachment of articular surface into the region of the distal tibiofibular syndesmosis in **A**, **B**, and **D** is indicated with heavy stipple.

marked) mobility of the shoulder and forearm also imply a relative freedom of forelimb mobility that might be expressed in behaviors such as climbing and (as illustrated by Fleagle, 1988) bridging. The closest analogs among living primates are the relatively generalized types such as *Cebus, Cacajao,* and *Alouatta.* It should be noted, however, that in its overall morphology *Pliopithecus* differs from each of these genera, just as they also differ from each other.

In the hindlimb, features of the hip, ankle, and foot again suggest some form of generalized arborealism. There are features of the hip and ankle region whereby *Pliopithecus* resembles hindlimb suspensory taxa such as *Ateles* and hylobatids. There are equally strong resemblances to the nonsuspensory cercopithecids, however, so the total morphological pattern does not suggest any suspensory use of the hindlimb more sophisticated than the bridging mentioned above for the forelimb. The overall impression is of a relatively long-limbed quadruped, perhaps moving rather deliberately through the canopy and using some climbing and bridging behavior.

SMALL APES

A number of hominoids, mostly from the Early and Middle Miocene of East Africa, will be considered here together under the collective name of small apes. This grouping is made because, on the

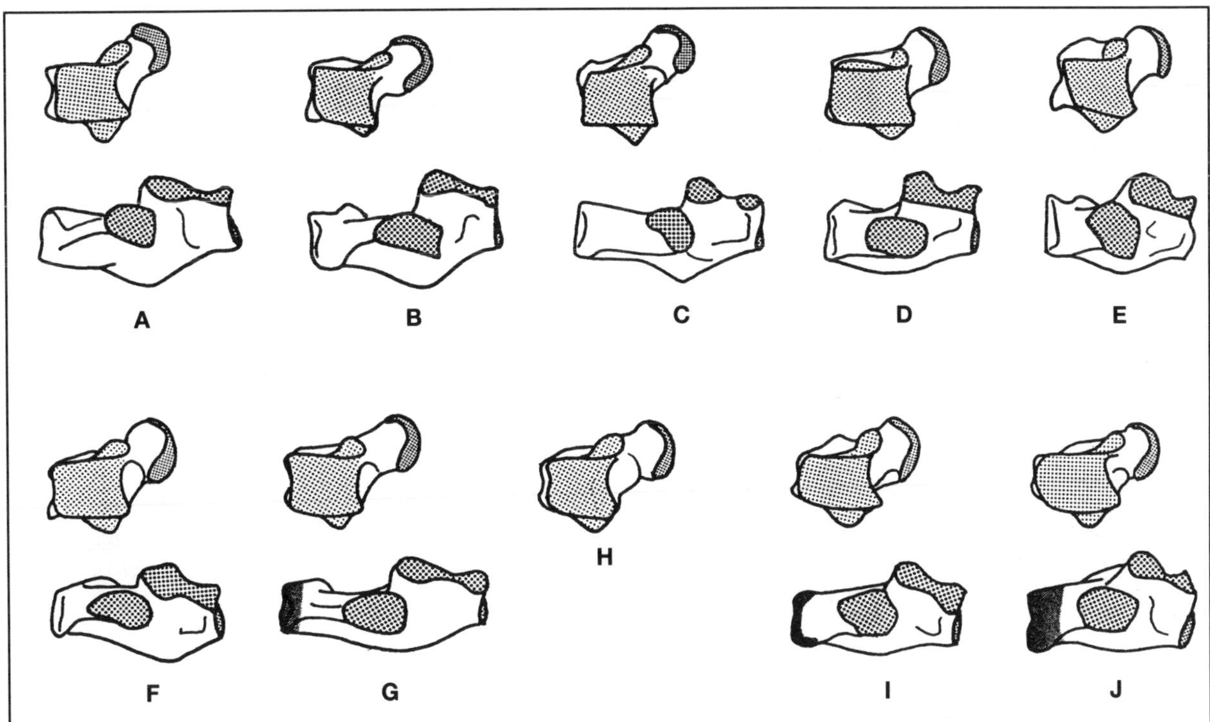

FIGURE 11.11. Right tali and calcanei, dorsal view, drawn to the same proximodistal length of the talus. A: *Cebus;* B: *Alouatta;* C: *Cercopithecus;* D: *Hylobates;* E: *Pan;* F: *Pliopithecus;* G: small ape; H: *Turkanapithecus;* I: *Proconsul;* and J: *Sivapithecus.* Talar articular surfaces for the tibiotalar (talocrural) joint are lightly stippled. Calcaneal articular surfaces for the subtalar (talocalcaneal) joints are indicated with coarse stipple. Articular surfaces on the talus and calcaneus for the midtarsal joint are indicated with coarse dark stipple. Speculative reconstructions of missing parts of G, I, and J are indicated with heavy stipple.

basis of the currently available specimens, postcranial morphology is similar in the species concerned. If and when more complete material becomes available this may well prove to be an oversimplification. The name "small apes" is not an ideal one. It was originally used when small apes were considered to be relatives of the living hylobatids, or lesser apes (e.g., Andrews and Simons, 1977). It is retained here only as a matter of convenience. The various small apes in fact vary in weight from approximately 3.5 kg to 10 kg (Fleagle, 1988). Small apes include species of *Dendropithecus, Kalepithecus, Limnopithecus, Micropithecus,* and *Simiolus* from East Africa, and *Dionysopithecus* from Asia. Until recently the only known associated postcranial specimens were of *Dendropithecus* (Le Gros Clark and Thomas, 1951; also discussed by Ferembach, 1958, and Fleagle, 1983). Currently, associated postcranial bones are also known for *Simiolus,* and a number of tentative attributions can be made for a relatively large number of unassociated specimens (Harrison, 1982, who also described many of these specimens for the first time). The following account is based mainly on *Dendropithecus* and *Simiolus* specimens.

Based on estimates of body weight (Fleagle, 1988) and limb-bone lengths (Le Gros Clark and Thomas, 1951), *Dendropithecus* is similar to *Pliopithecus* in the elongation of its limbs, but with slightly shorter hindlimbs (Jungers, 1984 and pers. comm.), quite like an *Ateles* monkey. The only axial skeletal specimen known for small apes is a lower thoracic vertebral body, possibly of *Micropithecus* (Leakey and Walker, 1985; Rose et al., 1992). This is kidney-shaped in cranial or caudal view, and is quite deep posteroanteriorly (more so than in *Pliopithecus).* In these features it most resembles the lower thoracic vertebrae of suspensory anthropoids.

The whole of the small ape humerus (Figs. 11.2F and 11.3F) can be reconstructed from various small ape specimens (Le Gros Clark and Thomas, 1951;

Harrison, 1982; Fleagle, 1983; Gebo et al., 1988; Rose, 1988, 1989). The humeral head region, including the tuberosities, is similar to that of *Pliopithecus,* but with a slightly more expanded articular surface. As for *Pliopithecus,* it most resembles the generalized platyrrhine condition. The humeral shaft is as gracile as that of *Pliopithecus* (Aiello, 1981) and is slightly retroflexed proximally. Although the bicipital groove is slightly narrower than in *Pliopithecus* (Fig. 11.2E and F), it is not as narrow as in suspensory anthropoids such as *Ateles* (Fig. 11.2C), where the groove undercuts the lesser tuberosity and is overhung by the deltopectoral crest. The deltoid plane is relatively flat and bounded by distinct deltopectoral and deltotriceps crests. This, in combination with distinct medial and posterior borders, gives the proximal shaft a diamond-shaped cross section. All of these features are similar to those of the humeri of platyrrhines such as *Cebus* (Fig. 11.2A) and *Alouatta.* The presence of a well-defined posterior border is, however, more characteristic of cercopithecids.

Distally the shaft is more anteroposteriorly compressed than in *Pliopithecus* and, as in some platyrrhines and most catarrhines, does not have an entepicondylar foramen. Otherwise, the distal shaft and epicondylar regions are quite similar to those in *Pliopithecus.* These similarities extend to the distal articular region (Fig. 11.4E and F), but some features are more emphasized than in *Pliopithecus.* Thus the trochlea is more expanded medially, to produce a more pronounced conical shape, is bounded by a better defined lateral keel, and is more concave posteriorly. These features are matched by similarly emphasized features on the trochlear notch of the ulna. The humeral articular surface for the radius is quite similar to that in *Pliopithecus,* but the corresponding radial features are more strikingly emphasized (Rose, 1988). Thus the radial head is markedly oval in outline, due to the presence of a pronounced lateral lip, there is a pronounced tilt to the head, and the radial neck is markedly compressed. In addition, the articular surface on the side of the head, for the proximal radioulnar joint, is very extensive on the anteromedial part of the head. Other features, such as the relative length of the neck and the orientation of the bicipital tuberosity, are similar to those of *Pliopithecus,* as are the surfaces for the distal radioulnar joints and the presence of a strong crest for the insertion of m. brachioradialis. The shafts of both the radius and ulna are quite gracile and mediolaterally compressed (Fig. 11.5F).

A number of bones of the hand are known from small apes (Harrison, 1982; Rose et al., 1992). A lunate and capitate are both most similar to those of cercopithecids, suggesting only modest mobility at the antebrachiocarpal and midcarpal joints. The first metacarpal exhibits a unique mixture of features that suggests limited mobility except for a well-developed flexion-grasping capability. The second metacarpal and proximal and middle phalanges resemble those of arboreal quadrupedal anthropoids.

Conclusions concerning function in the small ape forelimb are similar to those for *Pliopithecus.* Differences, which are relatively minor, include the possibility of more extensive movements at the glenohumeral joint, and greater stability at the joints of the elbow, especially in the semiflexed and fully pronated position. As with *Pliopithecus,* a predominantly quadrupedal locomotor pattern is to be inferred, with climbing and possibly some suspension as additional activities.

In the hindlimb, a partial hip bone is similar to that of *Pliopithecus* (Fig. 11.6D and E). It resembles platyrrhine hip bones (Fig. 11.6A) in having a relatively long, gracile, and anteroposteriorly compressed ilium between the acetabulum and the auricular area. This similarity extends to the morphology of the articular surface of the acetabulum.

The small ape femur and patella are quite similar to those of *Pliopithecus,* but the femur differs in being straighter and in having a more gracile neck and shaft (Figs. 11.7G and 11.8G), a trochanteric fossa that is less buttressed posteriorly, a more laterally directed lesser trochanter, and a distal end that is less expanded mediolaterally. The small ape femur most resembles those of pitheciine platyrrhines such as *Cacajao* (Figs. 11.7A and 11.8A), that combine above-branch quadrupedalism with hindlimb-suspensory activities. Although small apes do not share with pitheciines those foot bone features that are associated with suspension (Jenkins and McClearn, 1984; Fleagle and Meldrum, 1988), the femoral similarities may indicate a capability for a more limited type of hindlimb suspension in small apes.

The small ape fibula is only known from uninformative fragments, but the tibia is known for the shaft and distal end (Fig. 11.9G). The shaft is more compressed mediolaterally than in *Pliopithecus* (Fig. 11.9A). The distal end (Fig. 11.10F) is even more cercopithecid-like (Fig. 11.10C) than in *Pliopithecus.* In particular, the anterior part of the malleolus is hooked to fit tightly with the cup-shaped depression on the talus.

In the foot the small ape talus is similar to that of *Pliopithecus* (Fig. 11.11F and G), except that the trochlear surface is not as proximodistally elongated, making its overall morphology quite cercopithecid-like (Fig. 11.11C). The calcaneus is slightly more elongated proximodistally than in *Pliopithecus,* and in its overall shape is most like the calcanei of cercopithecids. However, it shares with *Pliopithecus* the presence of a medial tubercle on the calcaneal tuberosity (Sarmiento, 1983) which, as mentioned above, is indicative of a climbing and hindlimb-suspensory capability. The proximal facet for the talus is also somewhat elongated, as in *Pliopithecus* and large platyrrhines (Fig. 11.11B). Unlike *Pliopithecus* the distal facet, for the calcaneocuboid joint, is relatively narrow and deep. This feature is matched by the presence of a protuberant peg on the proximal surface of the cuboid. The depth of the calcaneocuboid joint most resembles the condition in large living hominoids and is related to considerable mobility combined with good stability in the midtarsal region. This is associated with the use of the foot in climbing (Langdon, 1986). The navicular, metatarsals, and phalanges most resemble those of arboreal nonhominoid anthropoids.

The evidence of the hindlimb reinforces the conclusions reached above for the forelimb concerning possible positional behavior in small apes. Again the pattern is similar to that of *Pliopithecus,* but with indications, especially from the femur and the foot, for climbing and some forms of suspension being important adjuncts to a predominantly quadrupedal positional repertoire.

PROCONSUL AND *AFROPITHECUS*

These genera are from the Early and Middle Miocene of East Africa. Postcranially, *Proconsul* is the best known of all the Miocene hominoids (Le Gros Clark and Leakey, 1951; Napier and Davis, 1959; Walker and Pickford, 1983; Beard et al., 1986; Walker and Teaford, 1988, 1989; Ward, 1991). Many additional specimens (Walker and Teaford, 1988) are currently being studied. There are probably four species of *Proconsul* represented by the known material, but the three currently named species will be referred to here. These are *P. africanus, P. nyanzae,* and *P. major,* with body weights of approximately 9 kg, 31 kg, and 50 kg respectively (Fleagle, 1988; Ruff et al., 1989), but with generally similar morphology despite these size differences. *Afropithecus turkanensis* postcranial bones (Leakey et al., 1988a) are very similar in size and morphology to those of *P. nyanzae.* The following account is based on specimens from both genera, which will all be referred to as *Proconsul* specimens for the sake of convenience.

In its intermembral proportions (96.4) *Proconsul* is similar to *Pliopithecus.* The limbs are, however, relatively shorter than those of *Pliopithecus* and the small apes. In these respects *Proconsul* resembles cercopithecids (Jungers, 1985; Ruff et al., 1989). In the axial skeleton the lumbar vertebrae and sacrum also resemble those of cercopithecids, as do some rib fragments (Ward, 1991). This implies that the trunk is relatively long and mediolaterally narrow, not short and broad as in hominoids. *Proconsul* differs from monkeys in that there are probably only six, rather than seven vertebrae in the lumbar column. *Proconsul* resembles hominoids in that the sacrum is relatively narrow, the sacroiliac joint is relatively small, and the tail (Ward et al., 1991) is reduced or absent.

In the forelimb the scapula (Fig. 11.1F) most resembles colobines and large platyrrhines (Fig. 11.1A and B) in the morphology of the groove for the origin of m. teres minor and of the root of the spine. There are no specimens of the humeral head region, but the humeral shaft is similar to those of cercopithecids (Figs. 11.2B and G, and 11.3B and G). Thus the proximal shaft is retroflexed, there is a relatively broad bicipital groove, the anterolaterally facing deltoid plane is flat, with well-defined deltopectoral and deltotriceps crests, and the medial epicondyle is directed posteromedially. These similarities with cercopithecids do not extend to the distal articular surface. The distal humerus of *Proconsul* has received considerable attention, and its morphology has been interpreted in a number of ways (see e.g., Morbeck, 1975, 1976; Feldesman, 1982 and references therein). The trochlea and the corresponding surfaces on the ulna are similar to those of small apes (Fig. 11.4F and G), but the trochlea is separated from the surface for the radius by a more pronounced lateral keel. The capitulum is more inflated than in *Pliopithecus* (Fig. 11.4E) and small apes and is separated from the trochlea by a narrow, recessed zona conoidea. The radial head bears a correspondingly reduced lateral lip, and the surface for articulation with the zona conoidea extends onto the side of the head. In these features of the humeroradial joint *Proconsul* resembles hominoids (Fig. 11.4D). Functionally this results in good stabilization of the radial head during pronation-supination movements, which are probably greater in amplitude than those possible in the forearm of *Pliopithecus* and small apes. A stable position of the elbow in semiflexion and full pronation is still present.

The radial and ulnar shafts are more robust than in *Pliopithecus* and the small apes (Fig. 11.5E, F, and H). The olecranon process of the *P. nyanzae* ulna points proximoposteriorly as in terrestrial cercopithecines (Jolly, 1967), rather than proximally as in arboreal anthropoids. While this may indicate some terrestriality in *P. nyanzae* (Harrison, 1982; Fleagle, 1983), this feature has also been interpreted in terms of the overhead use of the forelimb during arboreal activities (Rose, 1983). The bones of the hand are generally similar to those of quadrupedal arboreal nonhominoids (e.g., McHenry and Corruccini, 1983). In particular, the ulnar styloid process articulates with both the pisiform and triquetral in the wrist (Beard et al., 1986—see references therein for the extensive literature on this region). Some features relating to the mobility (especially ulnar deviation) of the midcarpal joint (Sarmiento, 1988; Lewis, 1989) and a relatively mobile trapezium–first metacarpal joint of the thumb (Rafferty, 1990; Rose, 1992) are more hominoid-like, and relate to a climbing capability and enhanced grasping.

In the hindlimb the hip bone (Fig. 11.6F) is generally like those of nonhominoids (Ward, 1991). Thus the ilia are relatively short, unexpanded, and laterally facing. The pubis is relatively short, so that the pelvis is relatively narrow mediolaterally. This is consistent with the axial skeletal features that imply a relatively narrow, nonhominoid type of trunk. As mentioned previously, however, the sacroiliac joint is like those of hominoids, implying that the ilium is more parallel to the vertebral column than in nonhominoids. In addition, the ischium is uniquely long, providing a long lever arm for the action of the hamstring muscles. The acetabular articular surface also has features suggesting the possibility of a hominoid-like degree of abduction at the hip joint. This combination of features in the pelvis again suggests a predominantly quadrupedal positional repertoire, with climbing as an important additional activity. The ischium lacks the flat, flared morphology found in anthropoids that have ischial callosities (Fig. 11.6B). This suggests sitting behavior similar to that of larger platyrrhines (Rose, 1974a, 1974b).

The femur is relatively shorter and more robust than those of *Pliopithecus* and small apes (Figs. 11.7F, G, and I). This is also true of the neck, which bears a tubercle of the neck on its posterior surface. The trochanteric fossa approximates the hominoid condition in having a relatively well-developed posterior buttress. The distal end resembles those of hominoids and large platyrrhines in being mediolaterally expanded (Fig. 11.7B, E, and I) with a splayed medial condyle (Fig. 11.10B, D, and H). The tibia is quite similar to that of *Pliopithecus* (Fig. 11.9F and H). The morphology of the talus, and the form of the tibiotalar joint resemble those of *Pliopithecus* and small apes (Figs. 11.10E, F, and H, 11.11F, G, and I). The fibula is more hominoid-like. It has a robust shaft; distally it is expanded anteroposteriorly; and the articular surface articulating with the lateral side of the talus faces mediodistally rather than medially. This may indicate a greater abduction-adduction capability at the talocrural joint than is indicated for the previously described fossil species (Barnett and Napier, 1952; Langdon, 1986).

The rest of the foot is generally similar to that of arboreal nonhominoids, (Conroy and Rose, 1983; Langdon, 1986; Lewis, 1989). The calcaneus is less elongated than in small apes, and does not bear a medial tubercle on the calcaneal tuberosity (Fig. 11.11G and I). Otherwise it is similar to the small ape calcaneus, particularly in the morphology of the calcaneocuboid joint. The forefoot is relatively narrow mediolaterally, and transversely arched to the degree seen in nonhominoids. The hallux is well developed and features related to the passage of the tendon of m. flexor hallucis longus on the fibula, proximal talus, and calcaneus indicate a powerful grasping function. This is also indicated by a degree of torsion of the first and second metatarsals similar to that found in African apes.

The overall implications for positional behavior in the *Proconsul* postcranium are rather different than for *Pliopithecus* and small apes. Overall body and limb proportions are most similar to those of arboreal cercopithecids and generalized platyrrhines, as are many details of postcranial morphology. These features are strong indicators of a basically quadrupedal, arboreal positional repertoire. In a number of regions, however, more hominoid-like features point to the importance of relatively deliberate but effective climbing behavior. This is evident in features of the humeroradial and radioulnar joints relating to stabilized, relatively extensive forearm rotation, in the midcarpal joint and thumb relating to effective midcarpal adduction and versatile grasping, in abduction at the hip, and in features related to intratarsal mobility and grasping in the foot. There are no indicators, in the fore- or the hindlimbs, of the suspensory behavior suggested for *Pliopithecus* and small apes.

TURKANAPITHECUS

A number of postcranial specimens are known for the species *Turkanapithecus kalakolensis* from the Early Miocene of East Africa (Leakey and Leakey, 1986; Leakey et al., 1988b). Most of these specimens are from a single individual with a body weight of approximately 10 kg (Fleagle, 1988). Nothing definitive can be said about body or limb proportions. However, the evidence of the femur and ulna suggests that the degree of limb elongation may be intermediate between that of *Pliopithecus* and *Proconsul* (i.e., only slightly elongated compared to most quadrupedal anthropoids), and that intermembral proportions may be similar to those of *Pliopithecus* and *Proconsul* (i.e., approximately 95).

In the forelimb the trochlear notch of the ulna is similar to those of small apes and *Proconsul* (Figs. 11.5F, G, and H, and 11.4F, G, and I). There is a relatively large, anterolaterally facing radial notch. The olecranon process for the insertion of m. triceps brachii is directed proximally, as in arboreal anthropoids, and is intermediate in length between the abbreviated process of hominoids (Fig. 11.4D) and the longer process of other anthropoids. The ulnar shaft is relatively robust, as in *Proconsul* (Fig. 11.5G and H). The insertion area for m. brachialis is relatively broad, as in hominoids. These features make possible rapid extension and powerful flexion at the elbow, as occurs during the reaching and pulling movements involved in climbing and hoisting. The rest of the ulna, including a partial styloid process, is fragmentary, but similar to the *Proconsul* ulna in its discernible morphology.

The radius exhibits a number of features that do not occur in the same combination in any other extant or fossil anthropoid. The radial head is oval, but not markedly so, and has a correspondingly modest lateral lip (Fig. 11.4I). The surface for the proximal radioulnar joint extends most of the way round the side of the head and is relatively constant in its proximodistal depth. These features suggest relatively extensive pronation-supination movements of the forearm, but with a radiohumeral joint that is most stable in full pronation. This is indicative of quadrupedal, climbing, and possibly suspensory activities. The distal shaft is quite compressed anteroposteriorly, and laterally there is a well-developed crest for the insertion of m. brachioradialis, as in *Pliopithecus* and small apes (Fig. 11.5E, F, and G). The ulnocarpal surface is quite deeply socketed and faces somewhat distoanteriorly. This combination of features suggests a working position of the forearm in midpronation. This might occur during climbing, or during quadrupedalism on small branches, when the branch is grasped from the side rather than from the top. The first, fourth, and fifth metacarpals and the phalanges are similar to those of *Proconsul.*

In the hindlimb the robusticity of the femur is intermediate between that of *Pliopithecus* and *Proconsul* (Fig. 11.7F, H, and I) and there is a high bicondylar angle. In the proximal femur, the angulation of the neck and the projection of the head proximal to the greater trochanter are similar to those of the femora described so far. The greater trochanter does not flare laterally as it does in *Pliopithecus* and small apes (Fig. 11.7F, G, and H), however, and the trochanteric fossa is not buttressed posteriorly as it is in *Proconsul.* The lesser trochanter is distinctive in pointing more posteriorly than laterally. The shaft is anteroposteriorly flattened, with a relatively broad linea-aspera area on its posterior surfaces for the insertion of the adductor muscles. Distally, the articular surfaces are less expanded mediolaterally than in *Pliopithecus* or *Proconsul.* The *Turkanapithecus* femur is extremely similar to that of *Alouatta,* an animal whose locomotor repertoire is characterized by slow quadrupedalism, climbing, and occasional suspension (Mendel, 1976; Fleagle and Mittermeier, 1978; Schön Ybarra, 1984; Cant, 1986). The thigh is frequently held in partial flexion, abduction, and lateral rotation, with partial flexion at the knee joint (Grand, 1968; Schön Ybarra and Schön, 1988). A talus and cuboid do not differ significantly from those of small apes or *Proconsul* (Fig. 11.11G, H, and I).

From the limited evidence currently available, *Turkanapithecus* appears to be an arboreal, basically quadrupedal animal, but possibly with a better developed climbing capability than *Proconsul,* and probably less suspensory than *Pliopithecus* or small apes. Although particularly close morphological similarities are only evident in the femur, *Alouatta* is probably the best analogue among extant anthropoids.

SIVAPITHECUS

Sivapithecus is from the Middle and Late Miocene of Asia. Several species are known, but comparable postcranial morphology does not indicate marked differences among them. *Sivapithecus* resembles *Proconsul* in this respect. The known postcranial material is mostly from *S. parvada* and *S. indicus,* with body weights of approximately 69 kg and 50 kg respectively (Kelley, 1988). Some postcranial specimens of related taxa are also known (e.g., Morbeck, 1983), which are similar to *Sivapithecus* in comparable morphology. The following account is summarized from Pilbeam et al. (1980, 1990), Rose (1984, 1986, and 1989), and Spoor et al. (1992).

The known postcranial material is not extensive enough to gain any clear idea of relative limb elongation or limb proportions. In the forelimb the morphology of the humeral head is not known, but the shaft resembles those of cercopithecids even more than does the *Proconsul* humeral shaft (Figs. 11.2B and H, and 3B and H). Thus, in addition to its retroflexion, the proximal shaft also inclines medially toward the region of the head. As in *Proconsul,* there is a well-defined, flat, anterolaterally facing deltoid plane (Figs. 11.2G and H, and 11.3G and H). This morphology contrasts with that of living hominoids, which have a straight, cylindrical humeral shaft that bears a curved deltoid plane (Figs. 11.2D and 11.3D). The distal humerus is distinctly like those of large living hominoids, particularly African apes (Fig. 11.4D and H). Thus the relatively broad trochlea is spool-shaped, but not to the marked degree seen in *Pongo*. This morphology, together with corresponding ulnar features, are discussed in the next section for *Oreopithecus*. The lateral trochlear keel is well defined, and separated from the globular capitulum by a narrow, recessed zona conoidea, as in *Proconsul* (Fig. 11.5G and H).

Different regions of the *Sivapithecus* humerus thus show resemblances to two behaviorly very different groups of living catarrhines, cercopithecids and large hominoids. This is even more marked than is the case for *Proconsul.* Although some combination of quadrupedalism and, especially, climbing is implied by this morphology, its particular characteristics are evidently different from those of any living anthropoid.

The forearm is only known from an incomplete juvenile radius. In comparable morphology this is quite similar to the radius of *Proconsul.* In the hand the capitate bears a moderately inflated, rounded head, to the extent seen in *Ateles*. The hamate is relatively broad mediolaterally, and has a small hamulus and relatively small area for articulation with the triquetral. Although the capitate features suggest relatively extensive movements at the midcarpal joint, the hamate features suggest limited extension and ulnar deviation, and load transmission mostly through the ulnar side of the hand. The midcarpal region lacks those features found in suspensory and in knuckle-walking hominoids. The overall morphology is relatively similar to that of generalized platyrrhines. The distal capitate and hamate, for articulation with the medial three metacarpals, bear irregularly curved concave surfaces, suggesting a relatively immobile, stable carpometacarpal region, as found in great apes. A partial first metacarpal indicates relatively extensive movement of the thumb, as in *Proconsul.* A pollicial proximal phalanx has an anteriorly facing, broad distal articular surface, suggesting effective thumb use in semiflexion, for both manipulation and locomotor grasping. The manual phalanges have proximal joint surfaces most similar to those of nonhominoid anthropoids, but the robusticity and degree of curvature are most similar to those of climbing and suspensory living hominoids. The forearm and hand thus show the same sort of combination of features that is found in the arm, and reinforce the conclusion reached for that region.

In the hindlimb the femur is only known from a number of fragments (Figs. 11.7J and 11.8J). The head has a relatively extensive covering of articular surface that encroaches onto the neck anteriorly and posteriorly, as in hylobatids and some *Pongo* specimens. The neck and proximal shaft regions are markedly flattened anteroposteriorly, to a greater degree even than in *Pongo*. The neck is proximally inclined as in the other fossil hominoids described so far. The trochanteric fossa is quite large and has only a modestly developed posterior buttress. These features suggest a relatively mobile hip, especially in terms of abduction. This is best interpreted in terms of climbing and, possibly, suspensory activities. In the foot the talus, calcaneus, and most features of the cuboid are quite similar to those of *Proconsul* (Fig. 11.11I and J). In the forefoot, however, the intermediate and lateral cuneiforms are African ape-like in their proportions and most of their articulations. There is, however, probably a well-developed transverse arch to the forefoot, as in nonhominoid anthropoids. The hallux is also African ape-like. It is large, robust, with an extensive articular area on the metatarsal head, and with an expanded ungual tuberosity on the terminal phalanx. As with the humerus, the foot exhibits a combination of features not matched by any living anthropoid. The forefoot morphology and the presence of a transverse arch suggest talocrural and subtalar movement, and the distribution of load onto the forefoot most similar to that of nonhominoid arboreal quadrupeds. The other features of the forefoot and of the hallux suggest a functional pattern associated with the secure hafting of the digits onto the tarsus and powerful grasping between the hallux and the other digits characteristic of climbing hominoids.

Although the same positional activities must be evoked for *Sivapithecus* as have been mentioned for other Miocene hominoids—quadrupedalism, climbing, and perhaps some suspension—the morphological pattern differs from the other fossil hominoids considered above. *Sivapithecus* resembles *Proconsul* in a number of respects, but possesses more features in common with living hominoids, particularly African apes. This may be a reflection of differences in the proportional contribution of the different activities to the positional repertoire. It more probably reflects particulars of the style of performance of, especially, the climbing and any suspensory components of the repertoire.

OREOPITHECUS

Oreopithecus bambolii is from the late Miocene of Europe. Of all the Miocene hominoids considered here, *Oreopithecus* is the only one that closely resembles hominoids of modern aspect. Body weight is estimated at about 33 kg (Stern and Jungers, 1985), as in female common chimpanzees. The intermembral index is about 120, as are the humerofemoral and radiotibial indices (Harrison, 1986). These are most similar to the proportions in *Hylobates*. For an animal of this size, the forelimbs are quite long and the hindlimbs slightly short compared to African apes (Stern and Jungers, 1985). Sarmiento (1987) links these particular proportional features to climbing and forelimb suspensory activities.

The following descriptions and interpretations are largely based on the studies of Harrison (1986, 1991), Szalay and Langdon (1986), and especially those of Sarmiento (1987 and 1988). Earlier and other recent studies are cited where relevant. Almost all of the known *Oreopithecus* specimens have been deformed during the fossilization process and are therefore not illustrated here. In the axial skeleton the lumbar region consists of five craniocaudally short vertebrae (Schultz, 1961; Straus, 1963) and the thoracolumbar vertebral spines all point caudally. These features are most similar to those of hylobatids and suspensory platyrrhines. The sacrum is long, consisting of six vertebrae, and there is no tail, as in hominoids (Hürzeler, 1958; Schultz, 1960). The ribs have an acute costal angle, so that the thorax is broad and shallow, as in hominoids (Schultz, 1960). This broadening of the thorax, in combination with long forelimbs, is advantageous for climbing large diameter supports.

In the forelimb the clavicle is long, as would be expected from the broad thorax. The humerus has a large, probably medially facing head, with greater and lesser trochanters that delimit a well-developed bicipital groove, and a deltoid plane that extends distally on the shaft (Schultz, 1960; Straus, 1963). These are features found in extant hominoids, and relate to considerable shoulder mobility and powerful abduction of the arm at the shoulder. These similarities extend to the elbow and forearm region. The humeral trochlea is markedly spool-shaped and extends posteriorly to a deep olecranon fossa. On the ulna these features are matched by a trochlear notch that faces anteroproximally and is strongly convex mediolaterally. The olecranon process is short. This results in highly stabilized flexion-extension movement that extends beyond 180°, and in the locking of the ulna onto the humerus in extension, which allows axial rotatory movements of the upper limb as a whole. Powerful flexion, especially with the forearm in the mid-pronated position, is indicated by the relatively distal placement of the bicipital tuberosity on the radius and the presence of a medially directed medial epicondyle that is well separated from the trochlea.

The surface for articulation with the radius has a well-defined, recessed zona conoidea articulating with a corresponding surface on the side of the radial head. The radial notch of the ulna is well defined and markedly concave, and the ulnar surface for the distal radioulnar joint is extensive. These features imply a greater amplitude of pronation-supination movement of the forearm than in *Proconsul.* The forelimb is thus highly mobile, but with joints that are well stabilized throughout their extensive movements. Such a forelimb is capable of making the reaching, embracing, and pulling movements associated with climbing and suspensory behaviors.

The hand is even more elongated than in extant hominoids, and the fingers are as elongated and curved as in *Pongo* (Susman, 1985). This hooklike configuration is clearly that of a climbing and suspensory animal. Within the wrist the lunate is relatively narrow compared to the scaphoid, suggesting predominantly radial transmission of force through the wrist. The form of the distal ulnar articular surface suggests that there was an intra-articular meniscus present. A medially facing facet on the hamate indicates limited contact between the ulnar styloid and the triquetrum. A distally directed hamulus on the hamate indicates that the pisiform is similarly oriented, and divorced from the ulnocarpal joint. These features are similar to the hylobatid condition and are linked to the increased excursion at the radioulnar joints, extensive abduction-adduction movements at the radiocarpal joint, and powerful flexion at the wrist.

In the lower limb, the hip bone is like those of large extant hominoids, with an expanded, dorsally facing ilium. The ischium does not support a callosity (Schultz, 1960). The femur has a high bicondylar angle and a negative torsion angle, both characteristic of climbers. The proximal femur has all the features mentioned in previous sections that relate to a high degree of mobility at the hip joint. Numerous extant hominoid-like features in the foot point to its use as a mobile, powerfully grasping organ. The talus has a relatively low, wedged trochlea, which allows for some abduction-adduction movement conjunct to plantarflexion-dorsiflexion. A highly divergent neck is an indicator of the shortness of the tarsus and of a divergent, grasping hallux. The calcaneus is short and broad, with a well-developed medial tubercle on the calcaneal tuberosity, and talocalcaneal joint surfaces that allow considerable inversion-eversion movement. Although the peg-and-socket morphology of the calcaneocuboid joint is similar to that described above for other Miocene catarrhines, its degree of expression in *Oreopithecus* suggests considerable midtarsal mobility. The other tarsal bones, together with the metatarsals, are proximodistally short and robust. There is a large distal facet on the medial cuneiform, and the foot axis passes through the second metatarsal. These features indicate the importance of hallucial grasping.

Although *Oreopithecus* does not show consistent similarities with any one particular living hominoid, its overall morphology is clearly that of a hominoid of modern aspect. Equally clearly, this morphology is that of a highly specialized climbing and suspensory animal.

DISCUSSION

As mentioned at the beginning of this chapter, the interpretations of Miocene hominoid functional anatomy and positional behavior given above are necessarily vague and leave many questions unanswered. More definitive interpretations will undoubtedly result from the collection of more extensive fossil evidence. Of equal importance, they will result from a more detailed understanding of the range of positional activities used by living primates, the habitat structures within which they are used, and the functional morphology that makes their performance possible. It is vital to be able to analyze compromise morphologies (Rose, 1991) in terms of a particular positional repertoire and the relative importance of its component activities. It is equally important to differentiate, on the basis of morphological evidence, the different varieties of particular activities, such as climbing and suspension, and to tell which types or components of arboreal habitats particular fossil taxa are most likely to have exploited.

A number of Miocene hominoids—such as *Proconsul major, P. nyanzae,* and *Sivapithecus parvada*—fall within the range of body size that in living primates is associated with some use of terrestrial habitats. In the case of living knuckle-walking apes, digitigrade baboons, and bipedal humans, there are obvious morphological complexes that are indicators of the use of terrestrial locomotor modes. No such complexes have been demonstrated convincingly for any Miocene hominoids. Such a demonstration—for known or yet-to-be-found Miocene hominoids—will be essential for an understanding of their paleoecology, and of hominid and African-ape evolution.

Many new specimens of Miocene hominoids have been collected in recent years, and it is apparent that they were a radiation that was extensive in both space and time. It is, however, strikingly evident that for the most part Miocene hominoids show no clear close relationships with, and are quite unlike, living hominoids. Indeed, it is tempting to think of extant hominoids as atypical, specialized offshoots from the more numerous and more generalized Miocene types. As well as filling in the above-mentioned gaps in our knowledge, future research may lead to a more complete understanding of the evolutionary history of these atypical offshoots.

Acknowledgments: I would like to thank DL Gebo for inviting me to prepare this chapter. The research reported here was supported by NSF grant BNS-9004502.

LITERATURE CITED

Aiello LC (1981) Locomotion in the Miocene Hominoidea. In CB Stringer (ed.): Aspects of Human Evolution. London: Taylor and Francis, pp. 63–97.

Andrews P, and Simons E (1977) A new African Miocene gibbon-like genus, *Dendropithecus* (Hominoidea, Primates) with distinctive postcranial adaptations: Its significance to origin of Hylobatidae. Folia Primatol. *28:*161–169.

Ankel F (1965) Der Canalis Sacralis als Indikator für die Lange der Primaten. Folia Primatol. *3:*263–276.

——— (1972) Vertebral morphology of fossil and extant primates. In RH Tuttle (ed.): The Functional and Evolutionary Biology of Primates. Chicago: Aldine, pp. 223–240.

Barnett CH, and Napier JR (1952) The axis of rotation at the ankle joint in man: Its influence upon the form of the talus and the mobility of the fibula. J. Anat. *86:*1–9.

Beard KC, Teaford MF, and Walker A. (1986) New wrist bones of *Proconsul africanus* and *P. nyanzae* from Rusinga Island, Kenya. Folia Primatol. *47:*97–118.

Begun DR (1987) Catarrhine phalanges from the Late Miocene (Vallesian) of Rudabánya, Hungary. J. Hum. Evol. *17:*413–438.

——— (1989) A large Pliopithecine molar from Germany and some notes on the Pliopithecinae. Folia Primatol. *52:*156–166.

Cant JGH (1986) Locomotion and feeding postures of spider and howling monkeys: Field study and evolutionary interpretation. Folia Primatol. *46:*1–14.

Ciochon RJ, and Corruccini RS (1977) The phenetic position of *Pliopithecus* and its phylogenetic relationship to the Hominoidea. Syst. Zool. *26:*290–299.

Conroy GC, and Rose MD (1983) The evolution of the primate foot from the earliest primates to the Miocene hominoids. Foot and Ankle *3:*342–364.

Corruccini RS, Ciochon RL, and McHenry HM (1975) Osteometric shape relationships in the wrist joint of some anthropoids. Folia Primatol. *24:*250–274.

——— (1976) The postcranium of Miocene hominoids: Were dryopithecines merely "dental apes"? Primates. *17:*205–223.

Feldesman MR (1982) Morphometric analysis of the distal humerus of some Cenozoic catarrhines: The late divergence hypothesis revisited. Am. J. Phys. Anthrop. *59:*173–195.

Ferembach D (1958) Les limnopithèques du Kenya. Ann. Paléontol. *44:*149–249.

Fleagle JG (1983) Locomotor adaptations of Oligocene and Miocene hominoids and their phyletic implications. In RL Ciochon and RS Corruccini (eds.): New Interpretations of Ape and Human Ancestry. New York: Plenum Press, pp. 301–324.

——— (1988) Primate Adaptation and Evolution. New York: Academic Press.

Fleagle JG, and Meldrum DJ (1988) Locomotor behavior and skeletal morphology of two sympatric pitheciine monkeys, *Pithecia pithecia* and *Chiropotes satanas*. Am. J. Primatol. *16:*227–249.

Fleagle JG, and Mittermeier RA (1978) Locomotor behavior, body size, and comparative ecology of seven Surinam monkeys. Am. J. Phys. Anthrop. *52:*301–314.

Gebo DL (1989) Locomotor and phylogenetic considerations in anthropoid evolution. J. Hum. Evol. *18:*201–233.

Gebo DL, Beard KC, Teaford MF, Walker A, Larson SG, Jungers WL, and Fleagle JG (1988) A hominoid proximal humerus from the Early Miocene of Rusinga Island, Kenya. J. Hum. Evol. *17:*393–401.

Grand TI (1968) The functional anatomy of the lower limb of the howler monkey (*Alouatta caraya).* Am. J. Phys. Anthrop. *28:*163–182.

Harrison T (1982) Small Bodied Apes from the Miocene of East Africa. Ph.D. dissertation, University of London.

——— (1986) A reassessment of the phylogenetic relationships of *Oreopithecus bambolii* Gervais. J. Hum. Evol. *15:*541–583.

——— (1987) The phylogenetic relationships of the early catarrhine primates: A review of the current evidence. J. Hum. Evol. *16:*41–80.

——— (1988) A taxonomic revision of the small catarrhine primates from the Early Miocene of East Africa. Folia Primatol. *50:*59–108.

——— (1991) The implications of *Oreopithecus bambolii* for the origins of bipedalism. In Y Coppens and B Senut (eds.): Origine(s) de la bipédie chez les hominidés. Cah. Paléoanth. Paris: CNRS, pp. 235–244.

Hürzeler J (1958) *Oreopithecus bambolii* Gervais, a preliminary report. Verh. Naturf. Ges. Basel *69:*1–48.

Jenkins FA, Jr., and McClearn D (1984) Mechanisms of hind foot reversal in climbing mammals. J. Morph. *182:*197–219.

Jolly CJ (1967) The evolution of baboons. In H Vagtborg (ed.): The Baboon in Medical Research. Austin: University of Texas Press, pp. 23–50.

Jungers WL (1984) Aspects of size and scaling in primate biology with special reference to the locomotor skeleton. Yrbk. Phys. Anthrop. *27:*73–97.

——— (1985) Body size and scaling of limb proportions in primates. In WL Jungers (ed.): Size and Scaling in Primate Biology. New York: Plenum Press, pp. 345–381.

Kelley J (1988) A new large species of *Sivapithecus* from the Siwaliks of Pakistan. J. Hum. Evol. *17:*305–324.

Langdon JH (1986) Functional morphology of the Miocene hominoid foot. Contrib. Primatol. *22:*1–225.

Leakey RE, and Leakey MG (1986) A second new Miocene hominoid from Kenya. Nature *324:*146–148.

Leakey RE, Leakey MG, and Walker AC (1988a) Morphology of *Afropithecus turkanensis* from Kenya. Am. J. Phys. Anthrop. *76:*289–307.

——— (1988b) Morphology of *Turkanapithecus kalakolensis* from Kenya. Am. J. Phys. Anthrop. *76:*277–288.

Leakey REF, and Walker A (1985) New higher primates from the early Miocene of Buluk, Kenya. Nature *318:*173–175.

Le Gros Clark WE, and Leakey LSB (1951) The Miocene Hominoidea of East Africa. Fossil Mammals of Africa (Br. Mus. Nat. Hist.) *1:*1–117.

Le Gros Clark WE, and Thomas DP (1951) Associated jaws and limb bones of *Limnopithecus macinnesi.* Fossil Mammals of Africa (Br. Mus. Nat. Hist.) *3:*1–27.

Lewis OJ (1989) Functional Morphology of the Evolving Hand and Foot. Oxford: Clarendon Press.

McHenry HM, and Corruccini RS (1983) The wrist of *Proconsul africanus* and the origin of hominoid postcranial adaptations. In RL Ciochon and RS Corruccini (eds.): New Interpretations of Ape and Human Ancestry. New York: Plenum Press, pp. 353–367.

Meldrum DJ, and Pan Y (1988) Manual proximal phalanx of *Laccopithecus robustus* from the Latest Miocene site of Lufeng. J. Hum. Evol. *18:*719–731.

Mendel F (1976) Postural and locomotor behavior of *Alouatta palliata* on various substrates. Folia Primatol. *26:*36–53.

Morbeck ME (1975) *Dryopithecus africanus* forelimb. J. Hum. Evol. *4:*39–46.

——— (1976) Problems in reconstruction of fossil anatomy and locomotor behavior: The *Dryopithecus* elbow complex. J. Hum. Evol. *5:*223–233.

——— (1983) Miocene hominoid discoveries from Rudabánya: Implications from the postcranial skeleton. In RL Ciochon and RS Corruccini (eds.): New Interpretations of Ape and Human Ancestry. New York: Plenum Press, pp. 369–404.

Napier JR, and Davis PR (1959) The forelimb skeleton and associated remains of *Proconsul africanus.* Fossil Mammals of Africa (Br. Mus. Nat. Hist.) *16:*1–69.

Pilbeam DR, Rose MD, Badgley C, and Lipschutz B (1980) Miocene hominoids from Pakistan. Postilla *181:*1–94.

Pilbeam D, Rose MD, Barry JC, and Ibrahim-Shah SM (1990) New *Sivapithecus* humeri from Pakistan and the relationship of *Sivapithecus* and *Pongo.* Nature *348:*237–239.

Rafferty, KL (1990) The Functional and Phylogenetic Significance of the Carpometacarpal Joint of the Thumb in Anthropoid Primates. M.A. thesis, New York University, New York.

Robertson M (1985) A comparison of pronation-supination mobility in *Proconsul* and *Pliopithecus vindobonensis.* Am. J. Phys. Anthrop. *66:*219.

Rose MD (1974a) Ischial callosities and ischial tuberosities. Am. J. Phys. Anthrop. *40:*375–384.

——— (1974b) Postural adaptations in New and Old World monkeys. In FA Jenkins, Jr. (ed.): Primate Locomotion. New York: Academic Press, pp. 201–222.

——— (1983) Miocene hominoid postcranial morphology: Monkey-like, ape-like, neither, or both? In RL Ciochon and RS Corruccini (eds.): New Interpretations of Ape and Human Ancestry. New York: Plenum Press, pp. 405–417.

——— (1984) Hominoid postcranial specimens from the Middle Miocene Chinji Formation, Pakistan. J. Hum. Evol. *13:*503–516.

——— (1986) Further hominoid postcranial specimens from the Late Miocene Nagri Formation of Pakistan. J. Hum. Evol. *15:*333–367.

——— (1988) Another look at the anthropoid elbow. J. Hum. Evol. *17:*193–224.

——— (1989) New postcranial specimens of catarrhines from the Middle Miocene Chinji Formation, Pakistan: Descriptions and a discussion of proximal humeral functional morphology in anthropoids. J. Hum. Evol. *18:*131–162.

——— (1991) The process of bipedalization in hominids. In Y Coppens and B Senut (eds.): Origine(s) de la bipédie chez les hominidés. Cah. Paléoanth. Paris: CNRS, pp. 37–48.

——— (1992) Kinematics of the trapezium-1st metacarpal joint in extant anthropoids and Miocene hominoids. J. Hum. Evol. *22:*255–266.

Rose MD, Leakey MG, Leakey REF, and Walker AC (1992) Postcranial specimens of *Simiolus enjiessi* and other small apes from the Early Miocene of Lake Turkana, Kenya. J. Hum. Evol. *22:*171–237.

Ruff CB, Walker A., and Teaford MF (1989) Body mass, sexual dimorphism and femoral proportions of *Proconsul* from Rusinga and Mfangano Islands, Kenya. J. Hum. Evol. *18:*515–536.

Sarmiento EE (1983) The significance of the heel process in anthropoids. Int. J. Primatol. *4:*127–152.

——— (1985) Functional Differences in the Skeleton of Wild and Captive Orang-Utans and their Adaptive Significance. Ph.D. dissertation, New York University.

——— (1987) The phylogenetic position of *Oreopithecus* and its significance in the origin of the Hominoidea. Am. Mus. Novitates *2881:*1–44.

——— (1988) Anatomy of the hominoid wrist joint: Its evolutionary and functional implications. Int. J. Primatol. *9:*281–345.

Schön Ybarra MA (1984) Locomotion and postures of red howlers in a deciduous forest-savanna interface. Am. J. Phys. Anthrop. *63:*65–76.

Schön Ybarra MA, and Schön MA III (1988) Positional behavior and limb bone adaptations in red howling monkeys *(Alouatta seniculus).* Folia Primatol. *49:*70–89.

Schultz AH (1960) Einege Beobachtungen und Masse am Skelett von *Oreopithecus* im vergleich mit anderem catarrhinen Primaten. Z.

Morph. Anthrop. *50:*136–149.

——— (1961) Vertebral column and thorax. Primatologia *4:*1–66.

——— (1963) Relations between the lengths of the main parts of the foot skeleton in primates. Folia Primatol. *1:*150–171.

Spoor CF, Sondaar PY, and Hussain ST (1991) A hominoid hamate and first metacarpal from the Late Miocene Nagri Formation of Pakistan. J. Hum. Evol. *21:*413–424.

Stern JT, Jr., and Jungers WL (1985) Body size and proportions of the locomotor skeleton in *Oreopithecus bambolii.* Am. J. Phys. Anthrop. *66:*233.

Straus WL, Jr. (1963) The classification of *Oreopithecus.* In L Washburn (ed.): Classification and Human Evolution. New York: Viking Press, pp. 146–177.

Susman RL (1985) Functional morphology of the *Oreopithecus* hand. Am. J. Phys. Anthrop. *66:*235.

Szalay FS, and Delson E (1979) Evolutionary History of the Primates. New York: Academic Press.

Szalay FS, and Langdon JH (1986) The foot of *Oreopithecus:* An evolutionary assessment. J. Hum. Evol. *15:*585–621.

Walker AC, and Pickford M (1983) New postcranial fossils of *Proconsul africanus* and *Proconsul nyanzae.* In RL Ciochon and RS Corruccini (eds.): New Interpretations of Ape and Human Ancestry. New York: Plenum Press, pp. 325–351.

Walker A, and Teaford M (1988) The Kasawanga primate site: An Early Miocene hominoid site on Rusinga Island, Kenya. J. Hum. Evol. *17:*539–544.

——— (1989) The hunt for *Proconsul.* Sci. Amer. *260*(January):74–82.

Ward CV (1991) Functional Anatomy of the Lower Back and Pelvis of the Miocene Hominoid *Proconsul nyanzae* from Mfangano Island, Kenya. Ph.D. dissertation, Johns Hopkins University, Baltimore.

Ward CV, Walker A, and Teaford MF (1991) *Proconsul* did not have a tail. J. Hum. Evol. *21:*215–220.

Zapfe H (1960) Die Primatenfunde aus der Miozänen spaltenfüllung von Neudorf an der March (Devinská Nová Ves) Tschechoslowakei. Schweiz. Palaeontol. Abh. *78:*1–293.

CONTRIBUTORS

ROBERT L. ANEMONE is Assistant Professor of Anthropology at the State University of New York at Geneseo and Research Associate in the Department of Mammalogy at the American Museum of Natural History in New York City. A specialist in the functional anatomy of primate locomotor systems, his published work has appeared in such journals as the *American Journal of Physical Anthropology* and the *Journal of Human Evolution.* Currently, he is engaged in a comparative study of dental development in humans and chimpanzees and its implications for understanding human evolution. He is coeditor (with Linda Winkler) of the forthcoming volume *Recent Developments in the Study of Hominoid Ontogeny.*

MARIAN DAGOSTO is Assistant Professor of Cell, Molecular, and Structural Biology at Northwestern University Medical School. She has published on primate anatomy, evolution, and systematics in such journals as *American Journal of Physical Anthropology, International Journal of Primatology,* and *Journal of Human Evolution,* and has coedited *Primate Postcranial Adaptation and Evolution.* She is currently studying positional behavior of Malagasy lemurs.

DANIEL L. GEBO is Associate Professor in the Anthropology Department of Northern Illinois University and a Research Associate in the Zoology Department at the Field Museum of Natural History. He has published numerous articles on primate anatomy and evolution in such journals as the *American Journal of Physical Anthropology, Journal of Human Evolution,* and *Folia Primatologica.* Currently, he is engaged in completing several manuscripts on the relationships between positional behavior, body size, and habitat use in several species of Old World monkeys.

SUSAN G. LARSON is Associate Professor of Anatomy at the School of Medicine of the State University of New York at Stony Brook. Her research centers on the functional analysis of morphology, with a particular focus on the shoulder and forelimb of nonhuman primates. She has concentrated her efforts on the use of experimental techniques to test hypothesized relationships between form and function with a particular emphasis on the analysis of muscle function using the technique of electromyography. With collaborator Jack Stern, she has published several articles on shoulder muscle function in suspensory and quadrupedal primates in such journals as the *American Journal of Anatomy, Journal of Zoology, London, American Journal of Physical Anthropology,* and *Journal of Motor Behavior.* She is currently working on a morphometric analysis of a new set of functional traits of the scapula and humerus generated from her electromyographic studies, and an examination of functional convergence in the forelimbs of hominoids.

JEFF MELDRUM is Assistant Professor of Evolutionary Morphology in the Department of Cell, Molecular, and Structural Biology at Northwestern University School of Medicine. Specializing in the study of primate positional behavior, he has focused on the evolutionary history of the New World primates. He has participated in paleontological field projects to Argentina and Colombia and is engaged in the study of recently recovered fossil primates and the description of new taxa. Currently, he is applying the analysis of nucleotide sequences to the problem of platyrrhine cladogenesis.

MICHAEL D. ROSE is Professor in the Department of Anatomy, Cell Biology, and Injury Science at the New Jersey Medical School. He is a leading expert on Miocene "apes" and has pursued a variety of other topics including the origin of human bipedalism. He has also published a number of articles on primate limb anatomy and evolution in several journals including the *American Journal of Physical Anthropology* and the *Journal of American Evolution,* as well as contributing chapters in prominent volumes concerned with primate postcranial adaptation. He is currently studying positional behavior of ateline primates in Ecuador.

LIZA SHAPIRO is Assistant Professor of Anthropology at the University of Texas at Austin. Her research on the functional morphology and evolution of the primate spine incorporates several approaches including morphometrics, electromyography, and kinematics. She has studied anthropoid vertebral function extensively and is currently expanding this work to prosimians.

SHARON M. SWARTZ is Assistant Professor in the Ecology and Evolutionary Biology Program at Brown University. She has published a variety of articles on biomechanics, especially on primates, in such journals as the *American Journal of Physical Anthropology, International Journal of Primatology, Journal of Zoology,* and *Nature.* She is currently conducting strain gauge experiments on the bones and wing dynamics of Australian bats.

PAUL F. WHITEHEAD is a staff member of the Division of Vertebrate Zoology at the Peabody Museum of Natural History at Yale University. He has extensive primatological and paleoanthropological field experience in Kenya and Ethiopia and is a codiscoverer of the Bodo D'Ar hominid skull. He is editor of the forthcoming collection, *Primate Quadrupedalism,* published by the *Journal of Human Evolution.* He has been author or co-author of papers and reviews in *Nature, Systematic Zoology, American Scientist, Journal of Vertebrate Paleontology, Palaeogeography, Palaeoclimatology, Palaeoecology,* and *Explorers Journal.*

INDEX